SUSTAINABLE POWER GENERATION

SUSTAINABLE POWER GENERATION

Current Status, Future Challenges, and Perspectives

NIKOLAY BELYAKOV

ACADEMIC PRESS
An imprint of Elsevier

Academic Press is an imprint of Elsevier
125 London Wall, London EC2Y 5AS, United Kingdom
525 B Street, Suite 1650, San Diego, CA 92101, United States
50 Hampshire Street, 5th Floor, Cambridge, MA 02139, United States
The Boulevard, Langford Lane, Kidlington, Oxford OX5 1GB, United Kingdom

Notices

Knowledge and best practice in this field are constantly changing. As new research and experience broaden our understanding, changes in research methods, professional practices, or medical treatment may become necessary.

Practitioners and researchers must always rely on their own experience and knowledge in evaluating and using any information, methods, compounds, or experiments described herein. In using such information or methods they should be mindful of their own safety and the safety of others, including parties for whom they have a professional responsibility.

To the fullest extent of the law, neither the Publisher nor the authors, contributors, or editors, assume any liability for any injury and/or damage to persons or property as a matter of products liability, negligence or otherwise, or from any use or operation of any methods, products, instructions, or ideas contained in the material herein.

Library of Congress Cataloging-in-Publication Data
A catalog record for this book is available from the Library of Congress

British Library Cataloguing-in-Publication Data
A catalogue record for this book is available from the British Library

ISBN: 978-0-12-817012-0

Publisher: Joe Hayton
Acquisition Editor: Raquel Zanol
Development Editor: Michelle W. Fisher
Production Project Manager: Anitha Sivaraj
Designer: Alan Studholme

Typeset by VTeX

In memory of my father, Dr. Sergey Belyakov (1953–2018)

Contents

About the author

Nikolay Belyakov is an acting manager of sales development in a petrochemical holding SIBUR where he leads corporate projects in digitalization, sales, marketing, talent development and capability building, and business process implementation.

Since 2017, he is also a country contributor and peer-reviewer in REN21 Global Status Report on a voluntary basis.

Previously he worked as a global segment manager for thermal and renewable power within Hilti Corporation — worldwide construction solution and service provider. His area of responsibility ranged from the development of the company's corporate strategy in energy-related business to global sales. Earlier he was employed by GE in Energy and Oil & Gas businesses in various roles, including sales, project management, customer business and technology evaluation, thermal power plant engineering, and marketing.

He obtained his engineering degree from Bauman Moscow State Technical University (Russia) specializing in general engineering and applied mathematics, followed by a PhD in engineering from Peter the Great St. Petersburg Polytechnic University (Russia) and MBA from the University of St. Gallen (Switzerland). His main research interests include sustainable development, sustainability education, energy innovations and emerging technologies, and business modeling.

He is the author of a number of articles in applied mathematics and heat transfer, economics of education, industrial automation, as well as several textbooks for students.

Preface

The philosophy of continuous economic growth, increasing world population, and globalization of economies lead to a rapid deployment of new power generating capacities. In parallel, such a race is facing the challenges of the inevitable impact on the environment that is seen through the climate change, shortage of resources, volatility of fuel prices, and many other traces throughout the world. In this regard, the society requires a balanced approach to the development of all primary energy sources and their proper utilization within the industry. The business is therefore challenged from two sides: efficient energy production to maintain its returns and run business to keep pace with growth strategies, and compliance with the regulation driven by the environmental challenges.

In the face of the growing issues that put sustainable development into question, the energy sector is by far among the top priorities. Thus, the disruptive innovations in the energy industry are usually those which drive the business, make the largest impact, and therefore result in new types of equipment and materials for electricity and heat generation, transmission, distribution, and usage. The amount of information available on power generation topics is growing at a high pace, and such information-intense environment challenges the new generation of managers and leaders to find comprehensive, compact, and useful form of knowledge on the industry and its sustainable development into the future. Such a source of information shall have a balanced approach to cover general aspects of technology and engineering approaches, as well as managerial implications such as basis for investment decisions, pitfalls of energy project management, key influencing factors, and their link to economic, political, social, and environmental aspects.

The problem becomes even more important in the face of growing consumption of resources, most of which are finite and becoming more and more scarce. Unequally distributed, resources become a cornerstone in the economy development, political and social activities, and, moreover, in the overall stability of the environment. Growing importance of carefully managing these resources, while keeping in mind the amount of people living in complete energy poverty, has been a subject of concern as shown in numerous research and publications within the last decades. One of the important roles is played by nonprofit organizations, which aim at attracting attention of the public to these problems, though being significantly influenced by the politics. On the other hand, big multinational corporations are also aware of the issue and step forward on the way of sustainable development through breakthrough innovations, especially in the areas of "green" energy.

Power generation is certainly one of a few areas which have traditionally depended on the natural resources like fossil fuels or fresh water, and has been closely connected to all aspects of human activity. We can call it the prime mover of the progress and the overall sustainable development of the humankind. Hence, it is important to have an overwhelming and system-based picture of the global energy footprint: what are the resources used, how the industry and technology are coupled to our political, economic, and social life, and what is the impact we face and are expecting to face in the future. On top of this, global and overwhelming digitalization of the world requires flexible power which would be the growth factor for the 21st century.

With the boundaries of managerial competences expanding rapidly, present day leaders must necessarily be aware of the information and knowledge of subjects that continue to become more and more complex and important. In the general business focus on the sustainability and future development, however, their knowledge seldom progresses beyond the level of elementary statistics and some ideas of the forecasts regarding major energy trends, as well as the level of basic vision of the green energies being the driver of the future. More detailed discussions or the history and underlying problems and developments, especially those that made the present situation and the innovations happen, are usually left outside of the focus areas. This is hence one of the goals of the current book to close this gap in aligning the overall structure of sustainable electricity production technologies into a single and balanced picture.

Moreover, despite rapid implementation of new solutions, traditional power is still being considered as the backbone of the industry and will remain the basis for the overall development. By these means, the idea of this book is also to give an overview of traditional power for graduate and postgraduate business students, mainly those with strong managerial ambitions who would become the future leaders and hence would guide the world into the new era of sustainable development. We shall outline here enough basics for understanding the technological side and the processes that take place in power plants, however, focusing on the underlining economic, political, and social aspects of this exciting and rapidly developing industry. In the long run, new business model innovations, change management, big data analysis, IoT and many other emerging technologies shall become a new set of competences that power generation managers would require in the coming decades.

With this in mind, this book has two major aims: firstly, addressing the importance to the future manager of understanding the challenges of providing energy for sustainable development, and secondly, giving a structured and balanced overview of the available technologies, their advantages, sustainability impact, and economic aspects to develop the basics of the sustainability competence. Throughout, we approach this new competence from an interdisciplinary perspective, leading to the conclusion that there is no single recipe of success when dealing with sustainability as it is an extremely complicated topic.

This book targets a wide audience. To start with, it shall be mostly beneficial for energy, chemical, civil, electrical, and environmental engineers who can get an overview of technologies and their development to extend their knowledge outside of pure design topics. Next, the book is essential for those studying or practicing energy policy-making, energy planning, and social and political sciences as they would learn the underlying issue of sustainable energy systems and some of the approaches to mitigate them. Last but not least, the book would be helpful for NGOs working for promotion of renewable energy and sustainable development.

Essentially meaning all supporters, I want to especially acknowledge Dr. Svetlana Tokareva whose constant support, encouragement, and personal scientific success drove me and showed an excellent example of expertise, imagination and success.

Nikolay Belyakov
Moscow
January 2019

Acronyms

AC Alternating current
ACC Air-cooled condenser
AFC Alkaline fuel cells
AI Artificial intelligence
AIS Air-insulated switchgear
AQSC Air quality control system
Btu British thermal unit
BOO Build, own, operate
BOP Balance of plant
BWR Boiling water reactor
CAES Compressed air energy storage
CANDU Canadian deuterium uranium reactor
CAPEX Capital expenditures
CCGT Combined cycle gas turbine
CCPP Combined cycle power plant
CCS Carbon capture and storage
CEMS Continuous emissions monitoring system
CFB Circulating fluidized bed
CHP Combined heat and power
CO_x Carbon oxides
COD Commercial Operation Date
COP Conference of the Parties to the UNFCCC
CSP Concentrating solar power
DC Direct current
DCS Distributed control system
EGS Enhanced geothermal system
EIA US Energy Information Administration
EIC Energy Industries Council (https://www.the-eic.com/)
EPC Engineering, procurement and construction
FESS Flywheel energy storage system
FGD Flue gas desulfurization
GDP Gross domestic product
GHG Greenhouse gases
GIS Gas-insulated switchgear
GT Gas turbine
HAWT Horizontal axis wind turbine
HDR Hot dry rock
HMI Human–machine interface
HP High pressure
HV High voltage
HVAC Heating, ventilation and air conditioning
HRSG Heat recovery steam generator
IAEA International Atomic Energy Agency

IGV Inlet guide vanes
IIoT Industrial Internet-of-Things
IIR Industrial Info Resources (https://www.industrialinfo.com/)
IoT Internet-of-Things
IP Intermediate pressure
IPP Independent power producer
IR Inferred resources
IT Information technologies
J Joule
LED Light emitting diod
LNG Liquefied natural gas
LWGR Light-water graphite-moderated reactor
LP Low pressure
LSB Last stage bucket (blade)
LV Low voltage
MCFC Molten carbonate fuel cells
MNC Multinational corporation
MOX Mixed-oxides fuel
MV Medium voltage
NDT Nationally determined contribution within the Paris Agreement
NGL Natural gas liquids
NO_x Nitrogen oxides
NSSS Nuclear steam supply system
NTP Notice to proceed
NUG Non-utility generator
O&M Operation and maintenance
OECD Organization for Economic Cooperation and Development
OEM Original equipment manufacturer
OPEC Organization of the Petroleum Exporting Countries
OTEC Ocean thermal energy conversion
OWC Oscillating water column
OWSC Oscillating wave surge converter
PAFC Phosphoric acid fuel cells
PEMFC Polymer electrolyte membrane or proton exchange membrane fuel cell
PM Particulate matter
PMC Project management consultant
PPA Power purchase agreement
PRO Pressure retarded osmosis
PV Photo-voltaic
PWR Pressurized water reactor
RAR Reasonably assured resources
RBMK Russian abbreviation for the High Power Channel-type Reactor
RED Reversed electro dialysis
RFP Request for proposal
RFQ Request for quote
RO Reverse osmosis
ROI Return on investment
rpm Revolutions per minute

SC Supercritical
SCPP Simple cycle power plant
SDGs Sustainable development goals
SLD Single-line diagram
SOFC Solid oxide fuel cell
SO_x Sulfur oxides
SI International System of Units
ST Steam turbine
TIV Total investment value
toc Ton of coal equivalent
toe Ton of oil equivalent
TPES Total primary energy supply
TPP Thermal power plant
UNFCCC The United Nations Framework Convention on Climate Change
UPS Uninterruptible power supply
USC Ultrasupercritical
VAWT Vertical axis wind turbine
WCED World Commission on Environment and Development
WEC Wave energy converter

PART ONE

Introduction to energy and energy systems

Overview

Understanding what energy and power are, and how energy is derived from various sources and transformed to the most usable shape, is of vital importance for sustainable development. The first part opens up the discussion about fundamental energy sources, global planetary cycles, and basics of energy systems that function these days.

CHAPTER ONE

Concept of energy

Contents

1.1. What is energy

The term *energy* has been utilized in everyday life so widely that there is hardly anyone who would object to its understanding. And it is partly true: energy is an essential part of our daily activities, we "use" it every moment while doing any kind of activity. Energy is everywhere and runs our cars, powers our gadgets, brings light — these are a few basic examples we may ever think of. Notwithstanding the foregoing, the path of energy is by far more complicated and much more important to understand as it underlines deeper processes than, for example, pure charging of your cell phone. To reveal these paths and trends, we shall first start with the understanding of what energy really is.

1.1.1 Concept of energy

The concept of energy was first introduced in mechanics by English philosopher Sir Isaac Newton when he hypothesized about kinetic and potential energies. However, the emergence of energy as a unifying concept in physics was not adopted until the middle of the 19th century and is considered one of the major achievements of the century [1].

https://doi.org/10.1016/B978-0-12-817012-0.00010-4

By definition [2,3], *energy* of a system is the capacity of this system to do work on another system, for example, to pull a weight against some force like gravity or friction. Work W is the consequence of the expenditure of energy and is defined as the product of a force F, acting on an object, and the distance d that the object moves, i.e., $W = Fd$. In this example, force F is assumed constant over time of the movement, and acts in the same direction as the movement of the object.

Energy is a scalar quantity that cannot be observed directly, but can be recorded and evaluated by indirect measures. While the absolute value of the energy of a system is difficult to measure, the energy change is relatively easy to evaluate [1]. The gravity force would pull any object down so one would require energy to lift this weight up against that force from the initial height h_1 to the desired height h_2, so that

$$W = mg(h_2 - h_1).$$

This means that this weight would accumulate some energy when lifted to any height. When pulling a weight over the surface or pumping air inside a vessel, one would require energy to perform this work, while generating some heat.

1.1.2 Forms of energy

Energy can neither be created nor destroyed, it can only be converted from one form to another. This statement is called *the first law of thermodynamics* and defines internal energy as a state function and provides a formal statement of the conservation of energy [1].

Energy comes in various forms; however, in general the total energy of a system can be represented as a combination of two types of energy:

- *Potential energy* reflects the potential (or possibility) of a system to have motion or to do work. This energy can have different nature and may be due to the position of the objects relative to others, stresses within the object, electric charge, etc.
- *Kinetic energy* is determined by the movement of the object or the composite motion. It is usually defined as the work needed to accelerate an object of a given mass from the rest state to its current velocity. In classical mechanics this energy is proportional to the mass of the object and the square of velocity, i.e.,

$$E_k = \frac{1}{2}mv^2.$$

These categories of energy have been introduced as a convenient way to describe the system and its behavior, as well as energy it may derive or consume from various sources. While these two forms of energy are sufficient to describe all categories of energy, it is still convenient to refer to a particular combination of potential and kinetic energies based on the nature of the physical or chemical phenomena. Some of these forms of energy that an object can possess as a measurable property include:

- *Mechanical energy* as the sum of macroscopic kinetic and potential energies;
- *Electrical energy* as potential energy caused by or stored in electric fields;
- *Magnetic energy* as a form of potential energy caused by or stored in magnetic fields;
- *Gravitational energy* as a form of potential energy caused by or stored in gravitational fields;
- *Chemical energy* as the potential energy stored within the chemical bonds;
- *Electromagnetic* or *radiant energy*, which is a form of potential energy stored in the fields generated by the propagated electromagnetic radiation, including light;
- *Nuclear energy* that binds nucleons to form the atomic nuclei. It can be released during nuclear reactions;
- *Thermal* as a type of kinetic energy on the microscopic level due to the motion of particles.

The above given list is not exhaustive and may include some other specific forms of energy used in particular areas of technology, physics, or other disciplines.

1.1.3 Sources of energy

When we "use" energy in our activities, we do not really consume it but rather transfer or convert it into another form that would bring us a benefit. Lighting up a bulb with electricity would lead to the energy transfer into light allowing our eyes to see, as well as producing some heat. Heat is a part of energy transferred to the surroundings to warm them up, which may not be always beneficial. Such dissipation of energy, which is not used for the purpose, requires much bigger initial supply. Eventually, energy consumption by all our devices and processes, similarly to the mentioned bulb, dissipates energy and finally wastes it. This means that we shall find such sources of energy, which are naturally concentrated and easily accessible for further transfer. On the other hand, in order to minimize wasteful dissipation of energy (like heat from a bulb), we shall make sure that the transfer process from this source is efficient, i.e., the energy we get from the natural source is fully utilized without or with minimal waste.

These concentrated and accessible natural sources of energy are usually referred to as *primary energy sources*. Primary energy is extracted or captured directly from the environment and can be of three types:

- *Nonrenewable*, which comprises energy stored in fossil fuels like coal, natural gas, crude oil, etc., and mineral (nuclear) fuel;
- *Renewable* that can be recovered by the Earth and is therefore considered infinite (relative to human history). These include solar radiation and its derivatives like hydro power in the form of falling and flowing water, winds, biomass, tidal energy (caused by the orbital movement of the Moon), and heat of the Earth;
- *Waste* of human activity which, similar to the fossil fuels, incorporates some chemical energy.

Remark 1.1. In some literature, waste is considered as a part of renewable energy. In this book, we shall treat waste as a separate source of energy as it is fully linked to human activity.

Each source of primary energy has its own *resource potential*. This theoretical value defines the physical upper limit of the energy available from a certain source of energy [4]. This value is used when considering the amount of resources available (either physically or theoretically) for utilizing it to produce secondary energy. Depending on the current rate of the energy transformation, i.e., the speed of converting one category of energy into another, one can estimate the number of years of powering the world with this resource under the assumption of the constant rate of consumption.

1.2. How to measure energy and power

Before going into further details of the energy and its conversion, it is essential to understand the ways how to measure energy. Energy measurement units will be required not only for the energy supply and demand calculations, but also for the comparison of various technologies, economic evaluations, and contractual agreements, making investment decisions and calculating the returns, as well as for understanding the implications of the human activity on the climate and the environment.

There are two major ways to express the values of energy. Firstly, energy values can be expressed in basic units used in physics — *joules*. The joule is a standard unit introduced in the International System of Units (SI) and is used to measure energy, work, and amount of heat. By definition, one joule is a unit of energy that quantifies the amount of work done on a system when an applied force of one newton is required to move an object one meter. For example, the amount of chemical energy stored in a barrel of crude oil is approximately 6 GJ. In other words, six billion joules of energy can potentially be extracted from a barrel of oil and then transferred into another useful form [5].

Similar to joule, there are several other units used in practice. *British thermal unit* (Btu) is the energy unit in the English system, which defines the amount of heat necessary to raise the temperature of one pound of water one degree Fahrenheit. The same approach is used to define a *calorie* (cal) that quantifies the amount of energy in the metric system needed to raise the temperature of 1 gram of water at 15°C by 1°C [2]. Of course, no matter how you define one unit of energy in various systems, you can still compare them.

Even though joule, British thermal unit, and calorie quantify an amount of energy, they sometimes lack transparency. Therefore in the energy industry, energy units can be also expressed in terms of the type of fuel they quantify. The petroleum industry measures energy by *tons of oil equivalent* (toe) or *barrels of oil equivalent* (boe), which define the amount of energy stored in one ton or a barrel of crude oil, respectively. A standard barrel of oil contains 42 US gallons, or 159 liters. There are approximately 7.3 barrels of oil in a ton, so approximately 41.9 GJ of energy are contained in one ton of oil [5]. These units are extremely useful when comparing various types of fuels adjusted to one of the major energy sources — the oil (see Table 1.1).

Similar to oil, *tons of coal equivalent* (tce) is a measure that is used to describe the energy in coal. The amount of energy in a tce can vary because of different coal types (due to their chemical composition), but the value of approximately 29 GJ per ton is accepted as an international standard [5]. Its comparison to other units is also given in Table 1.1.

In nuclear physics and in nuclear reactor engineering, a unit named *electron-volt* (or eV) is utilized. By definition, it determines the amount of energy gained (or lost) by the charge of a single electron moving across an electric potential difference of one volt. It is not an SI unit, and one eV is equal to approximately 1.6×10^{-19} J.

Table 1.1 Major energy measurement units utilized in energy industry.

Name of unit	Symbol	Conversion
Ton of oil equivalent	toe	41.868 GJ 10 Gcal 39.683 MBtu 7.143 boe
Barrel of oil equivalent	boe	5.862 GJ 5.551 MBtu 0.14 toe
Ton of coal equivalent	tce	29.308 GJ 7 Gcal 27.778 MBtu 0.7 toe

The above industry-specific units are often used for quantifying the overall amount of energy contained, transported or consumed. However, the consumption is usually expressed not in absolute numbers but in terms of the amount per period of time, or rate. *Rates* are a way of expressing how much energy an entity or facility is producing or consuming in a given amount of time. The rate at which the primary energy is converted to the secondary energy is called *power* [5]. More precisely, power is the energy exchange in time [2]. The unit called *watt* (W), which gives rise of one joule of energy per second, is the unit from the SI system that describes this rate. For example, a generator with the power of 100 W is capable of converting mechanical energy into electrical at an average rate of 100 joules per second. A large power plant generates electricity (converts mechanical energy into electrical) at a rate of 500–700 MW or 500–700 MJ per second [5]. Alternatively, in English units, power is in Btu/hr, which equals to 0.293 W.

Therefore power expressed in watts defines the *capacity* of the system to convert energy, i.e., capability of converting unit energy per unit time. It means that it is only the characteristic of the system and its capability. Consider, for example, a rotating machine that has the capacity of 1 kW so that it is capable to convert 1 kJ of energy per second. Assuming that this machine has worked for one hour (or 3600 seconds), we derive the total converted energy equal to

$$1\ \text{kW}\cdot\text{h} = 1000\ \text{W}\cdot 3600\ \text{s} = 3.6\ \text{MJ}.$$

Table 1.2 Energy conversion rates for major units.

	GJ	Gcal	Mbtu	toe	tce	MWh
1 GJ	1	0.239	0.948	0.0239	0.034	0.278
1 Gcal	4.187	1	3.968	0.1	0.143	1.163
1 MBtu	1.055	0.252	1	0.025	0.036	0.293
1 toe	41.868	10	39.683	1	1.429	11.63
1 tce	29.308	7	27.778	0.7	1	8.141
1 MWh	3.6	0.86	3.412	0.086	0.123	1

Source: calculated with the IEA unit converter: https://www.iea.org/statistics/resources/unitconverter/.

This number is usually referred to as the *kilowatt-hour*, which is another common energy unit. It simply describes how many kilowatts of capacity are used or converted in one hour.

Remark 1.2. The correct symbol for the kilowatt-hour is either kWh or kW·h, however, the latter is rarely used due to unnecessary complexity (dot can be omitted and kWh can be used instead). The notation kW/h is not correct.

The amount of various measures of energy may lead to the difficulties of reading and comparing the data coming from multiple sources, especially dealing with international data. As all the calculations and analysis have to be carried out in one system of units to ensure we compare the units of the same nature, there is a list of conversion factors readily available in Table 1.2 based on the IEA unit converter.

1.3. Basic principles of energy conversion

A large number of energy conversion processes take place in nature, so that various forms of energy are available for humankind. In our activity we require these various forms of energy: mechanical to do work and power machines, heat for industrial need and heating, radiant for lighting, etc. Man is capable of performing a number of additional energy conversion processes by means of various devices invented during the history of mankind. Such devices may be classified according to the type of construction used, the underlined physical or chemical principle, or according to the forms of energy appearing before and after the action of the device [6].

Generally, we can classify these energy converters into two broad categories:

Table 1.3 Examples of major energy conversion processes and technologies listed according to the initial energy form (first column) and the converted energy form. Some energy forms can be converted into multiple forms via a set of processes.

Initial form	**Converted form**				
	Chemical	**Radiant**	**Electrical**	**Mechanical**	**Heat**
Nuclear		Radiation		Kinetic energy of particles	Nuclear fission in a reactor[a]
Chemical	Chemical reactions	Combustion	Fuel cell, battery discharge		Combustion
Radiant	Photolysis, photosynthesis		Photovoltaic cells		Absorber
Electrical	Electrolysis, battery charging	Lamp, laser	Transformers, rectifiers, inverters	Electric motor	Resistance
Mechanical			Electric generator, MHD	Turbine	Friction
Heat			Thermionic, thermoelectric generators	Thermodynamic engines	Heat exchanger

Source: Adapted from [6], based on [7].

Some of the minor processes like piesoelectric effect or the release of radiation during heating are omitted.

[a] Heating is caused by the radiation and the kinetic energy of the particles inside the reactor.

- Devices that convert one form of energy into another; and
- Devices which change the characteristics of the same type of energy.

For a number of energy forms, Table 1.3 lists some examples of energy conversion processes or devices currently in use or contemplated, organized according to the energy form emerging after the conversion. In several cases more than one energy form will emerge as a result of the action of the device, i.e., heat in addition to one of the other energy forms listed. Many devices also perform a number of energy conversion steps, rather than the single ones given in the table [6].

The energy conversion scheme is therefore chosen according to the available source of energy and the needs, i.e., which type of energy is finally required as the result. In many cases this scheme incorporates several processes which end up as an energy conversion chain.

The efficiency with which a given conversion process can be carried out, i.e., the ratio between the output of the desired energy form and the energy input, depends on the physical and chemical laws governing the process [6]. An increase of the steps within the energy conversion chain would evidently contribute to the efficiency losses.

The power generation industry is focusing on converting various forms of energy, mostly primary energy, into electricity and heat as the end product consumed by the society. This process, according to Table 1.3, may not be always completed in one stage. The process flow mostly depends on the current level of technology, as well as economic matters and sociopolitical grounds. In certain cases this conversion can be undertaken in several ways, and the choice is usually driven by economic benefits. Therefore, the energy system is all about the proper conversion of energy into electricity and/or heat with the minimum possible waste at reasonable cost.

1.3.1 Chemical energy

Chemical energy is stored in various sources, for instance, coal, natural gas, and oil, which together are referred to as fossil fuels, as well as in biomass (e.g., wood or gases produced out of biomass), hydrogen, and other products of human activity: waste products, industrial by-products, etc.

1.3.1.1 Conversion of chemical energy into heat

Chemical substances like hydrocarbon have a certain amount of energy stored in their structure. During chemical reactions with oxygen (oxidation), hydrocarbons react to produce carbon dioxide and water with the release of energy in the form of heat and radiation. Usually referred to as *burning*, these chemical reactions are the fundamental conversion process of chemical energy into heat and radiation. In general, supplying common air is enough to provide the required level of oxygen for combustion, therefore this process has been known for many years and became one of the most studied and established. In many power generation process combustion is considered as the first stage, as schematically represented in Fig. 1.1.

The efficiency of the process depends on many factors, for example:

- Composition of the burnable material that defines which part of the compound can undergo reaction with oxygen;

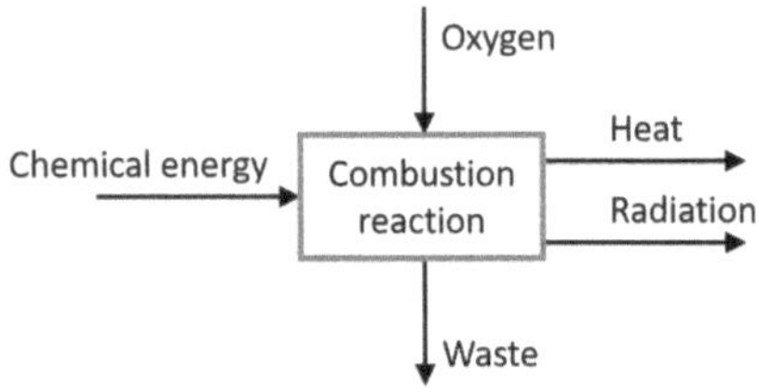

Figure 1.1 ***Conversion of chemical energy into heat.*** Chemical energy is converted into heat during the combustion process — an exothermal chemical reaction that requires a supply of oxygen and produces waste.

- Structure and shape of the material that defines which part of it would react: burning powder material will be more efficient than burning a single large piece that may not even be completely burnt;
- Concentration of oxygen within the incoming air. Pure oxygen would require less effort to maintain reaction;
- Amount of energy dissipated in the form of radiation.

1.3.1.2 Conversion of chemical energy into electricity

Apart from conversion into heat, chemical energy can be converted straight into electrical energy through an *electrochemical reaction*. This reaction takes place in an electrochemical cell — a device that works in two directions:

1. Generating electrical energy from chemical reaction, and
2. Facilitating chemical reaction through the introduction of electrical energy.

An electrochemical cell consists of two half-cells, separated by a semi-permeable membrane or by a salt bridge. We will consider an example of a typical galvanic cell with zinc and copper half-cells, as schematically represented in Fig. 1.2. Each half-cell consists of an electrode out of metal that is submerged in a solution:

- Zinc Zn half-cell has a solution of $ZnSO_4$,
- Copper Cu half-cell has a solution of Cu_2SO_4.

Both electrodes are connected by an external conductor. Because both metals are submerged into a solution, the chemical reaction would start and some of the cations (positively charged ions of metal) would dissolve in this solution through the reaction of oxidation, leaving negatively charged electrons in the metal. However, as metals differ, the speed of chemical reaction differs as well, and therefore one of the metals will have a "more negative" charge than the other (more electrons). In our case, Zn cations will dissolve

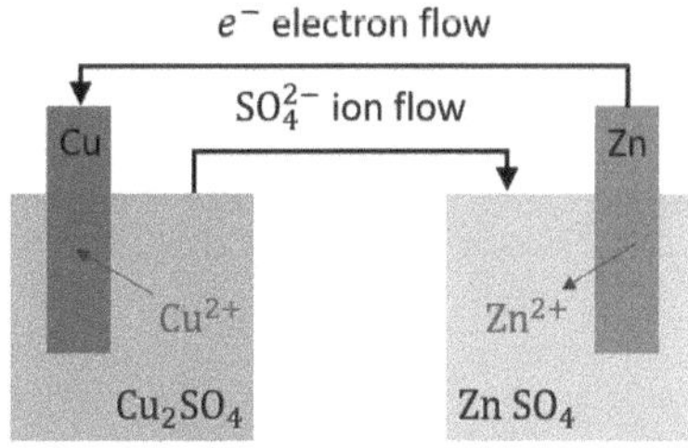

Figure 1.2 ***How a typical electrochemical cell works.*** **An electrochemical cell consists of two half-cells, separated by a semi-permeable membrane or by a salt bridge.**

faster, and therefore the excess of electrons from the zinc electrode would migrate through the external conductor to the copper one.

To compensate for this excess of electrons in the copper electrode, cations of Cu from the solution would be consumed through the chemical reaction called reduction $Cu^{2+} + 2e^- \rightarrow Cu$, leaving negatively charged anions SO_4^{2-} in the solution. Therefore, these anions would in turn migrate through the salt bridge to the other side to connect with Zn^{2+}. This would lead to the continuous flow of electrons through the external conductor or electric current, until all zinc dissolves in the solution or until all copper is absorbed from the solution through the reduction reaction.

A modern *battery* usually consists of one or more electrochemical cells connected in a required pattern. While they employ similar processes, the chemical composition differs.

The above described electrochemical reaction is limited in time and either has to be stopped to completely recycle the battery or recharge it applying electrical energy and initiate the reverse chemical reaction.

There is, however, a different opportunity which has been implemented in a device called a *fuel cell* — an electrochemical cell that converts the chemical energy from an externally supplied fuel into electricity through an electrochemical reaction. Contrary to batteries, fuel cells require continuous fuel and oxygen supply to maintain reaction.

While fuel cells come in many varieties, they still operate in the same way. Made out of anode and cathode separated with the electrolyte, two chemical reactions occur on the interfaces, as schematically represented in Fig. 1.3.

The anode, for example, made out of platinum, breaks down the hydrogen fuel into electrons and ions. This increases the amount of electrons in the anode and, being connected by an external conductor with the cathode, the excess electrons would flow through it. At the cathode the electrons

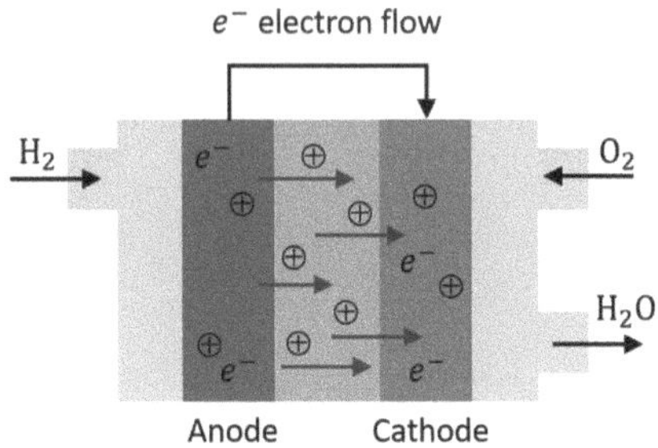

Figure 1.3 ***Schematic diagram of a hydrogen-powered fuel cell.*** Anode and cathode are separated with the electrolyte, and two chemical reactions occur on the interfaces.

and positively charged hydrogen ions that migrate through the electrolyte to compensate for the excess of electrons on the cathode would combine with oxygen to form water and be removed from the system to maintain the process.

As the whole process sounds pretty simple, it is still subject to multiple technological difficulties; however, some designs have already entered mass production. The implementation of fuel cells will be discussed in detail in Chapter 23.

1.3.2 Nuclear energy

Nuclear energy is the internal energy of the nuclear kernels, which is released under certain nuclear transformations. In nature, some heavy elements can emit radioactivity, when the naturally unstable atomic nuclei lose energy by emitting radiation like alpha particles, beta particles, neutrons or gamma rays or electrons. Therefore, a material containing such unstable nuclei is called radioactive. The same process can be initiated artificially so that the radioactive material will release energy in a controlled manner. Such man-made conversion process takes place within the special devices called nuclear reactors, where a specially treated amount of radioactive material can initiate and maintain a controllable nuclear fission reaction that would release large amounts of energy.

This nuclear energy, which comes in the form of kinetic energy of the fission products and fragments, and electromagnetic radiation, is converted within the reactor to heat as the particles and radiation collide with the atoms of the materials of the reactor and the medium that is used to absorb and collect energy and transport it out of the reactor. This process is schematically illustrated in Fig. 1.4.

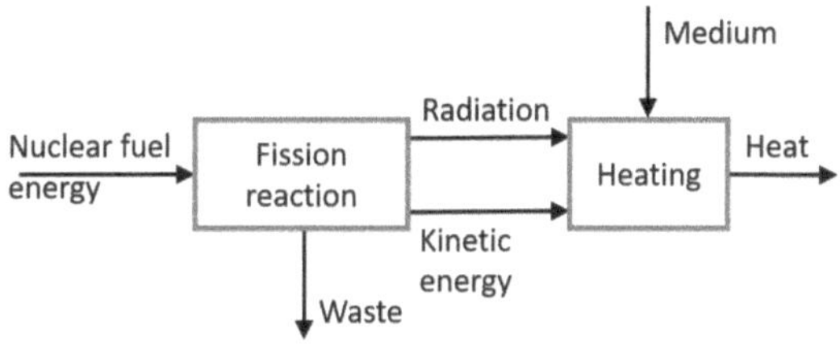

Figure 1.4 ***Conversion of nuclear energy.*** The energy of nuclear fuel is converted through the fission reaction into radiant energy and kinetic energy of the particles, which in turn are converted into heat within a reactor. During this process, a by-product of the nuclear reaction is waste.

Nuclear energy conversion is a complicated process that requires a lot of attention due to its nature and, consequently, high level of the implemented security: radiation is invisible and thus cannot be immediately felt by a human in case of an accident compared to heat or electric shock, for example.

1.3.3 Mechanical energy

Mechanical energy can be found in nature in multiple forms, for instance, one can mention:

- Pressurized hot gases or other fluids which have enough potential energy that can be turned into kinetic energy of the flow;
- Flowing water of the streams and rivers that have enough energy stored in the running water;
- Wind, or flowing air, which has the same logic as flowing water;
- Waves, tides, ocean streams, etc.

In order to capture this type of energy, special rotating devices called turbines are used. These turbines capture the mechanical energy of the flow and convert it to another type of mechanical energy in the form of rotation. Evidently, due to friction during rotation, some energy is dissipated in the form of heat. Moreover, turbines usually do not capture all mechanical energy of the flow which also negatively contributes to the overall process efficiency.

Next, rotation of the turbine can then be used in multiple ways (see Fig. 1.5):

- Run other machinery, for example, pumps, and do some work;
- Run a machine called electrical generator which converts mechanical energy of the rotation to the electrical energy. The physics of the pro-

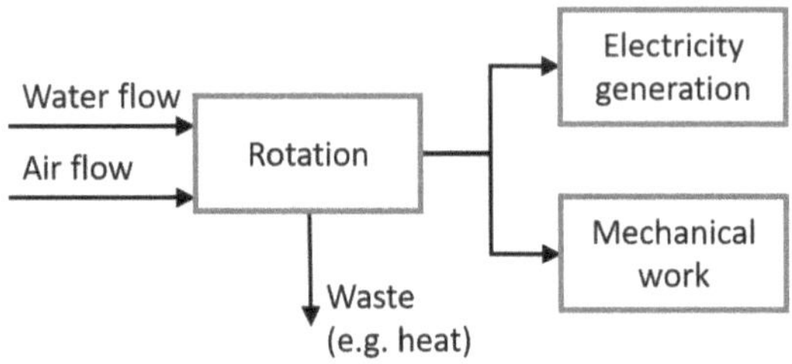

Figure 1.5 ***Conversion of mechanical energy.*** Conversion of mechanical energy of the fluid flow (e.g. water or air) is usually done through the mechanical energy of the rotating shaft, which can be further employed for doing work or converted into electrical energy.

cess involves rotating magnetic field that induces current, which is then transferred to the consumer.

It is important to mention that energy of the rotating equipment (wheel or shaft) is an important step within both conventional and emerging energy conversion chains. In many cases this is the only way to transform mechanical energy into electricity through coupling the shaft of the turbine to the electrical generator.

Moreover, mechanical energy can be also converted into potential energy (increase of pressure) and thermal energy (increase of temperature) of the medium when this medium (usually gas) is compressed within special equipment called compressor.

1.3.4 Radiant energy

One of the most common natural sources of radiant energy is the Sun: its electromagnetic radiation comes to the Earth and can be converted in several forms: thermal, chemical, and electrical. While the naturally ongoing energy conversion processes transform solar radiation into heat and chemical energy, humankind is focused on converting radiant energy into heat and electricity.

1.3.4.1 Natural conversion to heat

While the solar radiation propagates through the atmosphere and reaches the surface of the Earth, part of this energy is absorbed by the environment it goes through. The photons, or the electromagnetic waves, collide with the particles of the medium and therefore transfer their energy to the medium, which increases their temperature.

This manages the number of critical processes of the Earth. Unevenly heating up the atmosphere and the surface, due to the rotation of the planet, the inclination of the axis, and the ellipticity of its orbit, allows:

- Having comfortable conditions for the existence of life;
- Having variable temperature zones that lead to pressure differences, which further lead to winds and ocean currents;
- Maintaining the water cycle of the Earth.

1.3.4.2 Natural conversion to chemical energy

Natural conversion of radiant energy into chemical energy takes place during natural photochemical reactions. These reactions are vitally important for the whole existence of life on our planet, and the two main examples include:

- Conversion of the molecules of oxygen O_2 into ozone O_3 in the upper layers of the atmosphere;
- Photosynthesis — a process used by plants and some other organisms to convert radiant energy (simply light) and some other elements, for instance, water and carbon dioxide, into chemical energy stored in carbohydrate molecules, e.g., sugars. These molecules can be further used as a source of energy in energy conversion chains.

1.3.4.3 Conversion into electricity

Conversion of radiant energy into electricity takes place within the devices called photovoltaic (also known as PV) or solar cells. These devices use the physical phenomenon known as photovoltaic effect, when light is absorbed by the material, causing excitation of an electron or other charge carrier to a higher-energy state.

The photons of the light hit the solar cell and are absorbed by semiconducting material, e.g., silicon, as schematically illustrated in Fig. 1.6. They excite electrons by giving them photon's energy and "knock" them from their orbits within atoms to a higher-energy state, where they are free to move around the semiconducting material. As electron leaves the network of bonds, there is a positively charged "hole" left there, which can be filled by a neighboring electron letting the hole propagate through the material. In other words, the absorbed photon creates a pair of a free electron and a hole.

We can simplify the semiconducting material to have a layer of n-type silicon (negative charge with high concentration of free electrons) which is put in close contact with the area of p-type silicon (positive charge with a

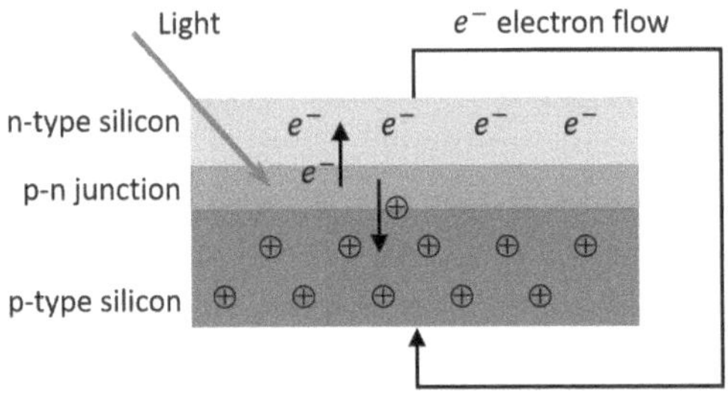

Figure 1.6 ***Conversion of radiant energy into electricity.*** Special semiconducting material absorbs radiant energy, which excites electrons and pulls them through the outer conductor to create current.

much higher concentration of holes) so that a diffusion of electrons from the n-type to p-type occurs, and holes diffuse in the opposite direction. When electrons diffuse across the p–n junction, they recombine with the holes, so that the interface layer of the n-area has the positive charge, and the interface layer of p-area becomes negative. They will create the electric field called depletion zone or energy barrier.

While the outer n-type area is exposed to the light, photons are continuously absorbed by the n-type of silicon and excite electrons to make them "free". Therefore the electron and hole separate because by the positive space charge within the depletion zone an electron will be pulled into the n-layer, while the hole would go to the p-area. In other words, the excess of electrons cannot flow through the depletion zone, as there is an energy barrier they cannot penetrate through. Therefore, these electrons have to move to the positively charged p-area (with excess of holes) through the outer conductor causing an electric current to flow. This phenomenon is called photovoltaic effect.

1.3.4.4 Conversion to heat

Conversion of radiant energy into heat has the same nature as that occurring in the atmosphere and is also similar to the nuclear energy conversion process. The electromagnetic radiation collides with the atoms of a material and increases their energy, which results in a temperature increase.

However, to increase the efficiency, the radiant energy is collected and guided with the help of mirrors, which allow concentrating a larger amount of radiation in one place to heat up the required material, for instance, evaporate fluid.

1.3.5 Heat

Thermal energy, or heat, is contained in the medium, which can be solid, liquid or gaseous and can be further transferred to another medium through the heat conduction process. We shall name three major processes:

- Conduction or diffusion takes place, when two objects are in close contact, so that the thermal energy flows from the object with a higher temperature to that with lower.
- Convection is the heat transfer between the object and the fluid, which flows around the object and either absorbs its energy (if the object has higher temperature) or increases it.
- Radiation is the transfer of energy by electromagnetic waves, similar to nuclear or radiant energy.

In most power applications a medium (either liquid or gaseous) is required to transport this energy and transfer it to another medium. While both of them are not likely to mix, the process takes place in devices called heat exchanges: for instance, the heating fluid flows inside a tube and heats it up, while the other heat absorbing fluid flows around that tube.

Heating up liquid or gas, i.e., increasing its temperature, would lead to the increase of pressure (if the volume stays constant), that is, $PV \sim T$, so that thermal energy is transformed into mechanical energy due to the pressure of that fluid.

1.4. Electricity as the energy carrier

When we convert energy from one form to another for purposes of convenience to meet some desired end use, we can consider some intermediate and easy-to-handle energy form to serve as the *energy carrier*.

Most of the energy conversion processes implemented nowadays are aimed to deliver electric energy. This energy is the category of energy that is transported via the motion of electric charges, e.g., electrons. It is usually produced by electromagnetic induction in an electric generator that converts mechanical energy of rotation to create moving electromagnetic field, or via some other less popular energy conversion processes discussed earlier: electrochemical reactions in fuel cells or photovoltaic process. All these processes basically create a difference in electrical voltage along a conductor which in turn causes electrons to flow and form electric current.

Technologically, conversion of other forms of energy into electricity is a real-time process, meaning that electricity can be produced for the required level of consumption but is difficult to be stored in large quantities at the present level of development. Except in comparatively small amounts, technologies for storing electricity are cumbersome and often costly. Rather, electricity can be stored via conversion to mechanical work, which can then be stored as potential energy [7]. Therefore, managing a continuous balance of electricity generation and its consumption is vitally important and becomes one of the primary tasks of an energy system.

Despite the difficulty to store electric energy, it still has a number of extremely valuable advantages that put it first in terms of efficiency and usability. For example:

- Electricity can be converted from all primary and secondary energy sources, including renewable ones;
- Electricity can be transported rather easily and over long distances. While it is weightless, it can be easily transmitted and distributed to all types of consumer through a system of electric conductors;
- Electricity itself is clean and does not damage the environment on the global scale even in case of accidents;
- There exists a variety of established and flexible technologies to change parameters of electricity for various use: direct and alternating current, frequency, voltage, current, etc.;
- Electric energy is easy to convert into other forms of energy via established and efficient technologies, for example, electric motors of various sizes and capacities;
- Direct use of electricity allowed multiple applications with electronics being the most significant.

It is not much of an exaggeration to say that electric energy, or simply electricity, has become an overwhelming source to provide energy for humankind activities. While it plays now and would most probably continue playing a major role as the most flexible and convenient energy carrier, it still does require further research and development in the way to efficiently store it in larger quantities, which are required for industrial purposes.

1.5. Motivation and structure of this book

Energy is the fundamental driver of human development. Without it, we would not be able to feed ourselves, protect against severe climate conditions, travel, extract natural resources, use them and manufacture goods. Basically all human activities require various forms of energy, and this trend would only grow along the way of humankind development.

This therefore poses a number of questions, though obvious but still very important, answering which would help us to ensure the ongoing development of humankind:

- What are the accessible sources of energy we can consume nowadays and in the coming future? Is there enough energy for us and the next generations, if we continue to increase our rate of consumption?
- What are the major types of energy we can utilize in our everyday activity, including living, industry and transportation? Why electricity is one of the most useful for us in terms of efficient production, transmission and consumption?
- What is the level of technology we have at the moment to generate electricity? What is the level of efficiency and what are the ways to increase it?
- Evidently, energy conversion drives the overall development. But does it bring any harm to the humankind or the environment? How can we mitigate these challenges and ensure we do not do irreversible damage?
- What will be our future? How long can we maintain the situation with the current energy conversion approach as we are aiming to grow?

Having this list of questions in mind, this book is therefore organized in seven parts. The first one, starting with the introductory basics discussed in the current chapter, will continue by focusing on the detailed discussion of the three main forms of energy readily available on the Earth. Secondly, we discover major planetary cycles which impact the energy resources available on the Earth, with their estimate and forecast. Furthermore, we pay attention to the energy system and its components, mostly concentrating on power generation and its role for humankind.

The second part is fully devoted to a complicated topic of sustainable development — the approach we have to decide on to maintain our growth without compromising on the environment and development. It first starts with the definition and principles of sustainability, its boundaries and limitations, followed by the sustainable energy system and the impact of traditional energy systems of the environment and human life.

Parts three to five cover conventional and established technologies of converting fossil fuels, energy of water, and nuclear energy into electricity. Thus, we cover in detail historical development, technology, major equipment, and project execution in detail, as it serves three goals: (a) discussing history of the technological development and providing a fact base, required for understanding emerging technologies; (b) understanding the impact of most widespread technologies, the ways to mitigate the negative consequences and at which cost; and (c) showing that these technologies still have to be considered as a bridge towards sustainable development.

Consequently, the sixth part focuses on the emerging technologies, which are usually considered as the major drivers of a sustainable energy system. These include conversion of wind, solar, biomass, geothermal energy, and the energy production from waste. Separately, we discuss some other forms of energy like tidal, wave, fuel cell technology, etc., and the potential of their conversion.

And finally, in the seventh part we shall be able to consolidate the complete picture, reviewing various technologies on the way towards sustainable development. Followed by the neighboring areas like energy storage, transmission and distribution, we finalize this book with the open discussion of the sustainable energy system of the future: key challenges, growth drivers, and choice of options.

References

[1] Dincer I, Rosen M. Exergy: energy, environment and sustainable development. Elsevier Science. ISBN 9780080970905, 2012.

[2] Demirel Y. Energy: production, conversion, storage, conservation, and coupling. Green energy and technology. London: Springer. ISBN 9781447123729, 2016.

[3] Dunlap R. Sustainable energy. Cengage Learning. ISBN 9781305156210, 2014. 672 pp.

[4] Teske S, European Renewable Energy Council, Greenpeace International. Energy [r]evolution. A sustainable world energy outlook 2015. Greenpeace International; 2015.

[5] Moan J, Smith Z. Energy use worldwide: a reference handbook. Contemporary world issues. ABC-CLIO; 2007.

[6] Sørensen B. Renewable energy conversion, transmission, and storage. Elsevier Science. ISBN 9780080559049, 2007.

[7] Tester J, Drake E, Driscoll M, Golay M, Peters W. Sustainable energy: choosing among options. MIT Press. ISBN 9780262304634, 2012.

CHAPTER TWO

Evaluation of energy resources

Contents

2.1. Fundamental energy sources

The formation and evolution of the Solar system dates back to roughly 4.8 billion years when a gravitational collapse of a part of the molecular cloud made it possible to form a new star — the Sun. While most of the matter collected in the center to shape the star, the rest flattened into the protoplanetary disc to become the rest of the Solar system.

The Earth is among those Solar system objects that originated from the gravitational collapse, which is estimated to have happened about 4.5 billion years ago. Due to the differences in density, heavier elements (e.g., those containing iron) tended to propagate towards the kernel because of gravitational forces of that still extremely hot planet which could melt metals and minerals. This is assumed to result in the multilayer inner structure of the planet, which led to the formation of the magnetic field that protects

https://doi.org/10.1016/B978-0-12-817012-0.00011-6

the planet from solar radiation. The atmosphere at those times contained hydrogen, helium, and a large amount of carbon dioxide.

The unique combination of various conditions — solar radiation, protective magnetic field, concentration of carbon dioxide that protected the water on the surface — made it possible for the organic molecules to appear, while the development of photosynthesis mechanism filled the atmosphere with oxygen, hence providing more favorable conditions for the development of life during billions of years.

Importantly, the Earth has always been exposed to the natural source of energy — the Sun. However, apart from the solar energy, the Earth itself possess energy that propagates in the form of heat from the inside of the planet, which is still extremely hot. And also the gravitational forces of the Solar system do impact the planet and result in some natural effects. Having said that, these three sources of energy — solar radiation, heat from the Earth's core, and the gravitational interaction — can be considered as *fundamental energy sources* that would stay forever within the scale of humankind history.

While gravitational energy and Earth's geothermal energy can already be considered as a source of "free" and readily available energy, solar radiation can be understood in two ways:

1. It is the source of solar energy, which can be captured and used; and
2. It is the initiator to many other sources of energy, where the conversion takes place naturally through major planetary cycles (water and atmospheric circulation) and life processes (especially the biomass, which is due to the natural cycles as well, especially of carbon and nitrogen).

As a consequence of the solar radiation, which has been the "feeder" of organic life, especially plants, millions of years have converted this biomass to what is now known as fossil fuels — hydrocarbons stored in the Earth's crust. Therefore they can be well considered as "fossil" solar energy, still being the result of the fundamental energy source.

Keeping this logic in mind, we shall draw our attention first to the fundamental sources of energy — solar, geothermal, and tidal, and then go through the major cycles to understand how the planet lives under the impact of the fundamental energy. And finally, we consider fossil energy and minerals which can initiate the nuclear decay process with considerable energy release.

2.1.1 Solar radiation

The Sun is the nearest star, and it is agreed by the scientists to be a big plasma ball composed primarily of more than 90% hydrogen, 8% of helium, and small amounts of other elements. While the temperature is high, kinetic energy of nuclei and electrons is large, and the nuclear fusion process takes place. The Sun is a star with an estimated age of 4.5×10^9 years and will most likely continue for another 5×10^9 years before starting the next phase of evolution, the burning of helium. At that point, the Sun will expand and become larger than the orbit of the Earth [1,2].

The Sun makes radiation by converting mass into energy through the reaction of smaller elements to make heavier ones, or, in other words, fusion occurs. When the lighter hydrogen nuclei (or protons, ^1H) are fused together to make a helium nucleus (sometimes called an alpha particle, ^4He), a slight amount of mass is lost which is converted to energy. The mass loss is incredibly small for each reaction, of order 5×10^{-28} kg, however, it produces a huge amount of energy, 6.5×10^{14} J/kg [1].

Most of the reactions occur in the core which has a radius of about one-quarter of the Sun's external radius, yet contains 40% of its mass. The core provides 90% of the energy making it the hottest part, due to its extreme conditions, with a temperature estimated to be over 10^7 K [1].

The high temperature dissociates hydrogen into its individual components, ^1H, and an electron, e^-, accounting for why it is nuclei reacting rather than atoms. Furthermore, it provides enough kinetic energy for two protons to get close enough allowing strong interactions to hold them together. Subsequently, when they successfully stick together, one of the protons changes into a neutron.

The sequence of several reactions is called the proton–proton cycle and can be written down in the simplified form [1] as

$$\begin{array}{ccccccc} 4\,^1H & + & 2\,e^- & \rightarrow & ^4He & + & 2\,\nu + 6\,\gamma \\ 6.693 \times 10^{-27}\ \text{kg} & + & 1.8 \times 10^{-30}\ \text{kg} & \rightarrow & 6.645 \times 10^{-27}\ \text{kg} & + & 0\ \text{kg} + 0\ \text{kg} \end{array}$$

Summing the mass over the complete reaction, one finds a very small mass loss of -4.8×10^{-29} kg, which is converted into energy and eventually directed out of the Sun. Even though this mass is small, according to the relation between mass and energy, $E = mc^2$, each reaction produces 4.5×10^{-12} J of energy, which on a per mass of reactants basis is approximately 6.5×10^{14} J/kg.

The Sun has an internal control system that regulates the fusion reactions [1], and its energy is transferred to the outer surface of the Sun by

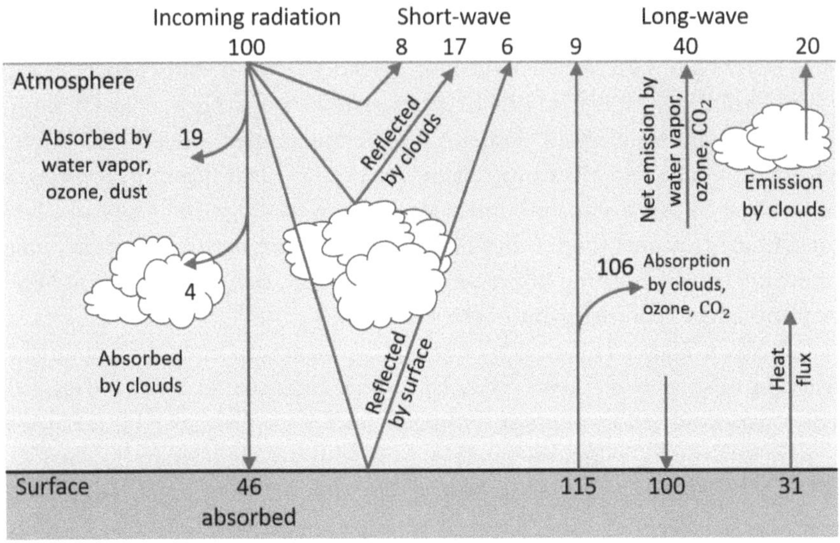

Figure 2.1 ***Global solar radiation balance.*** Schematic diagram of the mean global energy balance, in % of solar radiation. *Source: Based on [1,5,4].*

convection, and in the region beyond 0.7 of the Sun radius convection dominates the heat transfer. From the outer surface of the Sun, energy is radiated into space in the form of heat (electromagnetic waves) and light (photons) [3].

The distribution of the solar energy flux that intercepts the Earth is a strong function of the wavelength of the incident light. Because of variations in the absorption and reflection characteristics of different molecules contained in the Earth's atmosphere, the distribution changes from the top of the atmosphere to the Earth's surface. For example, most of the short wavelength with $\lambda < 0.4$ micron, or ultraviolet radiation, is absorbed by oxygen, ozone, and nitrogen in the upper layers of the atmosphere, while H_2O and CO_2 capture a good portion of the longer wavelength radiation with $\lambda > 0.6$ micron in the visible and infrared region [4].

This energy flux, or total *insolation* that strikes the top of the Earth's atmosphere, is referred to as the *solar constant* and has the value of 1367 W/m^2 [1,3,4]. However, this energy does not go directly to the surface, as schematically represented in Fig. 2.1. A certain portion of the short wavelet energy is reflected by air, clouds, and surface, but the largest portion of energy is absorbed by the clouds, atmosphere (water vapor, ozone layer, and dust), and the ocean and land surface. Long wavelet radiation goes directly through the atmosphere to be absorbed by the surface.

Much of the radiant energy that is emitted by our planet to the outer space comes from surface of our planet (partly due to the natural heat from the core), and emitted back by the atmosphere and the clouds.

Therefore, depending on the time of the day, season, and local weather conditions, the amount of insolation that actually reaches the surface will vary from essentially 0 to about 1050 W/m^2. In other words, while the estimated maximum solar flux at the surface is about 1 kilowatt per square meter, the annual average for a given point can be as low as 0.1–0.3 kilowatts per square meter, depending on location [6]. Even with all losses, the amount of solar energy that reaches the Earth and can potentially be captured is significant and ranges from as low as 1575 EJ[1] to 49837 EJ [6].

The solar energy potential that is available on the surface is more than sufficient to meet current and projected energy uses for the next hundred years. Thus the contribution of solar energy to global energy supplies will not be limited by resource availability. Rather, three major factors will determine the extent to which solar energy may be used in the long run [6]: the availability of

- Efficient and low-cost technologies to convert solar energy into electricity or other energy carrier;
- Effective energy storage technologies;
- High efficiency end-use technologies powered by electricity.

2.1.2 Geothermal energy

The structure of the Earth is studied by the science named physical geology. According to the current level of knowledge and the results of the research, the *core*, *mantle*, and *crust* constitute the three main layers of the Earth [7] as schematically shown in Fig. 2.2. These three layers can be further subdivided into the following layers [7]:

- *Continental crust* (0–75 km), also called the *lithosphere*, is the outer layer that forms the surface of the Earth. Depending on the location, the depth of the crust may vary, and the temperature of the lower boundary is expected at the level of approximately 1100°C.
- *Oceanic crust* (0–10 km) covers the ocean floor and is usually much thinner than the continental crust.

[1] 1 Exajoule equals to 10^{18} joules.

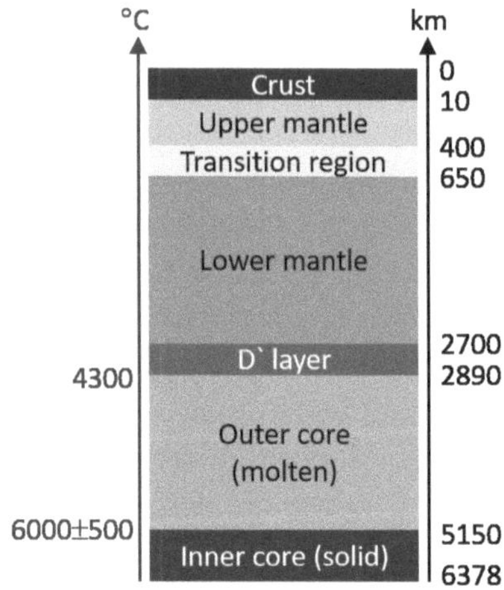

Figure 2.2 ***The interior structure of the Earth.*** The interior structure of the Earth consists of several layers, with the approximate depths being only indicative numbers due to the relative nature of research.

- *Upper mantle* (10–400 km) is still considered to be solid with partially molten areas.
- *Transition region* (400–650 km) is also called *mesosphere* or sometimes the *fertile layer*.
- *Lower mantle* (650–2890 km) is probably composed of silicon, magnesium, and oxygen with some amounts of metals.
- *D' layer* (2700–2890), called a d-prime, is a thin layer that might differ from the mantle chemically as seismic data suggests.
- *Outer core* (2890–5150 km) is very hot and composed of electrically conducting liquid made mainly of iron and nickel. This layer combines with Earth's rotation to create a dynamo effect that generates electric currents. This allows the Earth to have its magnetic field. Its temperature is assumed to range from 4300°C at the outer surface up to 6100°C near the inner core.
- *Inner core* (5150–6378 km) is made of solid iron and nickel and is suspended in the molten outer core, unattached to the mantle. Its surface temperature is estimated at 6000 ± 500°C.

Remark 2.1. Various sources suggest different thicknesses (depths) of layers and their temperatures. Evidently, the present research results can give only indicative values derived from indirect research methodologies like studies of the volcanoes, earthquakes, wave penetration, etc.

Even though the most evident tool to investigate the inner structure of the Earth is drilling, it is still limited to a few kilometers below due to increased geothermal gradient within the Earth's crust which defines the temperature increase. In simple words it means that with each kilometer of depth the temperature of the Earth increases by approximately 25–31°C [7, 8], forming the temperature gradient. The heat is mostly coming from the mantle, where it is thought to be primarily caused by the radioactive decay and release of energy by the disintegration of radioactive isotopes, primarily thorium Th-232, uranium U-235 and U-238, and potassium K-40 [7]. These processes can run because of the extreme pressures inside the mantle. Another source of the Earth's inner heat is the friction between layers that form it.

2.1.3 Tidal energy

Due to the gravitational forces, the Sun, planets, and their moons interact with each other in a way, so that the Solar system is stable at least on the scale of humankind history. The gravitational forces from the Sun, the Moon, and other planets impact mostly the Earth, with the most visible effect on the hydrosphere, which results in tides. The tidal systems that move the oceans are regular, reliable, and highly complex gravitational, centrifugal, and resonance-driven systems [9] and should not be confused with other oceanic flows such as ocean streams, which rely primarily on the solar energy and uneven heating of the surface.

The Moon is responsible for the largest proportion of the tidal forces applied to the Earth but it is not independent in its influence, since not only does the Sun also affect the tidal forces, but the Moon's own tidal influence is modified by the Sun's gravitational field. These cycles are themselves modified by the influence of the Sun, and occasionally other planets within our Solar system, depending on their relative position [9].

Tides are the rise and fall of sea levels, which are best seen at the shoreline. The timing and the amplitude of tides at the shoreline are influenced by:

- Alignment of the Sun and Moon as their major drivers;
- Pattern of tidal waves in the deep ocean;
- Amphidromic systems of the oceans — the pattern of the ocean movement around the points of zero tidal wave amplitude (amphidromic point) due to Coriolis effect (deflection of an object due to the Coriolis force which is an inertial force that seems to act on objects that are

in motion within a frame of reference that rotates with respect to an inertial frame);

- Bathymetric and topographical dimensions, which are the fundamental to the characteristics of the flow behavior.

Therefore, the tides may have various patterns: semi-diurnal (every 12 hours) with approximately equal high and low tides, diurnal (every 24 hours) with one high and one low tide, or with two uneven tides. This information is known and readily available for almost any point of the coastline, where tides are significant and may cause damage.

According to the report [10], the global estimates for global tidal stream energy potential is at approximately 3 TW, with 1 TW located in relatively shallow waters. However, due to geographical, technical, and environmental constraints, only a fraction of this could be captured in practical terms. In practice, suitable locations need mean spring peak tidal currents that are faster than 2–2.5 m/s to offer an energy density that allows for an economically viable project. Still, major tidal streams have been identified along the coastlines of every continent, making it a global, albeit site specific, resource.

2.2. Major planetary cycles

The Earth can be considered as a generally balanced system, which includes several *zones*:

- *Geosphere* is the solid part of the Earth;
- *Hydrosphere*, which comprises liquid water covering a major part of the surface of the planet;
- *Atmosphere* is the gaseous mantle of the Earth, which consists of several layers and preserves the planet's temperature regime;
- *Magnetosphere* is the region above the upper layers of the atmosphere that is defined by the extent of the Earth's magnetic field in space;
- *Pedosphere*, or soil, covers the lithosphere;
- *Cryosphere* comprises the frozen water (ice, glaciers);
- *Biosphere* is the worldwide sum of all ecosystems or the zone of life;
- *Anthroposphere* is that part of the environment that is made or modified by humans for use in human activities and human habitats.

The impact of the solar energy on these zones has made it possible to maintain life on the planet, which has been managed by a number of fundamental natural cycles: water, carbon, nitrogen, oxygen, and phosphor. These cycles, especially those of water, carbon, and nitrogen, are also im-

Table 2.1 Estimation of global water resources.

	Total water, %	Freshwater, %
Oceans and seas	96.5	
Glaciers and permanent snow	1.74	68.7
Saline groundwater	0.93	
Fresh groundwater	0.76	30.1
Ground ice and permafrost	0.022	0.86
Fresh lakes and rivers	0.007	0.26
Soil moisture and swamp water	0.002	0.08
Saline lakes	0.006	
Water captured in atmosphere	0.001	0.037
Water captured in biosphere	0.0001	0.003

Source: Aggregated and rounded from [2,4].

portant in the sense of power generation, and we further consider some of them in detail.

2.2.1 Water cycle

Water covers more than 70% of the Earth with around 96% of it concentrated in surface waters like oceans, seas, rivers, and lakes, excluding surface and floating ice. Only 2.5% is freshwater and most of that is in the polar ice caps. Still, the amount of freshwater from precipitation on the surface, much of which is stored water in lakes, reservoirs, and underground, is huge, with roughly 1250 km^3 accessible for human use (see Table 2.1) [2]. It was mostly due to water that made it possible to have life on our planet millions of years ago, as the first organic life appeared in the water. Being an important part of the life balance on Earth, water as a resource is extremely valuable and has to be carefully considered for our sustainable development.

The water cycle, as illustrated in Fig. 2.3, is a biogeochemical cycle and is due to the Sun radiation that heats up the Earth's crust and the water of the hydrosphere, which actually keeps the following cycle running:

- Being heated with solar radiation, the surface waters like oceans, seas, lakes, and rivers evaporate, and the water vapor flows upwards to the atmosphere.
- In the atmosphere, the vapor condenses into clouds, which are being transported by the atmospheric flows (also caused by the uneven heating with solar radiation).
- The clouds then initiate precipitation, e.g., rain or snow, which bring the water back to the surface waters or to the ground. Some of the

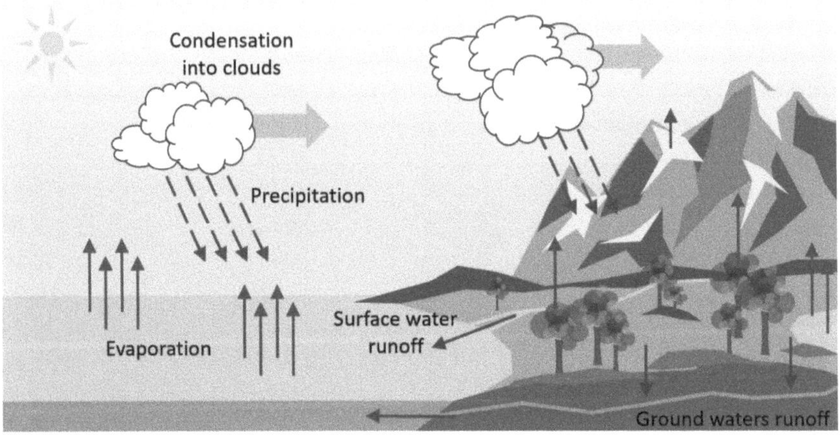

Figure 2.3 ***Simplified water cycle.*** Water cycle is driven by solar energy and provides ongoing circulation.

precipitation which is intercepted by biomass (e.g., stays on leaves) evaporates directly back to the atmosphere.

- The runoff incorporates a variety of ways the surface waters flow back to the larger reservoirs and evaporate, including seeping into the ground and underground flows.
- Part of the surface water infiltrates into the ground and becomes soil moisture or ground waters that feed the subsurface flows.
- Some water is directly consumed by the biosphere, namely the plants, that utilize it for photosynthesis to produce hydrocarbons. Later, this water comes back to the cycle through limited evaporation and after chemical decay of the dead biomass.

2.2.2 Carbon cycle

Carbon, chemically denoted as C, is the chemical element, which is the fundamental component of all organic matter on the Earth and hence life. The human body structure is based on it, and other animal and plant biomass consists predominantly of carbon [11]. Plants on land and algae in the oceans assimilate carbon in the form of carbon dioxide CO_2 from the atmosphere or hydrosphere, and transform it through the biochemical reaction called photosynthesis into energy-rich molecules such as sugars and starches, which initiates the natural cycle of carbon, which constantly changes its state through the metabolism of organisms and by natural chemical processes [11].

Therefore, the *carbon cycle* of the Earth is the biochemical exchange between the major reservoirs: atmosphere, soil (pedosphere), ocean (hydrosphere), burial in sediments (which are source of fossil fuels), and crust (lithosphere, where carbon is stored in rocks or limestone) [2].

The steps of carbon cycle are relatively simple and well-studied:

1. Green plants on land and algae in the seas and oceans assimilate carbon dioxide to use it during the photosynthesis reaction — a process to convert solar radiation into chemical energy stored in carbohydrate molecules (sugar), which are synthesized out of water and carbon dioxide:

$$6CO_2 + 6H_2O \xrightarrow{\text{Light}} C_6H_{12}O_6 + 6H_2O$$

Typically, oxygen is released as a by-product.

2. Biomass stored in plants is consumed by other organisms and follows the natural food chains. The absolute majority of the living creatures (including plants) have to consume oxygen to extract chemical energy stored in carbohydrate molecules and therefore release carbon dioxide as a byproduct of their *breathing* into the atmosphere.
3. Some carbon is buried in the soil and further in the lithosphere after death of plants or organisms. There, this dead organic matter can accumulate and stay for longer periods of time (under some circumstances dead biomass can turn into fossil fuels), or be converted by microorganisms into carbon dioxide to be released back to the atmosphere.
4. Carbon dioxide also dissolves directly from the atmosphere into the hydrosphere and, due to the air–sea gas exchange, is released back to the atmosphere.
5. Dissolved carbon dioxide can be consumed by the phytoplankton, which is basically using it as a product to convert solar radiation into organic matter and release oxygen. Dissolved oxygen is then consumed by sea living beings, which release carbon dioxide back and basically follow the same carbon cycle as plants and animals do.
6. Similar to land, some carbon is buried as dead matter on the ocean bed, which accumulates it.

According to the estimates [11,2], the atmosphere contains roughly 750 gigatons[2] (Gt) of carbon, soil and biomass comprise 1500–2300 Gt and 550 Gt, respectively, and the ocean around 38 000 Gt, which is more than 10 times as much carbon as the terrestrial biosphere, that is, all plant and

[2] 1 gigaton equals to 1 billion tons.

the underlying soils on our planet [11]. However, the Earth's crust itself is estimated to contain 10^8 Gt or 100 Pt of carbon.

While naturally the exchange between these reservoirs is fairly balanced, the rates differ. This may be influenced by the eruption of volcanoes, which may bring significant amount of carbon to the atmosphere and may lead to certain disbalance. However, the major impact on the carbon cycle is due to the human activity, which contributes to the increase of the atmospheric CO_2 by:

- Burning fossil fuels, which reservoir is estimated at 4000–10 000 Gt;
- Continuing deforestation (e.g., destroying rainforests);
- Damaging the soil through aggressive agricultural practices;
- Increasing production of materials (mainly cement).

The major absorbers of carbon (in the form of CO_2) are the oceans and biomass. The oceans absorb substantial amounts of carbon dioxide, and thereby consume a large portion of this greenhouse gas, which is released by human activity. This does not mean, however, that the problem can be ignored, because this process takes centuries and cannot prevent the consequences of climate change. Furthermore, it cannot be predicted how the marine biosphere will react to the uptake of additional CO_2 [11]. Biomass, similarly, is being damaged by the human activity through deforestation, and therefore cannot increase its capacity in absorbing carbon dioxide from the atmosphere.

With the CO_2 emissions from burning fossil fuels, cement production and changes in the land use at the level of more than 9 Gt per year (or 32.3 billion metric tons of CO_2 in 2012) [12], it brings imbalance to the carbon cycle, which leads to the increased amount of carbon dioxide in the atmosphere and, in response, to the increase of the Earth's temperature.

Remark 2.2. The ratio of carbon dioxide to carbon is 3.66, and so an emission of 1000 tons of CO_2 is equivalent to adding 366 tons of C to the atmosphere [11].

2.2.3 Nitrogen cycle

Gaseous nitrogen N_2 is the major component of the Earth's atmosphere, composing 78%, and is very stable. While nitrogen, together with carbon, oxygen, and hydrogen, is the "building brick" of organic matter, its capture from the atmosphere is an important step in the overall planetary life cycle. This process, also knows as *nitrogen fixation*, is naturally conducted by the nitrogen-fixing bacteria or atmospheric lightnings, while nitrogen is then used in a number of natural biochemical processes. Importantly, the biological growth is often limited by the availability of fixed nitrogen when climate and water supplies are suitable [4].

The freely living and symbiotic bacteria (living within the roots of the plants) play a dominant role in fixing natural nitrogen so that it can be further used in the food chains. With the investigation of the artificial nitrogen fixation, it is possible to produce nitrogen-based fertilizers. They are used to enrich depleted soil and increase agricultural productivity, however, overuse of fertilizers can poison soils and wash into groundwaters, spreading further contamination [4].

Apart from gaseous nitrogen N_2, the atmosphere also contains small quantities of N_2O (around 1.4 Gt) and NO_x (approximately 0.0006 Gt, including NO and NO_2). NO_x is a generic term for the nitrogen oxides: nitric oxide (NO) and nitrogen dioxide (NO_2). In natural quantities these substances do not directly affect the Earth's radiative balance, but they catalyze tropospheric ozone O_3 formation through a sequence of chemical reactions. However, anthropogenic emissions of additional NO_x can influence the environment.

NO_x is one of the major emissions from the combustion of fossil fuels, and the amounts of these compounds are regulated by the governments. The dominant sink of excess NO_x is through the atmospheric oxidation of NO_2 to form nitric acid HNO_3, which then comes through precipitation as acid rain.

2.2.4 Global wind circulation

By definition, *wind* is the flow of gases on a larger, planetary scale, which is caused by the difference in the atmospheric pressure. When there is a difference between the atmospheric pressure, air will move from the area with higher pressure to the lower pressure area trying to equalize them, which results in winds of various speeds. The pressure difference is due to uneven heating of the atmosphere by solar radiation and the Earth's rotation.

The lower layers of the air near the surface are heated up by the solar radiation, decrease in density and rise. Due to the symmetry of the equator, the colder air flows from the north and the south to the equator. As the air goes up and then poleward, it cools, becomes more dense and therefore descends at about 30th parallels, creating a high-pressure area. The descended air then flows towards the equator to close the loop which is called *Hadley cell*. Moreover, the poleward movement of the warm air in the upper part

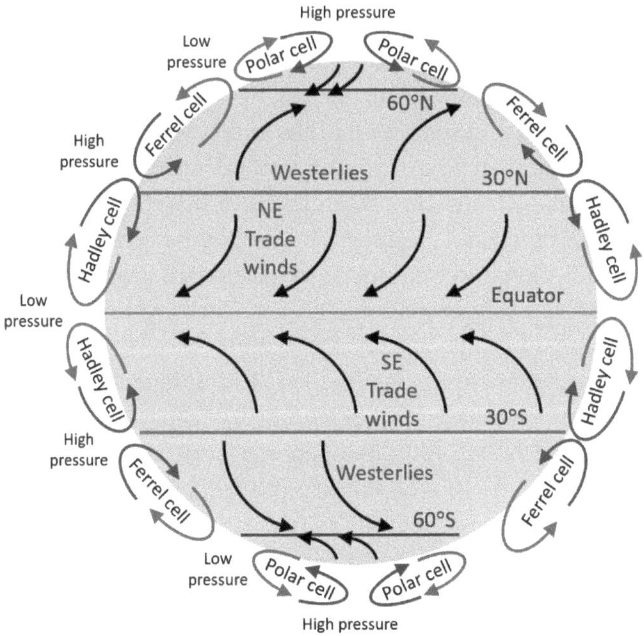

Figure 2.4 ***Global wind circulation schematic.*** Global wind circulation patter is formed by a number of atmospheric circulation cells and global wind systems, which are driven by uneven heating of the Sun radiation.

of the atmosphere deviates toward the east, caused by the Coriolis acceleration. These winds that flow to the west at the ground level in the *Hadley cell* are called the *trade winds* as illustrated in Fig. 2.4.

Secondly, near the ground level, air currents also flow from the poles towards the equator. Near the 60th parallels, the winds are heated by the solar radiation (which is limited in the pole areas) and rise up on the way to the South. In the upper layers they tend to flow back to the colder poles to close the circulation, which is called *polar cell*.

Finally, while the hot air rises at the 60th latitudes and the cold air descends at the 30th latitudes, a third circulation cell is formed in the area between the 60th and 30th parallels, which is called *Ferrel cell*. The western winds caused by this circulation are called *westerlies*.

2.2.5 Ocean currents

Ocean currents are movements of seawater generated by various forces: global wind flow pattern, Coriolis effect of the rotating Earth, temperature of water and salinity differences, cabbeling of water flows, surface heat, and freshwater fluxes. Their strength and direction is influenced by the tides, wave interaction, shoreline configuration, depth contours, and mutual interaction of the currents.

Most of the ocean currents move in broad circulatory patterns, called gyres, which flow around the subtropical heights. Because of the Coriolis effect, and a direction of flow around the cell of high pressure, oceanic gyres follow a clockwise direction in the Northern Hemisphere and counterclockwise in the Southern. Most surface currents do not cross the Equator, where the impact of the Coriolis effect is minimal [13].

The movement of the surface ocean currents influenced by wind is fairly intuitive. The trade winds drive currents near the equator in a westward flow of the ocean called the Equatorial Current. At the western margins of ocean basins along the eastern coastlines, its warm tropical waters are deflected poleward. As warm waters flow into higher latitudes, they move through regions of cooler waters flowing equatorward from cold currents [13]. Here the terms cold and warm mean that the current temperature differs from the adjacent water the current is flowing through.

In general, warm currents flow poleward along east coasts of continents as they carry tropical waters into the cooler waters of higher latitudes, like the Gulf Stream or Brazil Current. Cool currents, in turn, flow equatorward along west coasts of continents, e.g., the California Current or the Peru (Humboldt) Current. While the general circulation patterns of ocean currents are consistent throughout the year, their positions can respond to the seasonal changes in atmospheric heating and circulation [13].

Till recently the deep ocean waters were assumed to be static [14]. Contrary to the surface currents, in the deep ocean, the predominant driving force is the difference in density, caused by salinity and temperature variations, which gives the name *thermohaline circulation*, also known as the *ocean's conveyor belt* (see Fig. 2.5).

The mechanism of this water mix is assumed as follows [15–17]. The global conveyor belt is a combination of colder deep water currents and surface warmer currents [15]. It is assumed to start in the seas near Greenland and Norway, where the warm waters of the surface Gulf Stream release heat to the cold atmosphere, while part of the water freezes, leaving salt behind. This leads to the fact that the colder and denser water sinks to the ocean floor. The force of the sinking cold water pushes it to the south towards Antarctica, in a slow-moving cold deep water current. When it reaches

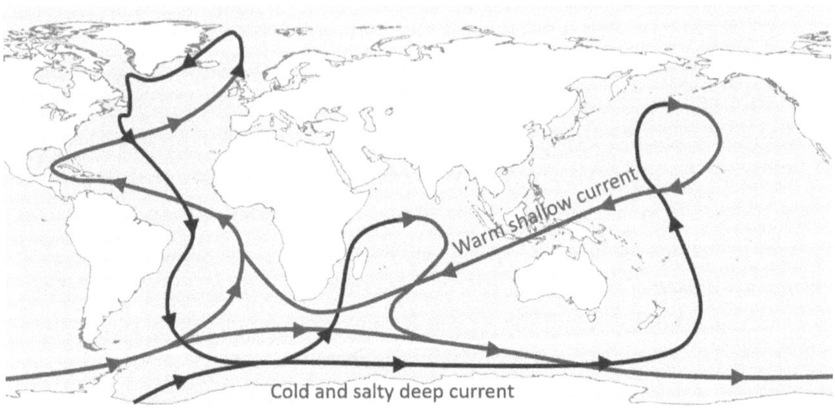

Figure 2.5 ***The global conveyor belt on a continuous-ocean map.*** The combination of colder deep water currents and surface warmer currents provides continuous global circulation of oceans.

Antarctica, the water flows east with the Antarctic Circumpolar Current, a massive and powerful current that circles the continent.

Parts of the Antarctic Circumpolar Current flow northward the Indian and Pacific Oceans. As the deep, cold waters travel through the oceans, they mix with warmer water, hence the water rises creating slow upwelling. Then these waters follow the surface currents: in the Pacific, they flow through the Indonesian islands into the Indian Ocean, around southern Africa, and back into the Atlantic Ocean, where they feed the Gulf Stream and therefore close the global cycle.

2.3. Biomass resources

Biomass is any living mass, both animal and vegetable, existing on the Earth, including its waste [18]. Consisting mostly of vegetation like trees, plants, and grass, biomass is one of the major components in the food chains and its contribution is twofold:

1. It provides chemical energy to the primary consumers, or herbivores — animals anatomically and physiologically adopted to eat vegetation. As these animals are heterotrophs, they consume organic carbon (chemical energy from biomass) and convert into other forms of energy for their living. Further, secondary or tertiary consumers in turn follow the food chain with the decomposers closing the cycle.

2. It consumes carbon dioxide, which is the by-product of the consumers, for its own energy conversion, and in most cases releases oxygen that animals require to convert energy.

Therefore, biomass can be considered as solar energy stored in chemical bonds of carbon–hydrogen as a result of metabolic activity or photosynthesis of different organisms [18]. It is important to emphasize that biomass requires surface land, soil, and water to maintain not only the natural cycles, but also the growing demand of humankind.

2.3.1 Land cover and use

The surface of the Earth can support only a limited amount of biomass, so that its growth rate and therefore energy production is also subject to many limitations: soil depletion, amount of fresh water available for growth, climatic conditions, and consumption rate. The greatest competing use for land comes from the demand for food, animal feed, and pasture. If technological improvements increased crop yields, or population decreased, or diets changed and the consumption of meat was reduced, then, at least in theory, surplus agricultural land would become available [19].

With forests covering the largest share of up to 40% of the Earth's surface (excluding snow, ice, and glaciers), there has been a net decrease of roughly 1.7% between 1999 and 2005, at an annual rate of change of 0.11% [20].

Major regional differences were found in the net rates of forest area change; only Asia and North America experienced gains in forest area, and all other regions saw net losses. South America had the highest net forest loss, losing some 3.3 Mha annually between 1990 and 2005. Africa had the second highest net forest loss of 1.6 Mha annually (for the same period). Europe, including the Russian Federation, had a net loss of 0.5 Mha annually and Oceania lost just under 0.1 million hectares annually. North America experienced a net gain in forest area of 0.2 Mha annually, while Asia had a net gain of 1.4 Mha annually between 1990 and 2005 [20].

There has been major deforestation due to conversion to farmland and pastures in the tropics, as well as inefficient wood production. In the Amazon Basin in Brazil, the estimated forest lost from 1970 to 2014 is around 76 Mha [21]. On the other hand, there were significant gains in forest area in the boreal (0.9 Mha annually) and subtropical (1.1 Mha annually) regions between 1990 and 2005.

Since the land is essentially fixed, the data on land use and cover, and how it changes over time, are important for understanding the dynamics.

Table 2.2 Estimated land cover and use for surface area of the Earth.

Type of land use	**Surface**	
	Gha	**%**
Agricultural	4900	36.3
— arable	1400	10.4
— permanent plantations	150	1.1
— pasture and meadows	3350	24.8
Forests	4000	39.6
Urban	600	4.4
Other	4000	29.6
Total	13 500	100

Source: Aggregated and rounded from [19,21].

As the population had grown, the amount of land converted to crops has grown accordingly [21].

In addition to the existing agricultural area, other types of land might be converted to either agricultural or energy crop use in the future. These areas include areas of marginal and degraded land, deforested and forested areas, and extensive grasslands, e.g., in Africa and Brazil [19]. Humans have transformed the Earth for permanent fields, plantations, and used conversion of grasslands and even forests to pastures for animals (see Table 2.2). These days, arable land plus permanent plantations for growing fruit, tea, coffee, cacao, and palms take around 11.5% of the ice-free surface of the Earth and the pastures and grasslands used by humans comprise another 24.8% of the area. The area for modern energy infrastructure: mines, refineries, power plants, etc., is estimated at 7.5 megahectares [21,19].

2.3.2 Biomass energy potential and its conversion

About 95 TW, or roughly 0.05% of the solar power that reaches the Earth (approx. 170 000 TW), is considered to be absorbed to produce photosynthesis [18].

The Earth's land plants assimilate about 10 Pmol, or 10×10^{15} mol, of atmospheric CO_2 each year, storing (at least temporarily) nearly 5000 EJ per year in sucrose, starch, and other carbohydrates. For marine organisms the total may be 7–8 Pmol of CO_2 per year. Those are values of gross primary production (GPP) or total photosynthesis. Annual global net primary production (NPP), defined as GPP minus the respiration of the photosynthetic organisms, may be nearly 5 Pmol of carbon on land (equivalent to as much as 2000 EJ a year) and perhaps 4 Pmol of carbon in the ocean. NPP is critical to present life on earth because it is the organic matter (and associated energy) potentially available to all non-photosynthetic organisms for use in support of their growth, maintenance, and reproduction. NPP also contains the bioenergy potentially available to society [22].

Actual solar energy conversion by photosynthesis and subsequent plant growth (biomass production) is less than 5% for productive plant communities. This low level of efficiency of photosynthesis is due to the nature of the reactions and the nature of the plants as they have to spend energy for living processes like aspiration and growth [22].

The chemical energy stored in biomass varies between 8 GJ/t for green wood and 16 GJ/t for dry wood, which are very low compared to some other energy sources. This energy of the biomass can be converted into other forms of energy in two major ways:

- By direct oxidation, or burning, it can be converted into heat and then further to electricity;
- By converting into other form of chemical energy: biogas and biofuels.

In many cases this process can be combined with food or other types of production, when biomass is first used within the initial process, while the waste product is then converted into fuel.

2.4. Valuation and forecast of fossil fuels

Fossil fuels belong to the primary energy sources. They are hydrocarbons in solid, liquid or gaseous form, formed from the remains of dead biomass by exposure to heat and pressure and preserved in sedimentary rocks over millions of years. Reacting with oxygen, fossil fuels would release certain amount of heat calculated per unit mass or volume of the fuel. In power plant, oxygen from the ambient air is used to maintain the combustion process.

From the economic point of view, the term *power generating fossil fuel* is used. It stands for any type of fossil fuel which is economically feasible for production of heat energy in large quantities. This assumes not only fuel cost itself with the associated cost of its extraction, refinement, and transportation, but also the overall fuel reserve, specification of the combustion process (amount of heat released and waste produced), influence on the equipment (wear, corrosion), cost of fuel treatment before combustion, greenhouse gases and solid particles emission levels, etc.

Power generating fossil fuels for power generation are almost readily available in nature, and their extraction process has been established during the last several decades. Most widely used fossil fuels for thermal power generation are obtained as solid, liquid, and gaseous matter. All of them consist of a large number of complex compounds of five major chemical elements [23]: carbon C, hydrogen H, oxygen O, nitrogen N, and sulfur S. Besides, they are normally combined with additional matter and moisture. This quantity of combustible part defines the quantity of heat produced by the complete combustion of a mass unit of fuel (e.g., 1 kg) and characterizes its heating value:

- *Higher heating value* (HHV) is the heat which includes the heat used for condensing all water and vapor produced within the combustion process;
- *Lower heating value* (LHV) is the heat of combustion excluding the heat used for condensing all water and vapor.

In different parts of the world, efficiencies are expressed on either heating value basis. Since LHV does not take into account the energy used to vaporize the water, efficiencies expressed in LHV are usually higher than HHV. The result is that, for virtually identical plant performance, efficiency may differ depending on the heating value used for calculations.

The properties of fossil fuels are extremely important to know, as they establish the basis for further understanding of the amount of primary energy available for conversion into electricity and heat. Such data plays a dominant role in economic evaluation of a power generating project like construction of a power plant and its future operation to decide on the profitability and payback period.

2.4.1 Solid fuels

Solid fossil fuels have been developed during millions of years through the process of partial biomass decomposition that had accumulated during the previous geological periods. They basically originate from the organic matter, namely trees and other plants, that were growing in a swampy type of the environment. This organic matter died and got buried as the years passed and turned into the matter called *peat*. Over the geological times peat was covered by rocks and stones, being exposed to high pressure, and therefore got buried deeper and deeper so that its temperature increased. Under such conditions, peat was eventually transformed into *lignite* and further into other forms of *coal*. And if you look at coal, you can see little parts of plant material still left inside. Due to various conditions, this process

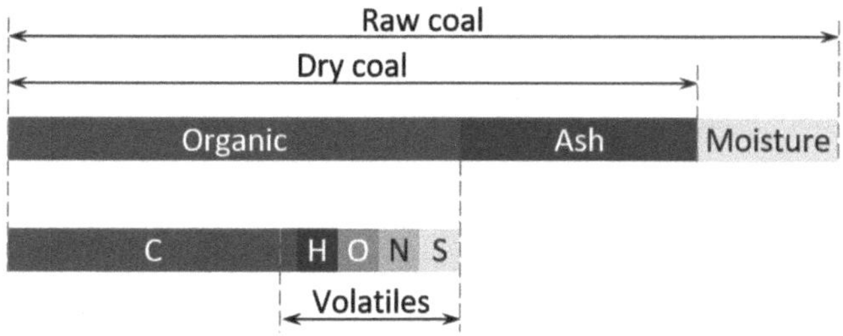

Figure 2.6 ***Schematic composition of coal fuel.*** Raw coal usually consists of dry coal and moisture. Dry coal, in turn, has a portion of ash and organic matter (carbon and volatiles like hydrogen, oxygen, nitrogen, and sulfur). Its exact composition depends on the origin of the fuel.

has gone through several stages leading to a different amount of moisture and volatiles' content compared to the burnable part of the fuel. By these means, there are several types of solid fossil fuels of various aging conditions available in nature that can be used in power generation:

- *Peat* can be considered as a precursor of coal from the earlier stage of its development. It is used as a major fossil fuel in some regions due to its low cost.
- *Brown coal* or *lignite* is the lowest rank of coal (formed from naturally compressed peat) which would burn relatively easily with a big amount of ash. It is one of the most common fuels for thermal power. Due to its high moisture content, it is not economically feasible to transport it and hence it is normally burned in power plants near the mines.
- *Bituminous coal* or *black coal* is of higher quality than lignite as it has lower ash content and hence can be burnt with a lower impact on the equipment.
- *Anthracite* is the highest grade of coal with the highest carbon content of more than 91.8% [24].

All of these solid fossil fuels are usually referred to as *coal*. Fig. 2.6 shows a general composition of a solid fossil fuel [24]. The raw fuel, besides the combustible organic substance, contains ash and moisture. Different types of coal are commonly differentiated according to their content of ash, moisture, and volatile components, and according to the characteristics of the combustible part [24]. It is the organic content that defines the capability of coal to release thermal energy during the chemical reaction; however, some of it is still wasted as a volatile component together with other substances.

Coal analysis is usually carried out in several steps. After heating it up to more than 100°C, both free moisture (located on the surface of the coal particles) and inherent moisture (bound inside the grain) evaporate, and the dry "moisture-free" solid fuel remains. When heating it further up to 850–900°C without oxygen access, it would release the volatile components in the form of vapor and gases, e.g., CO, H_2, CO_2, etc. [25]. The remaining part is called *char*. The combustible fraction of the char is named *fixed carbon*, while the rest is called *ash*. Note that the content of the fixed carbon is not the same as the C content of the fuel which, besides the fixed carbon, also includes the carbon in the volatile matter like carbon dioxide [24].

It is always better to burn the richest and the densest form of coal because it provides more energy per unit volume. But bituminous coal is perhaps the largest share of coal that is utilized for electricity generation around the world due to its abundance and proximity to the users.

2.4.1.1 Production of coal

There are two major methods of extracting coal from the ground, which mostly depend on the geology of the region, depth and quality of the coal seam, and some other environmental factors. While coal is mined, there are two versions of this process:

1. *Surface* or *strip mining* is economically feasible when the coal seams are near the surface and allow for the open pit extraction. This means that the surface is opened with the special heavy excavation machinery (draglines, shovels, and trucks) to get coal from the pit and transport it out to the processing and loading facility. Such mines can occupy big areas and damage soil, waters, and overall environment around, which would require a lot of effort for restoration and revegetation.
2. *Deep underground mining* is used for coal seams which are deep and hence not accessible through the open pit. There are various methods of underground mining using machinery, in many cases remotely controlled. However, this type of mining is by far more dangerous for the workers, for instance, due to unexpected behavior of the rock, methane release (leading to explosion) and coal dust in closed space.

Remark 2.3. Since the beginning of mining on an industrial scale, coal mining has been a very dirty and dangerous business with the health effects on miners and the neighboring community. Coal mines are usually huge and really scar the Earth and leave behind some very devastating environmental problems. These days, another approach for mining has been introduced, the *mountain top mining*: instead of tunneling, mining companies simply remove the mountaintop to get down to the coal seam. After scraping it off, they mine the coal like in an open pit. It requires huge displacement of earth and people in the area, causing devastating impact on the environment.

After being extracted from the ground, coal contains a lot of impurities like dirt, soil, and pieces of rock, and comes in fragments of different sizes. This coal is called *run-of mine* and has to be treated before transportation. This treatment is conducted at the *coal preparation plant*. This facility washes the coal to separate from contamination, and then crushes it into smaller graded pieces.

The fine graded coal is then transported via a conveyor or by a truck to the loading facility for further delivery to the consumer for final pretreatment and use mostly by rail or barge. For international transportation, vessels appear to be the cheapest means. Coal transportation can be quite expensive — in some instances it accounts for up to 70% of the delivered cost of coal [26].

2.4.1.2 Coal reserves

Coal is the only fossil fuel that is readily available throughout the whole world and distributed uniformly enough to support domestic power generation. According to [27], the total *proved reserves of coal* (that can be recovered in the future from known deposits under existing economic and operating conditions) are equal to 891.5 billion tons with the regional share presented in Fig. 2.7. Europe and Eurasia account for more than 34% of world coal reserves claiming the first place, with North America being second, while Africa and the Middle East region having less than 4% of coal reserves. The least amount of coal reserves is located in Central and Southern America. However, even these relatively small numbers can be enough for ensuring certain power security of local economy in case of need, though requiring the respective economies to search for other ways of power generation sources.

Summarizing, the total amount of coal reserves would last at least the next 110 years with the assumption of the same rate of its production. The top ten coal producers include such emerging countries as China (3748 Mt), India (668 Mt), Indonesia (471 Mt), Russia (334 Mt) and South Africa (253 Mt), with the total world coal production reaching a record level of 8104.8 Mt in 2014 (according to the BP Statistical Review). This

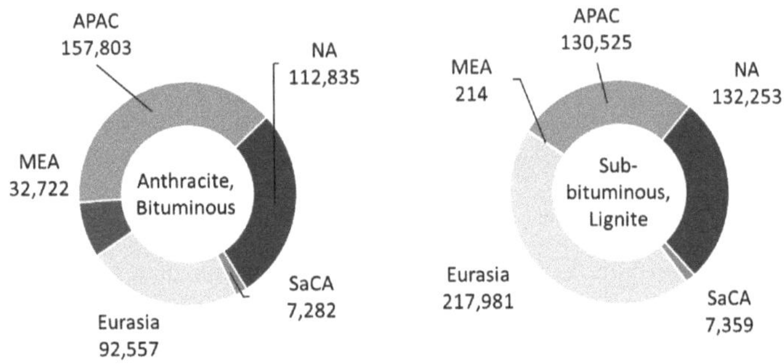

Figure 2.7 ***Total proved reserves of coal at the end of 2014.*** Total proven reserves of coal are rather evenly distributed between the regions: North America (NA), South and Central America (SaCA), Europe and Eurasia (Eurasia), Middle East and Africa (MEA), and Asia Pacific (APAC). *Source: Based on information from [27].*

gives us a signal that coal production is not declining despite its negative influence on the environment even in developed countries like the USA (being number 2 with 916 Mt), Australia (491 Mt), Germany (187 Mt), and Poland (137 Mt).

The growing amount of coal production is due to the fact that coal is the major fuel used for steel production and generating electricity worldwide. In 2013 coal was used to generate over 40% of the world's electricity [29], and in some countries coal has been the primary fuel for power generation, for instance, in Mongolia (95%), South Africa (93%), Poland (83%), China (81%), India (71%), or Australia (69%). Even countries highly developed countries like Israel or the UK continue to use coal for power generation at a high rate (with 61% and 39%, respectively). With the average efficiency of as low as 30%, currently operating coal-fired power plants utilize approx. 480 g of coal to generate a kilowatt-hour of energy that is delivered to the grid and sold, which would lead to the release of 1116 g of carbon dioxide emissions [30].

The price of coal is subject to its quality and usage (i.e., for power generation or steel production), as well as to market conditions. Being treated as a commodity, its price has shown certain volatility averaging to 69–98 US$ per tonne (for more details refer to multiple reports like BP Statistical Review [27]). However, the world coal market can be considered much more competitive compared to oil and gas as certain coal reserves are present on any continent in almost any country, and coal production is carried out in all world regions.

The most comprehensive and up-to-date information on reserves, production and economics of coal is consolidated and published by World Coal Association[3] on a regular basis. It is worth mentioning that most of the coal contracts are concluded for a long-term supply and therefore are less influenced by low to medium volatility of the markets.

2.4.2 Crude oil

There are many types of liquid fuel utilized in thermal power plants. These can be primary sources of energy like heavy or light residual oil, and crude oil, or secondary sources like various types of distillates, kerosene or jet fuel, and some others. Most of the liquid fuels used in traditional thermal power plants are products of crude oil refinement and have a general name *fuel oil* — basically a mix of primary and secondary energy.

Crude oil or *petroleum* is a naturally occurring liquid that contains various hydrocarbons and other compounds. Contrary to coal, oil forms in a very different way. Going back in time to the ancient continents and ocean, consider the phytoplankton and other marine organisms, which used to live at the continental shelf areas close to the shores but at relatively deep waters. When they died, they fell down to the bottom of the ocean. If they were buried fast enough in clay and silty material, the organic matter was then trapped in such clay rich mud, which would much later turn into rock, usually at the edges of the continents.

After being buried in mud for a long time, this organic-rich matter turned into special waxy substance named *kerogen*. Then, exposed to temperatures caused by pressure of the above layers of rocks, it would undergo the thermal maturation and eventually turn first into oil and, if heated up further, into natural gas. Thus thermal maturation would tell you about the type of hydrocarbon you are going to produce: very heavy viscous oil, light oil, or even oil with some of natural gas liquids or natural gas. *Natural gas liquids*, or *NGL*, are hydrocarbons that occur in a gaseous form. After the deposit is depressurized and they are brought up to the surface, they turn into a liquid and are available as liquid hydrocarbons.

Generally, crude oil contains a number of dissolved gases, especially lower hydrocarbons like methane, ethane, propane, and butane, while the others (pentane and higher) are in the form of liquids due to high pressure and hence are a part of the liquid. Other major organic compounds of crude oil are cycloalkanes, aromatics, and asphaltics.

[3] `https://www.worldcoal.org/`.

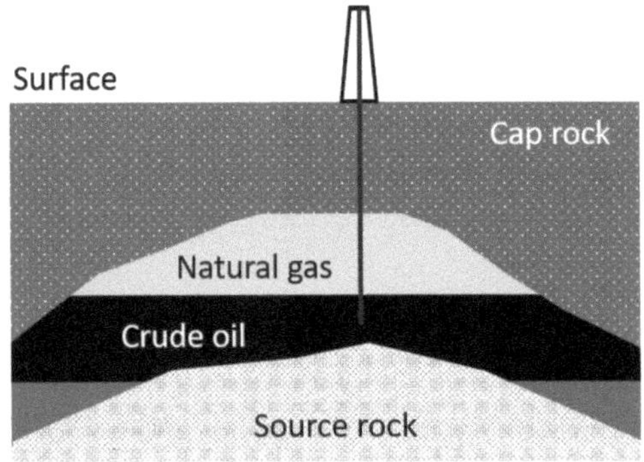

Figure 2.8 ***Oil reserve geology.*** Easily extractable oil is usually found in reservoirs — places where hydrocarbons have naturally migrated out of the source rock up into reservoir rock.

2.4.2.1 Production and processing of crude oil

Oil can be found in many source rocks, but it does not allow us to immediately produce it. Source rock is the organic matter trapped in mud, which provides very low *permeability* — the ease with which the fluid can move through the rock. If the permeability of the source rock is too low, you cannot collect and get hydrocarbons out — they simply would not flow through the rock to the well. Therefore traditionally people have searched for *reservoirs* — places where hydrocarbons have naturally migrated out of the source rock up into reservoir rock, as shown schematically in Fig. 2.8. This type of rock is much more permeable than the source rock. Due to its lower density compared to water and solids, oil would migrate upwards the reservoir, which is also sometimes called the *source rock*, until reaching the impermeable layer or *cap rock* though still trying to move up to the surface. Being separated from the water, the crude oil layer is still highly impacted by internal water flows which may move the oil horizontally for many kilometers. If trapped, an oil reserve may have the top filled with natural *oil-associate gas* (see Section 2.4.3).

Multiple configurations of reservoir are possible, and they depend on many circumstances, including the timeline and nature of the lithosphere formations, seismic activities, location, etc. In order to find reservoirs suitable for oil extraction, seismic data is usually used: the seismic waves go down and bounce off the different strata. Reconstructing the image of the subsurface, we can try to identify where these reservoirs are.

Oil rigs are used to bore the wells into the reserve and extract crude oil to the surface. While the oil is pressurized, in the beginning it would automatically flow up to the surface. However, during extraction, the pressure goes down and the process will require artificial methods of oil lifting: using compressors to pump the oil up or using an external source of high-pressure gas, which is pumped down the reserve to lift the oil — a widely used method called *gas lift*. Contrary to gas, pumping down the water can increase the pressure of the reserve but would lead to lower efficiency of oil production: the oil layer would contain too much water.

After extraction, crude oil is still mixed with water, associate gas and other contaminants (e.g., sands) and therefore goes first to a separator to remove gases and water before transportation. There are three major means of oil transport:[4]

- through the *pipeline* as the cheapest and the fastest means. Due to the pressure difference inside the pipeline, which is maintained by the compressor stations located along the way, oil flows with the speed of approximately three meters per second. Pipelines are not subject to weather conditions and can operate continuously without major maintenance.
- on a vessel called *oil tanker*, usually of medium to high deadweight to ensure economically efficient delivery. In certain cases the size of a vessel is limited by the capacity of the port and the depth of waters. Each vessel has a number of tanks: some of them for crude oil while others for water to balance the cargo. For the time being, there are more than 4000 oil tankers worldwide, most of them owned by independent transportation companies.
- by rail in tank cars. This form of transport is much more expensive compared to the other two and is normally used rarely and only for special needs or in case when other methods are unavailable.

The extracted oil is delivered to the *oil refinery*. The process of oil treatment has several stages. After separating mechanical particles and dissolved gases, oil is processed within the electrical desalination plant to remove dissolved salts and prevent corrosion of the equipment. Then the crude oil is separated into fractions by *fractional distillation* inside the rectification or fractionation column. The preheated up to 320–390°C crude oil is headed to the column mixed with *steam* as shown in Fig. 2.9. The vapors of light and heavy components condense on special distillation plates inside the column and therefore are separated into several components: gases, petrol (gasoline), kerosene, gasoil (diesel), and fuel oil (mazut).

[4] Based on information retrieved from Rosneft information web page: www.mirnefti.ru.

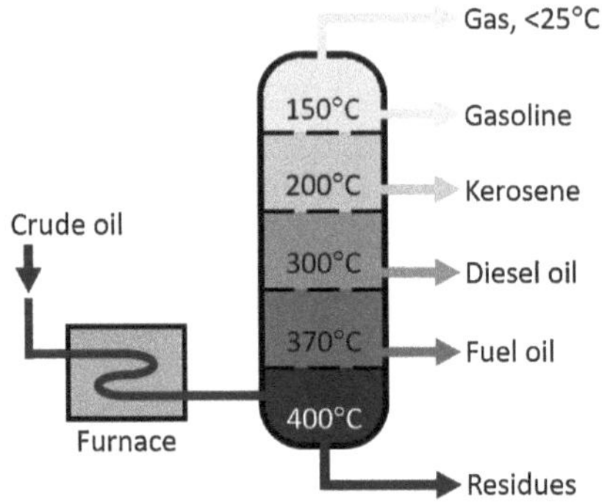

Figure 2.9 ***Schematic of fractional distillation column.*** Chemical engineering schematic of a continuous fractionating column: crude oil is heated within the furnace and separated into fractions of various temperatures, viscosities, and properties within the column into oil products. Here only major oil products are shown.

Due to the fact that oil is a source of a number of organic compounds that are used in petrochemical industry to produce multiple valuable materials and chemicals, as well as the source of fuels for vehicles, its use in power generation in power plants is limited to a few cases:

- *Diesel* is burnt in reciprocating engines to generate small power, usually for emergency needs. The emissions levels are quite high.
- *Fuel oil* is used inside the boiler to ignite the coal in many coal-fired power plants.
- *Mazut* is used in various gas-fired power plants as a second or emergency fuel;
- In a few cases, fuel oil is used as a primary fuel for base load or secondary fuel for peaking power plants;
- In oil production regions, oil is sometimes burnt in boilers to produce steam required for local needs.

2.4.2.2 Oil reserves

Contrary to coal, oil reserves are limited to a number of countries as shown in Table 2.3. According to the BP report [27], the total world proved reserves reached 1700.1 billion barrels at the end of 2014, sufficient to meet more than 52 years of global production (considering the present production rate). OPEC countries continue to hold the majority of the world's reserves, accounting for 71.6% of the global total. It is important to note that the proven reserves are a dynamic value that tends to change in either way over the years along with further exploration. Notedly, over the past

Table 2.3 Top countries with oil proved reserves of more than 5% of world share in 2014 in billion barrels (B barrel).

Country	B barrel	%
Venezuela	298.3	17.5
Saudi Arabia	267.0	15.7
Canada	172.9	10.2
Iran	157.8	9.8
Iraq	150.0	8.8
Russia	103.2	6.1
Kuwait	101.5	6.0
UAE	97.8	5.8
Total world	1700.1	100.0

Source: Adapted from [27].

Table 2.4 Top oil producing countries with the world share of more than 4% in 2014 in thousand barrels (T barrel) daily.

Country	T barrel	%
Saudi Arabia	11 505	12.9
Russia	10 838	12.7
USA	11 644	12.3
Canada	4292	5.0
China	4246	5.0
UAE	3712	4.0
Iran	3614	4.0
Total world	88 673	100.0

Source: Adapted from [27].

decade, the global proved reserves have increased by 24%, or by more than 330 billion barrels [27].

Contrary to reserves, the main oil market players are distributed in a different manner. Table 2.4 shows the top oil producers. According to statistics [27], world oil production has significantly outpaced consumption with growth mostly in the non-OPEC countries. This led to the critical drop of oil price in the period of 2014–2016.

2.4.3 Natural gas

There are multiple types of gaseous fuel which can be burnt in power plants. These are natural gas, oil-associated gas extracted during oil production, blast furnace and coke oven gases released in a metallurgical process, various synthetic gases coming out of the solid fuel gasification process, biogas, etc. While the first two belong to the primary energy sources, the others are usually considered as a product of the industry and therefore belong to the secondary energy category.

The most popular gaseous fuel from the primary energy category is the *natural gas* that is normally a mixture of several gaseous carbohydrates C_nH_{2n+2}, mostly methane CH_4. Similarly to oil, natural gas has been developed over the millions of years out of the biomass being exposed to intense heat and pressure turning first into kerogen. Heated up further, kerogen first undergoes the stage of oil and then turns into NGL, and further into natural gas. In other words, when you are producing natural gas from a reservoir, it is telling you that it came from an organic matter that was heated up much more than oil. Therefore natural gas can be found in the deep underground reservoirs alone or together with oil reserves as the process of its creation was basically the same. The latter, also known as *oil-associated gas*, can be burnt as *flare gas* (in case of low quality and quantity) like a type of waste or can be extracted for further needs.

Natural gas can be trapped under pressure within the reservoir inside the sedimentary rock from where it cannot easily migrate. Alternatively to the conventional reservoirs, gas can be located in the gas-rich shale layers (*shale gas*), coal seams (*coal-bed gas*), or within the sand layers. Shale layers are not easily permeable and therefore require special technology of production, which has not been economically feasible and widely available until the beginning of this century.

2.4.3.1 Natural gas production

From most of the conventional reservoirs, raw natural gas is extracted through the vertical *gas wells* more than a thousand meters deep. As the pressure of the gas down the well is much higher than the atmospheric, it naturally goes up to the well head, from where it is forwarded to the *gas processing plant* for extraction of higher hydrocarbons and purification process to remove contaminants like water, carbon dioxide CO_2, hydrogen sulfide SO_2, mercury, and a few others. Gas composition depends on the type and location of the gas field, as well as the geology of the area, however, the main component of natural gas is methane with approx. 70–98% of volume. Some of the contaminants, like SO_2 or higher hydrocarbons, retrieved from raw gas, can be further processed in a petrochemical plant and used in chemical production.

Contrary to this traditional method, shale gas extraction is subject to multiple difficulties: due to low permeability of the source rock, it is not possible to continuously extract gas from even a number of vertically drilled wells. The success here became possible thanks to the invention of the two breakthrough technologies:

- *Horizontal drilling* that allowed drilling the well along the shale layer;
- *Hydraulic fracturing* process, which uses pressurized liquid to pump into the well and further through the holes in the walls of the horizontal well to crack the source rock and increase its permeability along the well.

If the purified and dried natural gas meets the quality standards, then it is transported to the consumer via one of the two major ways:

- through the pipeline — a major means for gas delivery. The pressurized gas goes through the pipeline loosing its energy due to friction inside the pipe which is released in the form of heat, hence the gas would loose its pressure and would not arrive to the consumer. To avoid this, a number of compressor stations are located along the way to further compress the gas and cool it down to ensure delivery;
- in liquid state as a liquefied natural gas, or LNG, inside an *LNG carrier* — a special tank vessel. The liquefaction involves further removal of the contaminants (including heavy hydrocarbons) and condensing the gas into a liquid by cooling it down to minus 162°C.

The type of transportation depends on the geography of the gas production facility and location of the consumer.

When the gas is delivered to the power generating facility, which is normally done through the main pipeline either directly from the production facility or from the LNG unloading point, it has the *chemical composition* according to the supplier's standard. This composition, with an example shown in Table 2.5, is usually known upfront the power plant construction (or upgrade in case of fuel change) and is normally part of the contract as a reference case. It is used for defining the heating value of the gas fuel, calculating performances and designing equipment of the power plant. This knowledge is critical because the overall plant cost and performance guarantees of the power plant depend on the fuel and can be subject to penalties. However, the exact fuel composition can vary, especially if the gas is coming into one pipeline from several sources. However, the gas supplier is usually able to give the boundaries for each fuel component to ensure proper power plant operation over time.

Table 2.5 Possible natural gas composition delivered to the customer.

Component		% Vol.
Methane	CH_4	97.83
Ethane	C_2H_6	0.93
Propane	C_3H_8	0.29
Butane	C_4H_{10}	0.10
Pentane	C_5H_{12}	0.02
Nitrogen	N_2	0.74
Oxygen	O_2	0.01
Carbon dioxide	CO_2	0.08
Hydrogen	H_2	0.01

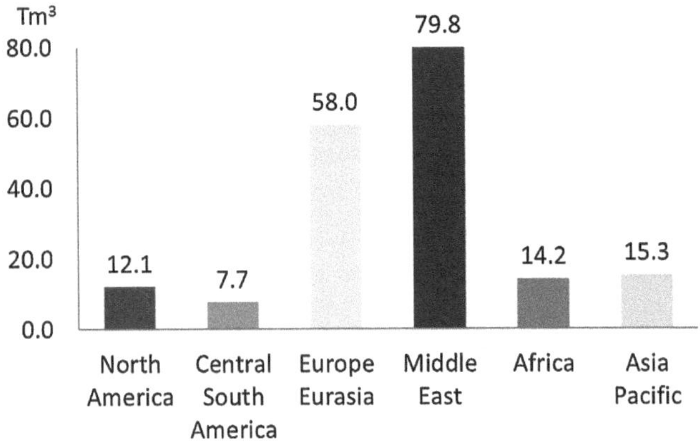

Figure 2.10 ***Estimated natural gas reserves per region in 2014.*** The Middle East and Eurasia have by far the largest share of natural gas reserves (figures in trillion cubic meters). *Source: Adapted from [27].*

2.4.3.2 Natural gas reserves

Being a non-renewable source of power, natural gas reserves are limited and located in a few key countries similar to oil reserves, as shown in Fig. 2.10. The total proven reserves of natural gas equal to 186.5 trillion cubic meters (Tm^3) and are basically split among a number of countries in Europe and Eurasia with 31% (mostly Russia and Turkmenistan), and Middle East (Iran and Qatar) with 42.7% being by far the biggest producers of natural gas, as shown in Table 2.6.

The total proved reserves of easily accessible natural gas are estimated to last for the next half of the century under the assumption of the current rate of production which already increased by 1.6% in 2014 [27]. Despite

Table 2.6 Top 10 countries with largest natural gas proved reserves in 2014.

Country	Reserves		Country	Reserves	
	Tm^3	%		Tm^3	%
Iran	34.0	18.2	Saudi Arabia	8.2	4.4
Russia	32.6	17.4	UAE	6.1	3.3
Qatar	24.5	13.1	Venezuela	5.6	3.0
Turkmenistan	17.5	9.3	Nigeria	5.1	2.7
US	9.8	5.2	Algeria	4.5	2.4
Total world	187.1	100			

Source: Adapted from [27].

the above conservative forecast and potential depletability of natural gas in the nearest future, the proven reserves are still an estimate of the known gas fields with the trend showing that they are growing in time and with new investigations. However, such findings cannot completely secure a natural gas supply for the infinite time period.

Remark 2.4. These reserves do not take shale gas into account. While shale gas is not yet considered as easily accessible and therefore requires higher investments into its production, it can become a very prolific source of natural gas and can therefore contribute to the green energy: contrary to other fossil fuels, burning natural gas would significantly decrease the amount of GHG emissions both directly and through increase of plant efficiency.

The annual production of natural gas, as shown in Table 2.6, has risen steadily over the last decades, resulting in more than 3460 billion cubic meters in total. With the US and Russia being the largest natural gas producers, they together extract more than one-third of total world natural gas annual production. The consumption of natural gas has been growing steadily during the last decade with a slight drop in 2009 due to the world economic crisis. With the world economy slowdown, the growth of consumption in 2014 was only 0.4% within the emerging markets like China, Algeria, and Brazil.

The main pipeline exporters are Russia (delivered 147.7 Bm^3 to Europe via pipelines in 2014), Norway (101.1 Bm^3), and Algeria (19.5 Bm^3) [31, 27] delivering gas to Western Europe, especially Germany, Italy, and Spain. The LNG exports and imports (see Table 2.7) have been historically held in two major trading regions: the Atlantic and the Pacific. While the Atlantic

Table 2.7 Top 10 LNG importers and exporters in 2014.

Exporters		Importers	
Country	**Bm^3**	**Country**	**Bm^3**
Qatar	103.4	Japan	120.6
Malaysia	33.9	South Korea	51.1
Australia	31.6	China	27.1
Nigeria	25.3	India	18.9
Indonesia	21.7	Taiwan	18.1
Trinidad & Tobago	19.3	Spain	15.5
Algeria	17.3	United Kingdom	11.3
Russia	14.5	Mexico	9.3
Oman	10.6	Brazil	7.9
Yemen	8.9	Turkey	7.3

Source: Adapted from [27].

is dominated by European importers (resulting in 52.1 Bm^3 in 2014 [27]), in the Pacific major importers of LNG are Japan, South Korea, and Taiwan. Recently, China and India have also increased their LNG imports within top 5 regional importers. In the Pacific, by far the biggest portion of gas is coming from Qatar (74.4 Bm^3), Nigeria (13.3 Bm^3), Australia (31.6 Bm^3), Malaysia (33.7 Bm^3), Indonesia (21.4 Bm^3), and Russia peaking up in recent years (14.5 Bm^3) with several international projects.

Pricing in world gas markets remains very fragmented, with prices driven more by local and regional factors than global dynamics. Gas prices in North America are largely set at trading hubs, while in Europe, natural gas is sold mainly via long-term contracts. In Asia and many emerging markets, the price of natural gas is often set via oil-linkages [32].

2.4.4 Comparison of fossil fuels

While fossil fuel is by far the biggest expense during a power plant operation, it has to be properly assessed in advance during the project feasibility study. Fuel heating value, or the amount of chemical energy stored in it, depends on many properties, including its chemical composition (carbon, hydrogen, sulfur, etc.), as well as moisture and ash content. The amount of energy retrieved from a unit of fuel (kilogram, liter or cubic meter) is important to understand while performing economic evaluation of the power plant.

Table 2.8 Higher heating values of some common fuels.

Fuel	Higher heating value	
	kJ/kg	**Btu/lb**
Anthracite	32 500–34 000	14 000–14 600
Bituminous coal	17 000–23 250	7300–10 000
Lignite	16 300	7000
Peat	13 800–20 500	5500–8800
LNG	53 600	23 050
	kJ/m^3	Btu/ft^3
Natural gas	35 000–43 000	950–1150
Methane	39 820	1050
Hydrogen	13 000	350
	kJ/l	Btu/gal
Crude oil	37 000–39 000	133 000–140 000
Heavy fuel oil	41 200–42 600	148 000–153 000
Kerosene	37 600	135 000
Diesel	36 300–39 000	130 000–140 000

Source: Based on [24,33].

Table 2.8 gives an overview and comparison of major fossil fuels, which can be utilized within a thermal power plant. Evidently, the best energy fuels are those with the highest calorific value, meaning that they release more energy per unit volume and, more importantly, reduce the fuel consumption. However, the fuel heating values have to be considered together with the following parameters:

- Availability and price of fuel;
- Amount of emissions produced by combustion, which directly influences the taxation or the amount and financial conditions of the investment;
- Cost of equipment required for burning the fuel and treating emissions, which is subject to the type of a thermal power plant.

The overall economic evaluation has to be carried out for the whole project lifecycle (at least 20–30 years of operation). However, the available fuel alternatives and their properties can be a starting point of the whole project evaluation with the associated costs partly represented in Table 2.9, as they define:

Table 2.9 Comparison of fossil fuels with regard to energy generation and emissions (average numbers for indication purposes).

Fuel	Efficiency	Fuel per kWh	CO_2 per kWh
Coal	30–45%	480 g	650–1200
Natural gas	40–62%	0.03 m^3	350–450
Oil	30–40%	0.27 l	450–520

Source: Aggregated from [30,34].

- A major portion of the operating costs, i.e., the fuel cost itself spent to deliver a kilowatt-hour of energy to the grid;
- The amount of GHG emissions produced for the generation of a kilowatt-hour of energy and therefore risks of taxation and carbon pricing;
- Exposure to various risks dealing with various types of fuel. For instance, the imports of LNG from other countries can be subject to foreign policies, volatile pricing, global structure of imports and exports, etc. Local coal mining is by far more stable unless targeted by international treaties regarding global climate change or ecological damages.

2.5. Nuclear fuel

Nuclear fuel is a fuel that is used in nuclear power plants to produce heat through the nuclear fission chain reaction. Contrary to fossil fuels, which undergo chemical reaction to convert their energy into heat, nuclear fuel goes through the special type of reaction and therefore requires different equipment, treatment, and level of safety. Depending on the nuclear reaction implemented in the reactor, there are several types of nuclear fuel:

- Natural uranium fuel that contains uranium-235 and uranium-238, capable of forming plutonium-239;
- Thorium fuel, which is not used presently due to its price, complicated fuel production and recycling processes, as well as certain abundance of normal uranium;
- Secondary, or reprocessed fuel which cannot be found in nature.

Traditional nuclear fuel is based on the natural uranium ^{235}U. This element is present in nature in the form of various types of uranium ores of different concentrations starting from 0.1% (poor) up to more than 1% (rich), and is approximately as common as tin or zinc [35]. Uranium identified resources consist of reasonably assured resources (RAR) and inferred

Table 2.10 Reasonably assured uranium resources (RAR) for the countries with significant share as of 2013 (rounded to nearest 100 tonnes).

Country	**Cost ranges**			
	< 40 $/kgU	**< 80 $/kgU**	**< 130 $/kgU**	**< 260 $/kgU**
Australia	–	–	1 174 000	1 208 000
United States	0	39 100	207 400	472 100
Canada	256 200	318 900	357 500	454 500
Kazakhstan	20 400	199 700	285 600	373 000
Niger	0	14 800	325 000	325 000
Namibia	0	0	248 200	296 500
Russia	0	11 800	216 500	261 900
South Africa	0	113 000	175 300	233 700
Brazil	137 300	155 100	155 100	155 100
Ukraine	0	42 700	84 800	141 400
China	1800	93 800	120 000	120 000
Mongolia	0	108 100	108 100	108 100
India	–	–	–	97 800
Uzbekistan	41 700	41 700	59 400	59 400
Total world	507 400	1 211 600	3 698 900	4 587 200

Source: Adapted from [36].

resources (IR) recoverable at a cost of less than 260 $/kg of uranium (or 100 $/lb of uranium oxide U_3O_8) [36]. Current estimates of RAR are presented in Table 2.10 for the top countries with significant share. RAR recoverable at costs less than 40 USD per 1 kg of uranium is the most economically attractive category as it allows for cheap extraction of uranium ore and hence higher margins on the market. The top five producing countries (Kazakhstan, Canada, Australia, Niger, and Namibia) retained their dominance accounting for 79% of world production in 2012 [36].

Significant uranium reserves are scattered throughout the world and therefore are not readily available to all countries. This is similar to natural gas or crude oil distribution and therefore influences the exports and prices as well. However, compared to other commodities, uranium does not trade on an open market, and buyers and sellers negotiate their supply contracts privately. Uranium prices are usually published per pound of the uranium oxide U_3O_8 equivalent in US dollars by independent market analysts and consultants. Most of the utilities have long-term delivery contracts, and the changes of the uranium price do not affect them significantly. Current price can be found within the information providers.

References

[1] Mackay M. Solar energy: an introduction. Oxford University Press. ISBN 9780199652105, 2015.
[2] Nelson V, Starcher K. Introduction to bioenergy. Energy and the environment. CRC Press. ISBN 9781498716987, 2016.
[3] Tiwari G, Tiwari A. Handbook of solar energy: theory, analysis and applications. energy systems in electrical engineering. Singapore: Springer. ISBN 9789811008078, 2016.
[4] Tester J, Drake E, Driscoll M, Golay M, Peters W. Sustainable energy: choosing among options. MIT Press. ISBN 9780262304634, 2012.
[5] Pitman A. The evolution of, and revolution in, land surface schemes designed for climate models. International Journal of Climatology 2003;23:479–510.
[6] United Nations Development Programme (UNDP), World Energy Council. World energy assessment: energy and the challenge of sustainability. New York: UNDP; 2000.
[7] Jain S. Fundamentals of physical geology. Springer Geology. Springer. ISBN 9788132215394, 2013.
[8] DiPippo R. Geothermal power plants: principles, applications, case studies and environmental impact. Elsevier Science. ISBN 9780081002902, 2015.
[9] Letcher T. Wind energy engineering: a handbook for onshore and offshore wind turbines. Elsevier Science. ISBN 9780128094297, 2017.
[10] World Energy Council. World energy resources 2016. Marine energy. London: World Energy Council. ISBN 9780946121588, 2016.
[11] The Future Ocean, International Ocean Institute. World ocean review 2010. Living with the oceans. Maribus. ISBN 978-3-86648-012-4, 2010.
[12] Information Administration (EIA). International energy outlook 2016. U.S. Energy Information Administration; 2016.
[13] Petersen J, Sack D, Gabler R. Fundamentals of physical geography. Cengage Learning. ISBN 9781285969718, 2014.
[14] Mills E. The fluid envelope of our planet: how the study of ocean currents became a science. University of Toronto Press, Scholarly Publishing Division. ISBN 9781442663602, 2011.
[15] Rahmstorf S. Thermohaline circulation: the current climate. Nature 2003;421:699.
[16] National Geographic. Ocean conveyor belt. URL: https://www.nationalgeographic.org/encyclopedia/ocean-conveyor-belt/.
[17] van Aken H. The oceanic thermohaline circulation: an introduction. Atmospheric and oceanographic sciences library. New York: Springer. ISBN 9780387480398, 2007.
[18] Guerrero-Lemus R, Martínez-Duart J. Renewable energies and CO2: cost analysis, environmental impacts and technological trends. 2012 edition. Lecture Notes in Energy. Springer. ISBN 9781447143857, 2013.
[19] Slade R, Saunders R, Gross R, Bauen A. Energy from biomass: the size of the global resource. London: Imperial College Centre for Energy Policy and Technology and UK Energy Research Centre; 2011.
[20] Lindquist E, D'Annunzio R, Gerrand A, MacDicken K, Achard F, Beuchle R, et al. Global forest land-use change 1990–2005. FAO forestry paper No. 169. Rome: Food and Agriculture Organization of the United Nations and European Commission Joint Research Centre; 2012.
[21] Nelson V. Wind energy: renewable energy and the environment. Second edition. Energy and the environment. CRC Press. ISBN 9781466581616, 2013.

[22] Tan DK, Amthor JS. Bioenergy. In: Dubinsky Z, editor. Photosynthesis. Rijeka: In-Tech; 2013 [chapter 12].
[23] Sarkar D. Thermal power plant: design and operation. Elsevier Science; 2015. 612 pp.
[24] Spliethoff H. Power generation from solid fuels. Springer; 2010.
[25] Trukhniy A, Izyumov M, Povarov O, Malyshenko S. Foundations of modern energy systems, vol. 1. MPEI Publishing House; 2010.
[26] World Coal Association (WCA). The coal resource. A comprehensive overview of coal. World Coal Association; 2009.
[27] BP. BP statistical review of world energy. BP; 2015.
[28] International Energy Agency (IEA). Coal information. Paris: International Energy Agency; 2015. 674 pp.
[29] WCA. Coal facts 2014. World Coal Association; 2014.
[30] World Coal Association (WCA). Coal and electricity generation, vol. 3. World Coal Association; 2012.
[31] Breeze P. Gas-turbine power generation. Elsevier; 2016. 98 pp.
[32] IGU. World LNG report. Norway: International Gas Union; 2015. 100 pp.
[33] Engineering tool box. URL: http://www.engineeringtoolbox.com, 2016.
[34] Teske S, European Renewable Energy Council, Greenpeace International. Energy [r]evolution. A sustainable world energy outlook 2015. Greenpeace International; 2015.
[35] Hore-Lacy I. Uranium for nuclear power: resources, mining and transformation to fuel. Elsevier Science. ISBN 9780081003336, 2016.
[36] OECD. Uranium 2014: resources, production and demand. OECD; 2014.

CHAPTER THREE

Energy system and basic electricity market

Contents

3.1. Fundamentals of energy system

There are various economically feasible ways of converting one form of energy into another, which are usually combined together into some energy conversion chains. Energy conversion technology and its efficiency depend on the current level of technological development of humankind, as well as the availability of resources and economic conditions. All together, these conversion chains and the respective technologies can be generally considered within one global energy system.

We can distinguish between the two energy systems: the natural and that which has been created and developed by humankind. The simplified *natural energy system*, as schematically represented in Fig. 3.1, has the following major energy conversion chains:

- Radiant energy from the Sun is converted into:
  - Chemical energy in the upper layers of the atmosphere to synthesize ozone;
  - Chemical energy during the photosynthesis process within the green plants and some other living beings. This energy is further converted into other chemical energy (food chain and further dead matter),

Sustainable Power Generation
https://doi.org/10.1016/B978-0-12-817012-0.00012-8

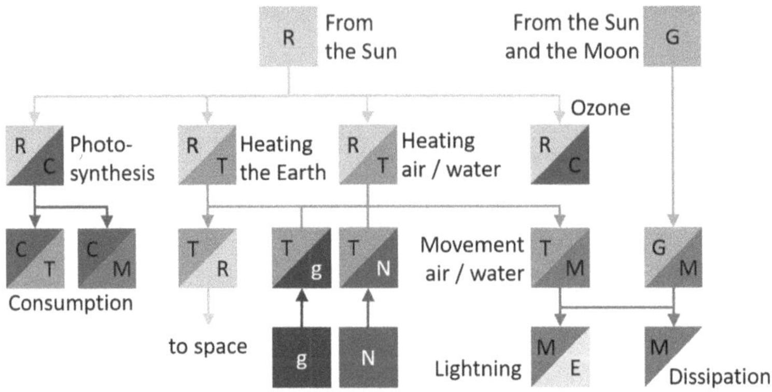

Figure 3.1 ***Schematic diagram of the simplified natural energy system.*** Energy conversion chains depend on the solar radiation. Here only main energy sources and principles are outlined to illustrate the complexity and logic of processes. R denotes radiant energy, C – chemical, T – thermal, G – gravitational, M – mechanical, E – electrical, g – geothermal, and N – nuclear.

heat and mechanical energy as the living matter consumes this energy.

- Heat within the atmosphere and hydrosphere to be further converted into mechanical energy of the moving air and water. Moreover, this mechanical energy can further be converted into heat (internal friction), electricity (lighting), and chemical energy (lighting produces ozone).
- Heat to warm up the surface of the Earth.

- Mechanical energy from the Sun and the Moon due to gravity is converted into mechanical energy of the moving waters which result in tides;
- Nuclear and geothermal energy of the Earth is converted into heat and, combined with the overall Earth's heat, is delivered to the surface (either directly like volcano eruption or through heat transfer), where it mostly converts heat into heat of the air or water, or into radiant energy.

Here some of the processes have been intentionally simplified or even omitted to focus mostly on the major energy conversion principles and the overall complexity of nature and its energy balance. Still, we can already see a number of natural mechanisms that have been observed by humankind for years and, consequently, multiple attempts have been taken to copy and utilize them. Therefore the current energy system employed by the human

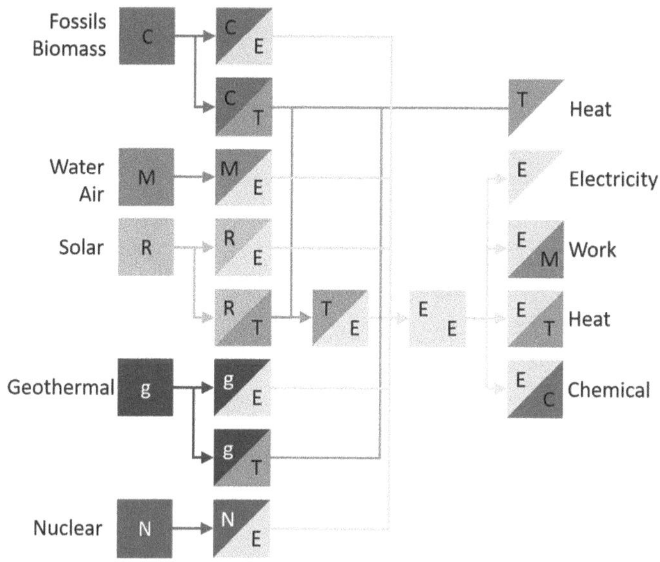

Figure 3.2 ***Schematic diagram of the simplified human energy system.*** **Energy supply paths illustrating the boundaries the energy industry: from primary energy sources like fossil fuels, water and air flows, solar radiation, geothermal heat, and nuclear fuel to electricity and heat supply. Here only main energy sources and principles are outlined to illustrate the complexity and logic of processes. R denotes radiant energy, C – chemical, T – thermal, M – mechanical, E – electrical, g – geothermal, and N – nuclear.**

civilization has evolved in a way to capture the easiest natural sources of energy with minimal effort.

The current *human energy system*, simplified and illustrated in Fig. 3.2, incorporates several traditional and commercially tested energy conversion processes, which here omit few interim stages and include two major energy conversion subsystems:

- Electrical, which uses chemical energy stored in plants or dead matter (eventually, fossil fuels), mechanical energy of the moving water and air, radiant energy of the Sun, thermal energy of the Earth, and nuclear energy. All these sources of energy are either directly or indirectly (with some interim stages, similar in many cases) are converted into electrical energy.
- Thermal energy, or simply heat, is usually converted from chemical energy of fossil fuels, from radiant energy of the Sun, and heat from the Earth.

Electricity and heat are then used in human activities, where they are converted into other forms of electricity and heat, as well as into radiant energy (for example, light), chemical energy or mechanical work (e.g., to run machines).

Evidently, the human energy system is based on the natural energy system, however, it is not fully integrated into it. Moreover, the energy system is also impacted by other areas of human activity and the environment: materials, processes, knowledge, and the overall impact are just a few to mention, though the interaction of the environment and energy system, as well as between the energy system and world economy, is extremely complicated and will only increase its complexity along the way of human development.

3.2. Energy industry

Historical development of the world economies has shaped the energy industry in the way we see it presently. It has mostly been directed by the economic efficiency of the energy production or, in simple words, to generate enough cheap energy and sell it to maintain economic development. The present *energy industry* in general comprises the following major components which basically represent the energy system:

- Extraction of primary energy;
- Conversion of primary energy into secondary energy: electricity, heat, fuels, etc.;
- Transmission of the secondary energy and its further distribution among the consumers.

However, to ensure proper functioning of this system, the energy industry has to supply a number of products and services from the neighboring systems through the cross-functional interaction value chains, which together form the global energy industry. These neighboring systems include, among others:

- Manufacturing of energy-related highly technological equipment which is used within the energy system;
- Construction of the facilities (i.e., supply of materials and services);
- Transportation of goods and materials;
- Educational services to supply qualified personnel;
- Other services including R&D, administration, regulations, management, etc.

Hence, according to [1], the energy supply sector comprises all energy extraction, conversion, storage, transmission, and distribution processes with the exception of those that use final energy to provide energy services in the end-use sectors (industry, transport, and building, as well as agriculture and forestry) with the support of neighboring industries and areas.

This covers not only the physical and technological aspects of the overall process, but also the economic side, including sale of energy, its pricing, and market relations. This industry is interconnected with all other branches and sectors of the economy, which are essentially bound to the energy either directly or indirectly. By direct connection we assume the processed energy consumption (transportation, utilities, as well as any electricity consumer), while indirect means supporting the energy process, for example, by supplying machinery, materials or equipment, training specialists, etc.

It is evident that all areas of our activity are closely connected to the energy industry and, therefore, contribute to the global energy chains. This makes the energy industry among the most complicated artificial systems the humankind has ever created and implemented on a large scale and, moreover, has used as a driving force for its ongoing development.

3.2.1 Power industry within energy industry

Within the energy industry, we can highlight the *power industry* or *power generation industry*, which can be named as a subsegment of the former. The power industry itself is a complex, real-time integration of technology, economics and business. Similarly to the parent structure, it inherits major connections to other sectors of the economy, however, it is limited to the electricity and partly heat — its generation and delivery to the final customers.

The power industry is closely connected to the world economy. This connection is implemented through a number of various energy receiving applications throughout the whole universe, which are fed from the modern energy industry. Consequently, this requires stability and reliability of the energy industry with the clear redundancy in case of emergency. This requirement is equally important for both electricity and heat supply. Therefore such a close connection requires that the power industry captures the pace of the economic growth of a specific country and even exceeds it in order to make sure that there is a certain reserve in energy supply for the future economic growth. In the worst case, the growing economy of a country would face power shortage, which may lead either towards dependency on the foreign supply, or to the economic slowdown. Moreover, in the unlikely case of a natural disaster or any "force majeure" situation, there shall be enough capacity to maintain the changes within the energy industry and therefore ensure the energy security of the nation.

The modern power system is responsible for the energy conversion process, when the end product is mostly electricity and, sometimes, heat. The power industry consumes primary energy from major sources to convert it to electricity and heat. Moreover, it can also utilize products of the energy industry, namely fuels (which can be considered as secondary energy). Moreover, the power system provides energy (electricity and heat) to the whole energy industry to enable its operation, therefore both are closely interconnected.

While the power system is in most cases global and extends over the borders of countries due to multiple reasons (cross-border trade, absence of local technologies, uneven distribution of primary energy, etc.), the energy system may still be localized within the region or country, though still exposed to limitation and therefore also going outside of these borders. This is especially seen nowadays due to global market liberalization, international trade and increasing complexity of technology: it becomes more and more difficult for any country to operate in a silo and own all necessary technologies. Moreover, international trade and globalization of large MNCs leads to cross-boundary relations also relevant to the power system.

3.2.2 Concept and types of power plants

The power industry establishes and maintains processes that support the energy supply paths. In order to handle the energy conversion chains and make sure that they are properly functioning, special facilities are used. These facilities are a combination of major equipment which is essential for the multistage energy conversion, supported by the additional systems and accessories. Their proper functioning is essential for both secondary energy production and for economic, safe and reliable operation.

Specially designed for converting various forms of energy into electricity (and sometimes heat), these facilities are called *power plants*. The type of a power plant and the choice of its equipment depend on the chosen energy conversion approach and therefore require thorough understanding of the following set of considerations:

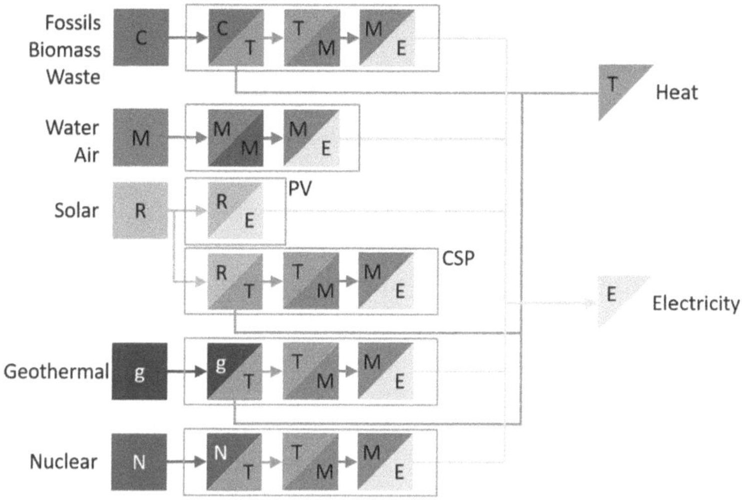

Figure 3.3 ***Major types of power plants within the power generating system.*** Major components of power generating system are based on energy conversion chains and define the types of power plants implemented within this system. R denotes radiant energy, C – chemical, T – thermal, M – mechanical, E – electrical, g – geothermal, and N – nuclear; PV stands for photovoltaic, and CSP – for concentrated solar power.

- Available primary energy (supply conditions, amount, properties, price);
- Required type of secondary energy (electrical, combined heat and power, mechanical or heat only);
- Operating profile and the underlying business model of the power plant;
- Required capacity and capability to evacuate power to the grid and deliver to the consumer;
- Local regulations and requirements.

Following the approach discussed in Chapter 1, we can represent the power system through the energy conversion chains as shown in Fig. 3.3. These energy conversion chains are executed in power plants of various types.

Historical development of the power industry has identified the major directions of the energy conversion and therefore has defined the key types of power generating plant. Therefore, all modern power plants can be split into two broad categories established throughout the last decades: traditional and non-traditional power plants.

3.2.2.1 Traditional power plants

Traditional or conventional power generation has been focusing on those types of primary energy and its conversion that are easiest to access and convert into electricity, mostly due to the well-established technology. These plants include:

- *Thermal power* technologies, which use chemical energy of the fossil fuels in various forms (coal, oil or natural gas). They assume that there is an exothermic chemical reaction of combustion that takes place within the power plant to release heat and utilize it further in the conversion process, either through hot gas or steam. Here, chemical energy is converted into thermal energy of hot gas or steam, which is in turn converted into mechanical energy of the rotation and further into electricity. These technologies and their economic advantages are discussed in Part 3 in Chapters 7–10.
- *Nuclear power* stations that have been developed since the middle of the 20th century are explored in detail in Part 4 within Chapters 11–13. These power plants use thermal energy released from the chain nuclear reaction to utilize it further in the energy conversion process.
- *Hydropower plants* that utilize mechanical energy of the flowing (or stored) water which is one of the oldest accessible sources of primary, and also renewable, energy. This energy is converted into mechanical energy of a rotating machine and further to electricity. Part 5 with Chapters 14 and 15 is devoted to these power plants.

Remark 3.1. In some literature thermal and nuclear power plants are not separated from each other and form one large segment of thermal power. This is mostly supported by the fact that a nuclear power plant still uses thermal energy to heat up water to produce steam, which has been a traditional approach in thermal power plants.

3.2.2.2 Emerging power plant technologies

Perceived abundance of fossil fuels and their seemingly low prices, established traditional technology, and lack of evidence of the climate changes resulted in the unquestionable economic advantage of traditional technologies on the global scale. While these conventional technologies still dominate the world and would continue to do so in the coming decades (see, for example, forecast in [2]), emerging technologies are becoming increasingly important and would be vital for the future development to supply sustainable energy.

Major emerging power generating technologies, which are the topic in Part 6, include:

- Wind parks that convert mechanical energy of the flowing air, namely winds, into the mechanical energy of rotation and further into electricity. Discussed in Chapter 16, these power plants utilize the old principle of wind mills.
- Solar power plants implement technologies that can convert solar radiation into electricity. Being abundant, solar radiation is considered as a promising source of renewable energy. Moreover, this energy can be also converted into heat and further used in a conventional process. Both approaches are described in Chapter 17.
- Biomass and energy from waste are becoming an alternative to fossil fuels due to the same approach: chemical energy is converted into heat (via combustion) and further into mechanical energy and electricity. Chapters 19 and 18 describe them.
- Geothermal power plants convert thermal energy into electricity similarly to conventional thermal power plants. This type of green and renewable energy has not yet reached its potential, which is approached in Chapter 20.
- Finally, a lot of energy stored in the oceans are converted by tidal, wave, and ocean stream energy technologies that convert this mechanical energy into electricity. Being renewable and green, these technologies are still at the start of their mass commercial implementation, as illustrated in Chapter 21.

The technologies mentioned above have been commercially implemented or at least have undergone the test phase to contribute to the global electricity generation to become a part of a contemporary or future power system. There are, of course, other developments, which may appear as technological and commercial breakthroughs in the next decades, but they fall out of the scope of our discussion.

3.3. Electric energy system beyond generation

Power systems are responsible not only for converting other types of energy like chemical, mechanical, nuclear, thermal or radiant into electricity (and sometimes heat), but also for transmitting this energy and distributing it safely and efficiently to consumers. Hence, power systems should include the following technical components to fulfill this function:

- Prepare electricity for transportation over long distances;
- Transmit electricity to the places, where consumers are located;
- Convert electricity into the required form to meet consumer's needs; and
- Control and manage the overall energy network to ensure its reliable operation.

3.3.1 Transmission and distribution of electricity

When the electricity is generated by the rotating machine called electrical generator, it has certain parameters like frequency or voltage. The frequency of the alternating current is historically subject to the region where the power system is located: while North, Central, and most of South America, part of Japan, Saudi Arabia and some other countries employ 60 Hz current frequency, others employ that of 50 Hz.

Ideally, power plants should be located near consumers, however, this is not always possible. For instance, large hydroelectric plants are situated in the areas where water energy is available, which is not always the nearest location to large industrial consumption. Therefore, there should be reliable means of electricity transportation, which would allow efficient transmission of energy.

At the power plant, electricity is generated at a relatively low voltage between about 2 kV and 30 kV, depending on the type and size of the unit. Here two major options are available for commercial implementation:

- To minimize electricity losses during transportation of electricity along multi-kilometer wires, the voltage is stepped up by the step-up transformer to a much higher voltage. This approach allows for lesser resistive losses over long distances in the wiring. This voltage depends on the country and may range from 100 kV up to more than 750 kV, varying by the transmission system.
- The alternative is to use DC current of higher voltage up to a few hundreds of kV, which is called High voltage DC (HVDC) and up to 1500 kV Ultra HVDC (UHVDC). The substation uses a special HVDC converter to convert AC current into DC current (rectify) and then back from DC to AC (invert).

HVDC lines have a number of important advantages over high-voltage AC lines, but only over long distances:

- While HVDC and UHVDC equipment is much more expensive, it generally has lower overall investment costs and much lower losses over

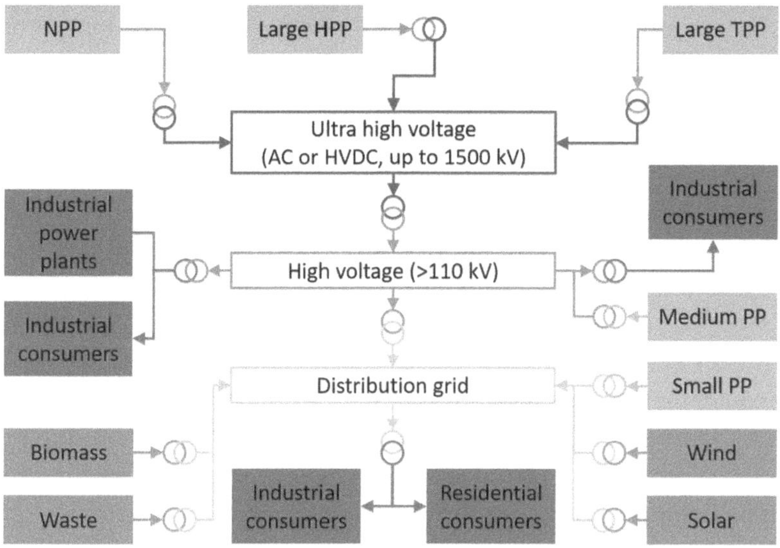

Figure 3.4 ***Simplified schematic representation of the grid.*** Larger generating capacities are usually connected to a major transmission grid formed by transmission lines and substations (HV and UHV or EHV), which is linked to a distribution grid of lower voltage that supplies power for industrial and residential consumers. Some smaller capacities can be directly connected to a distribution grid, so that the overall system management becomes even more complicated. PP – power plant, NPP – nuclear PP, HPP – hydroelectric PP, TPP – thermal PP.

equivalent AC transmission scheme, but only for long distance transmission. The break-even distance is estimated at around 600–800 km.

- HVDC can connect two AC networks without the need of synchronizing them (their direct connection requires synchronization of the two alternating currents), and provide a bridge between the two if the need arises.
- HVDC transmission line does not have a technical limit in length, contrary to AC lines. This is mostly important for underwater cable connections, where HVDC connection is the only viable alternative. Also, break-even distance for underwater cables is much smaller compared to overhead lines.

Remark 3.2. In urban areas, where overhead lines are not applicable, cables can be laid in underground tunnels.

When electricity is transmitted to the location of consumer, as schematically illustrated in Fig. 3.4, it shall be further converted to the applicable parameters. While major industrial customers may accept high-voltage electricity for their needs (and then further utilize it in their in-house energy systems), conventional customers are treated in a different way. Electricity from the line goes to the substation, where:

- AC high-voltage electricity goes through the step-down transformer, which converts it to medium and lower voltages, and then it is guided to the distribution network.
- HVDC electricity goes through inverters and then AC is treated in the same way as above.

A distribution network is usually operating with alternating current with medium voltage roughly between 2 and 35 kV electricity to distribution transformers located near consumers. These transformers again lower the voltage to the utilization voltage used by general purpose customers: lighting, industrial equipment or households.

3.3.2 Consumption of electricity

The overall power grid is a complicated system that has both in- and outflows of power and which has to be properly balanced in order to maintain its functionality. As shown in Fig. 3.4, a power grid binds together power producers and consumers. While various electricity generating facilities like coal or nuclear plants deliver various amounts of power to the grid, this power has to equal the consumption by industry, commercial, residential facilities or other types of customer.

Large utility-scale power generating facilities deliver power to the transmission grid, while the smaller units can be connected to a distribution system. Due to a large amount of production facilities of different types, and while the consumption of energy is not constant in time, there are different *operating profiles* of a power plant available in the market, which have to be properly balanced to manage the grid [3]:

- *Base load* plants, by definition, are developed to deliver the greatest hours of steady dispatch and provide reliable power supply to the grid. Here plants with the lowest variable operating costs and the highest efficiency are most likely to dispatch first.
- *Cyclic operation* plants are required to respond to demand fluctuations in the grid throughout both normal daily and weekly or monthly load cycles. In case the grid absorbs energy from renewable sources, flexibility characteristics like fast start and capacity ramping, the lowest possible

minimum load, or high part load efficiency, are valued by the operators as they can bring a much higher competitive advantage.

- *Peaking* plants are used to deliver cost-effective, reliable power to cover intermittent peak demands, which can be caused by the consumption profile or emergency. These plants typically dispatch less than 1000–1500 hours per year, so that the lowest capital costs of such units is the driver of the project.
- *Stand-by* power is usually required by industrial or commercial facilities to back up their grid connection in the event of a blackout. Stand-by power sources usually run a few hours per year during grid outages or in emergency of the main unit failure. As such, stand-by power offerings are mostly small gas turbines (e.g., based on aircraft engine technology) or reciprocating gas engines.
- *Ancillary services* are employed to balance short-term variations in electricity markets by dispatching resources within seconds or minutes to maintain grid stability and security. These services generally include frequency control, reactive power-voltage regulation, spinning reserves, and operating reserves.

The operating profile of a power plant can also change over its life cycle [3]. Plants initially constructed to deliver base load generation may decrease their dispatch over time and migrate towards cyclic operation as newer technologies are added to the grid, with even lower variable operating costs. In other cases, plants initially built as peak load can be well converted to combined cycle to improve their ROI.

Base load plants are intended to supply power consistently and reliably through the day and throughout the year, with breaks only for maintenance [4]. Hydropower plants are well suited for peak load because they can be switched on and off on a scale of minutes. Pumped storage plants are used for load shifting and frequency control. Load shifting refers to the use of pumped storage plant to compensate for the inflexibility of traditionally base load plants — large coal-fired and nuclear. These base load power plants operate most economically if their output is fixed at its optimum value. It is preferable to use the excess nighttime generation to pump water in a pumped storage scheme so that it may be released for generation during the peak demand.

Remark 3.3. It is important to mention that some technologies, for instance, wind and solar power plants, have variable dispatch of electricity due to natural conditions: wind speed may vary dramatically, and the clouds can also come unexpectedly. Therefore, these capacities have to be balanced by extra capacity available in the grid. These can be either peaking or semipeaking units that can ramp up quickly to deliver required power, or ancillary or stand-by plants. In either case, a thorough analysis of the grid has to be conducted to understand the limits of operation and to balance possible changes. This becomes even more critical in the phase of increasing share of renewables, which will be further discussed in Chapter 23.

3.4. Basics of electricity market

Electricity nowadays is treated as a commodity which still cannot be stored in large quantities. This commodity is bought, sold, and traded within the electricity market that involves a number of stakeholders, is regulated by the respective bodies, and is subject to market behavior.

Electricity market is a part of global economy from one side, and is an important component of the energy system. It is also a system that enables stakeholders to purchase a required amount of electricity through bids, sell electricity in required quantities and timeline, and conduct short-term trades on a short notice, depending on the local structure of the market. While electricity markets and their behavior differ in details in countries, they still have a number of common features and behavior patterns.

3.4.1 Major stakeholders

To describe how the electricity market works, it is useful to introduce various stakeholders that play major roles in these markets following the definitions from [5] and schematically shown in Fig. 3.5(A). Depending on the market and its state of development, not all of these stakeholders will be present in the power generation sector of a national economy. On the other hand, some companies can perform several functions.

Initially, there are essentially two major parties involved in a power market:

- *Generating companies* producing and selling energy (electricity and heat) via the physical dispatch;
- *Consumers* buying energy from the generators directly.

In a simple way, the physical power market connects a generating company to a consumer of power. The consumer is paying for the service

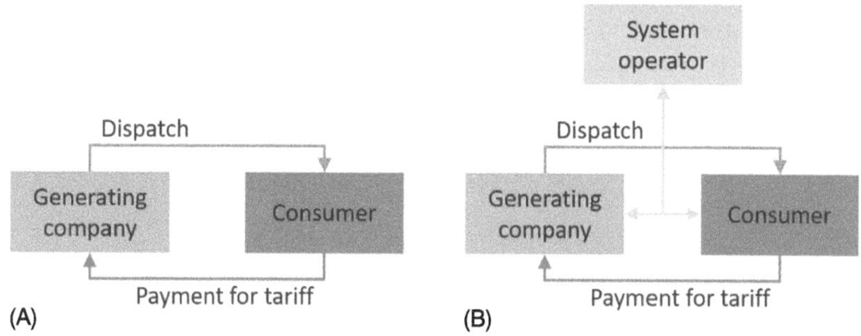

Figure 3.5 ***Simplified power market stakeholders.*** Power market stakeholders can have multiple forms of their interaction: (A) the ideal case of one generator and one consumer, and (B) the presence of the system operator.

according to the electric or heat tariff, while the generator delivers power based on the defined physical plan dispatch. However, this two-sided and simple market can only exist in an ideal and simplified world, when there are only one generator and one consumer, and the generator with always produce the required amount of energy which will always be consumed. In reality, this service is regulated by a system operator that gets ancillary services from the generator to allow for the dispatch control as schematically represented in Fig. 3.5(B). This way,

- *The system operator* has the primary responsibility of maintaining the security of the power system or basically a connection between the generator and the consumer. It shall be independent to not operate in favor of any market participant. It would normally own computing and communication facilities required to monitor and control the system.

In many cases, electricity (or literally kilowatt-hours) is sold and resold as a commodity a number of times before it is delivered to a final customer forming the *wholesale electricity market.* This type of market is usually open to any company, which has necessary approvals, can generate power and deliver it to the buyer. There can be multiple parties involved in-between that do not own any assets but can buy and resell electricity on the open market acting as traders.

We can split the power market players into three major groups. The first is formed by the asset owners and operators who physically operate and maintain the equipment and manage the whole system. They are:

- *Vertically integrated utilities* owning generating plants and T&D networks. In a traditionally regulated environment, such companies would have a monopoly for the production and supply of electricity and heat over a given area. Following the global liberalization of the electricity markets, its generation and T&D activities shall be separated.
- *Generating companies* producing and selling energy (electricity and heat). They may also sell services to the system operator to maintain the quality and security of the supply (e.g., frequency and voltage control). Depending on the size, such companies can own and operate a single plant or a number of plants of various technologies. They can coexist with vertically integrated utilities and are then called *independent power producers*.
- *Independent Power Producers* (IPPs) or *non-utility generators* (NUGs) which are private companies owning facilities to generate electricity and sell either to utilities, other resellers or straight to end users.
- *Transmission companies* owning transmission assets such as substations, overhead and underground (tunnel) lines with required equipment. They operate this equipment according to the instructions of the independent system operator.
- *Distribution companies* owning and operating distribution networks. Traditionally they have the monopoly to sell electricity to all consumers connected to their network. In a fully deregulated environment, the sale of energy to the consumers is decoupled from the operation, maintenance, and development to the network. Retailers then compete to perform this energy sale activity (one of the retailers can be a subsidiary of a local distribution company).

The second group deals with the energy market, operating and maintaining the process of buying and selling electricity and heat, and establishing economic relations among the stakeholders.

- *A market operator* typically runs a computer system that matches the bids and offers that buyers and sellers of energy have submitted. It also takes care of the settlements of the accepted bids: forwards the payments from buyers to sellers following delivery of energy.
- *Retailers* buy energy on the wholesale market and resell it to the consumers who are not able to participate in this market on their own. They do not have to own any assets, and their customers do not need to be connected to the network of the same distribution company.
- *Small consumers* buy electricity and heat from retailers and lease a connection to the grid from their local distribution company. Their partic-

ipation in the power market usually amounts to no more than choosing a retailer (if there is such an option).

- *Large consumers* often take an active role in the power market by buying their energy directly through the market (and not through retailers). Some of them offer their ability to control their load as a resource that the system operator can use to control the system (e.g., fast response to peak electricity demand). Large consumers can be directly connected to the transmission system.

We are now left with the third big player in the energy market that basically orchestrates the whole sector of economy:

- *The regulator* is the governmental body responsible for ensuring a fair and efficient operation of the power sector. It determines and approves the rules of the market, and investigates suspected cases of abuse of market power. The regulator can also set the prices for the products and services that are provided by monopolies (e.g., by vertically integrated utilities).

There are also some other stakeholders which can possibly impact the overall power industry directly, for instance, various NGOs. Moreover, in highly regulated environments like nuclear energy, there can be multiple regulators and international bodies involved in the process.

3.4.2 Models of market relationship

Due to many stakeholders, there usually exists certain competition within the electricity market, however, its level depends on the market regulation and how liberalized this market is.

Some authors [5] propose several major models of competition:

- *Monopoly*, as illustrated in Fig. 3.6(A), when there is a single vertically integrated utility in a region. This utility may own not only the generation and transmission part of the energy system, but also the distribution network. While these generation, transmission, and distribution parts can be represented as separate companies, in a monopoly scheme they are still integrated under one parent company. However, between the regions large-scale vertically integrated utilities may have bilateral energy trades.
- When a vertically integrated utility also acts as a purchasing agency, it allows independent power producers to connect to the network and sell their output to the agency, as schematically represented in Fig. 3.6(B). Further evolution of competition may result in a fact that there are

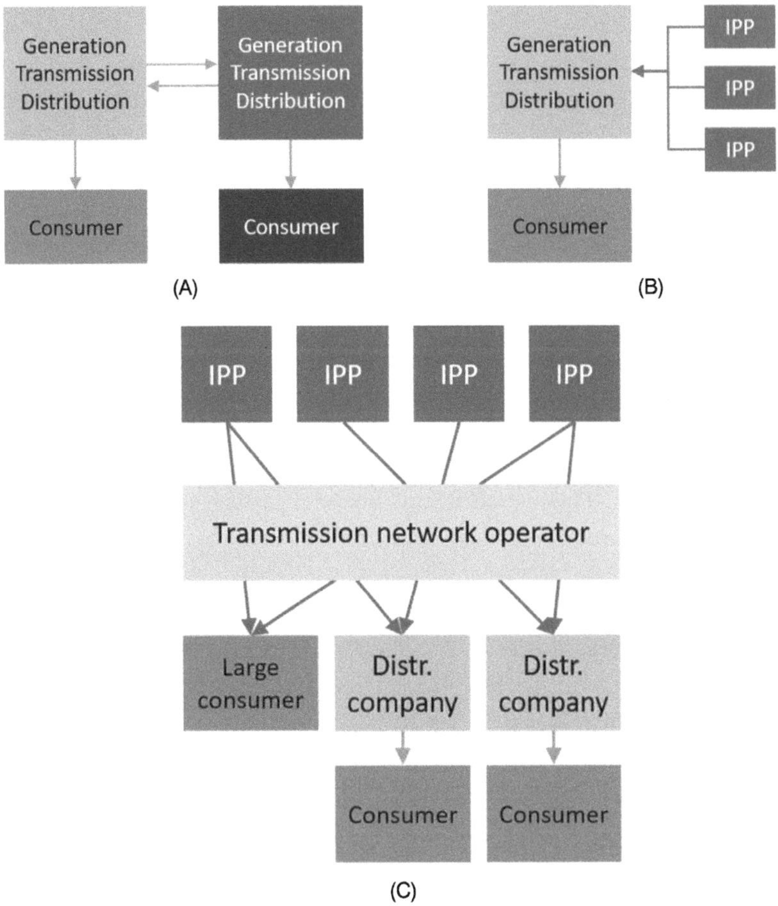

Figure 3.6 ***Power market models.*** Monopoly market model (A) and purchasing agency models: (B) vertically integrated utility purchasing from IPP, (C) wholesale market competition but still with centralized distribution system. Adapted with changes from [5].

no more vertically integrated utilities: instead, there are multiple IPPs who sell electricity to the wholesale purchasing agency which resell power further to disaggregated distribution companies and customers, which represents the basic wholesale market(s). However, the purchasing agency still has the monopoly position, so that the rates need to be regulated. This model at least introduces some competition into the generation market.

- *Wholesale competition* model assumes that no central organization is responsible for the provision of electrical energy. Instead, distribution companies purchase electricity consumed by their customers directly from generating companies, and these transactions take place in a wholesale market, as illustrated in Fig. 3.6(C). At the wholesale level, the only functions that remain centralized are the operation of the spot market, and the operation of the transmission network. At the retail level, the system remains centralized because each distribution company not only operates the distribution network in its area but also purchases electrical energy on behalf of the consumers located in its service territory.

This way, a wholesale electricity market appears only when competing generators offer their electricity to the market. Even more competition can be introduced within the fourth model:

- Retail competition model, finally, illustrates the ultimate form of competition in the energy market, in which all customers can choose their supplier. Because of the transaction costs, only large consumers choose to purchase energy directly in a wholesale market, while most of the medium and small consumers purchase it from the retailers, who in turn buy it on the wholesale market. In this model, the only monopoly functions are thus the provision and operation of the transmission and distribution networks. This model does not need price regulation, as even small consumers can change retailer when they are offered a better price.

While historically first electricity generating companies and the distribution networks appeared as private, most of them faced nationalization in the 20th century. When the liberalization of the market started to take place, in many countries the introduction of competition in the supply of electricity was accompanied by the privatization of some or all components of the industry, when they were sold by the government to private investors. Privatization is not, however, a prerequisite for the introduction of competition. None of the four basic models of competition descried above implies a certain form of ownership. Public utilities can, and in many instances do, compete with private companies [5].

It is also important to mention that liberalization of electricity tends to substantially benefit large consumers, mainly big industrial energy users, but benefits for small-size consumers compared with a public monopoly or a regulated private monopoly are questionable, as liberalization is suspected to pass on supply costs onto these consumers.

3.4.3 Types of energy contracts and markets

There are two basic ways to arrange electricity trade between generators and consumers:

1. They can sell electricity directly through bilateral trade. This trade may be conducted directly or involve a trade broker. Eventually, the system operator is also informed about the transaction.
2. Generators can sell electricity to an intermediary (e.g., retailer) who will then resell it to the consumer through mediated trade. In the most liberalized case, the pool market is used when all generators propose their capacities and prices to the market, and all consumers bid for the required amount of electricity. The system then automatically matches supply and demand.

While the bilateral trade offers fully customized transactions (as they are straightforward between two parties), mediated (centralized) market has lower transaction speed but much higher choice and flexibility for the stakeholders. However, the entire market can be a mix of markets: while for long-term energy contracts bilateral market is used, centralized market is implemented for shorter periods of time to match supply and demand in a more efficient way. Since the electricity demand is predictable but still rather volatile, electricity supply is subject to the market conditions and risks. As discussed earlier, there are several technologies available to fulfill electricity demand: base load, semipeaking, and peaking plants. Consequently, there are several types of trading to form electricity market and fill the needs for electricity.

Base load and predictable peak load electricity can be traded on a bilateral basis through various types of contracts, for example:

- Customized long-term contracts are very flexible contracts, which are negotiated between supplier and consumer (or retailer), and hence have specific conditions, pricing, etc. These are usually large transactions with high costs and large amounts of energy behind, negotiated for a long period of time.
- Forward and future markets, which usually operate from a year or more ahead. Forwards and futures are contracts for firm delivery of electricity at a certain time in the future for a price agreed upon today. Forwards are standardized contracts that can be further traded on power exchanges.
- Over the counter (OTC) trading is usually done using standard contracts with lower transaction costs for smaller amounts of electricity and shorter lead times.

For the flexible matching of supply and demand, several types of market are used:

- Day ahead market, which allows making deals between sellers and buyers for the delivery of power the following day, while the price is set and the trade is agreed at the moment a deal is closed. This trading is driven by the proper planning: for instance, a utility needs to assess how much energy it will need to buy tomorrow to meet the demand, and how much it is ready to pay for it on an hourly basis, and put a bid. The sellers, in turn, need to estimate how much energy they can deliver and at which price, which is reflected in the offer. These needs are reflected through orders entered by buyers and sellers into the market trading system.
- The intraday market supplements the day-ahead market and helps secure the necessary balance between supply and demand in the power market. At this market, buyers and sellers can trade volumes of electricity close to real time to bring the market to balance.

Remark 3.4. The intraday market is becoming increasingly important as more and more renewable sources of power with unstable dispatch are connected to the grid. This power, mostly coming from wind and solar, due to their unstable nature, is unpredictable and causes imbalances between day-ahead contracts (which are still a forecast) and real-time production of electricity. It is the role of intraday market and ancillary capacities to balance the market. Hence this type of market can be a key enabler to increase the share of renewable energy in the global energy mix.

3.5. Energy system safety as part of energy security

Ensuring the energy security of the country has been considered an important topic as shown by multiple research and business discussions resulting in a number of frameworks and models (see, for instance, [6–10]). The energy security concept has been evolving, from ensuring oil supply and dealing with the macroeconomic consequences of oil disruptions and/or price changes, to include many other concerns [9]. During the past few years, energy security topic has further risen up the global political agenda due to a number of important reasons [11]: rapid changes in oil and gas prices, highlighted awareness of political instabilities, and the increased impact of blackouts. It is therefore significantly important to understand the aspects of the energy security and its role within energy industry and overall economy.

Importantly, the energy security gap is commonly perceived by the public from the demand side perspective and therefore usually associated to insufficient supply or high prices of energy [9]. Therefore, the IEA defines energy security as the uninterrupted availability of energy sources at an affordable price [12]. On top of this, there are multiple dimensions of the energy security, which are derived from the numerous threats [11], including:

- Nonrenewable primary energy scarcity due to their possible depletion and unequal spread across the globe, which may lead to their price increase and other economic impacts;
- Potential external disruption to the primary energy supply due to political, social or other reasons, causing significant volatility of prices;
- Lack of domestic investments into the energy industry;
- Technology or system failure leading to blackouts and cut-offs. This may also include lack of investment into new build or innovative solutions and technologies leading to the increased average age of equipment and infrastructure;
- Political isolation causing a number of sanctions in either direction, mostly applicable in case of dependence on foreign energy sources or technology;
- Terroristic threat and regional instabilities;
- Global climate changes and natural disasters (e.g., earthquakes, tsunami, etc.) including other force majeure issues.

The whole energy system, being represented by interconnected and regulated generation and load capacities with the transmission network in between, is balanced in two ways:

1. The amount of energy consumed by the load part of the energy system shall be properly balanced by the generating part, i.e., by power plants;
2. As most of the electricity systems use alternating current, both generating and loading capacities shall operate within the predefined limits of the electric frequency, which has to be set and maintained for the whole grid. Frequency is considered among the most important characteristics of the quality of electricity. Stable frequency means that there is a balance between generation and consumption.

A risk of unpredictable disturbances exists on any network, for instance, unexpected outage of a power plant or part of the grid. This risk level can be set at least at the capacity of the largest generator (or a power plant in total) on the grid: in case of its failure, the system shall be able to maintain the power supply by introducing reserve capacities to the grid and providing overall energy security. Depending on the grid code of the country, there are various requirements to ensure energy security in terms of capacity and the speed of bringing it into operation. Generally, there are three major types of capacity required for these purposes [4]:

- *Spinning reserve* consisting of generators that are rotating synchronously with the grid but do not generate active power; it can respond within a short period of time to compensate for generation or transmission outages. These can be generators, which are rotating at the required frequency, but not connected and not synchronized with the grid.
- *Fast-start* or *supplementary reserve* consists of generators that are on standby but can be rapidly brought online within several minutes. These capacities are only used after all spinning reserve is online.
- *Replacement service* is related to those capacities, which can be put online for a longer period of time and not in an emergency mode. Such capacities are usually required to replace units, which undergo planned (or unplanned and emergency) maintenance, upgrades or retrofits of the capacities and the grid.

On the other hand, frequency control measures are used to keep the network frequency within the limits set out in the national grid code if a rapid and large loss of generation or an increase occurs. This requires sufficient operating reserves to be present on standby, and power plants may be required to declare up to 15% of their registered capacity as available for this purpose [4]. This means that a power plant shall be able to increase the frequency of the machine (rotation of the turbine and generator shaft), which in return shall lead to higher capacity demand.

Frequency response capabilities of the power plants are directly connected to their available extra capacity and flexibility and are usually divided into several time frames, which define the speed and the limits of frequency change or, in other words, how fast a power plant can change the rotation of the generator. This response of the network to a large and rapid frequency deviation is dealt within three stages [4]:

- *Primary response* is an initial automatic increase in power output intended to contain the frequency deviation over the first 30 seconds or so after the event and allow time for the secondary measures to be initiated. This response is usually managed through the special governor system of the turbine, and is aimed at keeping the frequency in the given boundaries. The capability of the system to make this response is called *primary reserve*.

- *Secondary response* also known as *re-dispatch* is the reaction of the energy system to bring the frequency back to its required level. It may take usually from 10 seconds up to tens of minutes and covers much bigger grid frequency instabilities. It is therefore the duty of the *secondary reserve* power plants to maintain the disbalance of the frequency till the time when the initial disturbance is fixed.
- *Tertiary response* is required when the problem cannot be fixed within the adequate period of time.

Moreover, while a large and united grid can provide certain stability as it incorporates multiple capacities and therefore can have greater flexibility, it also has to be capable of managing significant outages, which are unlikely to happen in case of natural disasters, technical issues, force majeure or any other unexpected accidents that may cause blackouts. In this sense, DC lines between separate energy systems can be of great importance as they can balance both systems in either way without the need of synchronizing frequencies. Other measures would include flexible capacities and proper management of the grid, especially in the face of distributed generating capacities introduced by small-size renewables.

References

[1] Intergovernmental Panel on Climate Change (IPCC). Climate change 2014: mitigation of climate change. Contribution of working group III to the fifth assessment report of the intergovernmental panel on climate change. Cambridge University Press; 2014.

[2] Energy Information Administration (EIA). International energy outlook 2016. U.S. Energy Information Administration; 2016.

[3] GE. Powering the world. Steam power systems product catalog. GE; 2015.

[4] Munoz-Hernandez G, Mansoor S, Jones D. Modelling and controlling hydropower plants. Advances in industrial control. London: Springer; 2012.

[5] Kirschen D, Strbac G. Fundamentals of power system economics. Wiley; 2004.

[6] Jewell J, Cherp A, Riahi K. Energy security under de-carbonization scenarios: an assessment framework and evaluation under different technology and policy choices. Energy Policy 2014;65:743–60.

[7] Mansson A, Johansson B, Nilsson LJ. Assessing energy security: an overview of commonly used methodologies. Energy 2014;73:1–14.

[8] Ranjan A, Hughes L. Energy security and the diversity of energy flows in an energy system. Energy 2014;73:137–44.

[9] Winzer C. Conceptualizing energy security. Energy Policy 2012;46:36–48. https://doi.org/10.1016/j.enpol.2012.02.067.

[10] Blum H, Legey LF. The challenging economics of energy security: ensuring energy benefits in support to sustainable development. Energy Economics 2012;34:1982–9. https://doi.org/10.1016/j.eneco.2012.08.013.

[11] Watson J, Scott A. New nuclear power in the UK: a strategy for energy security. Energy Policy 2009;37:5094–104.
[12] IEA. Energy supply security. Emergency response of IEA countries. International Energy Agency 2014.

PART TWO

Sustainable energy and power generation

Overview

The concept of sustainability is not an innovative concept of the 21st century, nor was it developed within the last couple of centuries. At its essence, sustainability is related to the basic needs of humankind to fulfill them continuously without the global ecosystem collapsing.

Energy in general is considered as one of the major enablers of development, although a major contributor to unsustainable future. With electricity gaining a much higher share, sustainable power generation becomes one of the key components of sustainable development.

In this part we focus on the approaches towards defining sustainability, its limitations, and challenges. Next, we address energy production and consumption patterns, as well as its impact on the environment.

CHAPTER FOUR

The system boundaries of sustainability

Contents

4.1. Introductory basics from systems theory

The system approach is considered to be among the most powerful tools for the analysis of complicated structures that are influenced by many parameters and which function in a complicated, sometimes even unpredictable, manner. Contrary to many other approaches of research and analysis, the system or systemic approach provides broadly applicable concepts and principles, which are consolidated and described within the general systems theory. Such an integrative approach is usually attributed to the names of W. R. Ashby, G. Bateson, L. von Bertalanffy, and some others [1], and further developed by many prominent scientists from multiple knowledge domains.

4.1.1 Definition of a system

The system approach, in general, considers the elements and interrelations between them, which all together form the whole, which may also be subject to some outer environment. Therefore, by definition, a *system* is an organized entity which is made up of interrelated parts or objects. These

https://doi.org/10.1016/B978-0-12-817012-0.00014-1

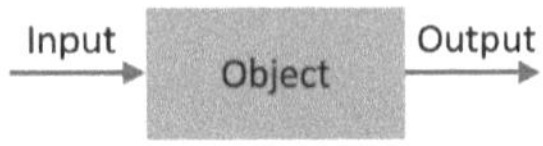

Figure 4.1 ***Basic definition of a system.*** A basic system consists of an object which transforms some input into some output.

objects can receive the inputs from the environment and other objects within the system, and transform inputs into outputs.

At the most basic level, a system can be represented by one object, which transforms inputs of the environment into some outputs, as represented graphically in Fig. 4.1.

For example, consider a typical coal-fired power plant. Under basic assumptions, we can consider it as a system that consists of one object — power plant itself, seen as a whole without detailed description of what is going on inside. The inputs of this system would be therefore coal as a fuel, air to maintain combustion process, water, and some others. The outputs will be electricity and heat, as well as waste produced by the system, e.g., carbon dioxide emitted to the atmosphere.

Such a graphical representation of a system is regulated by various standards of modeling, for instance, IDEF0 methodology, which stands for Integration Definition for Process Modeling. This methodology allows us to represent the systems through their process description and visualize interaction.

Depending on the goal of the observer, the system can be detailed further down to its objects. This can be done through:

- Considering the internal interactions and links among these objects within the system itself, or
- Consider each of these objects as a separate system called a *subsystem* of a parent system.

Both approaches together are illustrated in Fig. 4.2, where a system consists of multiple objects, which may in turn behave as smaller systems. Such a decomposition approach allows building a hierarchy of systems with the required level of detail. The concept of a subsystem also allows us to decompose the studied complicated system and treat its smaller objects separately as smaller systems using the same tools and methodologies.

4.1.2 Goal and limitations of a system

The implementation of the system approach starts with the definition of the goal of a system, which can be addressed in many ways. Throughout

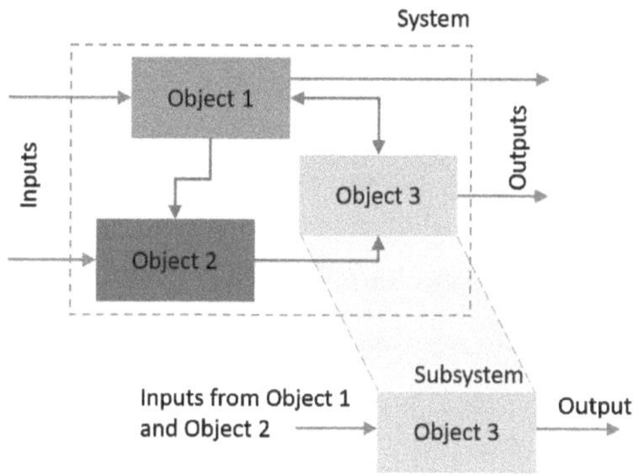

Figure 4.2 ***Concept of a subsystem.*** A system consisting of several objects, where one of them is studied as a subsystem.

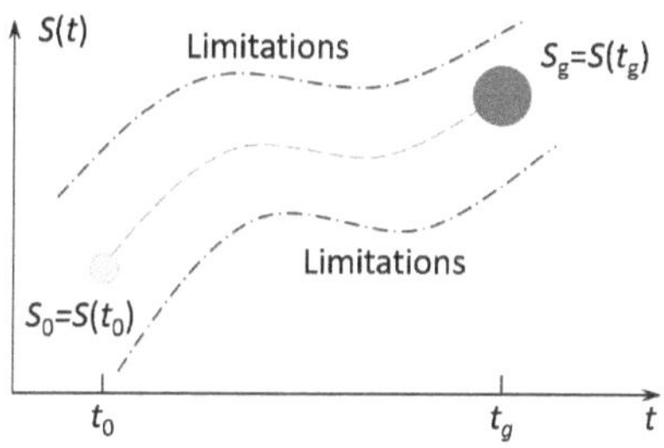

Figure 4.3 ***Schematic representation of the evolution of a system.*** State S_0 of the system at the initial time moment t_0 and at time moment t_g after its evolution when management is applied.

this book we will assume that the goal of a system is the definition of its desired state at the known time moment, and this state can be quantitatively (or sometimes qualitatively) and unambiguously described. Consider, for instance, an organization or an economy that has the strategic direction or a vision to act in the sustainable manner. This means that to achieve the goal the company has to be managed from the initial point of time t_0 to the agreed goal point t_g, while each time moment is characterized by a number of parameters to define the state of the system $S = S(t)$ and therefore assess it, as schematically represented in Fig. 4.3. This controlled evolution of the system from the initial state to its goal state is the responsibility of the management or the authorities, who are defining the activities according to the known and emerging limitations.

These limitations are by far the largest challenge for a manager and may include, but are not limited to, the following components:

- The initial state of the system, e.g., the way a given organization is set up when the management takes over the execution. This defines the so-called input parameters of the system, for example, for the company it will be the given setup, portfolio, processes, assets, liabilities, people, etc. These parameters shall allow for an unambiguous definition of this initial state of the system (in a simple way, in terms of money) and therefore give the possibility to set up the goal as the desired state of the system on a specific time scale. As the initial state is given and cannot be changed, we shall consider it as a limitation.
- Resources that are and will be available to achieve the goal. These can be materials for production, personnel, available assets, cash, etc. On a larger scale of a country or global economy, these can be further expanded to three major categories: human capital, natural resources (for material production), and energy for doing the work in a wide sense. Resources are by far the most complicated, and therefore most limiting, factor and shall be carefully assessed and observed.
- Environment, which usually stands for the natural habitat (like atmosphere or biosphere), as well as political, social, and cultural background developed and shaped by the humankind. While we live in a society, we have to follow certain rules and regulations which form the normative framework that includes policy makers, competitors, non-profit or non-government organizations, customers and suppliers, associations, and many other stakeholders that can limit (or sometimes broaden the boundaries of) the freedom of the system on its way towards the goal of sustainability.
- Managerial competences and skills that limit the activity of a person and hence impact the management process in general. As an example, one can mention educational background, socio-political views, personality, or other hard and soft skills that can impact the decision-making process in either way.
- Technology level, which is a limiting factor in that it may not allow achieving the goal at all or with desired efficiency. This is especially important for an energy transformation chain, where the efficiency levels of some of the cleaner technologies are still low, which leads to production of waste.

Most of the limiting factors are known and have been implemented in management practice, for example, in St. Gallen Management Model proposed in [2] and designed as an extensive tool for the in-depth examination of management. In the author's opinion, however, among these limiting

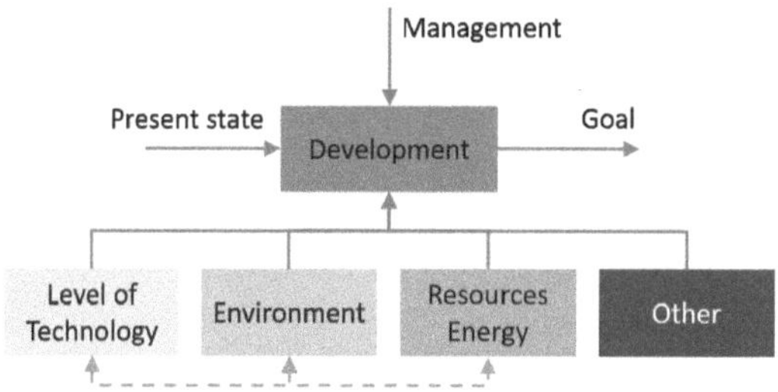

Figure 4.4 ***Schematic of the system with the limitations and management.*** Schematic diagram of the development process from the present state to the desired goal under management in the face of limitations: environment, level of technology, resources and energy, and others. Environment, in turn, is impacted both by resources and technological level.

factors, energy as a global resource plays the dominant role when we talk about the sustainability as an attribute of a goal on the global scale, i.e., the humankind. To support this, we shall represent the above mentioned model with the help of IDEF0 methodology that allows us to represent the actions and activities of the system under limitations. Fig. 4.4 shows a schematic representation of the system's activity which we name as "development": it is basically aimed at bringing the system from the present or initial state (the input) to the desired goal, however, omitting the discussions of the concept of the goal for sustainable development.

To carry out the system development process, management as the control activity shall be applied. Generally, management is also subject to a number of limitations, for instance, capacities, competences, people, external framework, etc. All limitations, formally represented in our model within four major blocks — environment, level of technology, resources and energy, and others — regulate and guide the system on its way towards the goal. As discussed in detail previously, energy as one of the most important resources for development is highly influenced by the technological level, that is, the efficiency of the present energy conversion principles. In turn, it has direct impact towards the environment through GHG emissions, solid waste production, etc.

This gives us confidence to state that a combination of energy value chain knowledge, technology and innovation capabilities, and the sustainable environment understanding, altogether referred to as energy for sustainable development, are among the most complicated and important factors of the sustainable development as they can question the overall attainability of the goal of the humankind.

4.2. Definitions and principles of sustainable development

The concept of sustainability is not an invention of the late 20th century. At its core, sustainability is related to the basic human need to maintain and nurture the conditions of which life depends. Sustainability therefore has been seen as the antonym to "collapse" [3].

While historically people subconsciously or from observations understood the limitations of the environment, for instance, in the amount of crops from the fields, timber production and the danger of deforestation, limits of fishing and bans for hunting during the breeding periods, these consequences were obvious and well-understood by one generation. Along the humankind development, when more and more tools had been developed and new technologies had been brought into life, the effects of their implementation became more vague.

From the emergence of industrialization at the end of the 18th century until the mid-20th century, for most people, development was largely about economic and social issues. Questions of survival and improving work conditions were more urgent than what we would today call environmental problems. In addition, new methods in agriculture and food industries improved food supplies and, in spite of greater opportunities for consumption, the population grew slowly or remained stable [4]. Moreover, the impact of industrialization on the environment has not been considered also due to its longer-term impact.

It was not until the 1960s that many scientists put nature and the environment back on the economic agenda, opening the discussion of the impact of such rapid, yet uncontrolled development and consumption, so that environmental protection became an issue of growing public concern [4]. This was partly impacted by a series of catastrophic accidents that started to pop up as a consequence of industrialization: smog, oil spills, chemical poisoning. This led to the fact that in 1972 the first United Nations Conference on the Human Environment took place in Stockholm. General understanding of the countries towards the need for a common outlook and principles to inspire and guide people of the world in the preservation and enhancement of the human environment resulted in reaching an agreement on measures to limit industrial pollution and protect the environment [5].

The final report from this conference [6] proclaimed a number of principles which may well be considered as a basis for further development of the humanity, including, among others:

- The natural resources of the Earth, including the air, water, land, flora, and fauna, and especially representative samples of natural ecosystems, must be safeguarded for the benefit of present and future generations through careful planning or management, as appropriate.
- The capacity of the Earth to produce vital renewable resources must be maintained and, wherever practicable, restored or improved.
- The non-renewable resources of the Earth must be employed in such a way as to guard against the danger of their future exhaustion and to ensure that benefits from such employment are shared by all mankind.
- The discharge of toxic substances or of other substances and the release of heat, in such quantities or concentrations as to exceed the capacity of the environment to render them harmless, must be halted in order to ensure that serious or irreversible damage is not inflicted upon ecosystems.
- Economic and social development is essential for ensuring a favorable living and working environment for man and for creating conditions on the Earth that are necessary for the improvement of the quality of life.

It has already been understood that rational planning constitutes an essential tool for reconciling any conflict between the needs of development and the need to protect and improve the environment. More importantly, it has also been understood that the developing and underdeveloped countries wanted to develop through the rapid industrialization, which usually does not take ecological considerations into account. Therefore, the conference members have agreed that the global community of developed countries shall support the other world members on their way with economic factors as well as ecological processes being properly taken into account.

Evidently, this desire from the developed world to ensure ecological compliance comes from the fact that these countries have already undergone the harsh process of industrialization, which caused severe damages to the environment. This way, the hypothetical impact of the industrialization of the developing countries might bring much bigger damages because of the relative scale of the developed part of the world and the rest.

The 1972 book *The Limits to Growth* [7] rekindled a centuries-old debate about whether continued industrial growth would result in serious, potentially irreparable harm to humans [8]. According to the authors of the report, "if the present growth trends in world population, industrialization, pollution, food production, and resource depletion continue unchanged, the limits to growth on this planet will be reached sometime within the next one hundred years. The most probable result will be a rather sudden and uncontrollable decline in both population and industrial capacity" [7]. However, according to their studies, it can be still possible to alter such uncontrolled growth trends and establish a set of conditions or ecological and economic stability that would allow for a sustainable future. This set was aimed to fulfill the basic material needs of every individual and provide an equal opportunity to realize individual human potential. However, action should have been taken as soon as possible.

Since those days, the popularity of sustainability has increased rapidly for several reasons, including concerns about population growth, increasing resource consumption, possible and already foreseen depletion of crucial resources, and the widespread deterioration of ecological conditions [9]. These were some of the reasons when in 1983, the United Nations appointed a World Commission on Environment and Development (WCED) chaired by the Norwegian Minister President GRO HARLEM BRUNDTLAND. The Brundtland Commission, as it came to be known, published its final report "Our Common Future" [10], providing what came to be the best-known definition of the concept of sustainable development [5]. The Brundtland Commission defined that to make development sustainable was "to ensure that it meets the needs of the present without compromising the ability of future generations to meet their own needs" [10].

However, the simplicity of this approach is deceptive, and obscures underlying complexities and contradictions [11]. However, the definition of the *needs* differs for various cultures and various stages of economic and political development, while it has been mutually agreed that sustainability was supposed to be for the whole humanity. Secondly, there is no clear vision of who and on which basis is going to decide on the preferable way of development.

Since the path-breaking deliberations of the Brundtland Commission, the expression *sustainable development* has been used in a variety of ways, depending on whether it is employed in an academic context, or that of planning, business or environmental policy [11]. Being broad enough, this concept possessed a number of strengths that made the most popular definition of sustainable development. First, it captures the essential elements of sustainability as the balance between the society and the environment for the sake of future generations. Second, its rather vague and generic nature allowed this definition to be adopted by various fields and activities and multiple purposes, so that sustainability has become a new and widely used buzz word. Multiple attempts have been undertaken to further develop the sustainability paradigm.

4.3. Sustainability paradigms and frameworks

From most of the definitions it has been clear that sustainability is about achieving the balance between human needs and environmental integrity, which is increasingly difficult in conditions of resource scarcity, and between intra-generational and inter-generational equity, which is only realizable when the first condition is met [9]. Meeting all human needs would require economic and social development, which has to be in line with nature. Therefore, sustainability has always been seen as having three major dimensions: economics, society, and the environment.

4.3.1 Triple bottom line

The triple bottom line term was coined in the early 1990s by business consultant John Elkington to describe economic, environmental, and social value of investment that may accrue outside a firm's financial bottom line [12]. This approach aims to more accurately value assets and leverage resources, so that capital is employed as efficiently and effectively as possible [13]. Therefore the triple bottom line approach expands the traditional reporting framework to take into account social and environmental performance in addition to the financial one.

With the rising popularity of sustainability in the business world, the political arena, and academia during the past two decades, TBL has been widely used as a conceptual standard for assessing and promoting corporate social responsibility in particular and sustainable development in general [9].

In [13], the authors define TBL or sustainable economic development as "programs, policies, or activities designed to create or retain jobs and wealth in ways that contribute to environmental, social, and economic well-being over time". According to the authors, it is distinct from economic growth or development, which may or may not contribute to overall well-being, including quality of life, fiscal health, resource stewardship, and resilience. This approach suggests that "economic systems exist to serve human well-being, that human and economic well-being are inextricably linked to environmental well-being, and thus, that human, environmental, and economic well-being must be considered in the design and evaluation of economic development efforts" [13].

This approach can be schematically represented as the intersection of the three dimensions: economy, society and the environment. From this perspective, one of the major challenges has been to understand the interrelation among these three dimensions according to the scale and timeline, if these dimensions can substitute or compensate each other, or if they can be further extended.

Further development of the sustainability concept led to the general understanding that a society is developing sustainably if it passes an undiminished stock of *capital* C from one generation to the next. In this context, capital comprises [14]:

- *Manufactured capital* C_m, including machines, buildings, infrastructure;
- *Human capital* C_h like knowledge, skills, benefits, values; and
- *Natural capital* C_n, e.g., minerals, air, water, soil, flora, fauna, ecological systems,

which in a simplified way can be mathematically represented as

$$C^{(t)} = C_m^{(t)} + C_h^{(t)} + C_n^{(t)}$$

for a considered time moment t.

Generally speaking, development is usually characterized by an increase in the stock of the manufactured and human capital over the period of time $[t, t+1]$, i.e.,

$$C_m^{(t+1)} \geq C_m^{(t)}, \quad C_h^{(t+1)} \geq C_h^{(t)}.$$

While it is still possible that these two types of capital can be depleted (either due to natural or artificial circumstances), economists are mostly focusing on the problem of depletion of the third part, namely the natural capital, which satisfies

$$C_n^{(t+1)} < C_n^{(t)}.$$

This happens due to the fact that the economic activities of the humankind have negatively influenced this natural capital over time, as the development is naturally characterized by the consumption of this capital for the sake of the development. This results in its continuous depletion, for instance, because of the extraction of natural resources, loss of top soil layers, accumulation of emissions in water and air, etc.

With these major categories of capital: manufactured C_m, human C_h, and natural C_n, and with what appears historically to be an increase in the first and second at the expense of the third, an immediate question arises whether this kind of development can be sustained. Will the well-being of future generations be maintained if some measure of aggregate capital *C* is constant, or are there limits to substitution that require minimum amounts of each type of capital to be passed from generation to generation [14]? Is there a possible way of such type and direction of development when the three parts of the capital either continue their growth unlimitedly or at least remain constant without damaging the others?

In defining the conditions for sustainable development, a differentiation is usually made between Solow–Hartwick, or the weak sustainability, and Holling, or the strong sustainability [11,14,15].

4.3.2 Weak sustainability

The economic growth theory considers sustainable development in the way of non-decreasing welfare over time, so that it is seen as a problem of managing a nation's portfolio of capital to maintain it at least at a constant level [16], with growth as the best case. However, persistent economic growth would lead to a violation of the ecological boundaries of the Earth [7]. Based on these findings, Robert Solow started a discussion about the extent to which optimal growth and a non-decreasing level of welfare will also be possible in the future [15]. This question has laid the foundation for what is now called the Solow–Hartwick sustainability concept [17].

This concept roots from the conclusion that it is impossible to reach a non-decreasing consumption path by using non-renewable resources [18, 15]. Natural resources are considered as finite, non-renewable, and, importantly, essential for production. This finite amount of resources would allow for a finite amount of output which cannot be sustained for an unlimited period of time, which indirectly sets the boundaries of growth.

However, Stiglitz [19] further develops this idea concluding that "the fact that there is a limited amount of natural resources and natural resources are necessary for production does not necessarily imply that the economy must eventually stagnate and then decline". His model of sustainable development points out technical progress as one of the ways to avoid the depletion of natural capital.

Therefore the concept of weak sustainability assumes that the savings from the extraction of non-renewable resources shall be reinvested in manufactured capital or human capital and that the latter are perfectly substitutable for natural capital. Furthermore, levels of capital are considered to be irrelevant, only changes matter. This leads to an interestingly strange fact that countries with a history of resource depletion and ecosystem damage may look sustainable [16,20].

The idea of trade-offs between the three types of capital is reasoned by the belief and trust in technology, which will permit future generations' well-being with the usage of less resources. Thus, current exploitation of resources is reinvested in the development of technologies that allow living with less material and energy inputs (and outputs). The basic assumption behind this mechanism is the possibility to decouple economic growth from resource usage [21].

Therefore one can conclude [15,22] that the concept of weak sustainability is a direct application of the Hartwick–Solow rule [20,23] in the case of non-renewable resources. A non-declining path of consumption can only be achieved if the capital stock does not decrease, meaning that the total capital stock as the sum of natural, human, and man-made capital remains at least constant. However, according to the concept, the composition of this total capital stock, which will be handed over to the next generations, is not relevant as long as the total stock does not decrease. This means that the decrease of the natural capital stock may be well compensated by the human or man-made stock increase.

Hence, the idea of the weak sustainability, in general, can be represented in the following way:

$$C^{(t+1)} \geq C^{(t)}, \tag{4.1}$$

without any specific condition on the change of C_m, C_h or C_n, and represents a pure economic sustainability approach.

4.3.3 Strong sustainability

The concept of sustainability has also been viewed in the sense of a closed system, whereas the Earth has been seen as a "spaceship". This idea, developed in detail by Boulding [24], grows out of the past open economy of illimitable resources, called a *cowboy economy* — "the cowboy being symbolic of the illimitable plains and also associated with reckless, exploitative, romantic, and violent behavior, which is characteristic of open societies" [24]. Similarly, he introduces the term *spaceman economy* of the future, in which the Earth would be considered as a single spaceship, without unlimited reservoirs of anything, either for extraction or for pollution, and in which, therefore, man must find his place in a cyclical ecological system which is capable of continuous reproduction of material form even though it cannot escape having inputs of energy.

While the weak sustainability tends more to the cowboy economy characterized by consumption and substitution of depletable natural capital by production, in a spaceship economy the essential measure of success is not production and consumption at all, but the nature, extent, quality, and complexity of the total capital stock, including in this the state of the human bodies and minds included in the system [24].

This philosophical work of Boulding [24] poses an important question of the closed system of the Earth, which is somewhat opposite to the open Earth vision of the past, and where boundaries are becoming more and more visible. These are the limits the humankind would be gradually facing on the way towards its development, and the stability and resilience of the system have to be understood and maintained. Moreover, the system is subject to the entropy law: in the context of entropy, every action or any process in nature must result in a deficit for the entire system [25]. This approach has been further developed by Georgescu-Roegen, who stated that "as long as there are abundant, easily accessible resources around, we might not really care how large this additional loss is... If there were not this entropic deficit, we would be able to convert work into heat, and, by reversing the process, to recuperate the entire initial amount of work — as in the imaginary world... In such a world, standard economics would reign supreme precisely because the Entropy Law would not work" [25].

This shifts the term sustainability from pure economic towards ecological domain, where Holling defines resilience of an ecological system [26] as the ability of to maintain the structure of a system in the case of disturbance to guarantee its functioning. The representatives of this approach concluded from the works of Boulding and Holling that without a comprehensive ecological system it is not conceivable that people on the Earth could work and act. Therefore they demanded that the economic system had to work within the scope of the reproductive and loading capacity of nature [15].

Therefore, this approach, usually called strong sustainability, is also known as Holling sustainability. It assumes that sustainable development is only achievable when the actions of the humankind are taken within the natural limitations of the global ecological system of the "spaceship Earth". Hence, strong sustainability requires maintaining both humanly created and natural capital intact separately, under the assumptions that they are compliments rather than substitutes in most production functions [15,27].

Compared to weak sustainability, strong sustainability, in contrast, is not so generous with trade-offs between the different aspects of sustainability.

There are variations of the extent to which humans are allowed to destroy the environment for their well-being. However, the main difference is that the economy is not seen as superior to the environment and that technological innovations will not provide the sole solution for the sustainability challenge [21].

Consequently, strong sustainability demands the same inequality (4.1), however, closed with two strict conditions:

$$\begin{aligned} &C^{(t+1)} \geq C^{(t)}, \\ &C_n^{(t+1)} \geq C_n^{(t)}, \quad C_m^{(t+1)} + C_h^{(t+1)} \geq C_m^{(t)} + C_h^{(t)}. \end{aligned} \tag{4.2}$$

This literally means that the decrease of the natural capital cannot be substituted by an increase of the human or manufactured capital.

Remark 4.1. Generally speaking, the approaches of weak and strong sustainability are not conceptually different. Weak sustainability can be seen as a necessary but not sufficient condition for strong sustainability, leading to the logical conclusion that strong sustainability is far more difficult to achieve and maintain from all perspectives. This is mostly due to the fact that weak sustainability demands keeping the aggregate capital as non-decreasing, while strong one requires natural capital to be non-decreasing as well.

Moreover, strong sustainability is purely in line with the TBL approach and represents a kind of a "win–win–win" situation, when all three dimensions complement each other.

4.4. Sustainability goals and metrics

Specifying the concept of sustainability and developing strategies for its implementation is an enormous challenge. And there are many different approaches that can be found in the literature about the goals, strategies, and instruments of sustainable development [5].

4.4.1 Sustainable development goals

The United Nations held a number of conferences to come up with a joint understanding of sustainability issues that are faced by the humankind. During these years, the concept of sustainability has been used in varied context, however, mostly implying the needs of the environment, also pointing out the global nature of these needs. A critical look at sustainable development discourses [11] suggested that the perceived need for global management of the environment stemmed, in part, from the assumption that it provided a way of correcting the anomalies of economic and trade policy.

One of the outcomes of Rio+20 Conference, "The Future We Want" [28], was the agreement by member states to launch a process to develop a set of sustainable development goals, or SDGs, which will take center stage at the post-2015 development agenda. These goals are aimed at transforming the current abominable conditions of poverty, health, education, employment, and emissions, among other issues, worldwide, especially in the least developed countries. These goals are designed to engage local, national, regional, and international players in solving the most chronic of economic, social, and environmental challenges the humanity have encountered through collaboration [29]. One of the major points proclaimed in the report [28] was that the member states *acknowledge the need to further mainstream sustainable development at all levels, integrating economic, social, and environmental aspects and recognizing their interlinkages, so as to achieve sustainable development in all its dimensions.*

In the 2015 Report "Transforming our World" [30], the United Nations finally announced a set of SDGs to guide world development until 2030. However, in stark contrast to the 1987 UN report "Our Common Future" [10], the reality of environmental limits and the potential drawbacks of ever-increasing economic growth have not been firmly placed on the sustainable development agenda [31].

These goals incorporate various goals covering all areas of global concern, including social issues like poverty, equality, education, growth; problems with resources (water, food, energy); and the environment we live in: climate, life on land and below water. This, evidently, corresponds to the Carter and Rodgers Triple Bottom Line Sustainability Model proposed in [12] and further developed in [32] which, at a broader level, consists of three components: the natural environment, society, and economic performance.

The Agenda includes a set of 17 Sustainable Development Goals which incorporate 169 individual targets and 232 indicators, ranging from poverty elimination, improving health and well-being, and quality of education to protecting and restoring life in the oceans or on the surface. According to [30], the SGDs seek to build on the Millennium Development Goals (which were mostly targeting poverty, global hunger and health) and complete what these did not achieve. Even though these goals are considered as a non-binding agreement, they do set an ambitious and transformational vision that would require global partnership to ensure its implementation. While each country has primary responsibility for its own economic and social development, they are expected to set their own priorities, allocate resources and build capacities to meet the targets. The implementation would first take place at the national level, including the generic stages of the policy-planning cycle, from prioritization of targets and indicators through to policy evaluation, decision-making, and implementation [33].

The publication of a comprehensive road map of goals and targets is already a great achievement on the way towards sustainable development to ensure simultaneous transformation of social, ecological, and economic spheres. However, in order to make this policy happen and lead to an anticipated result, these goals and respective targets have to be evaluated to ensure they rather complement than contradict each other. This question has been an important topic as it gained attention within the community and resulted in a number of research articles, which conduct thorough analysis (see, for instance, [33–36]).

While most of the policymakers and governments operate in silos [36], both on national and international level, various targets most probably be approached separately without overall coherent approach, so that potential synergies may be missed and unlikely trade-offs be exercised. Therefore policymakers require a comprehensive framework to approach these 17 SDGs based on deep understanding of their potential interactions. Several of such frameworks regarding the possible interactions between individual goals and the scale of their impact have been proposed by scholars [35,36].

Having said that, the sustainability goals, though aimed to cover major issues of the humankind on the way towards sustainable development, have to be considered carefully and integrally, with a number of precautions to be taken by a nation. Countries must interpret and rank the SDGs according to the resources, geography, level of development, governance, and national context instead of general knowledge to come up with reliable prioritization, as well as timescale. Another important point to mention is the exchange of information among nations, not only regarding the progress but also the basis for evaluation, preferably within a single platform which is unfortunately missing at the moment.

With this in mind, there have been several open questions posed regarding the 17 SDGs and respective targets:

- What is the overarching measure of sustainable development, which can provide an aggregated index to measure the progress towards global sustainability. According to the literature [37], an SDG Index was proposed, which approximated the performance for each goal and then aggregated the overall performance. The problem here lies in the absence of aggregated data for all 169 targets for all member countries. Being limited and not yet perfect, such an aggregated SGD Index still makes an attempt to measure the progress.

- Do nations possess a tool to properly assess the existing information, prioritize the sustainability goals and model scenarios to perform analysis and predict results? As recent research suggests [33], scenario analysis has been a method that is particularly well suited to simulate a long-term view so that it can be an invaluable tool for governments and policy-makers on the way towards SDG implementation. Various models have therefore been proposed and implemented, employing top-down and/or bottom-up approaches, as well as backcasting [33].
- How to ensure proper stakeholder engagement on national, regional and global level? While the SDGs have an overarching goal to fight poverty, climate change and other issues of the humankind, it requires efforts from all people and requires all members of the society to have an opportunity to participate in a decision-making process. On the national level, it may require limited consumption and societal changes. However, according to [38], the rise in income inequality seems to have leveled off during the last 10 years, however, many countries still remain pretty unequal. People in equitable countries emit less CO_2, produce less waste, and consume less water for personal use [38].

Despite ongoing criticism, the SDGs represent a much better and more comprehensive framework than the MDGs. Their scope is broader and includes important issues to address environmental challenges, social problems of inequality, governance, and many others. However, considerable efforts are required on both national and global level to identify SDGs most relevant to the local context and ensure they are properly achieved within the agreed timeline. Moreover, it has to be carefully observed and controlled that these goals do not lead to unexpected trade-offs.

4.4.2 Sustainability metrics

The definitions of goals should always imply the measure to quantify and estimate the progress and the impact of the activities towards that goal. This also applies to sustainability, which is though to be much more complicated to quantify properly.

Therefore, sustainability metrics are measures of sustainability, which attempt to quantify the current state and the progress of sustainable development. These metrics are usually influenced by various approaches, disciplines, political and social situation, and other factors, and there is no single approach how to quantify sustainability or the progress towards sustainable development. Because of a large number of systems related to or impacted by the humankind, even with regard to the power generation sector, many different indicators have to be implemented to track the progress.

It has been generally understood that the sustainability can only be achieved by integrating the different dimensions of economic and social development [5], so that the indicators of sustainable development to follow the same logical categorization into:

- Environmental, which include the levels of GHG emissions, fresh water supply and consumption, levels of pollution, biodiversity, generation of waste, and energy use (total and per capita production and supply of energy, electricity, and CO_2 emissions).
- Economic indicators, which usually include total and per capita GDP, GNP, national debts, investment rates and flows, and many others.
- Social indicators, which resemble the levels of poverty, education (level and access), and health (level and access), and may be defined in multiple ways.

The objectives of each of the 17 SDGs mostly have measurable targets to track the progress. As the SDGs themselves, the indicators also cannot be considered fully independent. In particular, energy is a major enabling force for many of the above indicators [39]. Reliable electricity supply is a prerequisite condition to increase the level of health and education, to provide access for information, and to develop the required infrastructure for economic development. The improved living conditions would then drive the longevity of life and impact the population level.

Similarly to other areas, there are several major categories of sustainability indicators [39]:

- Stock or total indicators, which are measures of total inventories, for instance, population, total proved fuel or material reserves, total land covered with forest or used for agriculture, etc.
- Flow indicators, which measure the changes of the values, for example, primary energy supplies, foreign investments, etc.
- Composite or multiattribute indicators, which use a combination of multiple individual indicators, e.g., SDG Index [37].

Stock or flow indicators are usually employed to track progress of an individual goal or metric, but may not be efficient in tracking the progress towards sustainability in general. Contrary to these simple indicators, composite or multiattribute indicators are much more complicated and hence difficult to valuate, and may require certain efforts to interpret results.

4.5. Concluding remarks

It has been generally agreed by the scholars, politicians and economists that sustainability comprises three major components: economic, social, and ecological. Major paradigms of sustainability do employ all three components, presenting the fundamental idea of growth as a red thread. However, development is mostly considered in the sense of building up economic or manufactured capital, which is relatively easy to estimate, calculate and understand how to achieve.

A far more difficult is the change of human capital, which is not that evident to evaluate. Moreover, setting up a clear goal of human capital is still an issue. Despite most of the scholars agreeing that the social side includes improved living conditions, namely longevity of life, availability of medical and educational services, these social goals cover only one portion of human capital development. Evidently, it is the most essential and emerging side, it may well be considered necessary but not fully sufficient. We might think of the possibility of a human to fulfill his/her ambition towards information in a general sense, including new knowledge, scientific search, ability to perform these activities to broaden the overall human capital, i.e., global information field.

Sustainable development is about the quality of life of the humankind. Created by humans and for humans, the concept of sustainable development focuses on providing the humankind an opportunity to increase the artificially created part of the world capital by not damaging the natural capital. This, according to the conservation laws, may seem unachievable, though on the scale of the humankind we do have an unlimited pool or resources from the outer space with the Sun taking center stage. This, however, is limited to the level of consumption, which shall be properly managed to minimize the probability of the "point of no return" when the consumption rate reaches the level at which the production of capital is not able to reproduce the natural one, even at the highest cost and under most urgency.

Finally, energy in all senses is one of the major components or critical points when we target sustainable development. Thanks to the Sun, we have the practical experience of life and theoretical potential of unlimited source of energy, at least on the human time scale.

References

[1] Dekkers R. Applied systems theory. Springer. ISBN 978-3-319-10845-2, 2015.

[2] Rüegg-Stürm J, Grand S. The St. Gallen management model [English translation of the second German edition]. Bern: Haupt Verlag. ISBN 978-3-258-07972-1, 2015.

[3] Grober U, Cunningham R. Sustainability: a cultural history; 2012.

[4] Heinrichs H, Martens P, Michelsen G, Wiek A, editors. Sustainability sciences: an introduction. Springer; 2016.

[5] Michelsen G, Adomßent M, Martens P, von Hauff M. Sustainable development — background and context. In: Heinrichs H, Martens P, Michelsen G, Wiek A, editors. Sustainability science. An introduction. Springer; 2016. p. 5–29 [chapter 2].

[6] United Nations. Report of the united nations conference on the human environment. UN; 1972.

[7] Meadows DH, Meadows DL, Randers J, Behrens III WW. The limits to growth. Universe Books; 1972. http://www.donellameadows.org/wp-content/userfiles/Limits-to-Growth-digital-scan-version.pdf.

[8] Marshall JD, Toffel MW. Framing the elusive concept of sustainability: a sustainability hierarchy. Environmental Science and Technology 2005;39:673–82. https://doi.org/10.1021/es040394k.

[9] Wu J. Landscape sustainability science: ecosystem services and human well-being in changing landscapes. Landscape Ecology 2013;28:999–1023.

[10] World Commission on Environment and Development (WCED). Our common future. Oxford: Oxford University Press; 1987.

[11] Redclift M. Sustainable development (1987–2005): an oxymoron comes of age 2005. Sustainable Development. 212–27. https://doi.org/10.1002/sd.281.

[12] Elkington J. Enter the triple bottom line. In: The triple bottom line: does it all add up? London: Earthscan; 2004.

[13] Hammer J, Pivo G. The triple bottom line and sustainable economic development theory and practice. Economic Development Quarterly 2016:1–12. https://doi.org/10.1177/0891242416674808.

[14] Victor PA, Hanna S, Kubursi AA. How strong is weak sustainability? In: Faucheux S, O'Connor M, van der Straaten J, editors. Sustainable development: concepts, rationalities and strategies. Economy & Environment, vol. 13. Springer; 1998. p. 195–210.

[15] Schlör H, Fischer W, Hake JF. The system boundaries of sustainability. Journal of Cleaner Production 2015;88:52–60.

[16] Ayres R, van den Bergh J, Gowdy J. Strong versus weak sustainability: economics, natural sciences, and "consilience". Environmental Ethics 2001.

[17] Solow RM. The economics of resources or the resources of economics. The American Economic Review 1974;64:1–14.

[18] Solow RM. Intergenerational equity and natural resources. Review of Economic Studies 1974;41. https://doi.org/10.2307/2296370.

[19] Stiglitz J. Growth with exhaustible natural resources: efficient and optimal growth paths. The Review of Economic Studies 1974;41:123–37. https://doi.org/10.2307/2296377.

[20] Hartwick J. Intergenerational equity and investing rents from exhaustible resources. American Economic Review 1977;66.

[21] Biely K, Maes D, Van Passel S. The idea of weak sustainability is illegitimate. Environment, Development and Sustainability 2018;20:223–32. https://doi.org/10.1007/s10668-016-9878-4.

[22] Gutés MC. The concept of weak sustainability. Ecological Economics June 1996;17(3):147–56. https://doi.org/10.1016/S0921-8009(96)80003-6.
[23] Solow RM. On the intergenerational allocation of natural resources. Scandinavian Journal of Economics 1986;88:141–9.
[24] Boulding KE. The economics of the coming spaceship earth. In: Environmental quality in a growing economy: essays from the sixth RFF forum, vol. 3; 2013. p. 1–20.
[25] Georgescu-Roegen N. Energy and economic myths. Southern Economic Journal 1975;41:347–81.
[26] Holling C. The resilience of terrestrial ecosystems: local surprise and global change. In: Clark W, Munn R, editors. Sustainable development of the biosphere. Cambridge, U.K.: Cambridge University Press; 1986. p. 292–317.
[27] Daly H, Cobb J. For the common good: redirecting the economy toward community, the environment, and a sustainable future. ISSR library. Beacon Press. ISBN 9780807047057, 1994.
[28] United Nations General Assembly. The future we want. UN; 2012.
[29] Rahdari A, Sepasi S, Moradi M. Achieving sustainability through Schumpeterian social entrepreneurship: the role of social enterprises. Journal of Cleaner Production 2016;137:347–60.
[30] United Nations (UN). Transforming our world: the 2030 agenda for sustainable development. Resolution adopted by the General Assembly on 25 September 2015, A/RES/70/1. New York: UN General Assembly; 2015.
[31] Holden E, Linnerud K, Banister D. The imperatives of sustainable development. Sustainable Development; 2016.
[32] Carter CR, Rogers DS. A framework of sustainable supply chain management: moving toward new theory. International Journal of Physical Distribution & Logistics Management 2008;38:360–87.
[33] Allen C, Metternicht G, Wiedmann T. Prioritising SDG targets: assessing baselines, gaps and interlinkages. Sustainability Science 2018:1–18. https://doi.org/10.1007/s11625-018-0596-8.
[34] Pradhan P, Costa L, Rybski D, Lucht W, Kropp J. A systematic study of sustainable development goal (SDG) interactions. Earth's Future 2017;5:1169–79. https://doi.org/10.1002/2017ef000632.
[35] Singh G, Cisneros-Montemayor A, Swartz W, Cheung W, Guy J, Kenny T, et al. A rapid assessment of co-benefits and trade-offs among sustainable development goals. Marine Policy 2018;93:223–31.
[36] Nilsson M, Griggs D, Visbeck M. Policy: map the interactions between sustainable development goals. Nature 2016;534:320–2.
[37] Wackernagel M, Hanscom L, Lin D. Making the sustainable development goals consistent with sustainability. Frontiers in Energy Research 2017;5. https://doi.org/10.3389/fenrg.2017.00018.
[38] Vandemoortele J. From simple-minded MDGs to muddle-headed SDGs. Development Studies Research 2018;5:83–9. https://doi.org/10.1080/21665095.2018.1479647.
[39] Tester J, Drake E, Driscoll M, Golay M, Peters W. Sustainable energy: choosing among options. MIT Press. ISBN 9780262304634, 2012.

CHAPTER FIVE

Sustainable energy development

Contents

5.1. Challenges for sustainable energy development

Reliable and sufficient energy supply is assumed as a key prerequisite for economic growth and cultural and social development in complex societies [1]. According to the cited work [1], many scholars have come to the conclusion that there exists a correlation between the degree of development of a country and its energy use. The increasing living standards would require higher energy consumption per capita, which would lead to a kind of a "vicious circle" to supply more and more power to maintain the required rate of development and living conditions.

Such growth reveals the challenges of sustainable development. With the world aiming to maintain economic growth and increase of population, it requires more and more energy to maintain this motion, which is challenged by multiple limitations of growth. Historically, these challenges have been mitigated in multiple ways, though leading to an unsustainable way of production and consumption.

Sustainable Power Generation
https://doi.org/10.1016/B978-0-12-817012-0.00015-3

5.1.1 Historical outlook of energy development

At the dawn of human society the major source of energy was wood. It was used for heating and cooking purposes, as well as for protecting the people against animals. During the years of development, since the fire had been adopted as a means of activity, the humanity has tried various types of burning material to get the benefit not only of the basic needs (heating, cooking and protecting themselves from enemies) but also producing more metals. Even though the easiest way to maintain the fire was to burn dried wood, it could not deliver that much energy for metal smelting. In some areas, however, people started to spot other organic matter like various types of coal, which would burn with much higher heat release. Later the portfolio of organic fuels has been broadened by crude oil and natural gas, and their products. These types of fuel have gradually become one of the major sources of primary energy due to their relative proximity to the consumer, certain ease of extraction and utilization, and continuously evolving technology.

The second major source of energy, readily available in nature, was the workforce of animals and people themselves. Without doubt we can admit that the invention of a wheel brought a new era in energy deployment. Further applied for building purposes and agriculture, and later contributed with the energy of water flows and wind, it established the major pillars of primary energy utilization.

Importantly, water and wind can be well named as first prime movers, the physical medium to perform work, used to power wind and water mills and old foundries. The flow of fluid (namely wind or water) was pushing the wheels, which were coupled, for example, to the grain grinders through the shafts, wheels, and belts. Here the potential energy of the fluid was converted into the mechanical energy of the rotating wheel and was used for performing any type of work.

Along with the new ways of primary energy utilization deployment, the society has faced the population growth. This required a massive increase in the production capacities and therefore consequentially led to the industrial revolution, which happened at the border of the 18–19th centuries. This revolution brought with it the shift towards machine work and consolidation of industries on a large scale. Global manufacturing production expanded considerably faster in the final 20 years of the 19th century, when the benefits of cheap steel were being fully felt, than in earlier periods [2]. Novel methods of cheap steel production made it possible to increase the transportation in-between countries and continents, and catalyzed other industries like machinery, chemicals, etc.

All this required massive natural and human resources and, more significantly, energy. Coal, being the major source of cheap and affordable primary energy, dominated the world. It was utilized for smelting and producing steel, generating steam and driving the steam engines. The latter became the major conversion tools: chemical energy derived from coal was transformed into mechanical work.

Along with water flows, steam became the world's biggest intermediate step in energy conversion. However, the next advancement in the energy utilization revolution belongs to electricity. There had been many years since the first experiments with electricity till its first commercial use in the second half of the 19th century. First electrical machines were powered with the direct current and were designed for various purposes: telegraph, mining, lighting, metal parts production. They were battery based and therefore very limited in capacity. The invention of the first DC generator made it possible to increase capacity and use it for lighting and powering some small machines. These units were initially powered by the steam driven reciprocating engines (which again burnt coal) that allowed for stable competition against gas lighting. Lastly, the birth of the synchronous generator and electrical motor became a starting point for modern power generation and made it cheaper and easier to generate and transfer electricity over longer distances.

The rapid deployment of steam driven technology, which resulted in the steam turbine, and the evolution of the alternating current made it possible to dramatically increase the electricity production, followed by the growth of coal consumption for electricity generating needs. Evidently, other types of fuel like crude oil (and its products) and natural gas became popular in the areas where their production turned out to be cheaper.

In parallel, the 20th century has brought to life another great innovation called gas turbine — a type of engine that first consumed kerosene, burnt it, and rotated the shaft with the help of the hot combustion gases. Initially designed for aviation, it has been applied to power generating needs and finally started a separate branch of technology presently known as heavy duty gas turbines. Both gas and steam turbines became the core machines for energy conversion: they both used the heat of the burnt fuel, either in the form of hot gas (in a gas turbine) or pressurized steam (in a steam turbine).

Finally, civil nuclear energy might be well named a supernova of the energy. The discovery of radioactivity by BECQUEREL in 1896 followed by a number of research works by prominent scientists like CURIE, RUTHERFORD, BOHR, FERMI, KURCHATOV and many others, established a solid basis of the nuclear physics and nuclear engineering. The dawn of a power nuclear reactor took place in the middle of the 20th century, with the civil technology originating from the military developments. This new approach to energy transformation made it possible to secure big capacities and stable output, yet paying a high price of potential risks. There were, however, no impediments to quick nuclear progress [3]. For one, fossil fuels were plentiful in the 1950s, so no immediate urgency in deploying civil nuclear energy was felt. Nuclear facilities and technical knowledge were under tight control of the governments, with many aspects kept secret because of the military connections. However, it finally appeared that nuclear power would be competitive to the fossil fuel alternatives in the longer run.

By the mid-20th century, the three major power generating technologies had formed the basis of the world economy consisting of thermal (mostly coal and later gas-fired), hydro, and nuclear power. The latter has been dramatically impacted by the series of accidents, for instance, Three Mile Island (1979), Chernobyl (1986), Fukushima Daiichi (2011) — the three most significant being publicly admitted. With the social impact being even more important than the environmental one, nuclear power turned out to be considered a "sleeping danger", becoming a great argument of the greens. Despite the efforts taken to increase its safety, nuclear power outperforms other rivals in terms of negative public opinion, especially after the recent accident that took place in Japan. However, nuclear power, potentially so dangerous, shall not be considered as an old technology that is going to fade away.

5.1.2 Exponential growth

Humankind consumption is closely linked, for instance, to the growth of population, rate of industrialization, and therefore characterized by food production, consumption of resources like fresh water, non-renewable ones (e.g., metals, fossil fuels), environmental impact, etc. Importantly, all these elements and their derivatives, which are closely linked to them, are increasing, while the amount of their increase follows a pattern called exponential growth.

Contrary to a simple linear growth, when a quantity increases by a constant amount in a constant period of time so that the process can be

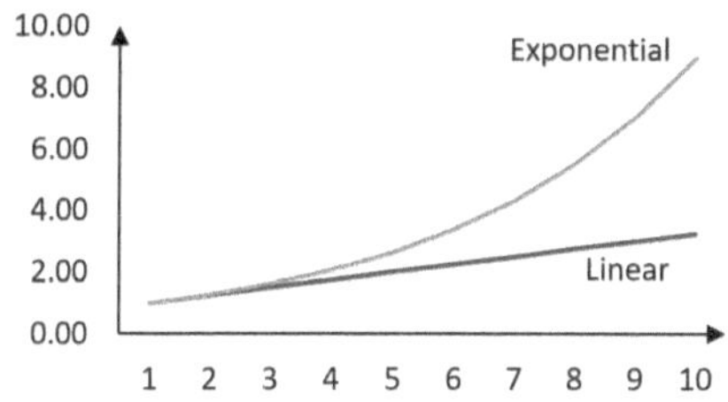

Figure 5.1 ***Linear and exponential growth over time t.*** Contrary to linear growth, when a quantity increases by a constant amount in a constant period of time, exponential growth means increases by a constant percentage of the whole in a constant time period.

depicted by a line on the growth over time t chart as illustrated in Fig. 5.1, exponential growth has a completely different patter: it means an increase by a constant percentage of the whole in a constant time period.

While the linear growth is easy to understand and therefore intuitively simple to forecast, exponential growth requires more effort and attention. One of the important definitions is the *doubling time* — the time it takes a quantity to double in size. Mathematically, the growth of an initial value x_0 at a time moment t can be represented as

$$x_t = x_0(1 + r)^t,$$

where r is the growth rate in a discrete time period.

In an ideal case or in a simplified system, exponential growth still has a relatively simple cause, so that it is mostly easy to calculate and predict. However, when many different quantities are growing simultaneously in a complicated system, and when all them are interrelated in a complicated way, the analysis of the causes of growth and of the future behavior of the system becomes very difficult indeed [4]. This is always the case when analyzing real systems: the growth of a population and its interrelation with industrialization, increase of pollution and its link to growing consumption, and so on. It is therefore important to study not only the quantities and their behavior, but the overall structure of the system and its dynamics.

With more than 7.5 B people on the Earth in 2017 [5], the UN estimates a population of nearly 9.8 B by 2050, which is about 30% higher than it is today, with an even more aggressive projection to a possible population of 11 B by the same time, so that the numbers may finally reach 16 B by 2100. Even though these forecasts depend on many circumstances, they clearly show that population would continue to grow, however, its exponential nature may well change to a peak around 2050 with a further decline, depending on a scenario. Nevertheless, the growth of the population would be mostly dominated by developing countries.

In developed and highly industrialized countries, the per capita energy use is relatively constant, so that the increase in energy use is largely governed by the growth of the population [6]. In developing countries the situation looks different: both the population growth and increase of energy use play important roles. This is especially relevant for the developing countries with large populations, like China and India. The same trend can be seen in emission levels: developed countries have a stable, though much higher, carbon dioxide emissions per person, which may be a benchmark for developing countries: their growth would drive consumption rate per person and in turn increase ecological footprint.

The question of predicting the population is an extremely complicated task, which is subject to a number of assumptions, and the data is usually derived from the governmental reports and undergo some statistical analysis. The magnitude of an error depends on the overall population and therefore varies from country to country. For some of them, like China, the error itself can be compared to the population of a small country, which introduces another uncertainty together with the variation of methods, absence of data in some areas, and so on.

The same problems may arise while estimating the possible growth rates and values of global data, including economic or social indicators, so that the forecasts and therefore scenario modeling becomes even more difficult, especially if it takes major limiting factors into account.

5.1.3 Limits to growth

If we consider our planet Earth as a closed system, at least at the present level of development and assuming the solar energy inflow as an input to this system, the continuous growth in such a system is not possible in the long run. The resources and energy the humankind consumes are taken from the environment and they do not disappear into nowhere. When their economic use is over, materials are recycled or, in most cases, are treated as waste and pollutants, while energy finally dissipates in the form of unused heat either within the system or outside into space. Therefore the economic subsystem of the global system Earth has the inflow of the energy and material streams from the planetary sources, which are finally led to the sinks [7], as schematically presented in Fig. 5.2. The rates at

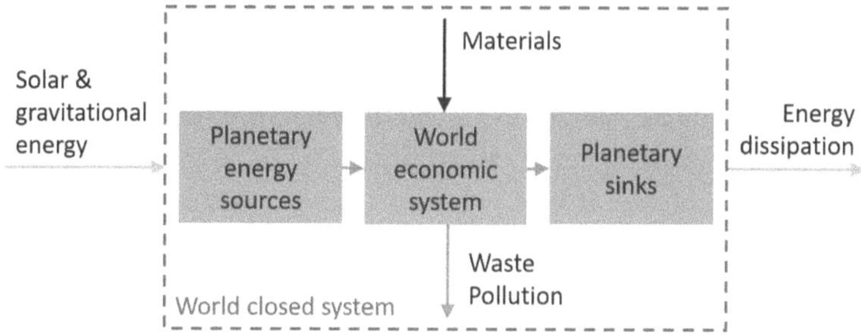

Figure 5.2 ***Economic subsystem within the global world system.*** Population and capital are sustained by flows of energy and material from the planetary sources and from the outside of the global closed system, and the world economic system initiates a larger sink of heat and waste. *Source: Based on [7].*

which the sources and sinks function are limited, and these limits may have complicated nature.

Growth is seen as necessary from economic and political standpoints, and it can only occur when all the current needs of the humankind are met and there is a certain surplus [8]. Growth, however, demands a higher level of consumption for basic operations with the loop feedback that generates even more growth. While growth is not vital for the very survival of the population, and in natural conditions growth is followed by a decline so that the population fluctuates around the balance point, it is in human nature to strive for it.

The required surplus of resources may also result in the increase of consumption beyond the basic needs and may not always result in growth at all but finally end up in prosperity or higher quality of life without any cumulative increase in the energy and material intensity of life [8]. Indeed, this surplus of resources is limited due to the nature of the closed system Earth, which is the major challenge of sustainability.

The surplus is the basis for the exponential growth of the population and increase of its living conditions to provide sustainable development of the society, which is always subject to a number of limitations: natural, technological, economic, political, and social. The growing population first drives the basic needs of humankind: production of food, consumption of fresh water, supply of material and energy to meet the demands, existence of the ecosystem of the planet to absorb waste, and moderate climate [7].

So far, there are more than 7 billion people on the Earth now, and roughly one in eight of us does not have enough to eat [9]. Moreover, the ongoing population growth is forecast to place new pressure on the global food system over the next decades. Much of the population increase will come from developing countries, where people are likely to be wealthier, and this will create a much higher demand for a more varied, high-quality diet requiring additional resources to produce [9]. Such a major population growth is expected in the poorest countries, where the standards of living are expected to rise rapidly, driving the demand in resources and energy.

The same trend is observed in the fresh water supply, especially in the face of the changing climatic conditions of some regions. While water is used not only for feeding or hygienic needs, much of it is consumed by agriculture for food production and industrial consumption. Water scarcity is emerging as a serous limiting factor in population growth and overall development of the humankind. Despite the agreement that there is a general human right for water, which has been proclaimed among the major global sustainability goals, it would still remain a problem and affect not only population directly, but may also have some effects on the environment. Importantly, water treatment solutions are highly capital and energy intensive, and their global implementation becomes a clear limiting factor for the growth of population.

The second category of requirements for growth consists of the social necessities of the humankind. Even if the system Earth is theoretically capable of supporting a much larger, more industrially developed population, the actual growth of the economy and the population will depend on such factors as peace, social stability, equity, security, education, openness to new ideas and admittance of mistakes, and many other social attributes [6]. While these factors are almost impossible to predict and quantify, they do plan an important role of a limiting factor. Their existence in the preferred form is a prerequisite condition for sustainable development, but will require extra effort and surplus of resources.

There are also a number of technological limitations that do not provide a viable opportunity to employ technologies that can provide sustainable energy because of low efficiency of conversion within the energy and material chains and increased waste. Suitable technologies may even exist and may enable us to fulfill the requirements of energy supply required for growth, but it does not ensure that they will be economically and environmentally viable [6]. This may be due to the lack of scientific knowledge, appropriate channels of its application to the real life problems, and economic and political will to implement such technologies which may be much more expensive and even not profitable at all.

According to [7], the current level of consumption shows that we are drawing on the world's resources faster than they can be restored. Even though major studies do suggest that the world in the year 2100 would still have enough resources for its development, the concern is more about the costs and sinks of these resources and their utilization. Such trend would for sure require constantly increasing amount of energy and capital to further maintain growth at the required level. Even though such a forecast cannot be proved and may only be projected with a certain probability, it still raises concerns regarding the limitations we have to face.

According to [7], there have been three simple rules suggested by Herman Daly that help to define the sustainable limits to material and energy throughput:

1. For a renewable resource, the sustainable rate of use can be no greater than the rate of regeneration of its source.
2. For a nonrenewable source the sustainable rate of use can be no greater than the rate at which a renewable resource, used sustainably, can be substituted for it.
3. For a pollutant the sustainable rate of emission can be no greater than the rate at which that pollutant can be recycled, absorbed, or rendered harmless in its sink.

These limitations are directly applicable to the power generation system on its way towards sustainable form. However, they are also driven by the time–cost dilemma: the economic feasibility to provide a complete solution and the timeline. Theoretically, the absolute sustainable solution is potentially feasible at the infinite cost and reasonable time, however, may not be exercised in reality due to social, political, and economic constraints. Similarly, maintaining the situation "as is" with some slow and cheap solutions may lead to sustainable but infinite future, which is not acceptable due to its timeless nature. Therefore the sustainable power generation system is all about the mitigation of limits and associated risks to develop and implement solution to provide the developing world with clean energy.

5.2. Global energy production and consumption

Global exponential economic growth and an ongoing increase of population require more and more energy. This is mostly critical on the way towards sustainable development, when one of the major goals is to provide equal and high quality of living which, in turn, is impossible to achieve without energy, especially if the form of fuels for transportation, production of materials and generation of electricity as one of the most convenient forms of energy utilized nowadays.

Global energy consumption has been marked by a strong increase especially in the past several decades [10]. One of the key determinants in the growth of energy demand is the overall economic growth as measured in gross domestic product (GDP). According to the analysis of Energy Information Administration (EIA), the world's GDP (expressed in purchasing power parity terms) is forecast to rise by 3.3% per year from 2012 to 2040. The fastest rates of growth are projected for the emerging, non-OECD countries, where combined GDP increases by 4.2% per year. In OECD countries, GDP grows at a much slower rate of 2.0% per year over the projection as a result of their more mature economies and slow or declining population growth trends. The strong projected economic growth rates in the non-OECD drive the fast-paced growth in future energy consumption among those nations [11].

Despite an ongoing push towards the increase of the process efficiencies, which is aimed to decrease the consumption rates and in turn increase the production to support growth, it would still require stable introduction of new capacities for primary and secondary energy production to support the increase of manufactured capital and eliminate the decrease of natural capital.

5.2.1 Energy production

The statistics regarding the primary energy production is readily available from many reports and data sources, for instance from the Key World Energy Statistics report published annually by International Energy Association [12,13]. According to this comprehensive and concise report, the world's total primary energy supply (TPES) has considerably evolved during the last decades as clearly visible from the pie diagrams represented in Fig. 5.3. These numbers, expressed in tons of oil equivalent, take into account production, imports and exports, as well as the stock changes of primary energy sources.

According to the historical data, represented in Fig. 5.3, the overall production of the primary energy has grown around 126% over the 40 years of economic development of approximately 2% per year on average. This growth, following the GDP development of around 5% on average,[1] has been outperforming the average growth rate of the world to ensure

[1] According to the World Bank data: `http://data.worldbank.org`.

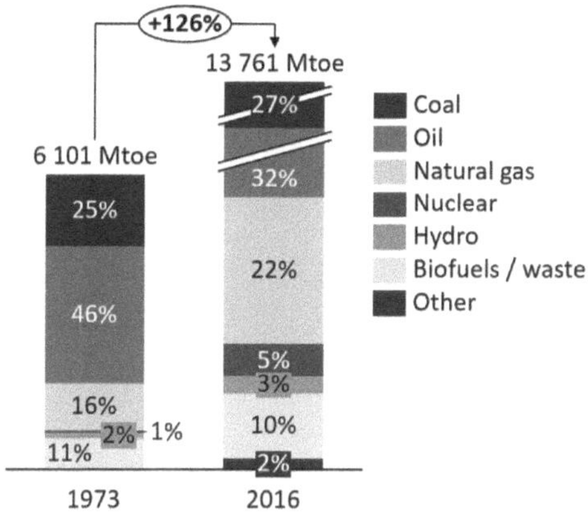

Figure 5.3 ***Total primary energy supply development from 1973 to 2016.*** For the given period the primary energy production grew 126%, with fossil energy occupying the largest share. *Source: [12].*

there is the energy resource for development. Evidently, massive growth of the energy production was mostly subject to the rapidly growing economies with booming GDP growth rate like China.

When looking into the details of each primary energy source, it is important to highlight that:

- Coal, natural gas, hydro, biofuel, and waste production have grown with the comparable rate as the overall primary energy fuel production (126%) with their shares being impacted by other sources;
- Oil growth rate has been underperforming the overall energy production growth mostly due to its decreasing importance as an energy fuel, increased efficiency in refining and consumption;
- Nuclear growth has been among the largest in the recent decades with roughly 1100% occupying the share of 4.9% of TPES in 2016;
- The largest growth has been delivered by other sources including modern renewables like wind and solar. With their growth of more than 3000% over the last 40 years, they have clearly shown that they may become a significant player in primary energy supply market in the coming decades.

The regional supply pattern has also changed significantly due to the shift of the global economic growth from OECD countries to the developing parts of the world. According to the statistics [12] represented in Fig. 5.4, China and Asia in general (excluding Asian OECD countries) correspond to the biggest growth in the world both in share and in total

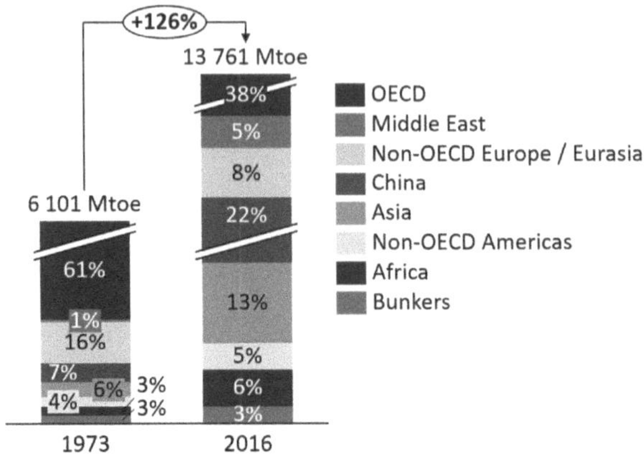

Figure 5.4 ***Total primary energy supply by region.*** Over the given period, the total primary energy supply has shifted from OECD countries to the developing countries in Asia, especially China.

values. OECD countries, subject to overall economic slowdown and more costly production, have lost about 30% of their primary energy production share.

It is also worth to highlight that both Middle East and Africa have claimed rapid development of the energy production which may be subject to booming exploration of natural resources, development of the economies through direct foreign investments, making them the second largest potential markets for the future.

The primary energy production is important to understand as it gives the input to the power industry which delivers secondary energy in the form of electricity.

5.2.2 Energy demand and consumption

Much of the world increase in energy demand occurs among the developing non-OECD nations, where strong economic growth and expanding populations lead the increase in world energy use [11]. Multiple forecasts and base scenario simulations predict that non-OECD demand for energy would rise by 71% from 2012 to 2040 [11], while the mature OECD economies would experience slower growth and therefore energy consumption increase, which may result in only 18% for the same period.

It took hundreds of thousands of years for the world population to grow to 1 billion [14], and since the second half of the 20th century, the world has experienced a radical increase in its population [15], which has still been

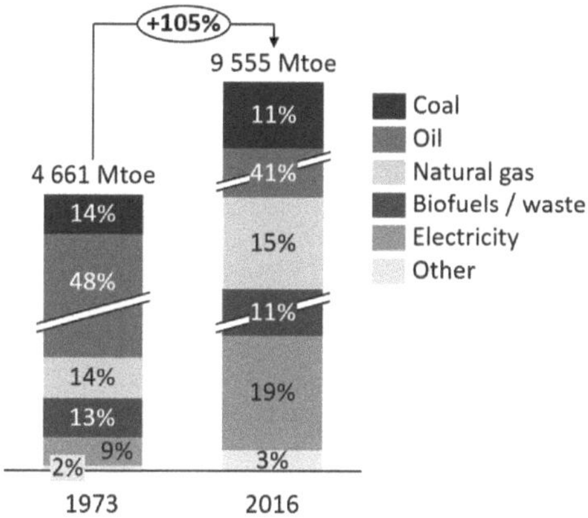

Figure 5.5 ***Total final consumption of fuel.*** 1973 and 2016 fuel shares of total final consumption. *Source: [12].*

the trend in the first half of the 21st century and would still drive the global growth. This in turn demands improvements in the living conditions (like housing, home appliances, fresh and clean water supply, hygiene), transportation (both private, like cars, and public, including buses, trains, taxis, planes, etc.), broadening the capacities to produce food, goods, and services for both domestic and foreign markets. Consequently, this growing demand in energy would require additional cheap and reliable supply.

Moreover, according to the International Energy Agency (IEA) [16], almost 85% of the 1.3 billion people without access to electricity are in rural areas. This means that it would require not only investments in electricity generating capacities, but also means of its transportation to these remote locations.

Economic growth, along with accompanying structural changes, strongly influences world energy consumption. As countries develop and living standards improve, energy demand grows rapidly [11]. According to International Energy outlook reports, published by the EIA, over the past 30 years, world economic growth has been led by the non-OECD countries, accompanied by strong growth in energy demand in the respective regions. From 1990 to 2012, real GDP grew by 4.9%/year in non-OECD countries, compared with 2.1%/year in OECD countries [11,16].

As shown in Fig. 5.5, world energy consumption has more than doubled from 4661 Mtoe to 9555 Mtoe during the last 40 years. While the oil and coal shares in the global energy consumption have decreased, these numbers are still relative and do not really mean the decrease of the importance of oil

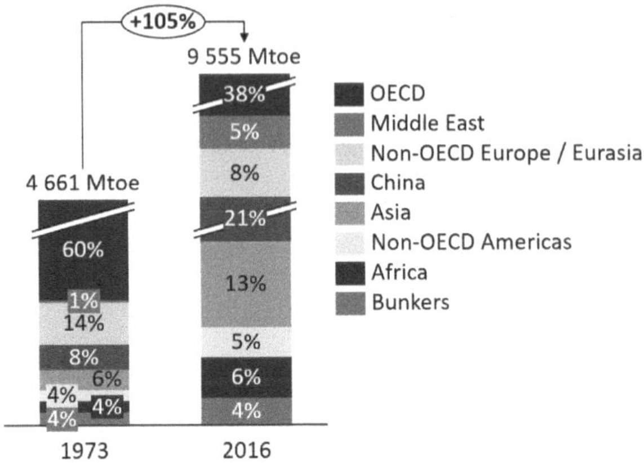

Figure 5.6 ***Regional shares of total final consumption.*** 1973 and 2016 regional shares of total final consumption. *Source: [12].*

and coal. The change in shares is mostly due to the massively growing share of electricity consumption — the major driver of the growing economies and technological development of the developed countries.

On the regional level, the OECD share of the energy consumption has been steadily declining due to strong economic growth of the emerging economies in Asia, mostly China (see Fig. 5.6). The latter has increased its share of the world energy consumption from 7.8% of the world's total to 20.7%, resulting in about 1978 Mtoe consumption in 2016. The same trend has been in place in the Middle East, where the energy consumption share has increased from 0.7% to more than 5% in 2014 with 487 Mtoe.

While energy is required literally everywhere, the total energy use has changed dramatically during the last decades. The overall consumption of three major sources of primary energy — coal, oil, and natural gas — has changed significantly:

- Industrial consumption has focused on the share of coal consumption, mostly due to the contribution of the steel industry. Similarly, the role of electricity became much more important, while the new highly technological industries require more and more electrical power;
- Transportation sector has increased its consumption of oil and therefore oil products, resulting in almost 2426 Mtoe of consumption in 2014 with the forecast to grow further. The share of electricity in transportation is still very small;

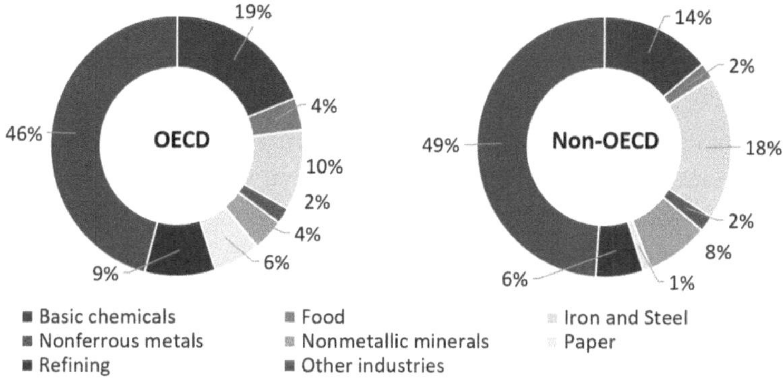

Figure 5.7 ***Energy consumption of major industries.*** Energy-intensive industry shares in % of total energy consumption for OECD and non-OECD (other industries include nonenergy-intensive manufacturing and nonmanufacturing). *Source: [11].*

- Energy consumed in the buildings' sector consists of residential and commercial end users and accounts for more than 20% of the total delivered energy consumed worldwide;
- In other segments of economy, which normally include agriculture, commercial and public services, etc., the share of electricity has won the race with 955 Mtoe of consumption in 2014, which is the largest portion of electricity use in the world economy. Such growth is mostly due to overwhelming growth of electrification, use of private electrical devices, IT development, etc.

According to [12,16], the total industry consumption of energy in 2014 was around 2751 Mtoe, which is roughly 29% of the total energy consumption of 9425 Mtoe in 2014, so that the industrial sector in total uses more delivered energy than any other end-use sector. It can be categorized by three distinct industry types: energy-intensive manufacturing, nonenergy-intensive manufacturing, and nonmanufacturing [11].

The bulk of energy-intensive industries include the iron, steel, metal, mineral and petrochemical industries [15], as well as food industry and production of nonmetallic minerals like cement, glass, etc. These are mostly segments that deal with production of raw materials and their refining, consuming large amounts of energy. In addition, the same industries emit large quantities of carbon dioxide (CO_2), related to both their energy consumption (combustion emissions) and their production processes (process emissions) [11].

Fig. 5.7 shows energy consumption shares of the energy-intensive industries compared with all industrial sector energy consumption in 2012 for the OECD and non-OECD, respectively [11]. According to the report, among the energy-intensive industries, the largest consumer of delivered energy is the basic chemicals industry, which in 2012 accounted for about 19% of total delivered energy consumption in the OECD industrial sector and about 14% in the non-OECD industrial sector. The iron and steel industry accounted for about 10% of energy consumption in OECD countries in 2012, while in the non-OECD countries its share was significantly larger and resulted in approx. 18%. Steel and metal production usually uses coal as the primary energy source for heat generation or electricity for within the electric arc furnaces to melt the metals.

Remark 5.1. It is important to mention that the energy-intensive industries are usually demanding in fresh water resources, for instance, steel and metal production or food industry. This close connection may require extra energy consumption in the future for fresh water production and its delivery to the consumers. While agriculture is not yet considered as energy-intensive industry, the growing population and possible trends towards biofuels may lead to the increase of its energy consumption and further water usage intensification.

Nonenergy-intensive manufacturing and nonmanufacturing industries therefore account for roughly half of the energy consumption in industrial sector for both OECD and non-OECD countries.

Energy use in the transportation sector includes the energy consumed in moving people and goods by road, rail, air, water, and pipeline [11]. Various oil refining products like petroleum and other liquid fuels (motor gasoline, diesel, and jet fuel) are the dominant sources of secondary energy for the transportation industry. It is important to mention that natural gas represents roughly 3% of the total energy consumption share, which is mainly due to the pipeline transportation, where natural gas is used as a major fuel for gas turbines to power big oil and gas compressors. However, the share of natural gas may increase due to larger introduction of gas-power engines for trucks, vessels, trains, etc.

Electricity remains a minor fuel for the world's transportation energy use, although its importance in passenger rail transportation remains high [11]. The growing importance of the electric vehicles is still not visible in the global transportation picture (see Fig. 5.8). It is still questionable whether electricity can become a significant source of energy for large scale means of transport on top for rail.

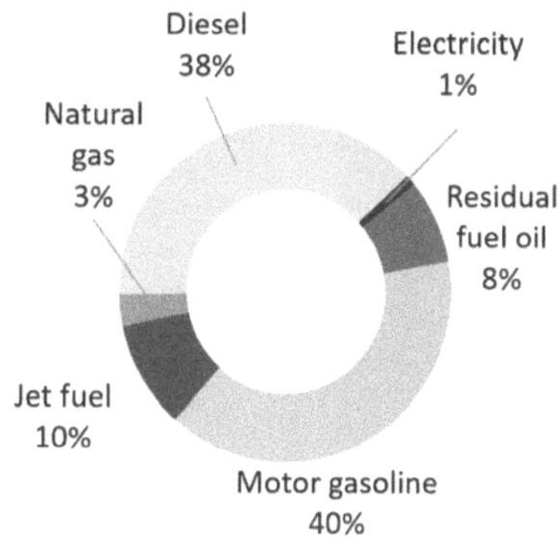

Figure 5.8 ***Energy consumption by transportation sector.*** World transportation sector energy consumption by energy source share. *Source: [11].*

Consequently, economic growth — as measured in gross domestic product — is a key determinant in the growth of energy demand. Currently the fastest growth rates have been witnessed in the emerging non-OECD countries, where the GDP growth rate is twice that in OECD countries [11]. This situation shall remain one of the key drivers of the overall energy production and consumption, especially within the industrial, transportation and services sectors, where the latter will be driven by the growing importance and worldwide collection and use of data.

5.2.3 Electricity generation

During the decades, the role of electricity has been increasing because it is an extremely versatile energy source, which can be generated from a wide range of primary energy sources and can easily be channeled to the point of use. Electricity generation uses some 40% of the world's total primary energy supply [17]. The growth of world electricity production has grown almost four times during the last four decades, and resulted in 24 973 TWh production in 2016 as illustrated in Fig. 5.9.

This growth has been the most significant in Asia and specifically China, with the latter accounting for about 25% of the world's electricity generation in 2016 compared to the modest share of less than 3% more than 40 years ago. This trend is mostly likely to continue, as the recent five years have shown stable electricity production both in the emerging regions in particular and in the world overall. Despite certain decline in some of the economies, there is still stable demand and therefore there will continue to be stable electricity production increase.

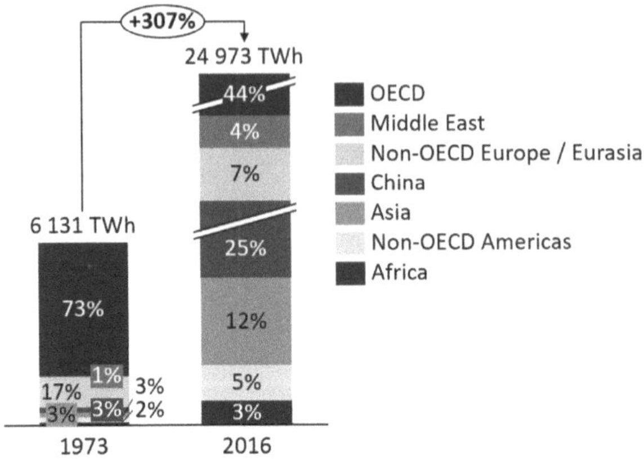

Figure 5.9 ***Regional shares of electricity generation.*** 1973 and 2016 regional shares of electricity generation. *Source: [12,13].*

It is important to mention that Europe and North America have decreased their growth rates (or even went negative in 2012) of their electricity generation due to the decline of economies and implementation of energy efficiency programs as well.

Consequently, the installed capacity growth has outperformed the electricity production figures especially in the face of global economic crisis times of 2008, which can be due to multiple reasons. For instance, the emerging markets can have lower efficiency of the installed capacities led by the tradeoff between the investment and production efficiency. On the other hand, the utilization factor — the share of time when the installed capacity is generating electricity — can partially explain the situation. Overall, the capacity growth has been following the production numbers pointing at Asia as the key emerging region.

Evidently, traditional power generating technologies occupy the central place in the capacity mix. As shown in [12,18], thermal power plants on fossil fuels form the basis of the world electricity generation. As stated in Fig. 5.10, the world electricity production in 2015 was mostly due to coal, natural gas and oil-fired power plants. While coal has been still the dominant source of electricity production with around 39% of market share, natural gas has become much more important with the current 23%. This has been mostly due to the emissions restrictions that have been imposed in developed countries and therefore led towards decline in coal-fired power plant construction, as well as greater flexibility of gas-fired generation. This

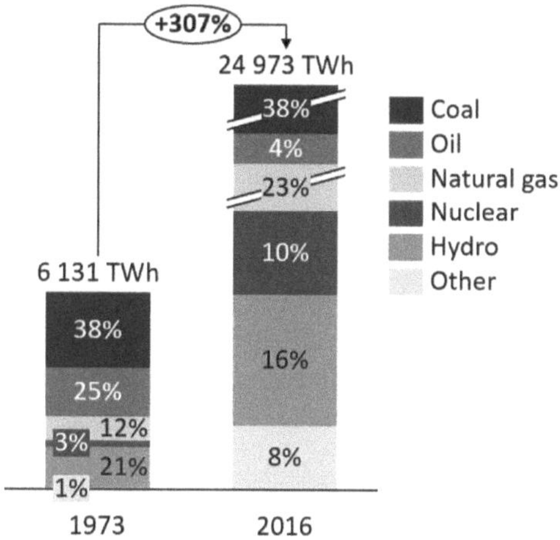

Figure 5.10 ***World electricity generation by source.*** The overall electricity generation has grown approximately four times within the last four decades and was 24 973 TWh in 2016. *Source: [12,13].*

is, however, true only in developed countries, while the emerging markets still invest in coal generation.

Nuclear energy, with its share of 10.6%, has been stable for several years and has not been significantly impacted by the recent Fukushima–Daiichi accident. Since the middle of the 20th century, nuclear power has developed to provide more than 10% of the world's electricity, however, economics, nuclear weapons proliferation concerns, waste management and safety issues are the major issues that can influence the use of nuclear power around the world. Despite rapid shutdown of nuclear plants in Germany, there are still a number of new units under construction worldwide, which would still keep the nuclear electricity generation share at the same or even higher level.

Other sources of energy occupy around 23.1% with the hydroenergy being by far the largest source of electricity from renewable primary energy source.

5.3. SDGs related to energy system

As we have seen, energy production and consumption are among the major indicators of development. Hence, the ongoing economic growth and population increase is expected to dramatically drive the energy demand. Thus, as sustainable development is about improving the quality of living for the humankind, both for current and future generations, energy production has to be in line with the global SDGs.

It is important to emphasize that energy is an integral part of all human activities, so that it is logical to expect energy be a part of most (not to say all) sustainable development goals, either directly or indirectly. For the purpose of our topic, the most interesting are the goals *directly* related to the energy system and environment. We highlight direct impact because most of these goals are interconnected with the energy system and visa versa, as the humankind is inextricably entwined with the environment. These goals therefore include:

- SDG 7: Ensure access to affordable, reliable, sustainable, and modern energy for all.
- SDG 12: Ensure sustainable consumption and production patterns, by implementing the 10-year framework of programs on sustainable consumption, production, and use of resources.
- SDG 13: Take urgent action to combat climate change and its impacts.

It does not mean that other goals are either irrelevant to the power generation system or less significant in general. All 17 SDGs are equally important for sustainable development and therefore have to be considered with equal attention.

5.3.1 SDG 7: ensure access to affordable, reliable, sustainable, and modern energy

This goal is the central one which is directly targeting the energy topic. The major reason of the UN to formulate this goal comes out of the fact that there are approximately 1 billion people who are functioning without electricity with around 50% of them are found in Sub-Saharan Africa alone. Unfortunately, this challenge is far from being solved and there needs to be more access to clean fuel and technology, and more progress needs to be made regarding integrating renewable energy into end-use applications in buildings, transport, and industry. Public and private investments in energy also need to be increased and there needs to be more focus on regulatory frameworks and innovative business models to transform the world's energy systems [19].

Moreover, the energy system is the dominant contributor to climate change, accounting for around 60% of total global greenhouse gas emissions.

This goal is supposed to have the following targets [20,19] by 2030:

- Ensure universal access to affordable, reliable, and modern energy services;
- Increase substantially the share of renewable energy in the global energy mix;
- Double the global rate of improvement in energy efficiency;
- Enhance international cooperation to facilitate access to clean energy research and technology, including renewable energy, energy efficiency, and advanced and cleaner fossil-fuel technology, and promote investment in energy infrastructure and clean energy technology;
- Expand infrastructure and upgrade technology for supplying modern and sustainable energy services for all in developing countries, in particular least developed countries, small island developing states, and land-locked developing countries, in accordance with their respective programs of support.

It is important to mention that these targets are mostly qualitative or do not fully comply with the SMART goal setting practice. This is mostly due to the fact that it takes an effort or may not be possible at all to quantify some of the mentioned targets. Indeed, the mentioned targets are important, but it may not be possible to track progress of their implementation.

These targets do not give any answer to the questions how to ensure access to affordable, reliable, and clean energy:

- What type of policies, incentives, regulations are required and which countries shall implement them?
- What is the required investment for such a transition? Even though UN states that the world needs to triple its investment in sustainable energy infrastructure per year, from around $400 billion now to $1.25 trillion by 2030, these numbers are rather indicative and may not fully reveal the situation.
- What is the roadmap towards this goal and each of the objectives?

Though these objectives require more attention and SMART approach on a regional or country level, they are for sure a significant step towards energy transition. It is all about the major goal of the energy system to provide affordable, reliable, sustainable, and modern energy, with the power generation industry being a significant part of it. This includes not only transition of the power generating system towards a completely green future, which may not occur within the nearest decades, but first a change in the global mindset to realize the change is vitally required.

5.3.2 SDG 12: ensure sustainable consumption and production patterns

Sustainable production and consumption patterns are the major components of the sustainable system Earth in general, and energy system throughput in particular. As discussed previously, energy and material production and their surplus are required for the growth and development, the material and energy consumption would increase consequently driving waste production.

Since sustainable consumption and production aims at "doing more and better with less", net welfare gains from economic activities can increase by reducing resource use, degradation and pollution along the whole life cycle, while increasing the quality of life [20,19]. Such an approach should be implemented to the whole throughput chain, which, applied to the power generation system, covers production of primary energy (e.g., mining and enrichment of fossil fuels), their transportation, consumption, and waste management. Additionally, there also needs to be a significant focus on operating this throughput, i.e., the change of the mindset of current and future leaders.

This goal has a number of qualitative targets, with those relevant to energy system listed below:

- Implement the 10-year framework of programs on sustainable consumption and production, all countries taking action, with developed countries taking the lead, taking into account the development and capabilities of developing countries;
- By 2030, achieve the sustainable management and efficient use of natural resources;
- By 2020, achieve the environmentally sound management of chemicals and all waste throughout their life cycle, in accordance with agreed international frameworks, and significantly reduce their release to air, water and soil in order to minimize their adverse impacts on human health and the environment;
- By 2030, substantially reduce waste generation through prevention, reduction, recycling and reuse;
- Encourage companies, especially large and transnational companies, to adopt sustainable practices and to integrate sustainability information into their reporting cycle;
- By 2030, ensure that people everywhere have the relevant information and awareness for sustainable development and lifestyles in harmony with nature;

- Support developing countries to strengthen their scientific and technological capacity to move towards more sustainable patterns of consumption and production;
- Rationalize inefficient fossil-fuel subsidies that encourage wasteful consumption.

Even though the goals will be difficult to measure, they reveal the importance of the efficient throughput, reasonable consumption and decrease of waste. It shall start from the individual's level to understand the importance of personal careful and reasonable consumption to minimize the waste. This includes reuse of material, decrease of plastics, energy saving technologies, careful food and fresh consumption, etc. On the industry level, companies should invest in energy saving technologies, increase their efficiency and design products keeping recycling topic in mind. Unfortunately, such an approach would lead to fewer economic benefits, which would again require the change of the system behavior and reevaluation of the concept of growth.

5.3.3 SDG 13: take urgent action to combat climate change and its impacts

Climate change is now affecting every country on every continent. It is disrupting national economies and affecting lives, costing people, communities, and countries dearly today and even more tomorrow. Weather patterns are changing, sea levels are rising, weather events are becoming more extreme, and greenhouse gas emissions are now at their highest levels in history. Without action, the world's average surface temperature is likely to surpass 3°C this century. The poorest and most vulnerable people are being affected the most [19].

To draw attention to the climate change problem and strengthen the global response to this threat, 175 parties have ratified the Paris agreement with the majority of them determining and communicating their contributions to combat the climate change.

Similarly to others, this goal has a set of qualitative targets related to power generation system:

- Integrate climate change measures into national policies, strategies and planning;
- Improve education, awareness-raising, and human and institutional capacity on climate change mitigation, adaptation, impact reduction, and early warning;

- Promote mechanisms for raising capacity for effective climate change-related planning and management in least developed countries and small island developing states.

These rather generic targets set up the framework for mobilizing funds and integrate the required measures into national policies and regulation, which shall be the first steps to enable the change. Secondly, there needs to be a change in the mindset to draw attention to the problem and increase awareness about the changes. Common people may not fully see the challenge and may not realize the scale, especially in a way that this change may not directly affect the current or next generation, but the problem would row in a few decades even if we immediately cut all emissions. The educational side of the topic is therefore among the most important.

Concerning the power generation industry, its contribution to both negative and positive impact on the climate is evident, so that their behavior, including optimization of the assets, investment decisions, and again change of the mindset, is the first priority on the way towards sustainable development, which is largely driven by the climate change challenge.

5.4. Energy system within the sustainable solution space

The sustainability system boundaries, according to [21], are formed by four major dimensions:

- The chosen paradigm of sustainability which shall be implemented to design or transform the future energy and power generation system. As discussed previously, the weak sustainability model would give the system more freedom for movement, while strong sustainability implies much stricter limitations within the natural boundaries of the global ecological system.
- The time frame of the energy system, which depends on the type of the energy system. Thus, the fossil solar age, which is characterized by the use of fossilized organic matter (solar energy stored in living organisms), is future-closed as it is limited and cannot be utilized forever. On the contrary, the active solar system is future-open, i.e., the solar radiation and its use are unlimited on the scale of the humankind. However, according to [21], "both systems would supply enough energy for the normal time frame of human development", though at different cost and effort, and with different impact.

- The expected world energy demand is the third limitation of the future energy system. The forecast of the demand depends on multiple factors like the expected population growth, level of living, condition of the environment, etc. As it continues to increase, the energy system in general and power generation system in particular shall be capable to meet this growth.

The climate change question can be sometimes considered as the fourth limiting factor which is treated separately due to its overwhelming nature and global impact. Evidently, the climate change limitation is closely connected with the chosen paradigm of sustainability, time frame of the future energy system, as well as energy demand. For instance, the transition towards the active solar energy system from the currently employed fossil solar energy system is not practically possible within the short period of time (due to technological, economic, and political constraints), so that the impact on the climate will still continue to exist even further also thanks to its inertia. Similarly, the growing energy demand requires implementation of cheap solutions, which are still fossil-energy related.

On top of these major dimensions, there are many limiting factors that can impact the transformation towards the sustainable power generation system:

- Available technological level to implement a sustainable solution, for example, designing, construction, commissioning, and operation of the new active solar technologies (like wind or PV) to effectively and efficiently replace existing and established conventional capacities, provide reliable sources of power.
- Economic feasibility of the sustainable power generation system, i.e., the feasibility of smooth transformation. Even if such a transformation were technologically feasible, it would be either too costly to implement within the reasonable time frame, or may require a tough step aside from the economic growth, accumulation of wealth or simple loss of money.
- Political feasibility of the transition which consists of the global and local political wills, development and implementation of policies, their ratification by governments, and minimization of the underlying interests that may impact the speed and direction of political forces.

Moreover, there may be another limitation within the sustainable solution space, which is closely connected to the human capital and social acceptance, that is, the perception of the humankind and its responsibility beyond the human life or beyond one generation. The idea here lies in the fact that the people are usually concerned about the living conditions of their own generation and, likely, about the conditions for the next one or two generations due to obvious biological reasons. The human life usually spans over 2–3 generations, at least for the most active age. Therefore, people are taking care about themselves and their children, sometimes grandchildren, with the limited concern either about the entire humankind of the current generation, or about the future generations.

The main difficulty here lies in creating just relationships between generations, which is a part of social interactions [21]. Since there are very limited tools to manage this process, it should be a political, governmental or other institutional obligation to establish and manage such kind of intergenerational bond. As per research [21], these institutions have a number of options to cope with this asymmetry:

- Investments into the sustainable development shall be financed by debts, which will put the obligations for their repayment to future generations. However, this poses a question of an unequal duty of current and future generations, as the current actions shall be clearly observed as serving sustainable future.
- The actions shall be financed from the taxes, which are payed by the current population, as they are responsible for the climate change. However, there is no clear definition of the level of this tax, as well as transparent system of calculating the tax accordingly: some people do contribute less to global warming than the others. This tax is much easier to calculate for companies, let's say, industries, however, individual payments may vary significantly.
- Finally, current generation shall invest in transferring knowledge and educating future generations along the way towards sustainable development.

Overall, sustainable development of the humankind in general and the development of the sustainable energy or power generation system is a complicated and versatile topic. Many attempts have been made to simplify the problem in order to start addressing it in a more constructive way, including the formulation of the three SDG related to power generation: develop responsible production and consumption, ensure affordable and reliable energy supply, and combat climate change. These goals are still subject to a number of limitations — political, economic, and social — and are governed by a number of important dimensions: the paradigm which would guide this development, available and targeted time frame, and expected demand.

Though the technological options and their availability, efficiency, and economics are only among the boundaries of sustainability, they do play a vital role. Therefore we further draw our focus to these technologies, paying special attention to their historical development and its reasons, efficiency gains, social acceptance, and economic valuations, and their possible role for the sustainable power generation system. This partly addresses the third option to deal with the sustainability asymmetry — knowledge transfer, not only for the future generations, but also horizontally to develop certain awareness of sustainable energy and power.

References

[1] Arto I, Capellán-Pérez I, Bueno G, Bermejo R. The energy requirements of a developed world 2016. Energy for Sustainable Development. 1–13. https://doi.org/10.1016/j.esd.2016.04.001.

[2] Marsh P. The new industrial revolution: consumers, globalization and the end of mass production. Yale University Press; 2012.

[3] Bodansky D. Nuclear energy: principles, practices, and prospects. New York: Springer. ISBN 9780387269313, 2007.

[4] Meadows DH, Meadows DL, Randers J, Behrens III WW. The limits to growth. New York: Universe Books; 1972. URL: http://www.donellameadows.org/wp-content/userfiles/Limits-to-Growth-digital-scan-version.pdf.

[5] United Nations Department of Economic and Social Affairs, Population Division. World population prospects: the 2017 revision, key findings and advance tables. Working Paper No. ESA/P/WP/248. 2017.

[6] Dunlap R. Sustainable energy. Cengage Learning. ISBN 9781305156210, 2014. 672 pp.

[7] Meadows D, Meadows D, Randers J. Limits to growth: the 30-year update. Earthscan. ISBN 9781849775861, 2012.

[8] Kreith F, Krumdieck S. Principles of sustainable energy systems. Second edition. Mechanical and aerospace engineering series. Taylor & Francis. ISBN 9781466556966, 2013.

[9] Rosenbaum A. Rapid population growth challenges global food society. URL: https://www.fcsi.org/foodservice-consultant/worldwide/rapid-population-growth-challenges-global-food-society/, September 29, 2014.

[10] Spliethoff H. Power generation from solid fuels. Springer; 2010.

[11] Information Administration (EIA). International energy outlook 2016. US Energy Information Administration; 2016.

[12] IEA. Key world energy statistics 2016. International Energy Agency; 2016.

[13] IEA. Key world energy statistics 2018. International Energy Agency; 2018.

[14] United Nations Population Fund. World population trends. URL: http://www.unfpa.org/world-population-trends.

[15] Petit V. The energy transition. An overview of the true challenge of the 21st century. Springer; 2017.

[16] IEA. IEA Sankey diagram. URL: http://www.iea.org/Sankey/, 2017.

[17] Hore-Lacy I. Nuclear energy in the 21st century: World Nuclear University Press. Elsevier Science. ISBN 9780080497532, 2010.

[18] WCA. Coal facts 2014. World Coal Association; 2014.
[19] United Nations. Sustainable development goals. URL: https://www.un.org/sustainabledevelopment/.
[20] United Nations (UN). Transforming our world: the 2030 agenda for sustainable development. Resolution adopted by the General Assembly on 25 September 2015, A/RES/70/1. New York: UN General Assembly; 2015.
[21] Schlör H, Fischer W, Hake JF. The system boundaries of sustainability. Journal of Cleaner Production 2015;88:52–60.

CHAPTER SIX

Power system and the environment

Contents

6.1. Introduction

The global energy system that provides modern society and economies with secondary energy for the development needs is closely connected to multiple spheres of the Earth's environment. Every step in the production, modification, and use of energy interacts with the environment and, consequently, modifies the Earth and its inhabitants. The resulting changes may be of various scale, timeline, and can be either beneficial or dangerous, by whatever environmental impact metric is most appropriate [1].

The major simplified patterns of resources' consumption include the following interactions between the major spheres and the energy system:

- Production of primary energy from the lithosphere, including mining for coal, production of oil and gas, mining for nuclear fuel, and drilling for geothermal energy. This type of interaction would also incorporate subsea drilling with offshore rigs, and various types of mining (open pit, underground, or in situ leaching).
- Consumption of water from the hydrosphere for process purposes, mostly for production and conversion processes. This includes processing desalinated water, water for cooling loops, steam for industrial needs, irrigation for biomass production, and so on.
- Use of oxygen (or the air) from the atmosphere for conversion purposes, e.g., burning fuel, and use of carbon dioxide for growing the biomass. While oxygen consumption for combustion reaction is involved in most of the processes where machinery with heat engines is used, we mostly focus on the conversion process as the most significant.

https://doi.org/10.1016/B978-0-12-817012-0.00016-5

- Biosphere (including the pedosphere) supplies primary energy in the form of wood, crops, or waste for energy production purposes.
- Finally, the Sun supplies solar radiation which is considered as the primary energy. While the Sun is not considered as a sphere, we can probably combine it with atmosphere for simplification purposes.

While this schematic description of interactions between the energy system and the environment and its spheres has been intentionally simplified and omits multiple bonds, including those between various spheres (which have been discussed in Chapter 2), it still reveals the complexity of interference between the natural system and the humankind created system. Unfortunately, this cannot be named as pure cooperation, as the energy system impacts the environment, and usually in a negative way.

6.2. Potential environmental threats

We elaborate now on the impact that the energy system has on the environment. Hence, Fig. 6.1 is a further development of the interaction

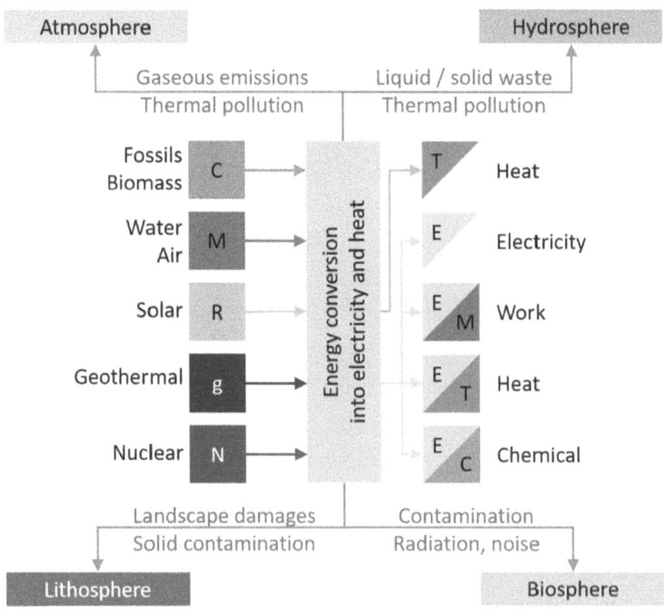

Figure 6.1 ***Impact of the energy system on the environment.*** The energy system releases solid, gaseous, and liquid waste, produces noise emissions, radiation, thermal pollution, and can have other potential impact.

model where the arrows outgoing from the energy conversion system depict its impact on the spheres of the environment. Thus, the impact on the lithosphere is mostly due to production of primary energy in the form of fossil or mineral fuels: open pit mines change the landscape and damage soil layers, while underground mining can lead to karst erosion, induced earthquakes, etc. Moreover, any type of construction impacts at least the top layer which is the most important for biomass growth, while using low permeable materials (like concrete or asphalt) decreases evaporation rate from lithosphere and therefore increases its heat.

Water used for industrial purposes is then contaminated either by the waste from production, or with secondary waste and process.

Gaseous emissions from the energy system processes are released to the atmosphere, and sometimes with little or no treatment. These include side releases of mining gases (for instance, methane), emissions from combustion reactions, production of materials for construction (a "good" example is cement manufacturing), as well as release of steam and waste heat.

Finally, biosphere is impacted (directly or indirectly) by construction, poisoning of soil, water, and air with the substances that damage health and therefore threaten the biodiversity and ecosystems.

Having said that, the types of impact from the energy system on the environment can be categorized in the following way:

- Energy systems release solid, liquid, and gaseous by-products of the energy production and conversion processes, which are called emissions, waste, or effluents.
- Noise emissions and vibration are produced by most of the components of the energy system, which imply motion. Rotating machines like turbines and generators, motors, wind mills, and any kind of machinery release sound waves. Depending on the frequency and intensity, they may result in a negative impact on human health.
- Using nuclear mineral fuel within the energy system may result in a release of ionizing radiation: electromagnetic waves, neutrons, electrons, or alpha particles. This radiation comes not only from the primary nuclear fuel, but also from the waste and the materials left after the nuclear power plant is shut down. It takes hundreds or even thousands of years for some materials to become safe for humans, therefore ionized radiation and its careful treatment is of high importance as it may cause cancer or radiation sickness.

- Electromagnetic radiation is emitted by all electric machines and electric lines which have flowing current. While the health impact has not yet been fully proven [1], it is still considered an issue.
- Waste heat and thermal pollution is always released during any thermal energy conversion process, e.g., combustion of fossil fuels. This thermal pollution can continuously increase the ambient air temperature, heat water causing damage to the ecosystem, or increase temperature of the human working and living environment leading to health issues.
- Other potential direct impact may include, for instance, a decrease of biodiversity through damaging the ecosystems, changing the migration patterns of animals and birds, breaks in the fish breeding paths (due to hydropower dams). Moreover, aesthetic damage is another possible impact of the interaction of the energy system and the environment [1].

While all of the mentioned threats are negatively influencing the environment, the largest impact still comes from gaseous waste and emissions which affect the environment on a global geographical and time scale.

There are multiple definitions of the air pollutants introduced by national or international agencies. For instance, in the US, the Clean Air Act requires EPA to set National Ambient Air Quality Standards (NAAQS) for six common air pollutants, also known as "criteria air pollutants" [2]:

- Carbon monoxide, CO, which is released during combustion process. The greatest sources of CO to outdoor air are cars, trucks, and other vehicles or machinery that burn fossil fuels. Breathing air with a high concentration of CO reduces the amount of oxygen that can be transported in the blood stream to critical organs like the heart and brain. Therefore, at very high concentrations, CO can cause death. While these levels are not likely to occur outdoors, they can still negatively impact human health.
- Ozone, O_3. While stratospheric or "good" ozone is vitally important to control the harmful ultraviolet radiation, tropospheric, ground level or "bad" ozone is created by chemical reactions between nitrogen oxides NO_x and volatile organic compounds in the presence of sunlight. This ozone is a harmful air pollutant, as it affects people and the environment.
- Lead, Pb. Sources of lead vary significantly, and may include combustion of leaded fuel (vehicle or aircraft), waste incinerators, utilities, lead–acid batteries, etc. Once taken into the body, lead distributes throughout the body in the blood and is accumulated in the bones, damaging major body functions.

- Nitrogen dioxide, NO_2, which is one of the gases from the broader group of nitrogen oxides NO_x, primarily getting into the air from burning fossil fuels in power plants, cars, trucks, etc. NO_2 along with other NO_x reacts with other chemicals in the air to form both particulate matter and ozone, form acid rain.
- Particulate matter (sometimes abbreviated as PM), being a mixture of solid particles and liquid droplets found in the air. They are usually distinguished by their size, e.g., 10 micrometers denoted as PM_{10}, or 2.5 micrometers as $PM_{2.5}$. These particles can consist of different chemicals, be released from various parts of the energy system including mining, construction of infrastructure, burning fossil fuels, managing waste, and can be inhaled directly and cause serious health issues.
- Sulfur dioxide, SO_2, as one of the major indicators for a larger group of sulfur dioxides SO_x. The largest source of SO_2 in the atmosphere is due to burning of fossil fuels by power plants and other industrial facilities. Sulfur oxides can have harmful effect when inhaled, or contribute to particulate matter pollution reacting with other atmospheric compounds. They can also cause acid rains.

According to [2], when sulfur dioxide SO_2 and nitrogen oxides NO_x are emitted into the atmosphere, they can react with atmospheric water, oxygen, and other chemicals to form sulfuric and nitric acids. These acids then mix with water and other materials before falling to the ground.

Apart from a limited portion of SO_2 and NO_x from natural sources like volcanoes, most of it comes from the burning of fossil fuels, that is, from power plants (approximately two-thirds), industrial processes, manufacturing and refineries, and transportation.

However, the most critical for the environment are greenhouse gases that trap heat in the atmosphere. While naturally these gases are important to control the temperature balance of the Earth and therefore provide conditions for living, in larger quantities these gases can contribute to an uncontrolled increase of atmospheric temperature that may result in irreversible damages to the global ecosystem and humankind.

These gases normally include:

- Carbon dioxide, CO_2, which enters the atmosphere through burning fossil fuels, solid waste, biomass, and is a result of certain chemical reactions. While it is absorbed from the atmosphere through the natural carbon cycle (see Chapter 2), its excess can significantly imbalance the system.

Table 6.1 Typical average lifetime values for some of the typical artificial greenhouse gas emissions.

Greenhouse gas	Lifetime, years
Carbon dioxide CO_2	several hundreds
Methane NH_4	9–12
Nitrous oxide N_2O	more than 110
Hydrofluorocarbons, e.g., CHF_3	up to 270
Perfluorocarbons, e.g., SF_6	up to 3200

Source: Based on [3] and various estimates.

- Methane, NH_4, is emitted during natural gas, oil, and coal production and transportation, agricultural practices, and by decay of organic waste.
- Nitrous oxide, N_2O, is emitted during agricultural and industrial activities, as well as during combustion of fossil fuels and solid waste.
- Fluorinated gases, like hydrofluorocarbons, perfluorocarbons, sulfur hexafluoride, and nitrogen trifluoride, are synthetic, powerful greenhouse gases that are emitted from a variety of industrial processes.

The effect of each of these gases on the environment depends on three main factors [2]:

- How much of these gases are in the atmosphere? Greenhouse gas concentrations are measured in parts per million (ppm), which is the ratio of one gas to another, e.g., molecules of CO_2 in the measured air volume. It can also be measured in %.
- How long do these GHG stay in the atmosphere? Each of these gases can remain in the atmosphere for different amounts of time, ranging from a few years to thousands of years. Table 6.1 shows averaged lifetime numbers for major GHG emissions.
- How strongly do they impact the atmosphere?

Remark 6.1. In order to provide a total amount of GHG emissions, a unified measure is used. CO_2 equivalent (CO_2e) is a measure of total greenhouse gas emissions or concentrations, converting all non-CO_2 gases to their CO_2 equivalent in warming impact [4].

6.3. Greenhouse effect

As mentioned in the previous section, gaseous emissions are the major contributor to the global climate change, which is mostly driven by the greenhouse impact. Long-term changes of the environment may well be not only in climatic changes, but also in an increased amount and severity of natural disasters, limited resources and problem of their quality, and health and safety issues. We briefly discuss here some of the challenges that are witnessed nowadays and can be connected to the negative impact of the global energy system.

The gaseous layer surrounding our planet and called *atmosphere* is held by gravitational forces. Due to the rotation of the Earth, radiation from the Sun and further water evaporation, as well as gravity of the Moon, it is continuously in motion. The Earth's atmosphere is a mixture of multiple gases of various concentrations called *air*. Roughly speaking, by volume, standard air contains 78% nitrogen, about 21% oxygen, 0.93% argon, 0.03% carbon dioxide, and trace amounts of other gases, including nitrous oxide. These ratios change very little with time or location. Most of the time, however, atmospheric air is not dry, and there is a moisture content of about 1% on average [5]. This moisture is present in the form of vapor, water droplets or ice crystals.

The Earth's atmosphere is responsible for the overall conditions at the surface: it provides favorable conditions for the organic life and, due to a proper composition of gases, balances the harmful solar radiation and protects the planet from being overheated or cooled down. The conditions of the atmosphere are characterized not only by the air composition, but also by the temperature of air, pressure, humidity, wind speed, precipitation, and other atmospheric events like thunderstorms, hurricanes, etc. These parameters, assessed at the current time moment, are called the *weather* and, measured over a significant period of time and averaged, define the *climate*. Therefore the Earth's climate is the vital environment for the entire living world and, therefore, has to be considered not only at the present stage of human development. The discussions about possible future climatic changes have reached all strata of our society and influence in many fields the political and economic action, both on a national and on the international scale [6].

The Sun's radiation goes through the transparent atmosphere down to the Earth and heats it up. While the atmosphere is represented by a mixture of various gases with different properties, some of them have a filtering effect [6]: they let the solar radiation pass through, while partly absorbing the infrared radiation emitted from the surface of the Earth, atmosphere itself, and the clouds. This leads to the heating up of the lower layers of the atmosphere and therefore creates the natural *greenhouse effect* — natural warming of the Earth. Therefore, naturally occurring greenhouse gases and clouds absorb radiation emitted by the Earth's surface and elsewhere in the atmosphere. Without the present composition of the Earth's atmosphere, a temperature of $-18°C$ would predominate of the Earth [6].

These transparent natural gases responsible for the heat absorption in the atmosphere are called *greenhouse gases* or *GHG*. Water vapor, carbon dioxide CO_2, methane NH_4, nitrous oxide N_2O, and ozone O_3 are the primary natural greenhouse gases in the Earth's atmosphere [5–7]. They are not toxic unless their concentration is very high, however, even a slight increase in their concentration may affect the climate and shift the greenhouse effect into a temperature increase. Naturally the composition of gases has been balanced by the overall Earth's ecosystem: for instance, an increase in the CO_2 concentration has been outbalanced by the growing vegetation and its consumption for the photosynthesis process.

This balance though has been affected by the development of the human activity, especially starting from the industrial age from the early 19th century. Rapid deployment of breakthrough technologies, increase of production and consumption of fossil fuels, rise of mining for ore to produce steel, and overall growth of manufacturing has led to tremendous and positive economic impact in the development, while causing an increase in harmful emissions. Accounting for greenhouse gas emissions has started to emerge as an issue due to the spotted changes in the thickness of ice fields, surfaces of glaciers, and some climatic abnormalities taking place worldwide. Therefore, a distinction has been made between the natural and balanced greenhouse gases and those produced by man, the so-called *anthropogenic* or *man-made greenhouse gases* [6]. These are several entirely human-made gases in the atmosphere, such as fluorinated gases, halocarbons, and many others. These greenhouse gases have been finally put into a global list of emissions to be controlled according to multiple intergovernmental documents like the Kyoto Protocol [8].

While power generation has been the largest source of greenhouse gases, it basically emits much more than just gases. Looking at the exhaust of a power plant, what one can see are soot particles or particulate matter — fine pieces of fly ash, which consist primarily of black and brown carbon coming out from fossil fuels (coal, diesel, and other oil products' exhaust). Being released into the atmosphere, these particles absorb sunlight directly, and convert that light into heat and therefore heat up the surrounding air. Black carbon absorbs all wavelengths of light, while brown absorbs shorter wavelengths (like blue and green) and reflects or transmits the remaining wavelength. But both carbons absorb sunlight and radiate heat around. On top of this, soot particles can cover the snow and ice surfaces and increase the light absorption, therefore causing their temperature increase.

Soot particles are much more powerful than greenhouse gases to impact the temperature of the atmosphere, but their mass is by far lower. The main reason for this is that soot particles are rapidly removed from the atmosphere over a period of several weeks and therefore can have an immediate effect on the climate situation and temperature, yet not sustainable enough. Greenhouse gases can stay in the atmosphere much longer: methane up to 12 years, CO_2 for at least 50–70 years before it comes out or is dissolved in oceans. The time scales of soot particles and gases differ a lot, but both types of emission are continuously contributing to a temperature increase.

Another contributor to the global climate warming is the *urban heat island effect* [9]: when we cover soil with concrete or asphalt, it prevents evaporation of water from the soil and stops a natural cooling process. Building a lot of structures worldwide basically leads to warming up the surface, but this contribution is relatively small compared to soot particles and greenhouse gases.

If we sum up the effect of the greenhouse gases, soot particles, and the urban heat island effects, it appears to be much higher than the net observed global warming. The reason is that there are other *reflective* or *cooling particles*, like SO_2, ammonia, or NO_x, that come from the power plant exhausts and increase reflectivity of the clouds. About half of actual global warming to date is being masked by cooling aerosol particles [9].

While it is important to decrease emissions of both soot and cooling particles, because all of them are causing severe health problems linked to air pollution, it may unmask half of the hidden global warming. The immediate action therefore will be to control first the soot particles (which can be done with much lower efforts through trapping these particles from the known carbon-burning sources). It can temporarily delay the global warming, bringing a very rapid feedback of the climate, as the soot particles do not stay long in the atmosphere.

Stopping the greenhouse gas emissions today will be a long-term measure, as they will still stay in the atmosphere for decades. So for a sustainable result we shall apply the joint approach: eliminate soot particle emissions, decrease the cooling particle emissions to eliminate health problems, and decrease all greenhouse gas emissions simultaneously. The only way to do so is to change the global energy infrastructure of the world, which is simply not possible in a short perspective.

6.4. Environmental consequences related to climate change

As greenhouse gases accumulate heat within the atmosphere and hydrosphere, the air temperatures will rise, and oceans will also get warmer, contributing to shrinking and thinning of the Arctic Sea ice cover, reduction in the volumes of glaciers worldwide, and decrease in the snow cover. Regarding the impact on snow cover, because the dark-colored surface of land and water reflects less sunlight than light-colored snow, warming is compounded by the albedo refractive effect. In other words, dark tundra and oceans absorb more sunlight than the reflective snow, accelerating energy uptake by, and warming of, these darker surfaces [10].

A warmer planet directly threatens planetary glaciers, particularly those of Greenland and Antarctica, where more than 2% of all the water on the Earth exists in the solid frozen state. As glaciers melt, they increase the amount of water in the oceans [10] which may lead to the rise of global ocean level and flood many coastal regions.

But greenhouse gases do not just affect the temperature of oceans, they also change their chemical composition. Seawater has been absorbing the majority of global carbon dioxide increase, which has directly increased the water acidity. Acidification of oceans alters the biochemical process of marine life and has been implicated in the collapse of major fisheries and disruption of the biodiverse coral reefs, both effects quite detrimental to human food security [10].

Moreover, extracting and burning fossil fuels and using other energy conversion chains, where heat is involved, inevitably leads to a release of excess waste heat to the environment. Adding more energy to the global complex system of the Earth, not only makes this system warmer, but also more unpredictable [10] in a sense of unexpected fluctuations in all spheres of the Earth. This has been recently observed by a number of natural disasters, some of them being extremely devastating: artificially induced earthquakes, tropical storms, tornadoes, droughts, floods, cyclones, forest fires or epidemic outbreaks.

For instance, on September 16, 2017, a new tropical storm originated in the area of Cape Verde. Undergoing extreme intensification as it approached the Antilles islands, it got its personal name *hurricane Maria*. It was classified at some point as a category 5 hurricane with likely catastrophic damages, before devastating Dominica [11].

After weakening slightly after crossing Dominica, storm Maria achieved its peak intensity over the Caribbean with wind speeds approaching

280 km/h, becoming one of the top ten most intense Atlantic hurricanes. At the time it stroke Puerto Rico, it had slightly weakened to category 4, and after crossing the country it moved towards Bahamas being a category 3 storm, and finally dissipated in early October.

In addition to its furious winds, the hurricane produced enormous amounts of rain on the islands over which it swept, causing serious flooding and mud slides [11].

While the amount of fatalities is disputed, it ranges from 112 to more than a thousand, with the damage estimated at more than 90 billion US dollars. It has damaged more than 85% of houses, causing severe food and fresh water problems [11,12].

One of the reasons of such devastating storms, which unfortunately was not the only one even in 2017, is the increased warming of some parts of the ocean and abnormal cooling of the others, which imbalances the system. Another reason is that larger temperature gradients of the atmosphere make the atmosphere more unstable. However, hurricanes are extremely complicated systems on their own, and no one can be sure that their growing intensity is due only to global warming.

Therefore, environmental consequences that are directly related to climate change include the following [13]:

- Increased temperature and frequency and/or duration of heat waves that lead to wildfires, damages of agriculture and severe health issues;
- Increased number of heavy precipitation events, which may cause health issues through flooding, directly injuring people and indirectly causing illnesses;
- Increased intensity and/or duration of droughts that put food and fresh water security at risk, especially in highly populated areas;
- Increased intense tropical cyclone activity due to changes in ocean temperature balance;
- Increased sea level.

Since heat waves, droughts, and floods can directly damage agriculture, the latter can also lead to significant issues with drinking water. Fertilizers and livestock life products that are accumulated in the agricultural areas can be transported by floods to the areas where water is taken for drinking. Despite major efforts to keep these areas closed and separated from water intakes, and to provide effective water treatment solutions, many areas are still far from the ideal situation. Moreover, unexpected floods caused by global climate change can take place in the areas where they have never been seen or expected. The germs and contaminants from the soil of agricultural areas can propagate through the sewage systems and spread infectious diseases, eliminating the efforts for sustained hygiene, hydration, and nutrition. Moreover, when heavy rainfalls combine with increased water temperatures and agricultural waste runoffs, algal blooms can arise and grow with amazing speed. These contaminate fisheries and exude toxins that are poisonous to humans in many ways, including liver and lung inflammation and neurologic diseases [10].

Natural disasters, food production, and fresh water supply are among major risks to take care of, which are mostly seen by the public, as other consequences may not be fully understood as being provoked by the climate change. For instance, the changes in the ecosystems can reduce biodiversity and cause [13]:

- Changes in the lifecycles of vectors, reservoirs, and pathogens. For example, in certain areas a temperature increase can lead to directional spreading of diseases like malaria or dengue, which are caused by migration of mosquitoes that spread to the areas of favorable living conditions.
- Impacts on diseases of wildlife and plants, which may result in complete extinction of species;
- Disruptions of the interactions among species and hence destruction of their habitats.

The forecasts of the global climate change have been of a major topic of interest for the last few decades, especially within developed countries. This activity is twofold: first, these countries account for the majority of greenhouse gas emissions, and secondly, the highest environmental impact is expected to take place in less developed areas with dense populations like Africa and Asia [13].

The discussion regarding scenarios and strategies of climate change mitigation will take place in Part 7.

References

[1] Tester J, Drake E, Driscoll M, Golay M, Peters W. Sustainable energy: choosing among options. MIT Press. ISBN 9780262304634, 2012.

[2] United States Environmental Protection Agency (EPA). Air topics. URL: https://www.epa.gov/environmental-topics/air-topics.

[3] IPCC. Climate change 2007: the physical science basis. Contribution of working group I to the fourth assessment report of the intergovernmental panel on climate change. Cambridge, United Kingdom and New York, NY, USA: Cambridge University Press.

[4] Harris J, Roach B. Environmental and natural resource economics: a contemporary approach. Taylor & Francis. ISBN 9781315448510, 2016.

[5] Tan Z. Air pollution and greenhouse gases: from basic concepts to engineering applications for air emission control. Green Energy and Technology. Springer. ISBN 9789812872128, 2014.
[6] Spliethoff H. Power generation from solid fuels. Springer; 2010.
[7] IPCC. Climate change 2013: the physical science basis: working group I contribution to the fifth assessment report of the intergovernmental panel on climate change. Cambridge University Press. ISBN 9781107057999, 2014.
[8] Nations U. Kyoto protocol to the United Nations framework convention on climate change. United Nations; 1998.
[9] Jacobson MZ. Review of solutions to global warming, air pollution, and energy security. Energy Environ Science 2009;2:148–73. https://doi.org/10.1039/B809990C.
[10] Lemery J, Auerbach P. Enviromedics: the impact of climate change on human health. Rowman & Littlefield Publishers. ISBN 9781442243194, 2017.
[11] Pasch RJ, Penny AB, Berg R. Hurricane Maria. 16–30 September, 2017; 2018.
[12] Barclay E, Fernandez Campbell A. Data shows hurricane deaths in Puerto Rico could be over 1,000. URL: https://www.vox.com/policy-and-politics/2017/11/29/16623926/puerto-rico-death-toll-hurricane-maria-count.
[13] Levy B, Patz J. Climate change and public health. Oxford University Press. ISBN 9780190202453, 2015.

PART THREE

Thermal power as a bridging technology towards sustainability

Overview

Traditional thermal power generating technologies from fossil fuels for several decades have played the dominant role in electricity generation. Despite the negative impact caused mainly by carbon dioxide, soot particles, methane emissions, and other waste products, these technologies are still needed to sustain economic progress now and for the foreseeable future due to their economic benefits, established technology and ease of implementation.

This part opens discussion about traditional power generating technologies with conventional thermal power plants based on steam and gas turbines. It outlines major technologies, their development and underlying challenges deep in detail as they can be considered the basis for emerging technologies.

CHAPTER SEVEN

Concept of a thermal power plant

Contents

7.1. Introduction

Traditional thermal power is based on a historically established and proven technology of burning fossil fuels — solid, liquid, or gaseous, when they release their chemical energy in the form of heat and radiation, justifying the name of the technology. Properties of the fuels, their availability, and cost in many cases define the approach that is used in a conventional thermal power station. Various plants have been built throughout the world, usually powered by two major kinds of machines: steam turbines and gas turbines. Recently, the combination of both has proven to be among the most efficient in delivering power out of natural gas.

Moreover, the overall plant configuration has been subject to multiple modifications in the race towards higher efficiency through eliminating energy waste in the process and equipment. The overall arrangement of plant's major systems evolved in the direction of modularization and following the same approach for all types of power plant: steam and/or gas turbine driven. All of them claim their own ways to design with some similarities, though still following the process of burning fuel to deliver heat.

Sustainable Power Generation
https://doi.org/10.1016/B978-0-12-817012-0.00018-9

7.1.1 Thermal power generation within global power mix

Thermal power generation from fossil fuels, namely coal, natural gas, and oil, accounts for more than 65% of the global electricity production [1,2].

Coal power is the most mature and common power generation technology in the world. With more than 2550 GW of installed capacity, it provides roughly 38% of world electricity generation, which proves the fact that it has served as a reliable, cost-effective, and durable baseload power source to utility scale networks [1,3–6]. Coal-fired power plants account for 8.5–10 trillion kWh of electricity production, which is expected to increase in the next decades [5].

The global coal power generation mix is dominated by Asia and especially China, as well as Africa and North America. China and India alone account for 69–75% of the projected worldwide increase in coal-fired generation, while the OECD nations continue to reduce their reliance on coal-fired electricity generation [3,5].

While natural gas-fired power generating technology accounted for roughly 21% of the global energy mix in 2016 [7], it is expected to grow for the next decades as well, however, with lower rate. In the United States, natural gas-fired generation is encouraged by low prices and favorable greenhouse gas emission characteristics compared to coal-power plants, however, there are some trends of revitalizing the coal industry and coal-fire generation. The fastest growing regions for natural gas-fired efficient technologies are Mexico, the Middle East, and some of the Asian countries, especially Japan, which imports LNG for these purposes.

Remark 7.1. In some countries, like Russia, many coal-fired power plants have been refurbished from solid fuel combustion to employ natural gas as a primary fuel and crude oil as a backup fuel. These power plants have proved to emit lower levels of CO_2, but provide the same or even higher efficiency range compared to those operating with coal. These plants shall not be mixed with modern gas turbine based combined cycle technology which can have much higher efficiency.

Fuel oil occupies roughly 4% of the global electricity generation mix and is mostly employed in the areas where oil is cheap enough to be burnt. Its use is supposed to steadily decline in the next decades, also due to the introduction of new energy policies.

7.1.2 Thermal power as a transition technology

Thermal power plants generate electricity with the economically feasible and historically established technology, which has been developed during the years of thermal cycle optimization, material research, and technology development. Still, these facilities burn fossil fuels, which has to deal with waste and releases large amounts of GHG. Environmental concerns notwithstanding, fossil-fired power generation would continue to play a significant and perhaps a dominating role, however, may not be considered as a sustainable technology in the long run. Still, it is impossible to replace this large share of base load generation in a reasonably short time: even though it may be technically feasible to replace all existing fossil-fired units with sustainable energy sources within few decades, such a conversion would almost certainly take longer due to the difficulty in implementing all necessary policies [8]. Since it may be much easier and cheaper to provide at least all newly introduced energy for the next 20 years from renewable sources, it would still stay as a favorable forecast rather than the real picture due to multiple factors, which shall be discussed in further detail in Part 7. Thus, thermal power is supposed to play a crucial role in the world power generation mix and would bridge the gap between the current power generation system and the future sustainable energy system.

Moreover, recent advancements in the development of more environment-friendly technologies, which at least eliminate some of the emissions to the atmosphere and increase efficiency, make it possible to replace older capacities with the newer technology at reasonable costs, hence maintaining certain balance during this transition period.

However, we shall clearly state that average coal-fired power plants are the dirtiest power generating facilities accounting for 650–1200 g of CO_2 per generated kWh [9,10]. This significant range is mostly due to two reasons:

1. There are several technologies in burning coal and converting its chemical energy into thermal energy, which differ in the physical parameters of the process like temperature and pressure, and therefore in different levels of efficiency, which finally result in certain decrease of emissions per kWh; and
2. Even though there are efficient technologies to eliminate some of the emissions, they are expensive to install and operate, so that it results in higher price of electricity for the final consumer. For this reason, some companies either do not install all the required systems or shut them

down during operation to decrease maintenance costs. This happens mostly in the regions with limited or more liberalized regulation.

Contrary to coal, which always incorporates multiple volatiles and hazardous components (for more details refer to Chapter 2), burning natural gas, which consists mostly of methane, is much cleaner in terms of CO_2 equivalent emissions with the levels almost twice lower than those for coal per kWh generated. Still, being relatively dirtier than most of the renewable energy sources, natural gas fired power generation can bring two major benefits:

1. Decrease the levels of emissions through cleaner combustion process and much higher efficiency of the thermal cycle with more than 62% for modern combined cycle configuration; and
2. Provide high capacity base load, which can also serve more flexibly in increasing and decreasing its output if required by the grid, which is extremely beneficial in the current power mix and may justify the higher cost of technology.

It is still debatable whether or not we shall shut down all thermal power plants. Despite obvious negative impact to the environment, the amount of energy these plants deliver cannot be replaced in a moment, and they would still be with us in the next decades, either newly built, upgraded or still operated.

According to this logic, we shall first draw our focus to the technological development of a steam power plant and go through the basics of the cycles and their optimization up to the modern level. This idea is twofold: first, we will see the possible areas of improvement to increase the level of friendliness of the thermal power, and secondly, we will understand the foundations of this established technology as it became fundamental for other types of power plants, including emerging sustainable technologies.

Next, we will outline key equipment areas and discuss in much detail overall plant configuration. The same approach will be applied to both steam and gas turbine driven power plants. With the short overview of the systems, we will get the idea of the common features of major types of plant, allowing for better understanding of the power generation business, as well as the EPC and OEM business approach. The logic here is obvious: due to the historic development, thermal power turns out to be most established, therefore providing a knowledge base for other power generating facilities, most of which have a lot in common and employ thermal energy conversion. This way, we may put much of the effort to understand this technology and further refer back to this discussion while investigating other power generating approaches.

7.2. Steam power plant

As one of the most historically established types of power plant, a steam power plant provides a reliable energy conversion technology. It is based on the Rankine thermodynamic cycle (for more details one can refer to any of the engineering books, for instance, [11–13] and many others). Its gradual optimization allowed coming up with the relatively efficient water and steam cycle supported by an established technology of the major equipment. Given that, we focus on the concept of a steam power plant in detail.

7.2.1 Steam turbine cycle

Consider a simplified cycle of a steam turbine unit as shown schematically in Fig. 7.1. To start with, fossil fuel is fed to the special equipment called *boiler* where it burns. To ensure that the combustion process goes continuously, it requires constant supply of oxygen from the air, which is taken from the ambient environment and which can also be considered as a resource. The chemical reaction of fuel combustion is exothermic and therefore releases heat and is accompanied by various combustion products — a mix of different gases and particles (that are released back to the environment as

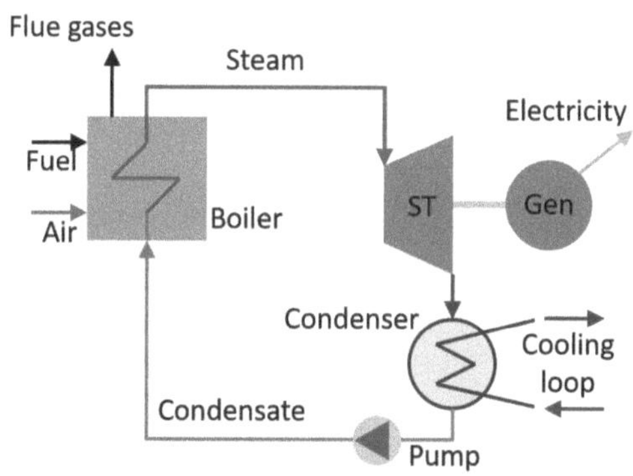

Figure 7.1 ***Simplified cycle of a steam turbine unit.*** Boiler burns fuel and generates steam that is fed to the steam turbine (ST), which rotates electric generator (Gen). Cold steam goes to the condenser and turns back into water, which is pumped back to the boiler to close the cycle.

flue gases). These hot gases possess enough energy, which is transferred to the water flowing inside the tubes called *heat exchangers* located within the boiler. This water evaporates, so that the pressure and temperature of the steam go up.

The superheated steam is fed to the inlet of the machine called *steam turbine* (ST). To ensure that the steam flows through and does not condense inside the turbine, the special vessel called *condenser* is located at the exhaust of the steam turbine. This device is kept under a pressure much lower than the atmospheric to make the steam go through the turbine straight into it and condense back into water. This so-called *vacuum* of the condenser requires constant supply of cold water to cool down the coming steam and maintain low pressure.

While flowing through the steam turbine to condenser, the steam cools down and expands. It gets kinetic energy during the flow and expansion, and hits the blades, which are mounted on the rotor, so that the rotor starts to spin. Therefore, the kinetic energy of the steam is converted into the mechanical energy of the rotating shaft. The rotor of the turbine is coupled to the rotor of the electrical *generator* (Gen): in its windings the electrical current is generated and then delivered to the grid.

On the other hand, exhaust steam releases its heat to the environment inside the condenser and therefore condenses back into water. Finally, this water is pressurized and fed with the *pump* back to the boiler where it receives heat from the burning fuel.

Major heat losses of a thermal power plant occur inside the condenser when the heat of the exhaust steam is taken away by cold water leading to around 50% of energy waste. Another part of energy is used for *auxiliary needs* of the process to provide its functionality. These include powering of special equipment like air ventilators, fuel supply, and water pumps for continuous operation and may result in 3–5% of energy losses. Certain part of energy cannot be transferred to the water inside the heat exchangers due to technological reasons and hence this part leaves the boiler through the exhaust stack into the atmosphere. These products are called *exhaust gases* and normally have the temperature range of 100–160°C. The amount of heat wasted to the atmosphere can be around 5–15% of the initial chemical energy of fuel and depends on the technology, type of fuel, and plant operating conditions. The temperature of the exhaust gases, as well as their chemical composition, is limited by the regulation to minimize environmental impact.

What is finally left is called the *net capacity* of the power plant. This amount of power can be then delivered to the grid and transferred to the consumer. The *net efficiency* of a thermal power plant is the ratio of the energy supplied to the grid within a certain period of time and the heat energy of the burnt fuel. The average efficiency of a steam power plant can be up to 40% with the losses due to the energy transformation process and the present state of the technology. In other words, *within the traditional steam power plant, we can use roughly 40% of energy stored in fossil fuel.*

7.2.2 Heat balance diagram and its optimization towards higher efficiency

A simplified steam power plant schematically shown in Fig. 7.1 employs a steam turbine, which generates only electricity with all the steam from the turbine going into the condenser. Such a power plant is therefore called a *condensing power plant*. In some cases, customers require not only electricity but also steam or hot water for district heating or industrial needs. A power plant that supplies not only electricity but also heat is called *cogenerating*. Cogeneration normally increases overall plant efficiency as it uses "waste" heat from the process.

In a cogenerating power plant, as shown schematically in Fig. 7.2, some steam is extracted from the steam turbine and guided to the consumer supplying heat for residential and/or industrial needs. This allows for less steam in a condenser and hence less waste of heat in the process. Then the cold water from the consumers is pumped back to the cycle.

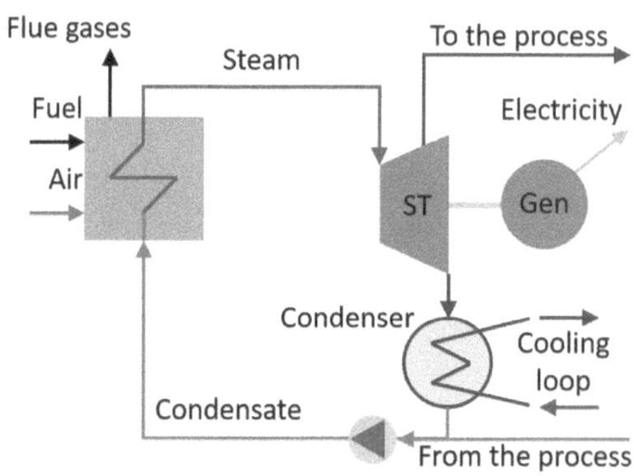

Figure 7.2 ***Simplified steam power plant with cogeneration.*** Schematic diagram of the process with cogeneration: steam is extracted from the steam turbine and delivered to some process (district heating, industrial use, etc.). Cold water is then delivered to the cycle.

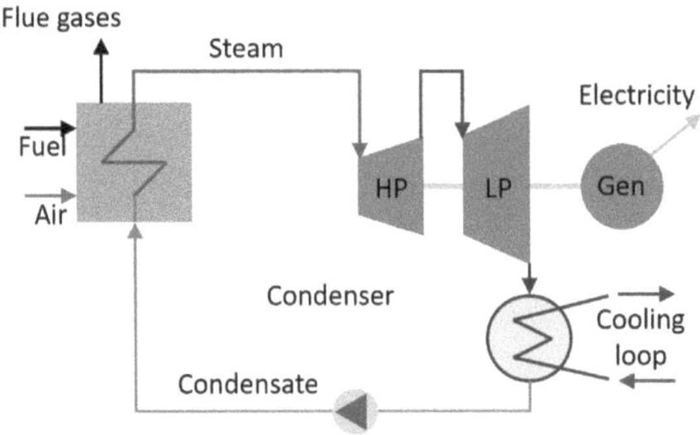

Figure 7.3 ***Simplified steam power plan with a two-pressure steam turbine.*** Schematic diagram of the process employing a two pressure steam turbine to increase efficiency.

This way, both types of steam power plant would have the following *major equipment* as shown in Fig. 7.2:

- steam generator, or boiler, which consumes water, air, and fuel to produce high pressure and high temperature steam;
- steam turbine with electrical generator;
- condenser;
- feedwater pump.

The above presented simplified process still has quite low efficiency due to multiple losses of energy. For instance, the hot high pressure steam goes through the steam turbine and loses some of its energy, but can remain hot enough to be reused and therefore guided to another steam turbine section called the *low pressure steam turbine* (LP ST) (see the schematic in Fig. 7.3). Such a configuration would allow to increase the plant efficiency by utilizing as much energy of the steam as possible.

On the other hand, the amount of energy from the burning fuel is high enough to not only evaporate water and heat up the initial steam for the high pressure steam turbine (HP ST), but also reheat the used steam and increase its temperature and pressure. This means the hot combustion products go through the first heat exchanger and evaporate the water, and then, instead of being released (wasted) into the atmosphere, they are guided through the second heat exchanger to reheat the steam. Fig. 7.4 shows the schematics of the respective steam reheat cycle: the cold steam from the HP ST is guided to another heat exchanger in the boiler called *reheater* where its temperature and pressure increase. Finally, this reheated steam is delivered to the LP ST and further to condenser. Such a cycle allows for better energy utilization and hence increased efficiency of the plant.

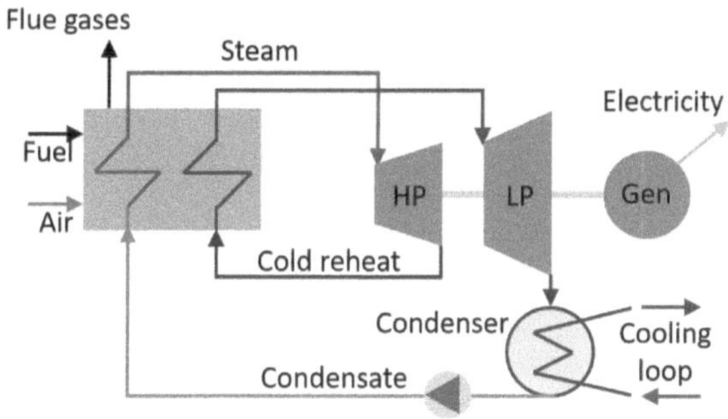

Figure 7.4 ***Simplified steam power plan with a two-pressure boiler with reheater.*** Schematic diagram of the process employing a two pressure boiler with reheater.

Assume now that we require to further increase the efficiency of the plant. One of the ways to do so is to optimize the burning process inside the boiler and provide additional heat exchangers to ensure we get the most out of the combustion process. This can possibly be delivered by increasing the temperature of the combustion products through better mixing the air and fuel, gas flow path optimization, furnace geometry correction, etc. The overall approach is only limited with two major factors:

1. Physical properties of the materials, which are capable to withstand very high temperatures and pressures, and
2. Economical factors or the cost of such technology to ensure that the generated net power is cheap enough.

By further increasing the pressure through the boiler engineering and optimization, we would produce steam of better quality and excess heat. This would allow us to go from two pressure (high and low) to three pressure cycle shown in Fig. 7.5. Here the steam with the highest energy is first admitted to the HP steam turbine and, after flowing through this section, it is delivered back to the boiler for reheating. After its temperature and pressure are recovered from the energy of the boiler, it is further fed to the IP steam turbine section and then further to LP.

The water that comes from the condenser back to the boiler is cold enough (due to the condensing conditions) and would require significant energy from the chemical reaction of the burning fuel to be evaporated. In order to save some energy and optimize the process, this water can be heated up and come to boiler already preheated. This is normally done through the special heat exchangers called *feedwater preheaters* or *economizers*. They use some of the hot steam extracted from the steam turbine and guided to this heater as schematically shown in Fig. 7.6. Depending on

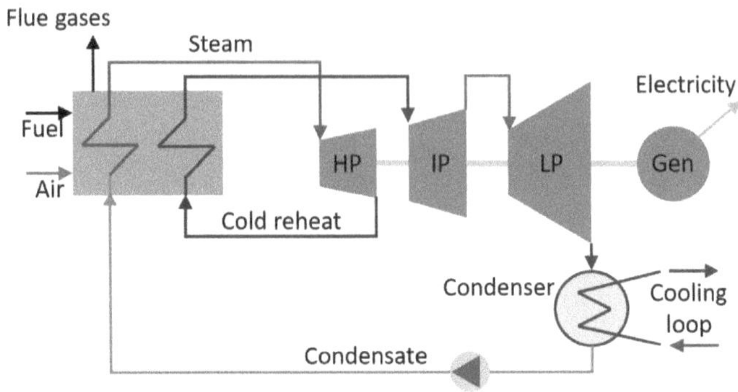

Figure 7.5 ***Simplified steam power plan with a three-pressure steam cycle.*** The three-pressure cycle allows employing a more efficient and larger steam turbine with three cylinders: high pressure (HP), intermediate pressure (IP) and low pressure (LP).

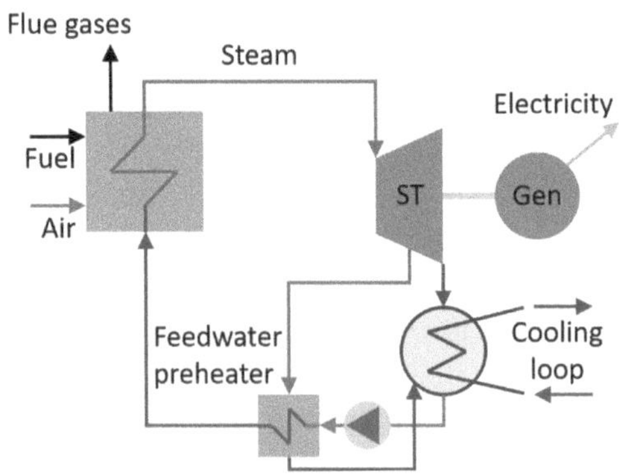

Figure 7.6 ***Simplified steam power plant with preheater.*** Employing a feedwater preheater within the schematic cycle of a steam power plant allows reusing waste heat to warm up water, so that it would require less energy of the fuel to evaporate it.

the configuration of the plant, there may be several preheaters to ensure that most of the energy is captured within the process and not dissipated as waste. After the preheater, colder steam is guided back to the process, for instance, into the condenser.

Despite proper insulation and the closed cycle, there are still some inevitable losses of steam and water inside the process. This can take place due to multiple leakages in the steam turbine, drainage within the pumps, tanks, and pipes, emergency situations or even quality issues which would require maintenance. To maintain a proper amount of water inside the cycle, it has to be properly prepared and constantly added to the process. This is normally done within the *water treatment plant* where water undergoes thorough cleaning to ensure that it does not damage the heat exchangers, pipes, steam turbine, and other equipment.

On the other hand, the condensed water may contain dissolved gases, especially oxygen, due to the vacuum inside this vessel as it would suck in not only steam from the steam turbine but also air from the outside, mixing it with water. This would require special equipment inside the cycle to *deaerate* the water and make sure it does not cause corrosion in the system. This equipment is called *deaerator* and serves not only for removing oxygen (and other gases) from the water but also as a storage for boiler feedwater. It means that extra water coming from the treatment plant goes first to the deaerator tank as shown in Fig. 7.7. Such a scheme would require a second *condensate pump* to guide water from the condenser to the deaerator.

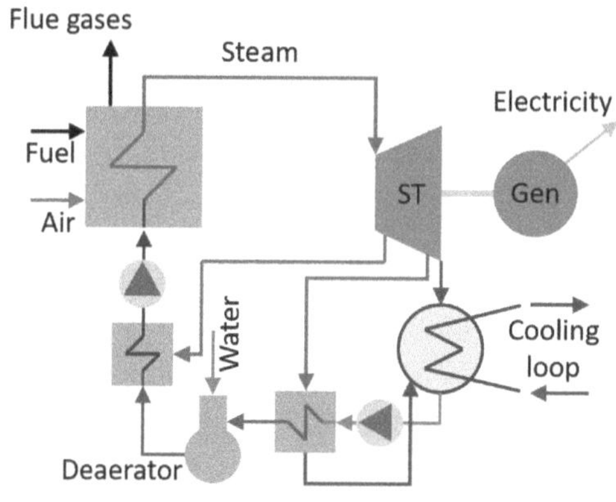

Figure 7.7 ***Simplified steam power plant with deaerator.*** Use of deaerator within the cycle of a steam power plant allows to remove oxygen that is inevitably sucked into the cycle and decrease corrosion of the system.

In spite of the historically developed optimization solutions, the efficiency of a fossil-fired condensing steam power plant does not normally exceed 40% and is subject to the technological process as shown above. However, it can be also influenced by the load regime of the plant or conditions of the equipment. This way, the optimization of the steam power plant requires not only efficient and modern equipment properly integrated into the plant, but mostly a well-designed and optimized heat balance. Fig. 7.8 shows a simplified diagram of a real three-pressure condensing steam power plant. The preparation and optimization of such a cycle shall be done by an application engineer within the special software on an iterative basis.

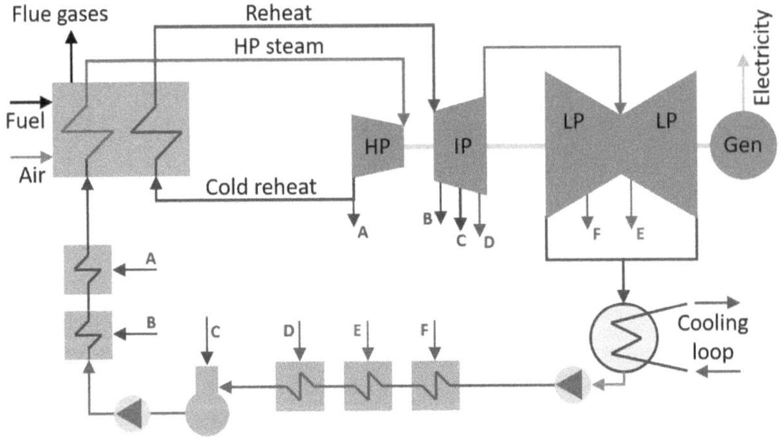

Figure 7.8 ***Simplified steam power plant with optimized cycle.*** A modern steam power plant with optimized cycle employs all of the discussed solutions: several pressure levels, cascades of feedwater preheaters, deaerator, etc.

Depending on the objective of the power plant and its required capacity (electrical and thermal), the application engineer would choose several pressures, type of boiler and auxiliary equipment (heat exchangers, pumps, condenser and cooling, etc.), and optimize the flows of water and steam.

The above mentioned diagram, together with the values of temperature, mass flows, and pressure, defines the overall *heat balance* of a power plant. Based on these calculations, the steam turbine application engineer would make the turbine configuration, leading to the fact that most of the steam turbines are adjusted or even specially designed for each project.

7.2.3 Support systems and equipment areas

The heat balance defines the configuration and the parameters of the water and steam cycle, performances and the major equipment with some auxiliaries. However, it does not give a complete picture of the overall plant configuration, equipment arrangement, and supporting systems to allow for proper functionality.

7.2.3.1 General process description and equipment areas

We now summarize the overall process of a fossil fuel fired steam power plant shown in Fig. 7.9. The process starts with the fuel supply from fuel handling system where it is properly prepared to ensure that the combustion process delivers most of the energy. The fuel is fed to the furnace of a boiler

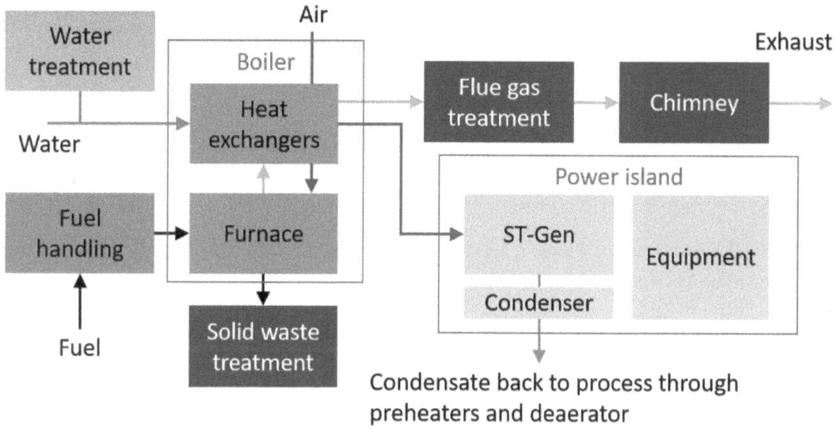

Figure 7.9 ***Simplified blocks of a steam power plant process.*** Simplified process to illustrate major steps within a steam turbine power plant (many steps and equipment units are omitted intentionally).

where it is ignited and burnt. The combustion gases follow the hot gas path of a steam generator (colored in yellow (light gray in print version)) transferring their energy to steam, water, and air through the system of heat exchangers. As these gases contain combustion products, they go through the flue gas treatment equipment before being released to the atmosphere through the chimney.

To maintain the burning process of the fuel, air is supplied to the furnace through the so called air path. To increase plant efficiency, the air is preheated with the hot flue gas inside the heat exchanger called *air preheater* and guided to the boiler by the fan.

In order to deliver steam and to minimize the damages inside the equipment, water shall be prepared within the water treatment station and then preheated with the hot flue gases inside the water preheater to optimize the efficiency of the plant. A number of heat exchangers inside the boiler transfer heat to the water to evaporate it and then heat the steam to increase its energy and decrease water content inside to supply hot steam to the steam turbine. The exhaust steam then goes to the condenser where it turned back into water. To ensure that the condenser is operating at the required capacity, the heat rejection loop is employed. The gathered condensate is then pumped back to the boiler through the feedwater preheaters.

The steam turbine is powering the electrical generator. The electricity is evacuated from the generator terminals through the main transformer into the grid. The overall control and operation of the plant is ensured via

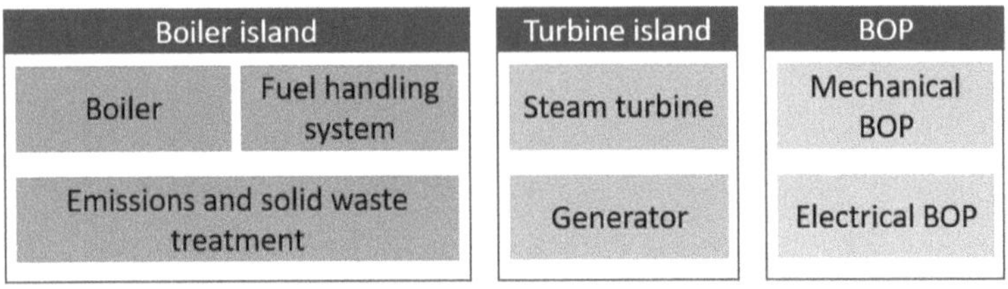

Figure 7.10 ***Equipment areas of a steam power plant.*** Simplified representation of the equipment areas of a fossil-fired steam power plant.

the control system. The description of the process, together with the heat balance diagram, gives an overview of how a power plant looks like.

The described process of a fossil-fired steam power plant allows for breaking down the facility into a number of parent systems. These equipment areas, as shown in Fig. 7.10, include:

- *Boiler island*, which supplies hot pressurized steam. Subject to definition, boiler island can consolidate all equipment around the boiler related to the steam production cycle, including fuel treatment and waste handling, i.e., everything required to transfer chemical energy of the fuel into the energy of hot pressurized steam, considering also the waste and its proper treatment.
- *Turbine island* that transforms the heat of the steam into the mechanical energy of the rotating shaft of the steam turbine coupled to the generator.
- *Mechanical balance of plant*, including all supporting systems like heat rejection, water preparation piping, etc.
- *Electrical balance of plant,* including the equipment that is responsible for transferring the power from the generator terminals to the grid, as well as for supplying power for all systems of the plant.

This simplified breakdown of the facility into the major areas or subsystems is helpful for two reasons. First, it generally corresponds to the scope of supply or the major manufacturers of equipment or companies that do engineering, procurement and construction (EPC) of a power plant. For example, there are companies that supply equipment only to the boiler island, or there are those, which act only in the steam turbine or generator market. Second, it gives the basis for comparing steam power plants to other power generating facilities and defining similarities not only in the smaller pieces of equipment but also in the major areas and systems. Finally, it allows for modularization of the power plant, which is mostly applicable for larger utilities.

7.2.3.2 Modular configuration of a steam power plant

The simplified process presented in Fig. 7.9 assumes one boiler feeding steam to one turbine. In many cases, however, the required capacity of a power plant cannot be fulfilled by only one boiler or only one steam turbine. This means that the overall configuration of both the boiler and power island can vary. Here several options are available:

- Several similar blocks or units, each of them having their own boiler island and turbine island. Such an option would require duplication of most of the systems, complicated routing of cables and pipes, and larger footprint, which leads to a higher cost of electricity generated at this plant. On the other hand, such a configuration may give certain flexibility, but practice shows this is usually not feasible economically.
- One boiler supplying steam for several steam turbines located in one power island, and each turbine employing its own electrical generator and steam condenser. Such a structure would optimize the boiler island equipment (e.g., single fuel and emissions treatment system) but would lead to complicated steam piping and sophisticated control system. However, from an economic point of view, this configuration can be favorable.
- Several boilers supplying steam to the main steam collector, from where it is distributed to a number of steam turbines. This configuration is applicable in large power plants or, for example, in case of adding new capacities to the existing plant. Depending on the output equipment arrangement, this can be a preferred option compared to the previous two.

7.3. Gas turbine powered energy

The development of gas turbines primarily for the aircraft propulsion has raised a question of their implementation for stationary power generation in a thermal power plant. Gas turbines have proved to be more flexible in operation and therefore could be used within a broader range of grid applications, i.e., both for constant load and when extra load is quickly required by the market. On top, combination of both technologies under one umbrella has enabled significantly increasing efficiency of a thermal power plant.

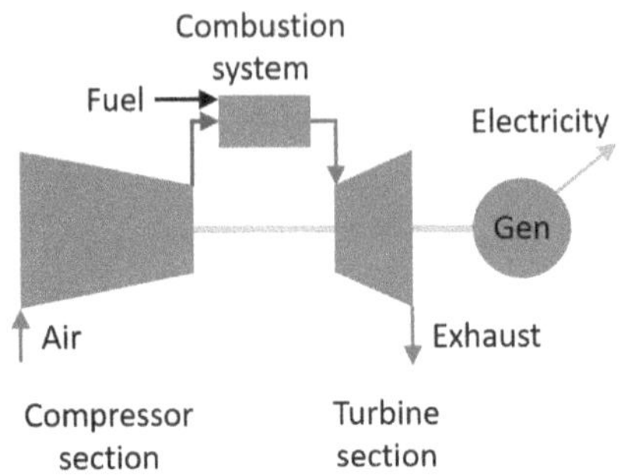

Figure 7.11 ***Simplified cycle of a gas turbine unit.*** Fuel is fed to the combustion system, and the air is compressed by compressor.

7.3.1 Gas turbine simple cycle

We will first focus on a simplified gas turbine cycle, which is schematically represented in Fig. 7.11. The idea of the process is based on the combustion of gaseous or liquid fuel and is fairly simple. The ambient air is taken from the atmosphere at the *inlet* of the compressor section of the turbine. The rotating blades of the compressor suck this air and compress it to a much higher pressure while pushing it through in the axial direction, so that its temperature increases. Hot pressurized air flows to the *combustion chamber* where it is mixed with the natural gas and burnt under constant pressure. The temperature of the combustion products, which are a mixture of various gases, rises up to 1200°C. Under this temperature and constant pressure, the hot combustion products are released to the turbine section where they tend to expand to the atmospheric pressure and therefore acquire kinetic energy. The energy of the hot exhaust gases is used to spin the shaft of the turbine, which is coupled to the shaft of the generator. After loosing some of their energy, the exhaust gases are finally released to the atmosphere.

The shaft of the turbine section is connected to the shaft of the compressor section, and the rotation of the former will spin compressor and maintain the constant air flow and its compression to maintain combustion. However, to start the process, there should be an external rotating machine to start the spinning of the shaft and start the flow of the air to the combustion. Then, part of the energy of the rotation that comes from the energy of the hot gases will be used for rotation and air compression.

Remark 7.2. A simple gas turbine cycle is a steady repetition of the *Brayton cycle*. In this simple cycle, combustion and exhaustion occur at constant pressure and compression, and expansion occurs continuously. This means that a gas turbine generates power continuously rather than in pulses like in an internal combustion engine.

This process, usually called an *open* or *simple cycle* of a gas turbine, uses hot gases as the main working medium. The environment, or simply the atmosphere, is considered as an infinite supply of cold ambient air fed to the compressor, while the energy is supplied to the system with the fuel. It is the chemical energy of the fuel that is transferred first into the energy of the hot pressurized gases (combustion products), and then into the mechanical energy of the rotating shaft. Part of this energy is used to rotate the compressor and therefore suck the air and compress it before delivering into the combustion chamber.

As the gas turbine cycle is exposed to the ambient conditions, its performance and flexibility characteristics depend on these conditions. Therefore the cycle is usually designed for particular conditions, and the performance calculations are conducted at both standard or ISO conditions and site average conditions.

Similarly to a steam turbine, gas turbine cycle has quite low efficiency in transferring the chemical energy of the fossil fuel into electricity compared to a theoretical limit. This is mostly due to multiple points in the cycles, where energy is wasted or dissipated within the environment. Concerning the efficiency of a gas turbine cycle, there are several sources of energy losses:

- Friction within the rotating equipment (compressor and turbine shaft);
- Exhaust gases of the gas turbine, which are released to the atmosphere. They are still hot enough, however, the energy is simply wasted.

A thermal power plant based on such a gas turbine cycle is called a *simple cycle power plant* and has efficiency around 30%. Among the approaches to increase efficiency of this type of power plant, one of the simplest and most evident ways is to recover the heat from the hot gas turbine exhaust and utilize it for other purposes. Here two major options are possible:

1. Use special heat recovery equipment to boil water and produce steam for industrial utility purposes. A power plant, which produces both electricity and heat, is called a *cogeneration plant* (Cogen) or *combined heat and power plant* (CHP).
2. Implement a heat recovery steam generator to supply steam to the steam turbine. Combining gas and steam turbine cycles together in one power plant allows for increasing the efficiency considerably. Such a power plant is called *combined cycle power plant* (CCPP).

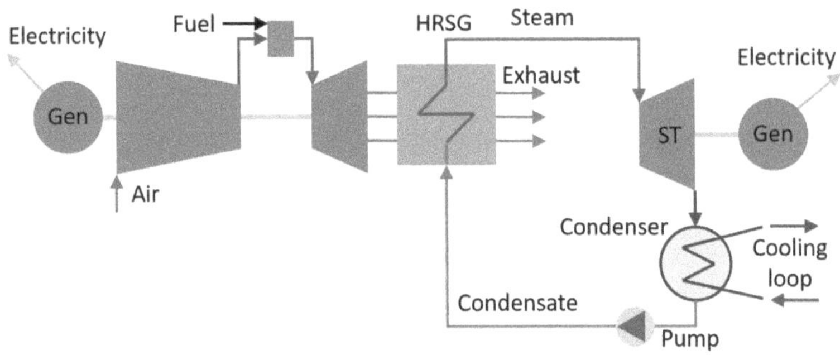

Figure 7.12 ***Simplified diagram of a combined cycle power plant.*** Hot exhaust gases from the gas turbine transfer their waste energy to the water within the heat recovery steam generator, which acts as a boiler. Stem is then fed to the steam turbine. Both turbines can transfer their mechanical energy into electricity.

7.3.2 Combined cycle as a way to increased efficiency

In a combined cycle power plant, two thermodynamic cycles are employed:

1. Brayton cycle for a gas turbine, the same as for a simple cycle power plant. This cycle is also known as *topping cycle.*
2. Rankine cycle of a steam turbine, where the energy for evaporation is taken not from the burning fuel but from the hot exhaust gases of a gas turbine. The steam turbine cycle is called *bottoming cycle.*

The implementation of this cooperation is fairly straightforward. Consider the simplified diagram of the combined cycle power plant schematically represented in Fig. 7.12. The gas turbine follows the simple cycle schematic with the only difference: the exhaust energy is fed to the heat recovery equipment or a special type of boiler called *heat recovery steam generator*, or *HRSG*. This unit transfers the energy of the hot gases to the water that circulates through the tubes of the heat exchangers and evaporates. The steam is admitted to the steam turbine where it expands and is released to the condenser.

The steam cycle is similar to that in a steam turbine power plant, thus being limited by the exhaust parameters of the gas turbine, namely temperature and flow. For instance, if the exhaust gases are hot enough (especially for bigger units), the steam turbine performance can be optimized in the same way as for a steam power plant, for instance:

- High pressure steam from the HRSG can be fed to the HP steam turbine section and, after being released, can be brought back to the boiler as cold reheat;

- This cold reheat can be further heated in a boiler's reheater section and, with recovered temperature, be fed to the IP section of the steam turbine;
- Finally, released from IP ST section, steam can be fed to the LP ST section.

However, even the largest available gas turbines still have a limited exhaust gas flow and temperature, which is definitely much lower than the temperature inside the conventional fossil-fired boiler. Therefore, in order to optimize the output of a combined cycle power plant, multiple approaches are used:

1. The above discussed combined cycle power plant with one gas turbine coupled to the HRSG and supplying steam to one steam turbine. Each turbine has its own generator (two in total in this case) coupled to a respective turbine shaft. Therefore such a configuration is called *multi-shaft* and shown in Fig. 7.13(A).
2. If we are not able to increase the output of a single gas turbine, we are still left with the option of using multiple gas turbines with their own

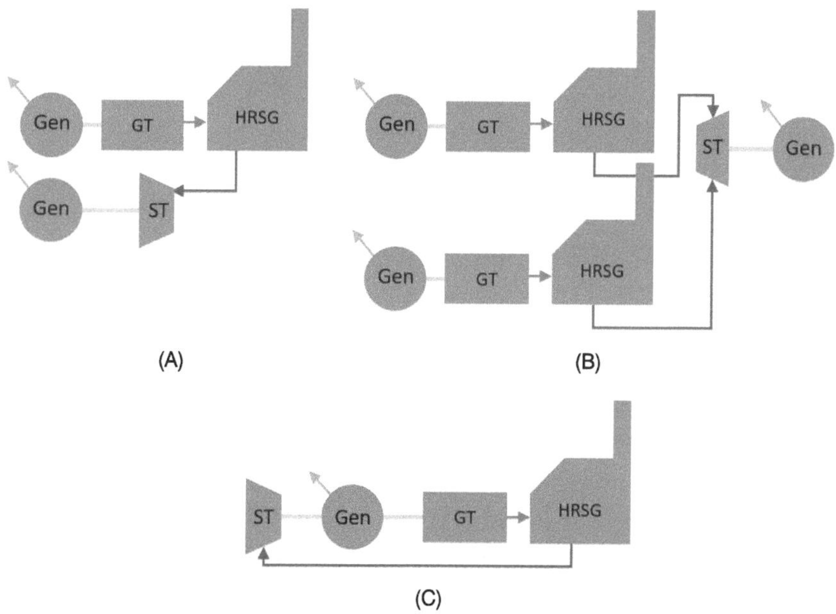

Figure 7.13 ***Combined cycle power plant configurations.*** Multi-shaft configuration (A); two gas turbines with their own HRSGs feeding steam to a single steam turbine (B); single-shaft configuration (C) with gas and steam turbines coupled to a single generator.

heat recovery steam generators. The steam is then collected from all of them and supplied to one large steam turbine to increase its output and efficiency. In practice, usually two or three gas turbines are combined with one steam turbine, as shown in Fig. 7.13(B). In some literature, such configurations are called, for example, "two on one" or "three on one" meaning that there are two or three gas turbines operating with one steam turbine, respectively.

3. If we want to increase the output of the plant, we can also use only one generator, which is coupled to both gas turbine and steam turbine. Such a configuration is called *single-shaft* and is represented in Fig. 7.13(C). It is normally implemented for higher capacities and base load operation at top levels of efficiency.

All of the above mentioned configurations of the combined cycle power plant can be implemented in either a single plant or in modular configuration. In general, configuration of a combined cycle power plant depends on many factors. To start with, the major defining factor an investor considers is the MW capacity of the plant and his capability to sell energy to a consumer. The capacity here covers not only electrical megawatts but also thermal power in the form of steam or hot water. The latter can bring much more profit compared to electricity, for instance, when the power plant produces high quality steam of required properties for industrial site, e.g., for refinery or chemical plant. Another example may be a cogeneration power plant where the major output parameters of the station are electricity and hot water for district heating needs.

The electrical efficiency of a combined cycle power plant (without considering possible heat production) is evidently higher that of a simple cycle or conventional coal-fired power plant. The incoming chemical energy of the fuel is used for the topping cycle, however, the major waste of the simple cycle — the exhaust — is reused in the process.

Remark 7.3. The term *combined cycle* is used to define a type of a power plant. In some cases it is appropriate to name it a *gas turbine combined cycle (GTCC) power plant*, specially pointing out that this power plant is based on a gas turbine cycle combined with steam turbine. Also the name *combined cycle gas turbine (CCGT) plant* is widely used. For simplicity, however, it is recommended to use the term *combined cycle power plant*, or *CCPP*, as it immediately assumes gas and steam turbines.

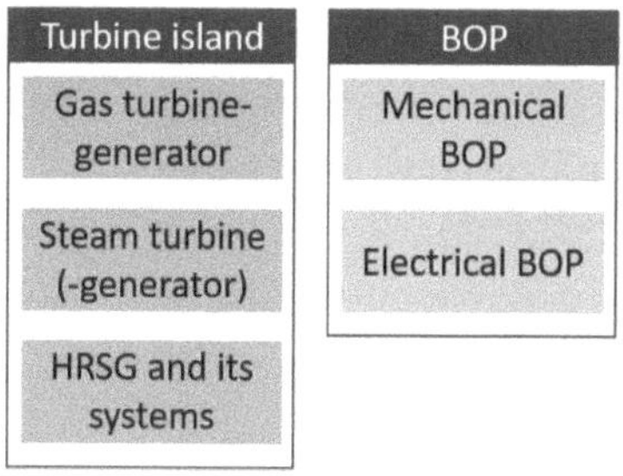

Figure 7.14 ***Equipment areas of a combined cycle power plant.*** Simplified representation of the major equipment areas of a CCPP includes turbine island with a topping cycle (gas turbine-generator) and a bottoming cycle (HRSG and steam turbine with a possible generator) supported by the BOP systems.

7.3.3 Major components and systems

The schematic process diagrams and heat balances discussed above define the key parameters of the gas and steam and do not fully specify all necessary equipment, systems and site layout. Similarly to the steam power plant, every simple and combined cycle power plant can be segmented into several major systems, with the heart being always gas turbines, steam turbines, and generators. A gas turbine driven power plant can be broken into the following major equipment areas (Fig. 7.14):

- *Turbine island* or *power island* that transforms the chemical energy of the fuel into the mechanical energy of the rotating shaft of the turbines coupled to the generators. It normally includes:
  - *Topping cycle* that incorporates a gas turbine with its generator and dedicated systems;
  - *Bottoming cycle*, which includes heat recovery steam generator, steam turbine with its generator (if applicable), condenser, essential valves and supporting systems (applicable only for combined cycle);
- *Mechanical balance of plant*, including all supporting systems like heat rejection, water preparation piping, etc.;
- *Electrical balance of plant* including the equipment that is responsible for transferring the power from the generator terminals to the grid, as well supplying power for all systems of the plant.

This breakdown allows for the customer to clearly see the potential scope of supply and define the configuration of a power plant. On top of this, such a modular approach allows customers to implement upgrades and power plant extension projects in a simpler way. The typical configurations can be derived one from the other and include:

1. Mechanical drive gas turbine, when a unit is used to power other machinery, for instance, a compressor for a gas pipeline or LNG facility;
2. Simple-cycle power plant with a gas turbine and generator operating in an open cycle;
3. Simple cycle with cogeneration mode, when the power island consists of a gas turbine, generator and heat recovery equipment;
4. Combined cycle power plant.

Depending on the engineering approach and historical trends, the breakdown into equipment areas may differ. Various market players prefer to have their own itemization, which does not necessarily coincide with the given here. However, in most cases the approach is similar to what has been discussed here and can be easily adopted to the market conditions.

References

[1] World Coal Association (WCA). Coal and electricity generation, vol. 3. World Coal Association; 2012.
[2] World Energy Council. World energy resources 2016. World Energy Council. ISBN 9780946121588, 2016.
[3] GE. Powering the world. Steam power systems product catalog. GE; 2015.
[4] IEA. Energy supply security. Emergency response of IEA countries. International Energy Agency 2014.
[5] Energy Information Administration (EIA). International energy outlook 2016. US Energy Information Administration; 2016.
[6] IEA. Key world energy statistics 2018. International Energy Agency; 2018.
[7] IEA. Key world energy statistics 2016. International Energy Agency 2016.
[8] Jacobson MZ, Delucchi MA. Providing all global energy with wind, water, and solar power, part I: technologies, energy resources, quantities and areas of infrastructure, and materials. Energy Policy 2011:1154–69. https://doi.org/10.1016/j.enpol.2010.11.040.
[9] Teske S, European Renewable Energy Council, Greenpeace International. Energy [r]evolution. A sustainable world energy outlook 2015. Greenpeace International; 2015.
[10] WCA. Coal facts 2014. World Coal Association; 2014.
[11] Gupta M. Power plant engineering. PHI Learning. ISBN 9788120346123, 2012.
[12] Kehlhofer R, Rukes B, Hannemann F, Stirnimann F. Combined-cycle gas and steam turbine power plants.
[13] Sarkar D. Thermal power plant: design and operation. Elsevier Science; 2015. 612 pp.

CHAPTER EIGHT

Efficient and clean combustion of fossil fuels within boiler island

Contents

8.1. Boiler island configuration

A *boiler island*, or a *boiler unit*, is a major part of a thermal power plant, which utilizes steam within a steam turbine. As discussed in the previous chapter, this system is basically transferring the chemical energy stored within the fossil fuel, into the energy of the hot pressurized steam, which is then used for various purposes. This energy transformation process can be separated into the following major steps:

1. Preparation of the fossil fuel for the combustion process;
2. Combustion of the fuel with the release of heat;
3. Evaporation of water to produce steam of the required parameters;
4. Process waste treatment and disposal.

Therefore the whole set of equipment, both mechanical and electrical, utilized for the steam production, together with the necessary supporting systems, is called *boiler island*. This set of equipment can be sometimes referred to as a *boiler unit* with the combination of some mechanical and

Sustainable Power Generation
https://doi.org/10.1016/B978-0-12-817012-0.00019-0

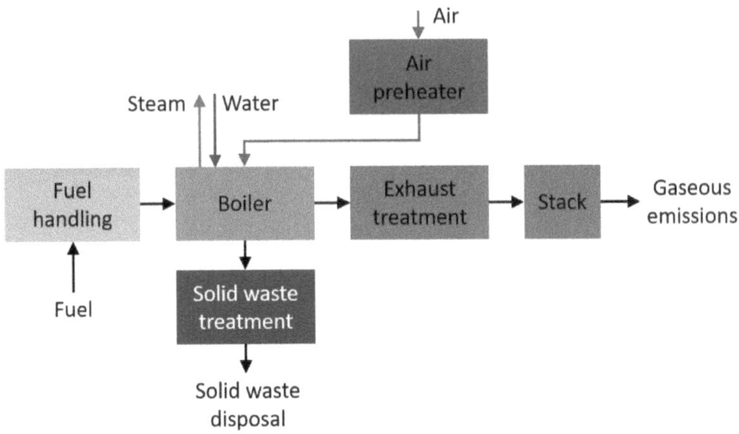

Figure 8.1 ***Simplified schematic of boiler island systems.*** The process starts with the fuel treatment system that delivers fuel to the boiler. Steam from the boiler is fed further to the process, and the waste from the combustion process goes through the exhaust and ash treatment systems.

electrical systems related to major circuits: fuel, water and steam, air, and gas.

The general concept of a boiler island is shown in Fig. 8.1. The energy conversion process starts with the delivery of the raw fuel to the power plant and its proper treatment within the fuel treatment system allowing for the optimized combustion process with the least possible energy waste. This fuel is directed to the boiler, the configuration of which depends on the type of fuel used on the power plant. The combustion process requires air to maintain the chemical reaction so that the preheated air is directed to the boiler as well. In order to minimize energy consumption during the chemical reaction, the air is preheated with the waste energy of the process, so that it allows directing the heat from the burning fuel straight to the water. As discussed in the previous chapter, the water is also preheated due to the same reasons of process optimization.

Finally, the evaporated water is further used within the steam turbine or for process needs, while the exhaust gases from the combustion area inside the boiler go through the waste treatment system to the exhaust chimney or stack and are released to the atmosphere. Note that the strict ecological regulation defines the allowable limits of the emissions that are met with the proper cleaning and filtration equipment. The solid waste in the form of ash goes through the ash handling to the disposal as well.

So the boiler island is a system of interconnected circuits, including:
1. Fuel handling;
2. Boiler unit with the furnace area, heat exchanger equipment, and supporting systems to evaporate water and deliver steam to the cycle;
3. Emissions and solid waste treatment.

In the next sections we consider these systems in detail.

8.2. Efficient fuel treatment

A steam power plant can utilize different types of fossil fuel like solid (coal, peat), liquid (crude or diesel oil), and natural gas, discussed in detail in Chapter 2. In some countries it is obligatory to ensure that a plant can operate on at least two types of fuel: major and backup. Traditionally, the most widespread fuel for steam power plants is coal due to its proximity, ease of production and lower cost, hence the coal-fired technology is one of the most established for steam power plants.

8.2.1 Coal handling

Generally, the coal handling components are determined by the design and requirements of the boiler [1] and the capacity of the thermal power plant.

Coal can be delivered by any means of transportation allowing for bigger volumes and weights depending on the location of site. The cheapest transportation is by water, however, this requires special facilities for vessels. Railroad is the preferred option where water transportation is impossible, and the rail tracks normally go directly to power plant site to ease the unloading process. In cold environments special heating compartments are required to warm up the coal carriages before unloading.

When the coal is unloaded via the *unloading facility* (e.g., cranes, rotary dumpers, etc.), it goes to the *treatment station* and passes the following stages:

- *Crushers* and *breakers* decrease the size of coal as it is delivered as bulk material and is not suitable for an efficient burning process;
- *Vibration* and *magnetic separators* sort out excess particles (like iron) and coal crushed below certain size of particles;
- *Heaters* and *dryers* dry out excess moisture and vapor which might lead to temperature decrease during combustion and therefore decrease efficiency;
- *Coal bunkers* for intermediate storage.

Then, subject to the type of a boiler, two options are possible:

1. Small coal particles are headed from the coal bunker directly to the boiler furnace for combustion. This normally applies to boilers with the stoker and grate type of combustion, where, for example, coal particles are bunt over the special grate.
2. The particles are headed to the coal pulverizer system to produce fine coal powder for further combustion.

When pulverized, the surface of coal exposed to the air during combustion increases dramatically and allows for more efficient process of heat release. This technology therefore became increasingly important though significantly more expensive. The pulverized coal handling system also depends on the overall configuration of the plant (arrangement of the boiler units) and can be:

- *Centralized*, where the fuel treatment is done on a centralized basis for all burners on the plant, and
- of a *unit* type, where each burner has its own pulverizer system.

Both approaches utilize the same process of fuel handling, including feeder systems to guide coal from the bunker, the mill to make fine coal powder, and the cyclone separator. However, in the centralized system the fine powder is first headed to the bunker and then distributed to separate burners at the required rate, while the unit system delivers coal powder from the cyclone separator straight to its dedicated burner.

The centralized system requires only one big mill and hence is more reliable, allowing for precise powder quality control with stable operation of equipment thus leading to the danger of powder explosion inside a big bunker. The unit systems provide simpler plant layout, but require variable operating profile for the mill to fulfill the coal consumption.

8.2.2 Liquid fuel treatment

Burning pure liquid fuel in a boiler is economically feasible only in the areas where this fuel is really cheap and easily accessible. In most cases liquid fuel is considered as a backup fuel together with natural gas due to its wide usage in chemical production. After crude oil refinement, the result of this process named *fuel oil* is delivered to a power plant by various means of transport, usually by rail. The technological process of liquid fuel treatment at the premises on site includes:

1. The unloading point, where the oil is poured from the railway tank cars;

2. Tanks for continuous and emergency supply of fuel, usually capable of storing fuel for a predefined period of time defined by the local regulation;
3. Pumps to increase the pressure for the transportation of oil through the facility pipe system;
4. Oil filtering to remove mechanical particles, water droplets or some unnecessary chemicals;
5. Oil heaters to maintain fuel temperature and proper viscosity for easier circulation and delivery within the system and further atomization and spraying inside the burners.

The heated oil circulates between the heater and tank to maintain stable temperature and viscosity regimes. Its temperature is usually kept at 60–80°C throughout the year with the help of the oil heaters. These heaters are fed either from the special auxiliary boiler (during startup or shutdown times) or from the main steam cycle as steam extraction. The fuel tanks are manufactured out of steel with anti-corrosion treatment, and can be located above or under the ground. As fuel oil is flammable, an active fire protection system should be installed around and on the tanks.

To ensure proper combustion of the liquid fuel, it is sprayed inside the combustion chamber of the boiler and atomized to the very fine mist before the burning process starts. This provides a larger fuel–air surface, shortens the time of evaporation of the droplets and hence increases the efficiency of the overall combustion. The atomization is carried out either by mechanical means or with the help of the pressurized steam (for heavy oils) or compressed air (for light oils). It is therefore essential to have atomized air or steam supply on site.

8.2.3 Natural gas treatment

Fuel gas is usually supplied from the main gas pipeline with given pressure and temperature. These parameters may vary within certain limitations defined in the supply agreement with the gas provider. The same is valid for the gas composition that states the mass and/or percentage composition of each component. This data is normally available before the power plant is erected to properly design the burners and the whole heat balance.

Fuel gas does not require any specific treatment before being burnt within the gas fired boiler. The gas processing plant consists of pressure and flow meters, filtration system against moisture and particles, safety shut-off and control valves. The gas pressure before combustion is normally lower than in the gas pipeline.

8.3. Boiler design and challenges

8.3.1 Introduction

Since the 19th century, steam generation technology providers have been working on optimizing and developing technology of producing steam for multiple purposes. The idea of a boiler, utilized first in steam reciprocating engines like steam locomotives, drives, or pumps, was then applied to a steam turbine. The major function of a boiler is therefore to convert chemical energy of the fuel into heat, evaporate water and supply steam of the required pressure, temperature and flow to the consumer, mostly a steam turbine.

The work on the boiler technology has been conducted in multiple directions. First, in order to increase its efficiency, the parameters of the supplied steam have to be improved. This means that the steam pressure and temperature shall go as high as possible to transfer the maximum amount of energy to the turbine. Despite theoretical possibility, there are certain limitations to steam conditions due to the properties of material, overall level of technology (not only for the boiler itself but for the whole water and steam system), as well as the overall optimization of the combustion process.

Combustion chemistry and dynamics was the second big challenge after the new material development. Contrary to liquid or gaseous fuels, which can be burnt without major efforts, solid fuels vary a lot in quality and combustible material share, therefore require not only proper pretreatment but also approach to fully burn them inside the furnace area of a boiler. This leads to a number of approaches of burning low quality fuels with enough efficiency.

In general, burning a fuel (especially of a lower quality) leads to various types of emissions, both solid and gaseous. This, in turn, attracted even more attention in the view of the temperature rise and the possibility of irreversible global warming. The amount of emissions released when burning any type of fuel has been a concern faced by the boiler manufacturers. Even though this problem has been addressed with the support of the external systems, which will be covered in detail in Section 8.4 of this chapter, the combustion process itself plays an important role. This was the third driver of boiler manufacturers to optimize and develop the design.

Finally, despite technological capabilities and advances in material studies, these manufacturing companies are still having the goal of earning money and in the end be profitable. Evidently, this leads to a certain tradeoff between the technology (and all the points mentioned above), on the one hand, and the cost, on the other. The existing technologies we discuss here are a reasonable balancing of the required efficiency, emissions level, fuel flexibility and cost, which shall be chosen according to the overall economic evaluation based on fuel availability, local regulation and financing options.

8.3.2 Traditional boiler schematics

An optimized heat balance diagram of a big steam power plant, schematically shown in Fig. 7.8, assumes that a boiler supplies two major types of steam: superheated and reheated. The former is normally fed to the HP steam turbine, where it transfers part of its energy into the mechanical energy of the turbine's rotating shaft, and then discharged back for reheating. After this process, reheated steam goes to the IP steam turbine. This means that a boiler in this case has at least a device to evaporate water and then two heat exchangers to superheat and reheat the steam. All these parts are enclosed within the boiler shell, together with the combustion area.

Consider a simplified schematic of a boiler water and steam cycle represented in Fig. 8.2. Cold water from the condenser is fed with the *feedwater pump* through the *economizer* or preheater, where it is preheated. Warm water enters the boiler, where the combustion reaction of the fuel–air mix takes place. Due to this chemical reaction, the hot combustion gases and radiation are used in the *evaporator* — special heat exchanger where the feedwater is turned into steam. This steam is then fed through the *superheater* — another type of heat exchanger, which also uses the energy of combustion and increases pressure and temperature of stem. The super-

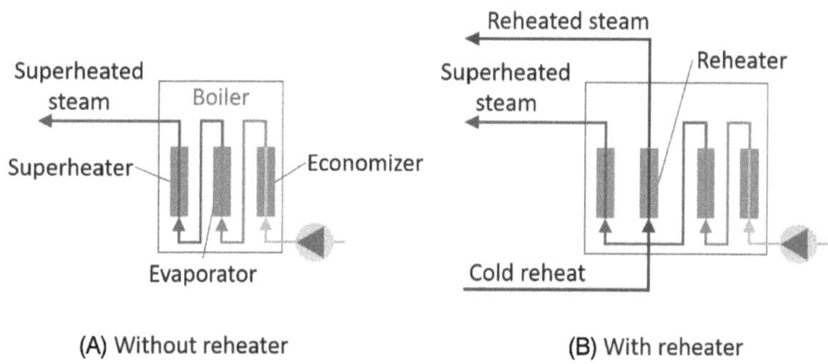

Figure 8.2 ***Simplified boiler diagrams.*** Schematic diagram of a water and steam cycle in a boiler without (A) and with (B) reheater.

heated steam goes straight to the HP steam turbine and, after being utilized there, returns back to the boiler as *cold reheat*. This "cold" steam is then heated up again within the heat exchanger named *reheater*. Finally, the *hot reheated* steam goes to the IP section of an ST and further through the cycle.

There is no single design of a boiler which covers the above described process. The overall configuration depends on the type of fuel, capacity of the power plant and the amount of steam required, properties of the materials used to manufacture the equipment, as well as types of the heat exchangers used in the process. We will consider here only major approaches, which significantly influence the economics of the power plant. These are:

- Configuration of the evaporator;
- Overall arrangement of heat exchangers inside the boiler, including hot gas ducts downstream;
- Types of fuel and conditions of fuel that is utilized, and their influence of the overall configuration.

8.3.3 Configuration of evaporator

Various boilers differ in the configuration of the evaporator, while the superheater and the economizer units are usually of the same type. The *evaporator* is a special type of equipment, which is designed to properly evaporate water into steam and separate this steam from possible water for further use. There are two major approaches: circulation and once-through. Depending on the type of the evaporator, the boilers are be named after them, e.g., "once-through steam generator".

In *circulation* steam generators (see Fig. 8.3(A)), water is heated to boiling temperature in the vertical tubes, forming steam bubbles. In the special vessel called *drum* mounted above the tubes, the rising water–steam mixture is separated: steam is further fed to the superheater, while the water flows back through the *downcomer* pipes to reenter the heating tubes. Complete evaporation is achieved only after several recirculations [2].

Water circulation can be achieved through two major technologies [3, 2]:

- *Natural circulation* assumes that water flows down from the drum due to gravity, while the rising mixture goes up because of the density difference between the falling water and the rising water–steam mix. This simple solution, shown in Fig. 8.3(A), does not consume extra energy and can utilize less purified water as all impurities will be accumulated in the drum;

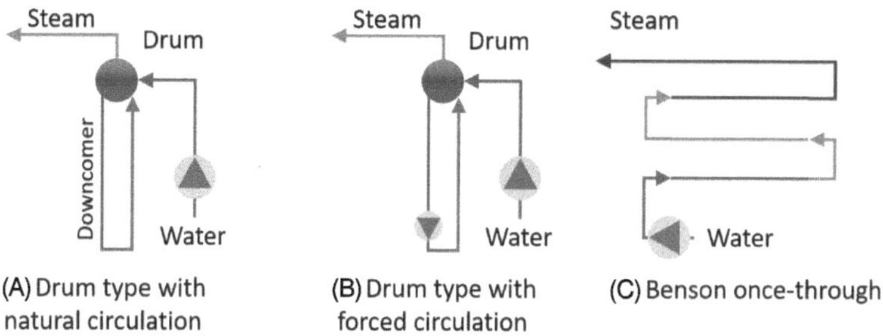

Figure 8.3 ***Comparison of evaporator approaches.*** Drum type with natural (A) and forced circulation (B), and Benson once-through (C).

- In *forced circulation*, as shown in Fig. 8.3(B), the overall process is driven by the *circulation pump*, which is usually installed at the bottom of the downcomer where the water pressure is the highest. Such a configuration requires extra auxiliary power, however, may result in less material in terms of boiler size.

In contrast, in *once-through* evaporator, also known as *Benson boiler*,[1] the water in the evaporator tubes is preheated, evaporated and partially superheated along the way without any circulation, as shown schematically in Fig. 8.3(C). Because steam leaves the evaporator tubes straight, this system does not need any separator for water–steam mixture.

One of the biggest advantages of a once-through boilers is that they are not limited in pressure as the drum boilers where circulation is linked to the difference in water and steam density. Thus the plant cycle can be increased by an increase of steam parameters above the critical point. The limitation on once-through boilers is mainly linked to the selected materials for pressure part and turbine design [4].

Therefore the configuration of an evaporator is subject to the pressure, and hence the type of a boiler and steam turbine. Depending on the steam parameters (pressure and temperature) boilers are usually classified into three major groups: subcritical, supercritical (SC) and ultrasupercritical (USC). The values of steam temperatures and pressures can be assumed as shown in Table 8.1.

Remark 8.1. The definition of supercritical and ultrasupercritical boiler pressure and temperature profiles may differ from one country to another [5].

[1] Mark Benson, a German engineer, was granted a patent for the supercritical drum-free boiler in 1922.

Table 8.1 Approximate pressure, temperature and efficiency ranges for subcritical, supercritical and ultrasupercritical pulverized coal power plant.

Type	Main steam pressure, MPa	Main steam temperature, °C	Reheat steam temperature, °C	Net efficiency, % (HHV basis)
Subcritical	< 22.1	up to 565	up to 565	33–39
Supercritical	22.1–25	540–580	540–580	38–42
Ultrasupercritical	> 25	> 580	> 580	> 42

Aggregated from: [4,5] and public materials of manufacturing companies.

In a subcritical boiler, the water is boiling to steam within the drum, where the mixture is separated, while in supercritical boilers the water can turn into steam inside the tube avoiding the common boiling process. It means, that *drum-type boilers can be only subcritical*, while the once-through boilers can be either subcritical or supercritical.

8.3.4 Efficient arrangement for various types of fuel

As coal has been the major fuel for boilers, multiple approaches have been utilized to optimize its combustion process. The major coal-based power generation technologies available today, or being under development, include:

- Pulverized coal combustion:
  - Single-pass (tower) boiler;
  - Two-pass boiler;
- Circulating fluidized bed combustion.

Oil and gas-fired boilers can have similar configuration as those running on pulverized coal. The major difference is in the fuel injection system: special nozzles are implemented to allow for optimized burning process. For liquid fuel, water or steam injection can be used to control the firing temperature. On top of this, compressed air supplied to the nozzle helps break the droplets of fuel into smaller ones for better combustion.

8.3.4.1 Pulverized coal combustion

In *single-pass* or *tower boilers*, the heating surfaces are mounted above the furnace, so that the flue gases only have to be redirected after the last heat exchanging surface. This helps minimize erosion, especially with high-ash coal types. Only after being cooled down to 400°C the flue gases are guided to the air preheater [2], where NOx reduction equipment can be installed (see Section 8.4).

Such arrangement has a number of advantages: straight flow path with less thermal stresses, no cooling requirement of the top part, and simplified assembly at site and smaller footprint. Moreover, the tower type can be efficiently fired with highly erosive high ash coal. Further, the straightforward tower design allows for easier cooling of the walls (spiral walls). However, the main disadvantage is its height compared to a two-pass boiler.

A *two-pass boiler* guides the hot gases from the furnace through the evaporator heat exchanger. Then the hot flue gases turn through a small horizontal pass downwards through another set of heat exchangers (part of superheater, reheater and economizer).

These boilers offer an optimized hot gas path through the adapted cross-section of the boiler. This allows for a compact boiler structure and faster construction: both passes can be erected simultaneously.

8.3.4.2 Fluidized bed combustion

The above mentioned types of boilers — two-pass and tower — were mostly designed to burn *pulverized coal* and sometimes liquid and gaseous fuels (usually for economical reasons). Preparation of fine coal powder requires complicated equipment, while the whole approach is normally used with better grades coal. These limitations were among the drivers to develop a new design of a boiler with the *fluidized bed combustion*.

The idea of this type of combustion is as follows. Assume there is a special grate in the furnace of a boiler which holds a layer of coal pieces. Feeding the air from the bottom of the grate and applying some preliminary heating would initiate combustion process on the surface of the fuel layer with the release of hot flue gases. If the air flow is further increased, the particles of the fuel would be exposed to the air force, which compensates the gravity, so that they become fluidized in the air flow. This would dramatically increase the thickness of the burning layer.

A further increase of the air flow would lead to bubbles inside this layer, increasing the thickness of the burning layer and the so-called *bubbling fluidized bed*. Finally, the maximum air flow would get the particles out of the layer before they are completely burnt. These unburnt particles of fuel will be flows away from the furnace up to the special cyclone separator, in which the combustion products are separated from the unburnt pieces and released downstream. The latter are fed back to the burning layer inside the furnace. This cyclic process, as schematically represented in Fig. 8.4, ensures

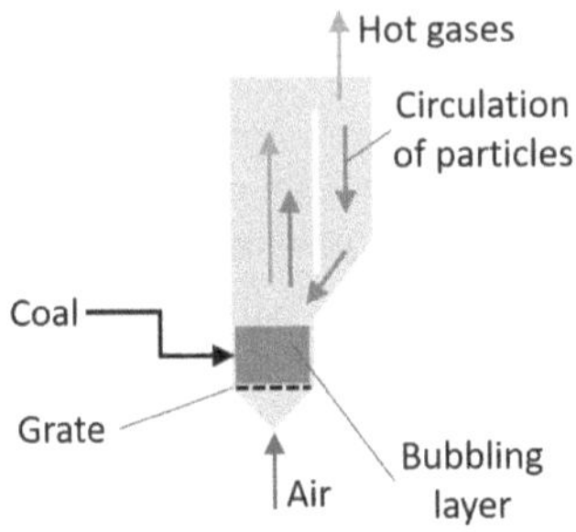

Figure 8.4 ***Schematic of the circulating fluidized bed combustion process.*** Coal is fed to the grate and, exposed to heat and induced air flow from the bottom, starts to burn and bubble. Unburnt particles recirculate within the hot gas flow and until final combustion.

proper circulation of the fuel particles till they get burnt completely. Such a combustion process is called *circulating fluidized bed combustion.*

There are several CFB boiler configurations used in modern fossil-fired power plants. CFB boilers can be both of a drum and once-through type depending on required steam parameters: subcritical or supercritical. However, one of the major advantages of this type of combustion is that the process takes place at temperatures up to 900°C, which are much lower compared to 2000°C of pulverized coal combustion. It means that the NO_x emissions are much lower. Moreover, continuous recirculation of fuel and additionally fed lime stone dramatically decrease the amount of SO_x emissions. Having said that, CFB technology does not require any big equipment for NO_x and SO_x emissions control, leading to smaller sizes of the boilers.

Moreover, CFB technology has a number of other advantages:

- It can burn various types of fuel, including those of poor quality (low grade coals, biofuel, municipal waste), some of them simultaneously;
- There is no special milling equipment required for fuel pulverization.

Unfortunately, these boilers are much more complicated and are subject to higher erosion of the heat exchanger plates due to their exposure to the circulating particles and dust. Auxiliary consumption to power the air fans is another disadvantage of this technology.

Remark 8.2. There is no clear answer which type of boiler is economically better. Evidently, the prices of various types of boiler shall not be directly compared, as a proper analysis of the whole power plant economics has to be conducted. This is due to multiple influencing factors: type of fuel, steam amount and parameters, additional necessary systems and equipment, footprint, erection time, and operating costs.

8.4. Emissions and process waste treatment

The process of burning any fossil fuel leads to a release of waste products in the form of solid particles (in case of burning solid fuels) and gaseous emissions backed with the impact on the materials (e.g., corrosion) and the environment. While thermal power has been the biggest sole producer of greenhouse gases [6], it has led to the strict regulations being enforced in many countries, which resulted in predefined acceptable levels of emissions. These regulations become even stricter year after year, and major equipment manufacturers invest more and more in waste treatment facilities.

Regulations normally require that a power plant has to be equipped with the *air quality control system*, or *AQCS*. This system is by far more important for coal-fired power plants: solid fuel is the dirtiest fuel in terms of gaseous and solid emissions. Contrary to coal burning facilities, gas-fired power plants emit much less greenhouse gases and therefore require less treatment and specific equipment. Generally, the AQCS represents a set of equipment implemented into the systems of a boiler island for any type of the fuel in order to target two major groups of pollutants:

- Solid ash from the furnace of a boiler, and
- Flue gas, including:
  - Solid particles in the flue gas, mostly represented by fly ash;
  - Sulphuric oxides SO_x;
  - Nitrogen oxides NO_x;
  - Carbon oxides CO_x;
  - Other pollutants including mercury, volatile organic components, extra water, or just heating the environment.

Depending on the chemical composition of fuel, various combinations of pollutants are released on site and therefore multiple approaches and systems are implemented to minimize the harmful impact. We now consider each system in detail.

8.4.1 Furnace solid ash treatment

Coal contains up to 30% of ash [3] and, depending on the capacity of the thermal power plant, ash is produced in significant amounts. The ash can be released either directly from the furnace (bottom ash) or from the flue gases at the exhaust. Despite advanced burning technologies and continuous optimization of the combustion process, coal still does not burn completely, and some particles leave the furnace, contributing to the ash content as well.

Bottom ash is formed in the furnace of a boiler. It is made from ash particles that are too large to be carried away in the flue gases so that they fall through the grates to the bottom of the furnace. On top of this, some of the ash particles melt under high temperatures of up to 1700°C within the furnace and therefore are accumulated on the walls and heat exchangers inside the boiler. This mass is sometimes called *clinker* and it builds up for a while before it is blown off by jets of air or simply falls down to the bottom. This is why it is called *bottom ash*. Finally, these hot pieces of ash are carried away by the ash handling equipment, which normally has two major functions:

1. Remove ash directly from the furnace;
2. Deliver ash to the fill where it is stored and then transported to the disposal facility or to a customer for further usage.

Due to high temperatures of combustion, bottom ash treatment can be challenging due to high temperature and hence huge corrosive impact on materials. There are several approaches of how to deliver ash from the bottom of the boiler:

- A *mechanical* system carries away the ash on the conveyor belt. Ash falls to a water tank where it is cooled before passing through a clinker grinder, which reduces the size of the pieces. Then, it is transported by the belt to ash bunker. This system is cheap and simple enough to be used continuously, however, it has quite low capacity and shorter lifetime.
- A *hydraulic* system uses a water channel to carry out the ash that falls from the bottom of the furnace. The slurry is then carried to the ash sumps, where the ash is separated from the water. Water is reused again in the process. This system has a major advantage: it does not produce any dust and is therefore much cleaner with the limited erosion of material (as no moving parts are exposed to ash). On top of this, it is capable of handling big amounts of ash and is hence suitable for large capacity boilers. However, special materials are required to manufacture water pumps and tanks.
- A *pneumatic* ash removal system utilizes high pressure air to collect ash particles. The bottom ash is collected within the bottom hopper and fed

through the crushers to optimize the particle size before feeding them to the air stream. Further, cyclone separators sort out ash and collect it in ash hoppers, while the clean air is discharged through exhaust. This system is normally dust free and quite flexible. On top of this, it is usually cheaper compared to the hydraulic one as no water is required. On the other hand, it has a higher noise level during operation and higher maintenance costs.

- A *steam jet ash handling* system is similar to the pneumatic one, but uses steam instead of air. This system, however, is exposed to much higher corrosion and has significantly lower capacity.

The choice of the system depends on the overall plant design and the applied conditions (e.g., available land, water, local regulation, budget, etc.) and is usually proposed by the plant designer or the supplier of the AQCS.

8.4.2 Flue gas handling

Solid particles that fall down to the bottom of the boiler are big enough to be sorted out and removed easily from the furnace. The flue gases are much more difficult to treat before they are released to the atmosphere. Due to strict regulation on greenhouse gas emissions and toxic substances, which negatively influence human health, exhaust gases of a coal fired plant have to be cleaned from multiple chemical substances, namely combustion products. These include both gaseous products and much smaller solid coal and combustion products particles that can be carried away by the air flow — the fly ash.

8.4.2.1 Fly ash

While the minor part of the ash is removed from the bottom of the furnace as solid mass, up to 80% of ash is still present in the flue gases [3]. Because of its high corrosivity, and therefore severe impact on the equipment and the environment, it is extremely important to remove fly ash particles from the exhaust gases downstream the furnace. This is critical especially for the CFB boilers, where the overall recirculation of fuel particles leads to much higher concentrations of dust in the flue gases.

Technologically, there are two major approaches to sort out fly ash from the flue gas: mechanical and electrostatic.

Mechanical ash collecting systems can be dry or wet and are based on various approaches:

- *Dry* systems, according to the name, do not need extra water and can be based on different physical approaches to separation methods:

- *Gravitational* that works on the principles of gravity: particles simply fall down because of their weight. Being simple, this system cannot remove fine particles.
- *Cyclone* uses centrifugal forces, which send particles towards the walls of the collector. Being efficient, they still require auxiliary power and large footprint.
- *Fabric filter* or *bag house* is basically a type of cloth filter that collects the dust. It usually contains a large number of filter elements placed in parallel rows. Special self-cleaning systems like pulse jets or simple shaking can be implemented to clean the filter cartridges and remove particles from the filter house.

- *Wet* systems employ water as part of the filtration process. In these systems, fly ash is collected from the air and removed by spraying water into the flue gas flow. Though being very efficient, they require a lot of water and are exposed to high levels of corrosion.

Electrostatic systems, usually called *electrostatic precipitators* (*ESP*), implement the following approach, as illustrated in Fig. 8.5: ash particles pick up a negative charge when going through the negatively charged grid, and then are attracted to the positively charged collecting plates. These plates are then vibrated to remove the ash particles, which fall into ash hoppers.

Similarly to bottom ash handling, the choice of a fly ash handling system depends on the designer and the requirements of the customer. However, the majority of the modern coal-fired power plants employ either bag filters (cheaper solution) or ESP system (much higher efficiency and low maintenance costs).

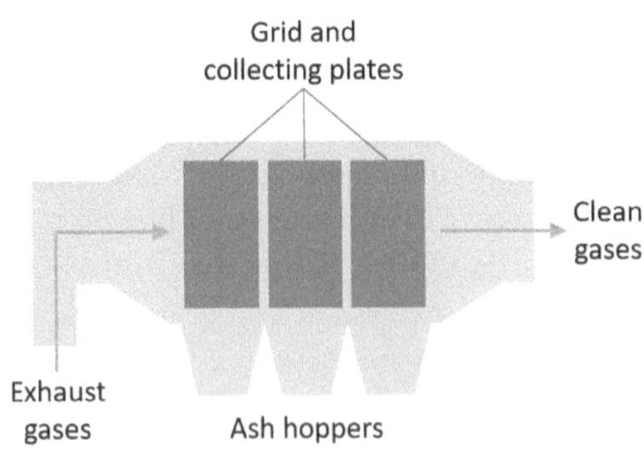

Figure 8.5 ***Schematic of the electrostatic precipitator.*** ESP absorbs negatively charged ash particles on the positively charged plates.

The collected fly ash is conveyed to the site storage, from where it can be disposed off and used for other purposes. Usually considered as waste product, ash can contain traces of precious metals, particles of unburnt fuel, and after separation can be used in agriculture as fertilizer or in construction industry, for instance, to produce concrete.

8.4.2.2 SO_x control

The sulfur content in coal can be up to several percent of the mass, and its combustion leads to significant SO_x release. The most common form is the SO_2 gas, which is extremely toxic for people and the environment: released in big amounts, it can cause acid rains (see Chapter 6 for more details).

There are various strategies to reduce SO_x emissions from the coal plants. In countries where emissions are controlled by legislation, it is not allowed to burn coal with high sulfur content. However, this measure can lead to the price increase of low-sulfur coal and therefore impact the electricity price for the end consumer. Therefore an alternative, and the most common approach adopted today, is to capture the sulfur from the flue gases that emerge from the furnace [7].

There are three major technologies implemented in the facilities that emit SO_x gases:

- A *wet scrubbing system* uses water carrying a special sorbent (limestone, lime, or calcium hydroxide) or simply pure seawater to capture the pollutant. The acid gases dissolve in water and then react with the sorbent material. The scrubbed gas is trapped by the chemical and finally collected in the recycle tank. When using limestone, this system delivers gypsum as a byproduct, which is finally separated from the water and can be sold.
- A *dry injection system* is used in the regions where water is scarce. The dry adsorbent (e.g., fine powder of limestone) is injected into the flue gas and the acid gas is absorbed in the reactor, while the air passes through an additional fabric filter.
- A *spray dryer system* injects an atomized chemical, usually a mix of limestone and water, that absorbs acid gases while the heat of the flue gases evaporates the droplets. The particles are then collected in an additional fabric filter.

The choice of the *flue gas desulfurization* (*FGD*) system depends on the capacity of the power plant, as well as capital and operating costs. In most cases water is not a big issue for coal-fired power plants (as they are normally situated near a proper water source), therefore wet scrubbing systems are usually the preferred solution for an FGD system as they prove to be the most efficient.

8.4.2.3 NO_x control

Even though low NO_x combustion systems can significantly decrease the quantities of the nitrogen oxides, they cannot reduce them completely to the levels required by legislation. Therefore additional systems are implemented to treat the flue gases. They are [7]:

- *Selective non-catalytic reduction* (*SNCR*), which involves the injection of ammonia into the flue gases right after they leave the furnace when their temperatures are around 870–1200°C. If the temperature is too high, ammonia will start to decompose thermally, and if it is too low the reaction rate will be too slow. The reaction between ammonia and NO_x will produce a mixture of nitrogen and water vapor.
- *Selective catalytic reduction* (*SCR*), which uses the injection of ammonia into the flue gases as well, but requires the mix to pass over a solid catalyst that facilitates the chemical reaction. This reaction takes place at much lower temperatures of around 300–400°C and has much higher efficiency and precise control of the ammonia injection.

The SNCR is effective only if the NO_x concentrations in the flue gases are already low. On top of this, it can contaminate flue ash with ammonia, which leads to its release to the atmosphere. Using feedback loops, the ammonia concentration within the final exhaust can be controlled, however, this brings additional complications and may not be suitable for large coal-fired power plants. The SCR technology is therefore a much more suitable solution for large plants, especially in combination with an efficient FGD system to remove all acidic gases that may occur due to the catalyst.

8.4.2.4 CO_2 and other emissions control

The excess amounts of CO_2 emissions from a thermal power are considered to be among the largest contributors into the greenhouse effect. Various approaches have been proposed to decrease its release from coal-fired plants, ending up in the commercially feasible (yet too expensive) *carbon capture and storage* (*CCS*) technology that can capture up to 90% of the carbon dioxide emissions produced from the fossil fuels. The overall process usually consists of three stages [8]:

1. *Carbon dioxide capture* implies a technology to extract CO_2 from the power generation process and prepare it for transportation. There are

several commercially available process technologies which can, in principle, be used for CO_2 capture from flue gases. However, comparative assessment studies have shown that absorption processes based on chemical solvents are currently the preferred option for post-combustion CO_2 capture.

2. *Transportation* of compressed CO_2 can be carried out via pipelines (like other gases) or on special vessels.
3. *Storage* or further *use* of CO_2 for other purposes. Carbon dioxide can be stored underground in geological storages, which can be depleted oil reservoirs, deep unused water reservoirs or other cavities suitable for gas injection. In some cases, CO_2 can be used in enhanced oil and gas recovery process.

This process is, however, not commercial attractive because of its complexity and costs. Quite high operating costs are usually associated with the leakages of the gas, and no-one can really guarantee that the underground storage would be safe enough for the coming centuries. However, the problem of CO_2 still remains of vital importance, and we can expect to see the changes in regulations related to the CCS technology.

There are some other emissions and substances that are considered as pollutants and therefore have to be controlled at a power plant level. These include:

- *Mercury*, which can accumulate inside a human body and cause poisoning and severe damage. Depending on the fuel, special sorbent like activated carbon is used for high removal efficiencies: it binds mercury that comes from the coal combustion and is collected in a special filter.
- *Water vapor* or *water* is also considered as a greenhouse gas, so that its discharge to the atmosphere is regulated in some countries (so-called zero fluid discharge). Most of the processes in a coal-fired boiler island involve water, and its use and release have to be strictly controlled.
- *Heat pollution* is also considered an important topic, especially in the face of the global warming. Exhaust gases that are released in the combustion and filtering process have to be cold enough not to increase the temperature of the atmosphere dramatically. On the other hand, the heat energy can be used in the process again (e.g., to preheat the air) to increase the plant efficiency.

To comply with the regulations and to ensure continuous monitoring and control of the exhaust gas emissions, the flue gas analyzer system should be implemented downstream of the flue gases in a boiler island, usually near the exhaust stack. The gas probes are usually located at the discharge with the control and analyzing equipment installed in a container nearby. The *continuous emission monitoring system* (*CEMS*) has to fulfill the requirements given by the local authorities and enable the authorities to perform a check of emission characteristics and limit values along the operation of a plant.

8.4.3 Exhaust chimney

Downstream of the flue gas treatment equipment, cleaned exhaust gases are finally released to the atmosphere. This is done through the *exhaust stack*, also called *flue-gas stack* or *exhaust chimney*. It guides hot gases up to the upper layers of the atmosphere. Depending on the local regulations, exhaust chimney can be up to several hundred meters high to dissipate low concentrations of exhaust gases further in the atmosphere. The height of the stack allows protecting people from breathing these gases, while the stronger winds hundreds of meters above dissolve these gases much more efficiently. The tallest stack of around 420 m was built in Kazakhstan at Ekibastuz coal-fired power plant.

The design of such tall structures is subject to multiple challenges. The overall structure has to be stable, so that strong winds, which flow around the chimney and may lead to vortex shedding, do not cause significant dangerous oscillations of the whole structure. Tall chimneys are usually built from reinforced concrete or masonry and erected along with the civil construction works. The material has to withstand high temperatures of the exhaust gases.

Stacks are equipped with lightning protection, aircraft lights, maintenance platforms and stairs, sometimes even with a special elevator.

8.5. Development of sustainable boiler technology

Since coal-fired boilers play a crucial role in the power generating technologies, especially in emerging markets where coal is the only cheap and accessible fuel for large energy conversion, they will still continue to play this dominant role for the next few decades. Therefore, it is vitally important to ensure that these technologies become more and more environment-friendly, or in simple words "cleaner".

Generally speaking, there are four major approaches to make coal-fired technology cleaner:

- Employ better fuel pretreatment, for instance, coal washing. While this can be a typical treatment procedure in developed countries, some emerging markets may override it to save costs.

- Optimize efficiency of coal-fired boiler technologies to the level up to 40% and above;
- Properly implement waste management technologies to control particulate matter, NO_x, SO_x, mercury and other trace elements;
- Use direct technologies to decrease CO_2 emissions, e.g., carbon capture sequestration (CCS).

High efficiency, low emission (HELE) coal-fired power plants are a key first step along a pathway to near-zero emissions from coal with carbon capture, use and storage (CCUS). HELE technologies are commercially available now and, if deployed, can reduce greenhouse gas emissions from the entire power sector by around 20% [9].

These HELE technologies are basically a group of diverse technologies developed to increase the efficiency of coal-fired boilers and power plants in general, therefore reducing carbon dioxide and other greenhouse gas emissions, as well as non-GHG emissions such as nitrogen oxides, sulphur dioxide and particulate matter [9].

Low efficiency of electricity generation literally means that more coal is required to produce a kWh of electricity. Consequently, as more coal is consumed, more carbon dioxide is hence released as a by-product. Upgrading and retrofitting existing boilers and introducing new boiler technologies that operate more efficiently would address the climate change problem in two ways:

- Increasing efficiency would directly lower CO_2 emissions per unit electricity generated, and
- Carbon capture technologies would be more appropriate for high-efficiency boiler options.

As discussed previously, the major efficiency gains are achieved through the increased temperatures and pressures of steam in the cycle, leading from subcritical boilers to superctritical and ultrasupercritical technologies. A switch from subcritical to current USC steam conditions raises efficiency by around 4 to 6 percentage points. Historically, the majority of pulverized coal-fired plants were based on subcritical steam-cycle technology, but supercritical technology is now widespread, largely due to improvements in boiler tube materials [10].

Moreover, advanced USC (AUSC) steam cycles are expected to continue the trend towards unit growth and efficiency gains, aiming at 400–760°C inlet steam temperatures compared to "just" > 580°C for USC technology [10]. While the planned efficiency of such plants is expected to be at the level of 50%, which is still lower than modern combined cycle plants, such a combustion technology would emit more than 25% less of CO_2 than a conventional subcritical boiler.

Evidently, the deployment of such technologies has some practical issues, mostly with special heat-resistant materials for heat exchanger tubes that can withstand much higher temperatures within the combustion area. This problem has probably been the number one challenge for energy engineers on the way towards advanced thermal boiler technologies.

Even using already existing USC technologies, which have been employed in several coal-fired power plants (e.g., in Germany), can dramatically impact the situation. According to some studies [9], moving the current average global efficiency rate of coal-fired power plants from 33% to 40% by deploying more advanced off-the-shelf technology could cut two gigatonnes of CO_2 emissions now, while allowing affordable energy for economic development and poverty reduction. Two gigatonnes of CO_2 is roughly equivalent to India's annual CO_2 emissions or running the Kyoto Protocol three times over.

The challenge here is all about economics: USC boilers are much more expensive. When the low quality coal is affordable (in other words, it is cheap), the tradeoff between emissions and electricity cost, and hence price, is usually done in the way towards economic gain, especially in the regions with weak to no climate-related regulation.

References

[1] Miller B. Clean coal engineering technology. Elsevier; 2010. 696 pp.

[2] Spliethoff H. Power generation from solid fuels. Springer; 2010.

[3] Gupta M. Power plant. PHI Learning. ISBN 9788120346123, 2012.

[4] Nielsen F, Danesi P, Radhakrishnan M. Modern boiler design. BWE; 2012.

[5] Nalbandian H, ICRCC Centre. Performance and risks of advanced pulverised coal plant. IEA coal research, clean coal centre. IEA Clean Coal Centre; 2008. 70 pp.

[6] International Energy Agency (IEA). CO2 emissions from fuel combustion. Paris: International Energy Agency; 2016.

[7] Breeze P. Coal-fired generation. Elsevier; 2015. 96 pp.

[8] IPCC. IPCC Special Report on Carbon Dioxide Capture and Storage. Prepared by Working Group III of the Intergovernmental Panel on Climate Change. Cambridge, United Kingdom and New York, NY, USA: Cambridge University Press; 442 pp.

[9] World coal association. URL: https://www.worldcoal.org/.

[10] Barnes I. Upgrading the efficiency of the world's coal fleet to reduce CO_2 emissions. Cornerstone 2015;3.

CHAPTER NINE

Power island and balance of plant

Contents

9.1. Power island configuration

A *power island* of a thermal power plant, which is sometimes called a *turbine island*, accommodates major rotating equipment, i.e., turbines and generators, which are responsible for converting energy into electricity via mechanical rotation. As discussed in Chapter 7, the power island is one of the major areas of a power plant. By default, rotating equipment is always considered as critical, meaning that its performance, vibration, speed, and temperature — basically all sorts of critical parameters — are closely monitored during operation. An emergency or accident within the rotating equipment can lead to significant damages for the whole plant and therefore commercial consequences: loss of profit, idle time, and so on. Note that these machines rotate with significant speed and possess significant momentum because of their weight, therefore it is simply impossible

Sustainable Power Generation
https://doi.org/10.1016/B978-0-12-817012-0.00020-7

to stop them in a moment. The operator has to consider some extra time even for emergency shutdown.

Depending on the type of a power plant, the configuration of the power island differs as follows:

- Steam power plant:
  - Steam turbine with condenser and their dedicated systems;
  - Electrical generator.
- Simple cycle power plant:
  - Gas turbine and its dedicated systems;
  - Exhaust heat recovery equipment (if applicable);
  - Electrical generator.
- Combined cycle power plant:
  - Gas turbine and its dedicated systems;
  - Exhaust heat recovery equipment;
  - Steam turbine with condenser and their dedicated systems;
  - Electrical generator (at least one).

Remark 9.1. Even though steam turbines for steam and combined cycle power plants may significantly differ, they do have the same underlying physical principles and therefore structure. Gas turbines for various plants differ much less and therefore can be assumed the same within the scope of this overview.

9.2. High efficiency steam turbine

The idea of using steam as driving power dates back to the first century BC when a Greek engineer Heron of Alexandria developed the first device to use steam power [1]. Multiple devices that used steam have been developed since those times. However, it was THOMAS NEWCOMEN (1664–1729) who first developed a practical steam engine with a piston and cylinder. This machine was initially aimed at pumping water in the mines, which was among the biggest challenges at that time since those areas had few streams large enough for water-wheels, and animals were not powerful enough [2]. During the industrial revolution, James Watt patented a modified version of the steam engine in 1781. This machine had a separate condenser, however, still used a cylinder–piston mechanism. These units were widely used to run machines in factories replacing the power of water flow and therefore enabling factories being placed far from the rivers.

Further development of steam engines found their applications to run steam locomotives. Toward the end of the 19th century, steam turbines were introduced in the field of steam power [1]. This new type of machine had fewer moving parts, could deliver higher power and proved to be more efficient. KARL G.P. DE LAVAL (1845–1913) contributed to a prototype of a modern impulse steam turbine with the special type of steam nozzle, while in 1884 Sir CHARLES A. PARSONS (1854–1931) officially patented a reactional steam turbine design. Because the turbine was generating rotary motion, it became particularly suitable for driving an electrical generator. Coupled together, the newly invented machines were able to supply cheap and abundant electricity. This initial decision to run an electrical generator already predefined the future of a steam turbine.

A number of other variations have been developed over time, and multiple engineers and scientists contributed to the ongoing innovation. The overall theoretical development, calculations and justification of the technology had been carried out by AUREL STODOLA (1859–1942), who is considered one of the founders of modern theory of steam turbines.

Further development has followed two major directions: increase of the unit output and increase of the steam parameters (temperature and pressure). For years, steam turbines have been an established technology for fossil-fired steam power plants. Continuous need for bigger capacity and efficiency was supported by new heat-resistant high-chromium-percentage steels that enabled steam turbines to reach pretty elevated steam temperatures [3]. Along with the development of SC and USC boilers, steam turbines have also evolved. Broadly speaking, big utility-size steam turbines for pure steam and combined cycle applications differ in size, number of pressures, casings and steam extractions, i.e., in overall configuration.

9.2.1 Configuration of steam turbines

A steam turbine configuration directly depends on the conditions of the steam that comes from the boiler. Following the boiler classifications, steam turbines must also match the steam parameters and therefore can be classified into three major groups: subcritical, critical, and supercritical, as shown in Table 8.1.

The configuration is usually understood in the sense of the arrangement of *casings* or *cylinders* — sections of a steam turbine related to one specific pressure, for instance, high or low. Generally, the number of cylinders depends on the number of steam reheaters used in the process. Consider several examples:

- If a boiler supplies only one type of steam, usually superheated, the steam turbine can have either one section (usually called an HP section)

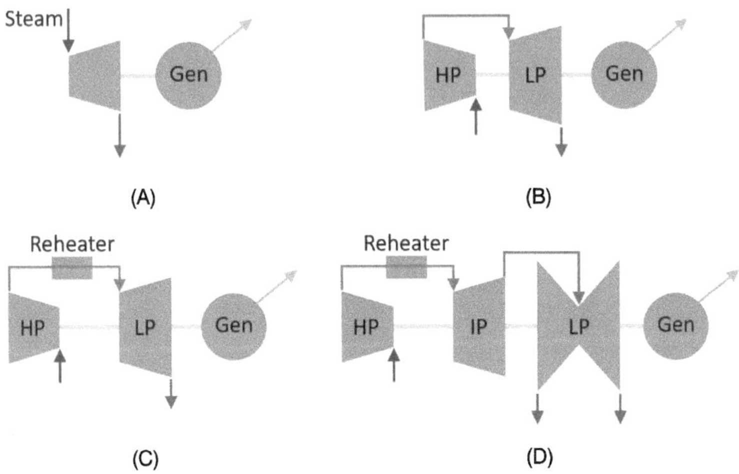

Figure 9.1 ***Steam turbine schematic configurations.*** Illustrative configuration of the steam turbine based on the type of steam supplied from the boiler (direction of steam flows within cylinders and the number, size, and configuration of cylinders are for indicative purposes only).

or two (HP and LP). While a single pressure turbine releases cooled steam straight into condenser, a two-pressure unit first uses steam energy within the first section, and then releases it straight to the second one, as illustrated in Fig. 9.1(A)–(B). This type of steam turbine is called *non-reheat*. Usually its capacity is up to 300 MW. These steam turbines are utilized in smaller-size conventional steam power plants or in combined-cycle mode.

- If a boiler provides two steam flows, they are usually called superheated steam and reheat. The superheated steam flows into the HP section, as shown in Fig. 9.1, and, after being discharged from its exhaust, is fed back to the boiler as cold reheat. Being reheated, it flows back to the steam turbine. This type of a steam turbine is usually called *reheat*. Similarly, it is used both in small to medium size conventional steam power plants and within the combined cycle. Here two options are possible:
  - Reheated steam flows to the LP section and then to the final exhaust (C). In this case the steam turbine will called *double pressure reheat*.
  - Reheat, being fed through the IP section, then goes to the final LP stage (one or several) as shown in figure (D). Such a configuration is usually referred to as *three pressure reheat*.

- In a few cases of big machines for large coal-fired power plants, there can be two reheaters. It is basically achieved by adding the second IP section to the steam turbine: superheated steam goes to the HP section, then being fed back into the boiler, where it is reheated for the first time. After going through the first IP section, steam goes back to the boiler, where it reheated again in the second reheater and then guided to the second IP section. Finally, the steam goes into the LP sections of the turbine and further to the exhaust. *Double reheat steam turbines* were of three pressure configuration (high, double intermediate, and low). Most double-reheat units were built around the 1960s and 1970s in Germany, the USA, and Japan. However, they fell out of favor in the late 1970s largely due to the introduction of nuclear power and combined cycle gas turbines, and also persistently low oil prices [4].

The absolute majority of the steam turbines are designed to accommodate all sections on one shaft, which is, in turn, connected to the generator shaft. In a very exceptional case, however, a few of the double pressure reheat steam turbines were designed in a *multishaft* configuration. It means that some of the sections are connected to one shaft with one generator, while the others are connected to the second generator. This allows increasing the overall output and decreasing technological problems arising due to very long and heavy shaft, its manufacturing, transportation, balancing, and operation.

On top of this, the steam flow direction plays an important role in overall stresses of the rotor and the whole unit. The steam flow through any of the sections has enough energy to force the unit along the direction of the flow and may cause problems with the rotor (or at least would require much stronger structures and supports). To avoid this, sections of the steam turbine are designed in the way to compensate the stresses of the flow as much as possible.

Looking at Fig. 9.1, the HP and LP (for a two-pressure unit) or HP and IP (in a three pressure unit) sections are often designed in a way to have the axial steam flows in the opposite directions. This is less critical for smaller units; however, for larger turbines, especially of a reheat type, it becomes much more important. Moreover, for the larger units employed in utility scale coal-fired power plants, also the LP sections are designed in a way to minimize this impact. Combined in pairs, they have the opposite direction axial flows of LP steam, as schematically shown in (D), which also increases the capacity due to much higher flow. Such a section configuration is called *double flow*.

The choice of a steam turbine configuration depends on the overall plant design, mainly its heat balance configuration — the amount and properties of steam implemented in the process. Steam turbines for coal-fired power plants can be designed for inlet steam conditions of 300 bar and 630°C (4351 psi and 1005°F), i.e., supercritical superheated steam conditions. These turbines can also have multiple provisions for the admission or extraction of steam for various process uses, which has to be considered within the heat balance to make sure that the steam flow through the steam turbine is sufficient for optimal efficiency. Therefore steam turbines are usually customized for a particular project and hence may require longer schedules, unless the standardized design has been employed.

This way, fossil utility steam turbines are a class of large, reheat units used almost exclusively for electric power generation and sometimes co-generation. Because of their large size, usually greater than 200 MW, these units utilize higher steam pressures up to modern ultrasupercritical, have several stages of regenerative heating, and incorporate other design features to maximize performance, reliability, and availability [5].

Steam turbine units for combined cycle applications are usually smaller and can be up to 300 MW capacity. Depending on the size and configuration of the plant, they can also be of a reheat type and have up to three pressures; however, the steam conditions from the HRSG are usually much worse than those from a conventional boiler: HRSG cannot deliver supercritical steam parameters, therefore steam pressures are usually up to around 185 bar (2680 psi) with the steam temperature of 600°C.

9.2.2 Major components

By definition, steam turbines transform the energy of the hot steam of into the mechanical energy of the rotating shaft. Even though the steam parameters may differ, all steam turbines used in modern coal and gas-fired power plants (both steam and combined cycle) have the same operating principle, and therefore the overall structure with major components. To emphasize again, the major difference lies mostly in the number and type of steam sections.

Keeping this logic in mind, consider a typical three-pressure steam turbine schematically illustrated in Fig. 9.2. Any steam turbine consists of stationary and moving (rotating) parts. All stationary parts of a turbine are usually called the *stator*. The stator of a steam turbine consists of lower and upper parts bolted together. Moving parts that rotate inside the casing as one assembly are named the *rotor*.

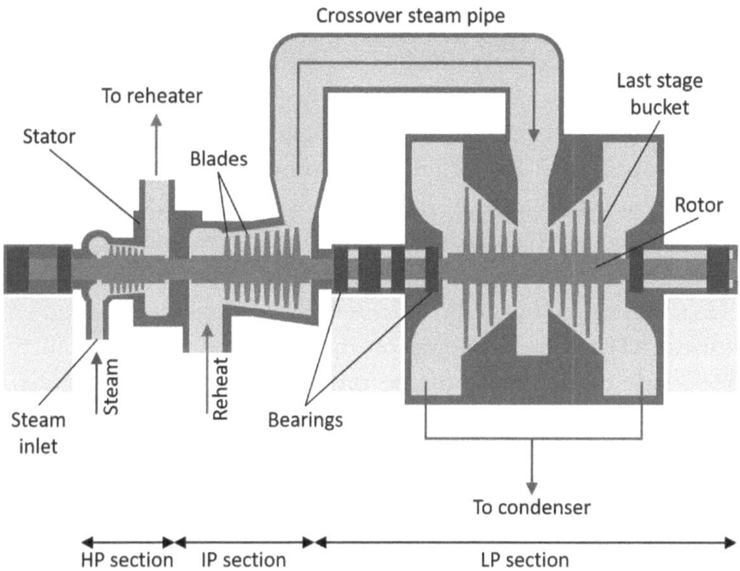

Figure 9.2 ***Steam turbine major components.*** A typical three-pressure steam turbine simplified schematic drawing (for indicative purposes only).

High pressure superheated steam is first admitted to the single flow high pressure section of a steam turbine through the steam inlet. This flow is controlled via the main steam stop and control valves. After going through the stages of the HP section, it expands and transfers part of its energy into mechanical work of the rotating shaft. Then, through the steam exhaust it is fed to the superheater, and is then further admitted to the IP section. Finally, the steam flows to the double flow LP section through the single crossover steam pipe. There, the steam further expands in two opposite directions and again does mechanical work to rotate the same shaft. Finally, the steam is released down the steam turbine through the exhaust discharge.

Parts of the rotor can consist of several discs holding a number of *buckets* or *blades*. On the stator, there are several stationary blades sometimes called *nozzles*. Each pair of buckets and nozzles defines a *stage* of a steam turbine. The size of the blades corresponds to the respective steam properties, mostly pressure: for an HP section all buckets and nozzles are much smaller compared to the LP section due to much bigger energies they are exposed to. One of the major parameters defining the steam turbine is the *last stage bucket* (*LSB*) of the LP section: it is normally done as long as possible in order to convert as much energy of steam as possible. Longer last-stage buckets can accommodate larger steam flows and loadings at relatively the same performance level by maintaining or reducing exhaust losses, without increasing the number of low pressure turbine flows [5].

The overall rotor of the steam turbine can be a connection of the separate rotors of HP and LP sections. The complete rotor is first assembled and balanced at the factory to eliminated vibrations and then put into the stator. It is supported by a number of bearings, which ensure smooth and safe rotation without material damage and with minimum frictional losses. This is ensured by the hydraulic oil system that supplies pressurized oil into the bearings. The rotor is sliding on the thin lube oil layer, and the oil is continuously pumped into the bearing to support it. Importantly, during the rotation, the oil temperature increases, so that its continuous in- and outflow is important for cooling purposes. To ensure secure operation, several special proximity sensors and temperature devices are installed inside the bearings for rotor position and temperature indication.

To prevent steam leakages from the steam turbine sections, special insulation called *journal bearings* is implemented to make sure the pressurized steam does not exit the cylinders (or at least this leakage is minimized), so that it does not decrease the overall efficiency.

A turning gear is an essential part supplied together with a steam turbine and is used to rotate the turbine-generator shaft slowly during preparation for startup and shutdown. This is required to prevent a rotor from being damaged: when a unit is shut down, the internal elements of the system continue to cool down for many hours. This gear is powered by an electric motor.

9.2.3 Accessories and auxiliary systems

To ensure safe and reliable steam turbine operation, various *accessories* and *auxiliary systems* are required. These systems are usually supplied as modules together with the unit by the turbine manufacturer in the *equipment package*, and can be split into mechanical, fluid, and electrical systems.

Mechanical and fluid systems are responsible for turbine proper rotation during all periods of its utilization, cooling, and safe and reliable steam supply. They may include:

- Hydraulic and lubricating oil system, which supplies oil from the oil reservoir to the bearings of the steam turbine. In order to maintain the required quality of oil, the oil module has a filtration system and a number of pumps to ensure the required reliability of the system through redundancy. This system is considered critical and therefore requires special attention during the design, construction, and operation of the plant: measurement devices of the oil system are connected to the overall control system and can initiate shutdown of the whole plant;

- Shaft sealing system, which is required to prevent steam leakage from the HP and IP sections of the turbine, as well as stop air ingress into the LP section. This system incorporates not only special packing rings inside the unit, but also valves and other equipment;
- Steam turbine enclosures and acoustic treatment to protect the personnel from noise and high temperatures of the stator;
- Various electrical, instrumentation, and control systems used for operation and control purposes. These devices usually measure many essential parameters, for example, speed, load, positions of valves, oil parameters, sealing system properties, vibration, rotor position, turbine stator thermal expansion, etc.;
- Steam valves.

The latter can be considered as major and inevitable parts of the steam turbine package. Depending on their functions, there are several types of valve:

- *Stop valve,* used with the primary function to protect a steam turbine against overspeed, or simply to shut off the steam supply into the steam turbine in case of emergency. These valves are either fully open or closed;
- *Control valve,* used for speed and load control, as well as providing the initial overspeed protection through regulating the steam flow into the unit;
- *Combined stop and control valve,* which is a combination of the above mentioned valves and can be used for multiple purposes;
- *Overload valve,* enabling operation of a steam turbine using the unit's reserve capacity. An overload valve is provided in parallel with the steam control valves but is connected to discharge steam to a lower pressure stage of the turbine. During operation of the turbine at less than full load, the bypass overload valve is maintained closed and the control valves are positioned to sustain a pre-selected load. As load on the turbine increases, the control valves are increasingly opened until they have all reached their fully opened position corresponding to a nominal rated capacity (100% load) of the turbine. Above the nominal rated capacity, the bypass overload valve is used to control load while the main control valves remain fully opened. In other words, this valve is used

to control the operation of the steam turbine at capacities greater than 100% where it is not possible to open the control valve even more.

The overall configuration of the power plant (boiler and steam turbine) defines the steam parameters and overall steam flows into the steam turbine. All of these flows have to be controlled and protected with a number of valves, for example, in case of a three-pressure steam turbine:

- HP inlet stop and control valves, which regulate the steam flow from the HP superheater to the HP section of the steam turbine. They also protect the unit from the excessive energy stored in the superheated steam contour.
- Reheat or IP stop and control valves protect the unit from overspeed and regulate the steam flow from the reheater of the boiler into the IP section.
- LP admission stop and control valve.

Depending on the size of the turbine and therefore a number of steam pipes that deliver steam to the unit, the number of valves may differ. In most cases, however, there is either one or two valves of the same type. For example, for the turbine implemented in big coal utilities, two combined reheat valves can be located on each side of the IP section of the turbine.

9.3. Modern heavy duty gas turbine

Historically, gas turbine types were defined by output, firing temperatures and pressure ratio. As advances were made in materials and cooling technologies, gas turbines were able to fire hotter, resulting in better efficiencies and higher outputs. Design changes in the compressor and turbine section were commonly needed, and thus when a manufacturer made improvements significant enough to increase output and efficiency, a new turbine class was born [6].

9.3.1 Classes of efficiency

There are several generations of heavy duty gas turbines installed and operated in thermal simple and combined cycle power plants. These are the turbines with the gross output from few dozens up to more than 500 MW. Traditionally, they are denoted with capital letters, for example, B, D, E, F, etc. Presently, gas turbine generations available in the market for utility use are mostly of D/E, F and G/H/J classes (or frames), the latter being the most advanced.

- Currently available *D* and *E-class gas turbines* are usually within the range of 75–100 MW, and can be considered as robust and reliable workforce for those applications, where superb efficiency is not required and where the fuel is either cheap or is of low quality. They usually utilize special combustion systems to minimize emissions and comply to the regulation, and at the same time deliver reliable power with average efficiency.
- *F-class gas turbines* are in the range of roughly 150–230 MW, being mostly designed for combined cycle applications as a further development of D and E class machines. They possess an optimized design of the air and hot gas flow path, as well as more efficient and therefore complicated combustion system. This, in return, brought certain limitations on the quality of fuel (mostly natural gas) utilized in these machines. However, increased efficiency pays back in the countries where regulations are strict and fuel prices are quite high.
- Recent developments in gas turbine technologies resulted in *G*, *H* and *J classes* — a new generation of units with the output of more than 250 MW, which have been installed since a couple of decades. These units show superb efficiency of more than 62% in a combined cycle mode — a breakthrough value delivered by a 9HA.01 gas turbine from GE in 2016 which has already been beaten by the same type of unit. These units employ state-of-the-art materials and technologies to provide the most efficient and clean gas-fired power.

Thus, all types and classes of heavy duty gas turbine used in power plants consist of three major sections: compressor, combustion system, and turbine section. On top of these, there are several essential systems and accessories required for their normal operation and which are supplied together with the turbine. These are:

- Air inlet system,
- Exhaust system,
- Fuel module;
- Lube and hydraulic oil module,
- Compressor washing.

9.3.2 Major components

We consider a typical heavy-duty gas turbine used in simple and combined cycle utility power plants. Fig. 9.3 shows an example simplified schematic of a modern gas turbine. A typical gas turbine has three major sections:

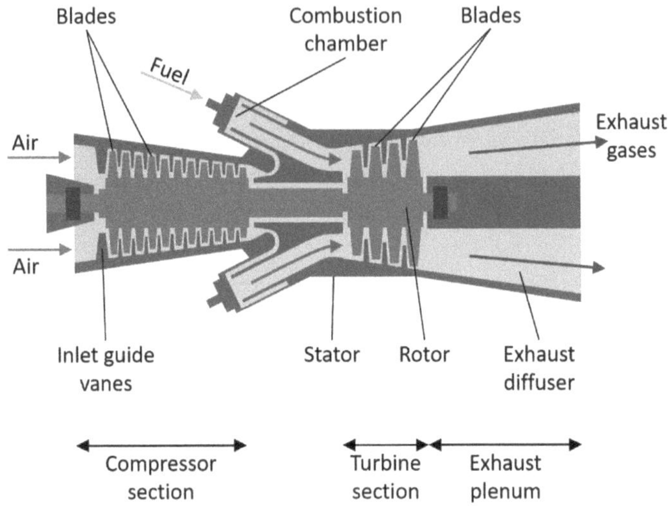

Figure 9.3 ***Gas turbine major components.*** Schematic simplified drawing of a typical gas turbine (for indicative purposes only).

compressor, combustion system, and turbine, which follow the energy conversion chain.

Clean and cold ambient air goes to the compressor section *1* through the inlet system. The air flow and its direction is controlled by the movable *inlet guide vanes (IGV)* — a set of special vanes operated by the gas turbine control system according to the operating regime. The compressor section is responsible for compressing this ambient air and feeding it to the combustion chamber. The compressor consists of the rotor with the rows of blades (or buckets) assembled and balanced at the factory. These buckets normally have a special complicated shape to ensure that the air is properly sucked into the machine and compressed to the required ratio. In-between these rotating rows of buckets, the stationary nozzles are located on the compressor casing for optimized air flow inside the compressor.

The compressed hot air is then guided into the *combustion chambers* arranged into the combustion section. Here the compressed air is properly mixed with the fuel (either gaseous or liquid) to prepare it for burning. To start the process, spark plugs are usually used, and then the flame is maintained continuously on its own to allow the ongoing process of combustion. Basically, the combustion chamber is used for converting the chemical energy of the fuel into the heat energy of combustion and then into the kinetic energy of the hot gas, which is then guided into the turbine section.

The turbine section is designed to take most of the energy of the hot gas flow and convert it into the mechanical energy of the rotating turbine rotor. The turbine section consists of several rows of moving buckets (mounted on the shaft) and stationary nozzles (sitting on the stator). The hot gas is guided through the rows in such a way that it gives the energy of the flow to the buckets, which have the shape of the wing and hence face the lifting force when exposed to the gas flow, while the nozzles guide the air flow from one row of buckets to another. This leads to the rotation of the shaft itself.

Finally, after being fed through the turbine section, hot gases are discharged through the *exhaust diffuser*, which is a part of the exhaust system.

Compressor and turbine rotors are assembled as one piece at the factory and altogether are called the *rotor of a gas turbine*. Proper balancing of the rotor eliminates vibration and prevents the machine from failure. A number of bearings lift the rotor on the thin layer of pressurized lube oil, which are located on the stationary part of the turbine — the stator. A bearing is an important part of the turbine which significantly reduces friction and basically allows rotor to rotate within the turbine. This requires a continuous supply of lubrication oil inside each bearing and its constant cooling as the whole part would heat up due to friction. Moreover, this oil should be pressurized in order to lift the heavy rotor and keep it away from the surface of the bearing.

9.3.3 Accessories and supporting systems

Gas turbine *accessories and supporting systems* are required for proper turbine operation. They are supplied together with the unit by the same OEM in the *equipment only package* and can be split into mechanical, fluid, and electrical systems, as shown in Fig. 9.4.

Mechanical systems cover the inlet and exhaust systems of the gas turbine, as well as the overall turbine assembly and may include:

- Air inlet system;
- Exhaust diffuser and downstream systems;
- Gas turbine enclosure (container) to protect personnel from the high temperature, noise and possible leakages of gas. It usually has ventilation, gas detection with alarm, and fire protection (compressed CO_2 or water mist).

Fluid systems include:

- Fuel system and steam or water injection module (if required);
- Lubricating and hydraulic oil;

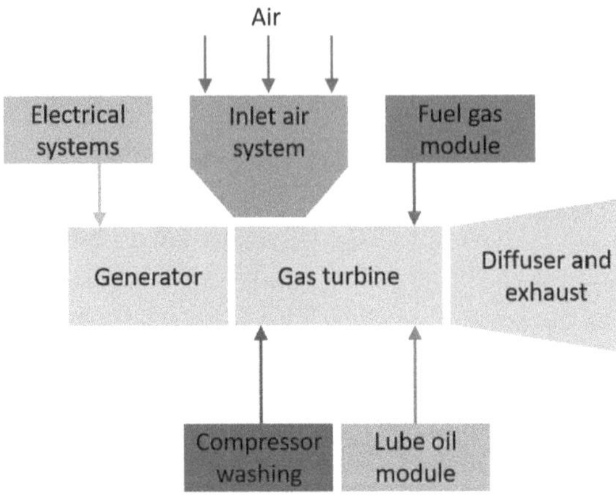

Figure 9.4 ***Essential systems supplied with a gas turbine.*** Gas turbine systems in the equipment only package include all necessary mechanical and electrical systems.

- Cooling of the turbine systems (stator frame, exhaust diffuser, buckets, combustion system, etc.) by air, water or steam depending on the type of a gas turbine;
- Compressor washing module.

Electrical and instrumentation systems are necessary for starting and operating the unit, as well as for monitoring and controlling all parameters. They include, among others,

- Various instrumentation, like vibration monitoring system, flame detectors, temperature and pressure probes, velocity measuring to make sure that the control system and the operator get all necessary information about the condition of the turbine.
- Gas turbine control system, usually in a special container, which is connected to the main control room of a power plant. This system is responsible for operating the unit during start-up, loading, operation, and shut-down. It reads data from the instrumentation and assesses the condition of the unit to maintain it safe. In case of emergency, this system can automatically issue a trip signal and stop the unit if the safety boundaries are reached.
- Starting means, usually supplied together with a generator. They provide electricity to allow the generator operate as a motor and start the rotation until the machine goes to firing speed (i.e., the compressor supplies enough air to the combustion system).

We now consider some of the systems in detail.

9.3.3.1 Air inlet system

Gas turbines should be able to successfully operate in areas with different ambient conditions and air quality (dust, moisture, salt, etc.). The information regarding the ambient conditions is important at the early stage of the projects and should include:

- Ambient temperature range (min, max, average);
- Air humidity range (min, max, average);
- Level of precipitation and snow and wind loads;
- Information about environment conditions (e.g., chemical plant nearby, salty seaside air) to define corrosion level;
- Amount of insects at peak season (high or low).

In order to adapt the serial unit to the environment, it is necessary to know as many details as possible about the environment and hence properly treat the air that comes to the compressor through the *air inlet system*. Such treatment provides protection against the contaminants, which may degrade gas turbine performance and therefore its life. On the other hand, this system also conditions the air to optimize gas turbine performance.

A failure of the air treatment system can lead to the following impacts:

- *Foreign object damage,* usually occurring in the early stages of compressor;
- *Erosion of material* that can reduce the aerodynamic performance of the unit and lead to serious turbine damages (up to break down);
- *Fouling of compressor blades,* which decreases the efficiency of compressor. The surfaces of compressor blades can be roughened by ingestion of substances, which adhere to the surfaces. This can significantly decrease the output of a turbine;
- *Cold corrosion* or *pitting of compressor components* caused by wet deposits of salt or acids, which damage the blades and can shorten their life;
- *Hot corrosion,* caused by ingestion of metals (like sodium, potassium, vanadium, or lead), which would combine with sulfur and oxygen during the chemical reaction of combustion and deposit on the surface of the hot gas path parts. Such deposits damage the protective film coating of the parts so that they oxidize very quickly.

Any gas turbine application typically needs more than one type of filter, and there are no "universal filters" that will serve all needs. Therefore, two- or three-stage filtration systems are used. Any arrangement of filters is subject to the ambient conditions and highly influenced by the environment the machine will be operating in. Therefore the inlet air system consists of the following equipment:

- Weather hoods to protect the equipment against precipitation;

- Trash, bird, and insect screen to protect the filters in certain areas against mechanical damage;
- Filter house with a set of filters against different sizes of particles that allows for protection of the machine for the guaranteed operation period;
- Coalescers and separators to remove liquid moisture: they catch the small water droplets inside the fibers within the coalescer, or through certain physical principles (momentum, gravity, and centrifugal forces) in separators;
- Air temperature treatment to optimize compressor and turbine performance;
- Inlet silencers to provide abatement of noise from compressor bellmouth due to high velocities of the airflow.

This equipment is connected to the gas turbine compressor with the air ducts. In case of a failure of the filters due to partial or complete obstruction with the dust which results in excessive pressure loss, special pressure relief protection called *implosion doors* is used. These doors shall automatically open in case of filter failure. With the doors open, the compressor will digest unfiltered air, but this will prevent damage to the filter housing structure due to high forces from increased pressure loss across the filter.

There are two major types of inlet filter house: *self-cleaning* and *static*. According to their names, the major difference is in the way the cleaning process of the filter is carried out. In static filters, the filtering panels shall be removed and cleaned once they are stuck with the particles taken out of the air, while the self-cleaning filter can do the cleaning itself with the compressed air blown in the opposite direction of the airflow. Depending on the type of system, they have various stages: static includes coalesce, pre-filter and fiber glass fine filter. Water droplet have to be caught as much as possible before the air reaches the final filtration stage.

Additional systems to increase performance of the machine and ensure its operability include:

- *Inlet bleed heating* used for distributing the hot air flow extracted from the compressor into the inlet to provide a uniform temperature rise at the compressor inlet and optimize its performance.
- *Evaporative cooler* to reduce the compressor inlet air to increase turbine efficiency. The water is injected into the airflow where it evaporates and absorbs energy from the air. Evaporative cooling is limited to low humidity conditions and ambient temperatures above approx. 15°C to completely avoid icing conditions inside the compressor and ensure

proper and efficient evaporation takes place. The evaporative cooler is located downstream the filter house to reduce its fouling and not to influence the pressure drop along the filters. Located after the filter, the water droplet screen protects the air from pure water content.

- *Chiller* to cool down the compressor inlet air. The chiller is a water-to-air heat exchanger that circulates cold water and cools down the flowing inlet air. It can be located in the filter house and does not affect the filtration process. Despite high energy consumption (to run the pump and cool down the water in the loop), it does not waste water compared to the evaporative cooler and can increase power output by up to 15%.
- *Anti-icing* for static filter to avoid clogging on the filters with ice when icing conditions appear (from around −5°C to +5°C). This can be done either by using hot air from the compressor (inlet bleed heating system) or with the special water-to-air heat exchanger (in case inlet bleed heating is used to preheat the air to optimize performance): a heating coil is installed in the inlet to heat up the air.

Remark 9.2. Instead of a special anti-icing system, self-cleaning filters use compressed air to remove ice together with the dirt.

9.3.3.2 Exhaust system

The exhaust system guides the hot exhaust gases from the turbine exit to the downstream equipment. For instance, this can be a silencer, bypass stack, heat recovery equipment, or exhaust stack. Depending on the technology, two major configurations are possible:

- Axial exhaust, where the hot gases from the turbine go axially. Such a configuration is the most efficient in utilizing the exhaust heat in heat recovery equipment to minimize losses. The generator is then connected at the compressor side ("cold end").
- Radial exhaust, when the gases change their direction and go to the side of the gas turbine. This configuration is applicable for the units where the generator is coupled at the side of the exhaust, or "hot end".

When the flue gases leave the exhaust system of the gas turbine, either radially or axially, to the heat recovery system, they normally pass through a set of ducts and expansion joints, which guide them properly to the HRSG inlet. In some cases, there can be a *bypass system* used for guiding hot gases straight to the atmosphere when the boiler has to be turned offline with the running gas turbine. This bypass system includes:

- Diverter damper to guide the hot exhaust gases;
- Bypass exhaust stack to release hot gases to the atmosphere;

In case there is no heat recovery system, i.e., the gas turbine is operating in an open, or simple cycle, mode, flue gases from the exhaust are guided to the exhaust stack through a silencer and, potentially, through the exhaust gases' filter system to further decrease NO_x and CO emissions.

9.3.3.3 Fuel module

Modern gas turbines are quite sensitive to the quality of the fuel and therefore require specific fuel treatment and precise control in order to maintain guaranteed levels of emissions and performance values. This is delivered through a combination of the state-of-the-art combustion and fuel treatment.

Normally, preliminary treatment of the fuel gas or liquid is done on the power plant basis outside of the turbine compartment at the terminal point, where the gas line is connected or where the liquid fuel is stored. This overall gas treatment is included either into the boiler island (for steam power plants) or into the *balance of plant* scope. The fuel module normally consists of control valves and fine filters, driven by the gas turbine control system. It distributes the fuel to the combustion system to optimize the amount of fuel and hence control the rotation of the shaft and the temperature at the exhaust of the combustion chamber for emissions control.

For some combustion systems, water or steam injection is required to maintain required levels of emissions, which are guaranteed through the automated control of the proper fuel–air–water (steam) mixing.

9.3.3.4 Hydraulic and lubricating oil system

The lubricating oil system is usually used for both the gas turbine and generator. Its main function is to take lube oil from the tank, pump it to high pressure and use in the turbine and generator hydraulic system. This system usually includes:

- Oil pumps, since usually the lube oil system relies on several redundant pumps to deliver oil from the reservoir to the systems that need lubrication and hydraulic support;
- Coolers to reduce oil temperature when it is pumped back from the turbine and generator. Coolers are oil-to-coolant heat exchangers where all oil is cooled down. The coolant is supplied from other BOP systems;
- Filters are used to clean return oil;

- Instrumentation and control equipment;
- Lube oil reservoir contains all necessary lube and hydraulic oil used within the process.

Normally, during the commissioning of a gas turbine, fresh oil fill is supplied. After all procedures and guarantee runs, used oil is drained from the system and replaced by fresh oil for further ordinary operation. Its further renewal is subject to manufacturer's recommendations.

9.3.3.5 Compressor washing

During the hours of operation, despite proper air filtration, the first stages of the compressor blades are exposed to some dirt, substances, insects, chemicals, or precipitation, coming from the environment. The problem becomes even worse in case the power plant is located in an industrial environment of a chemical plant or near the seaside exposed to salt.

For cleaning purposes, the compressor washing system is used on the operating gas turbine. Normally, the washing procedure can be done in two ways:

1. for regular washing, demineralized water is discharged through the special nozzles directly to the compressor and cleans up the blades during the normal operation of the power plant.
2. for serious cleaning, water is mixed with some chemical detergent and then injected; however, this procedure requires specific turbine operating conditions and takes more time.

9.4. Waste heat recovery equipment

The hot exhaust gases from the gas turbine can be up to 650°C and therefore have enough energy which can be further utilized in order to increase the overall plant efficiency. This heat is recovered inside a special device called a *heat recovery steam generator*, or *HRSG* for short. Due to the limitation of the gas turbine exhaust temperature, HRSG cannot supply steam of supercritical conditions, therefore this type of boiler can be of a subcritical type only.

9.4.1 Arrangement

Consider a combined cycle power plant that incorporates one gas turbine and one non-reheat steam turbine with the process diagram represented in Fig. 9.5 and discussed in Chapter 7. The hot exhaust gases from the gas

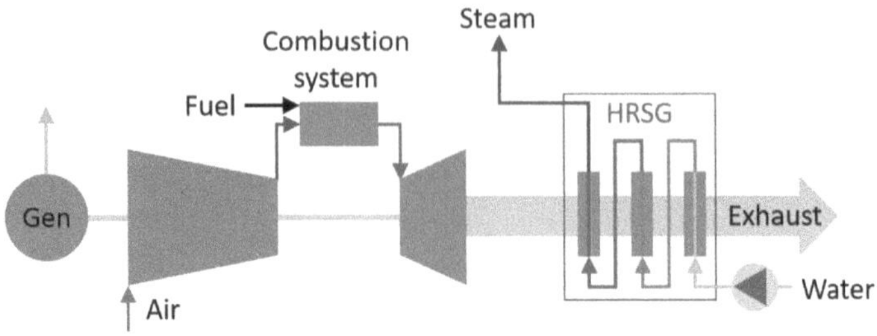

Figure 9.5 ***HRSG within the process diagram.*** **Simplified schematic process diagram of an HRSG coupled to the gas turbine. Its configuration is designed based on the turbine performance, steam requirements and overall plant configuration.**

turbine are fed to the *heat recovery steam generator*, which has the primary function to evaporate water and supply steam to the steam turbine. Similar to a conventional fossil-fired boiler, the cold water is preheated within the *economizer* and then guided to the *evaporator* — a heat exchanger exposed to the hot exhaust gas flow. Then, steam goes through a *superheater*, another heat exchanger that increases pressure and temperature of steam. This superheated high quality steam is then supplied to the steam turbine. From the turbine, the cold steam follows its common path through the condenser, preheaters, deaerator and several pumps to close the cycle.

In case of a reheat steam turbine, an HRSG would have another set of heat exchangers, which would heat up the cold reheat that comes back from the HP discharge. The reheater is incorporated in the process using the same logic as in a conventional boiler.

A heat recovery steam generator is usually designed for a particular gas turbine exhaust characteristics (which mostly depend on the type and composition of fuel, operating profile, and ambient conditions) and steam requirements. The heat balance diagram, calculated for the plant for various regimes and conditions, is used as basis for an HRSG design and construction. Normally, there are three major parameters to mention when considering the overall arrangement of an HRSG: direction of the flue gas flow and type of evaporator, which would define the overall components and size of the unit.

The *direction of the gas flow* in an HRSG depends on many circumstances, for instance:

- Amount of space available on site, i.e., the footprint of the HRSG;
- Direction of the GT exhaust flow;

- Overall plant configuration and required characteristics (e.g., steam conditions, exhaust temperature and flow, etc.).

Therefore, two major options are available:

1. *Horizontal flow*, which means vertical heat exchangers, natural circulation, less efficient heat exchange, light structure, larger footprint;
2. *Vertical flow*, which means horizontal heat exchangers, natural or forced circulation (extra pump), equal heat flux with efficient heat exchange, heavy structure (increases cost), smaller footprint.

Similar to the conventional boilers, there are two major *types of evaporator* that can be used in an HRSG:

- Steam drum, where the water is heated up within the tubes and evaporated, and the water–steam mixture is separated in the drum. This is an established and commonly used technology in an HRSG for all pressures of steam (as all of them are subcritical).
- Once-through Benson type can be implemented instead of the HP drum in both horizontal and vertical flow boilers. Eliminating an HP drum would increase boiler flexibility and significantly reduce startup time of the plant [7].

Therefore, there are four major configurations of the heat recovery steam generators being present in the market:

- Horizontal drum type, with lower installation costs and shorter installation period as parts of it can be installed at the same time, but requiring a larger site space. This type of HRSG is the most common and widely used.
- Vertical drum type, which is suitable when a smaller footprint is required, especially for repowering or plant extension. It usually has longer installation period and more associated costs as the modules are installed one after another.
- Horizontal Benson, which is suitable for the high temperature and high pressure HRSG. Implementation of the once-through evaporator in the HP section with the drums for LP and IP sections results in greater thermal flexibility, higher efficiency, and the ability to support various power cycling.
- Vertical Benson, which has a much lower footprint and is modularized. On top, it can also benefit significantly from the flexibility, allowing for fast start-up and shut-down capabilities.

As always, the choice of the suitable technology depends on the requirements and needs of the owner and designer of a power plant. There is no

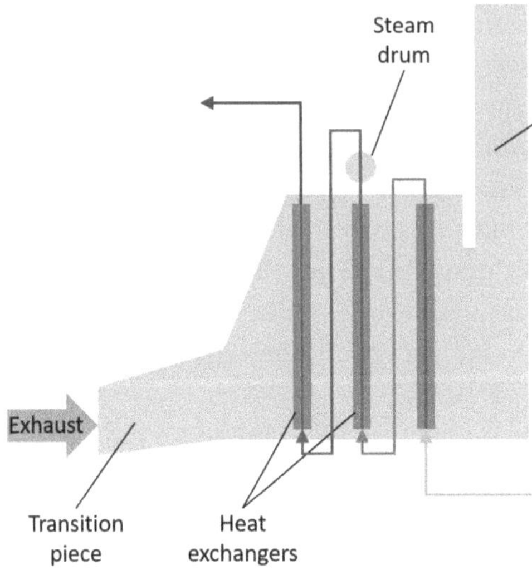

Figure 9.6 ***HRSG major components.*** Simplified schematic of a horizontal flow HRSG (for indicative purposes only).

single solution as each plant design is unique, otherwise there would have never been several solutions on the market at the same time.

9.4.2 Major components

There are multiple configurations of the HRSG components, but we will focus here on one of the most typical HRSG: horizontal drum-type with two pressure configurations. As illustrated in Fig. 9.6, an HRSG consists of the following major components:

- Transition piece and inlet duct, which are exposed to most extreme temperatures of the hot gases coming directly from the gas turbine;
- Overall steel structure and shell with thermal insulation;
- Evaporator tubes of the heat exchangers, mounted on the headers and located inside the shell across the hot gas path (in this case evaporator tubes are vertical);
- Steam drum, where the steam is separated from the boiling water circulating naturally through the tubes;
- Exhaust stack, coupled to the main structure of the HRSG through another expansion joint;
- Pumps, valves, and tanks to maintain the overall water and steam cycle properly. These can include recirculation pumps, feedwater pumps, safety shut-off and control valves, etc.;

- Emissions control equipment aimed at reducing gaseous emissions that arrive from the gas turbine all the way through the HRSG to the stack. Usually, an SCR is implemented to decrease NO_x concentration (refer to Section 8.4). However, these systems significantly impact both the performance of the HRSG and its cost, therefore it is extremely important to have a low emission combustion system within the gas turbine.

Similar to the conventional boiler, the exhaust temperature at the stack is subject to multiple factors, however, it is usually regulated by the local legislation. Normally, at the top of the stack, CEMS is installed to ensure monitoring of the NO_x and CO emissions levels.

9.5. Electrical generator

All modern generators are based on the phenomenon discovered by Michael Faraday at the beginning of the 19th century [8]. If a conductor passes through a magnetic field, this movement generates an electric current in the conductor. When rotating the coils within the poles of a permanent magnet, the system produces a pulsed DC current. After further development, which implied electromagnets (powered by an external DC source to generate a much more powerful magnetic field) and multiple poles with a series of commutators to automatically compensate the current pulsation, it became possible to manufacture bigger machines and produce a more stable current. These machines were named *dynamo-electric* or simply *dynamos*.

Initially, dynamos generated only direct current as it was used for lighting purposes and therefore was much more practical. It was the need to transport the current over longer distances to power the industry that made the advantages of the alternating current evident. The voltage of the AC supply could be easily changed with the transformer, while the wire transport of the electricity was much more efficient. Moreover, newly introduced *alternating current generators* or *alternators* and machines using this type of current helped overcome the complexity caused by a commutator.

9.5.1 Concept of modern generator

From the physics point of view, the design of a modern generator is defined by what is moving: the poles of the magnet or the coil or armature, where the current is generated. Here two options are possible:

1. The coil rotates inside the stationary magnetic field. The poles are located in the stator and generate the static magnetic field. When the

rotor moves, the coil starts its movement inside this magnetic field, so that the current is generated. This current has to be transferred for practical use. This is done through the special *brushes* or special *slip ring*, which touch the moving part.

2. The rotating magnetic field turns inside the fixed coil. The poles of the magnet are located on the rotor, which excites the magnetic field. When the rotor starts it rotation (powered by a turbine), the magnetic field starts moving, or, in other words, the fixed armature of the stator will be exposed to the moving magnetic field. This would generate current inside the coils of the stator (as the conductor would move through the magnetic field).

Generators with the revolving armature are not often used. The second design with the poles located on the rotor is more favorable as it does not require transferring large current of high voltage through the brushes or rings, which makes them much smaller and cheaper. On top, the stationary armature can be wound for any required voltage level.

The frequency of the generated current is determined by a number of poles of the rotor. These poles are always paired (one north and one south) to generate a magnetic field. One rotation of this two-pole magnet would generate an alternating current in the stator fixed coil of one period. The required number of periods per minute is determined by the grid and is called the *frequency*.

If the frequency of the grid is 50 Hz, it means that the generator has to deliver the same frequency of the current, i.e., 50 rotations per second, or 50 AC periods per second. Usually, the rotations of the shaft are measured in revolutions per minute, or *rpm* meaning that 50 Hz equals to 3000 rpm if we have only 2 poles. Adding another 2 poles would allow for twice more periods of generated current per minute so that the rotation speed shall be two times slower to maintain the same 50 Hz frequency of the grid. However, on a thermal power plant high speed turbines and generators are used. This means that for 50 Hz their speed will be 3000 rpm, while for 60 Hz it will be 3600 rpm with a two-pole generator.

Moreover, each stator has three independent coils which are located symmetrically 120° to each other. This allows for three phases of the current generated simultaneously with the same number of poles; moreover, these currents will have the same frequency and a shift of their phases.

Having said that, the basic function of the generator is to convert mechanical power delivered from the rotating shaft of the turbine into electrical power. Therefore a *generator* is a rotating mechanical energy converting machine by means of a rotating magnetic field produced by direct current in the winding of the rotor. This field generates a three-phase alternating current and voltage in the winding of the stator (armature). This stator winding is connected to terminals, which are in turn connected to the grid through transformers [9].

Remark 9.3. All generators within the same grid must generate the current of the same frequency or, in other words, generate the same number of periods per minute. This can be achieved only if all machines are synchronized before being plugged to the grid. This is why these machines are sometimes called *synchronous generators*.

9.5.2 Major components and accessories

Like any rotating machine, a generator consists of two major parts: the stator and the rotor. The *stator* is a stationary part of the generator with the space inside where the *rotor* spins. The stator accommodates the rotor, which is then coupled to the turbine. The rotor spins on the bearings, which receive a constant supply of lubricating and hydraulic oil. The windings of the stator, where the HV current is generated, are then coupled to the generator terminals and further to the bus bars.

Each generator is equipped with a special *excitation system* to generate and supply direct current in the rotor windings that creates the magnetic field. While this field rotates with the rotor, it generates HV current in the windings of the stator, and this current is further transferred to the consumers. There are two major approaches to the excitation system:

1. Separate excitation, when the current for excitation of the main generator coils is produced by a separate small generator, coupled to the rotor of the main generator. A small generator delivers an alternating current, which is transformed through the thyristor bridges into a direct current.
2. Self-excitation uses part of the generated energy to power the excitation system. To start the process, the initial excitation coil generates a short pulse of voltage (either from the plant auxiliary power supply or from the DC battery) in the windings of the rotor, which in turn generates a current in the stator. This immediately starts the whole process of self-excitation.

The above approaches require electrical connection between the spinning rotor and the stator within the excitation system, which is done through the so-called *brushes* or *spinning coils*. These brushes and slit rings are the weak point of the generators. Exposed to the friction and wear process, they have to be frequently maintained, cleaned and finally replaced.

This problem has been solved through the *brushless* type of the generator excitation system. It consists of the stator and the rotor parts as well. Contrary to the previous approaches, the stator part of the excitation system uses current to generate a magnetic field, which will be static. The rotor then contains the coils, which would spin in that magnetic field and therefore generate an alternating current. Being transformed through the thyristor bridges into DC, it is then fed to the coils of the main generator.

A generator, as any other rotating machine, still requires a number of supporting systems that are supplied together with the machine. These include:

- Lubrication and hydraulic oil system to ensure reliable and efficient rotation of the rotor within the bearings;
- Cooling system;
- Control and protection system to ensure safe and reliable operation of the generator at all regimes. These protections are connected to the plant control system through the generator control panel.

We now consider separately the generator cooling system. The energy losses during the generator operation lead to the heating of its elements. Even though the efficiency of modern generators is quite high, these losses are still present and may be up to 2.5%, which can be around 10 MW for an 800 MW unit. Such an amount of energy converted into heat would increase the temperature of the material, including the conducting copper parts and insulation.

The highest temperature of the generator parts is limited by the material properties, namely the winding insulation of the rotor and the stator. High temperatures may lead to a decrease of conductivity, loss of material strength, as well as to direct damages of the insulation and decreased lifetime of the machine. In order to maintain the temperature of the rotor and the stator within the required boundaries (defined by the technical regulations for various types of generators), a continuous heat rejection system is utilized. This system is usually called a *generator cooling system*.

Generally, three major cooling approaches are used [8]:

- *Air* is the cheapest cooling to implement, but it the least effective of the three for carrying out the heat. Usually this type of cooling is used when the generator capacity is up to 350 MW for simple and combined cycle applications.

- *Hydrogen* is 4–5 times better at cooling than air. A hydrogen-cooled generator itself can be significantly smaller, and therefore less expensive than an air-cooled one. However, hydrogen is explosive and would require proper sealing and detection systems. Moreover, a special gas heat exchanger is required to reject the heat of the gas. This type of cooling is suitable for 250–450 MW simple and combined cycle applications, as well as steam plants.
- *Water* cooling is implied for very large generators. Despite great capability of heat rejection, such a cooling system is much more complex to implement. Generally, water cooling is employed for cooling down the stator, while the rotor can be hydrogen-cooled. Such cooling is well suited for the highest power generators (more than 1 GW), in large fossil-fired steam power plants.

Remark 9.4. Hydrogen is a very flammable gas. Its flammability limits and its tendency to self-ignite or form explosive mixtures with air requires special safety measures. The generator enclosure must be purged first with inert gas and then replaced with air before shutdown, maintenance, and refilling it with hydrogen. The opposite sequence is used before startup. Usually carbon dioxide or nitrogen are used for this purpose, as they do not form combustible mixtures with hydrogen, are quite cheap and usually available from other power plant systems.

9.6. Balance of plant systems

A balance of plant systems is required to support the operation of major equipment within a thermal power plant. Here we give an overview of the plant balance and first discuss the mechanical systems, including heat rejection (cooling), water treatment, and others that are required for operation. Then, we focus on electrical systems, especially power transfer to the grid, control system and other auxiliaries.

9.6.1 Scope and importance

The *balance of plant* (*BOP*) for a thermal power plant covers all systems and equipment, which are required to support the process of power generation. In addition to the boiler and power islands, where the energy conversion process takes place, the balance of plant supports this process to ensure that it is efficient, safe, and reliable.

BOP is usually split into two major parts: mechanical and electrical. The key subsystems of the *mechanical BOP* typically include any kind of material handling, water treatment facilities, heat rejection system, exhaust chimney, waste treatment, and some others. *Electrical BOP* refers to the systems that cover the electrical side of the power plant, for instance, power transfer, electricity supply, control system, and protections.

9.6.2 Mechanical

Mechanical BOP comprises all systems related to proper operation of the boiler and turbine island, providing mechanical and fluid systems interconnection between the specific parts of the plant. Mechanical BOP usually includes the following systems:

- Heat rejection system to condense the steam from the steam turbine exhaust;
- Water treatment system;
- Material handling and fuel treatment system;
- Grid water heaters, which are installed if the plant is operating in a combined heat and power mode. This special type of heat exchanger uses steam from the boiler to transfer heat to the grid water supplied directly to consumers (for instance, for district heating purposes);
- Oil treatment system, which ensures that all equipment receives clean oil of certain properties, including not only major rotating equipment, but also various pumps, motors, as well as transformers. Used oil is flushed from the equipment and either treated (filtered) to recover as much oil as possible, or disposed. If required, fresh oil is supplied to the system from the main oil tank;
- Compressed air system, which is used to provide pressurized air to power valves, pumps, and other equipment, requiring high-pressure air for operation;
- Drainage system, used to remove excess water like precipitation, waste, and other liquids from various systems of the power plant;
- Blow down system, utilized for removing excess water and impurities from the boiler or HRSG to avoid operational issues;
- Emergency diesel generator or other power supply to run the emergency power supply in case of power loss. In case of a blackout, power is required to run lubrication oil pumps and continue rotating the rotors of the turbines during the cooling-down process. Also, an emergency diesel generator is used to run the minimum required control system functions, lighting, ventilation, etc.;

- Fire protection system for the whole power plant to control the overall operation, detect overheating, dangerous concentration of flammable gases and smoke. This system can automatically stop the plant and extinguish the fire through multiple sources like compressed CO_2 or water mist;
- Heating, ventilation, and air conditioning (HVAC), designed to remove excess heat from the enclosures and compartments where equipment is installed. The system can also protect potentially dangerous systems from explosion, especially in restricted areas exposed to potential leakages of fuel gas or hydrogen. Moreover, it maintains favorable operating conditions for the equipment and personnel;
- Civil and structural works that include all foundations, turbine halls, equipment buildings, structures and other site facilities like fencing, gates, camp, office buildings, etc.;
- Auxiliary boiler, used to supply steam for startup purposes and to run auxiliary steam turbines driving some of the water cycle pumps;
- Hydrogen system required to fill-in the leakages within the system (if applicable). Hydrogen is often produced on-site in electrolyzers, as this reduces the need for stored amount of compressed hydrogen and allows storage in lower pressure tanks, with associated safety benefits and lower costs;
- Nitrogen system, implemented mostly for purging gas pipes before maintenance. Nitrogen would remove explosive or flammable gases like hydrogen or fuel gas and therefore protect the overall system and personnel.

Remark 9.5. Depending on the approach, material handling and fuel treatment systems for fossil-fired (either coal, oil, or gas) steam power plants are included into the boiler island. These are fuel preparation and treatment, boiler waste handling and emissions reduction equipment. We use here the same approach and therefore discuss these systems in Chapter 8 related to the boiler island. However, gas stations for gas turbine driven power plants are considered as part of BOP and therefore are discussed separately. This is also due to the fact that contrary to steam boiler, fuel gas for gas turbines requires more specific treatment.

Let us now consider major mechanical BOP systems in detail.

9.6.2.1 Efficient heat rejection system

During plant operation, most of the equipment heats up and therefore requires timely and efficient cooling. The equipment subject to cooling includes all rotating machines (turbine, generator, pumps, motors), as well as all heat exchangers. Even though some of the equipment can be easily cooled down with the recirculating air, major heat exchangers require efficient and powerful heat rejection system to transport waste heat outside the plant. This is mostly relevant to the equipment used to condense steam released from the steam turbine exhaust, which will be the focus of this section.

The *heat removal system*, also referred to as a *cooling loop*, shall be designed to fully support the steam turbine requirements. This system is required for any steam turbine cycle, no matter whether it is a conventional steam or a combined cycle power plant. The type of the system is normally defined by:

- The capacity of the steam turbines, or simply the amount and parameters of steam going to the condenser;
- Capability of the plant to bypass all steam over the steam turbine straight into condenser in case of emergency and steam turbine shutdown;
- Amount of exhaust steam used in the industrial process;
- Availability of water;
- Commercial estimation.

In the majority of cases, steam from the exhaust is fed to the special vessel called condenser, where vacuum is maintained. This vessel is cooled down by external cooling loop or simply by circulating ambient air to ensure the steam is condensed and returned back to the cycle. In a few cases, however, all exhaust steam can be driven to the consumer so that no condenser is implemented. The steam is fed to the external loop (which is out of scope of the power plant) and returned back as condensate. This loop can also be considered as a heat removal system, as it still cools down the steam to return back water.

This way, there are several widely used options available on the market for the power plants that utilize steam turbine cycle:

- All steam from exhaust goes to an external process and the cycle returns a condensate, which is pumped back to the boiler by the condensate pump. Being less popular, this cycle is normally an external facility outside the plant (or may employ special type of a steam-to-water heat exchanger to heat up process water) and therefore will be out of our focus.
- Steam-to-water heat exchanger called *surface condenser* can be used to cool down steam from the turbine. Cold water is fed from the external

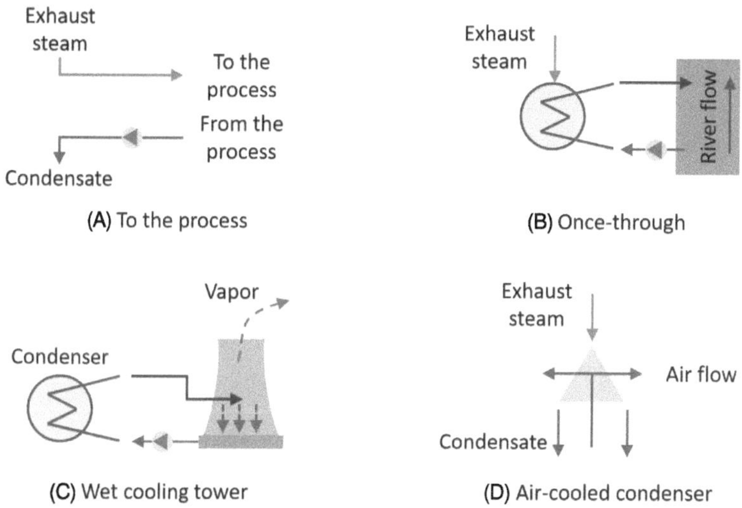

Figure 9.7 ***Heat removal system options.*** There are four major heat removal options available for thermal power plants that employ steam turbines with condensers: exhaust steam is used in external process loop (A) or the system employs once-through cooling (B), wet cooling tower (C) or air-cooled condenser (D).

cooling loop. In turn, to decrease the cooling water temperature, two major technologies are used (see Fig. 9.7):
- Once-through cooling;
- Cooling tower.

- Steam from the turbine exhaust can be condensed straight within a steam-to-air heat exchanger called *air-cooled condenser*. Exposed to colder ambient air, it would condense steam into condensate, which will be pumped back to the cycle.

We now consider these options in detail.

Surface condenser with cooling loop

A conventional surface condenser has been the most popular heat removal technology implemented in thermal power plants. Until recently, the local regulations paid much less attention both to water consumption and to thermal pollution, and such a technology has been among the cheapest and the most efficient, especially for large coal-fired steam power plants. Even today, this technology has been widely implemented in newly built power plants, where efficiency of cooling is critical.

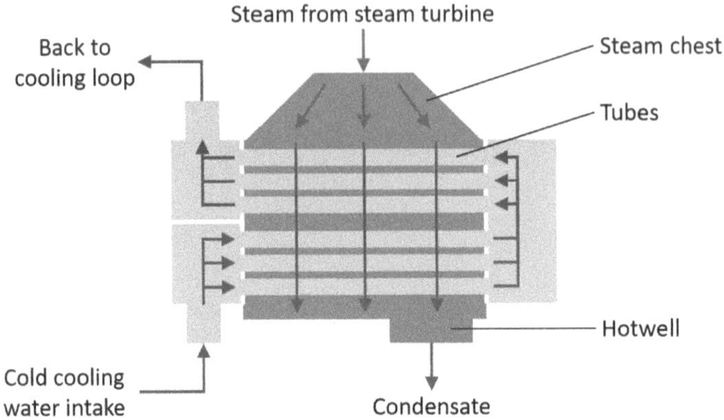

Figure 9.8 ***Surface condenser.*** Simplified schematic diagram of a surface condenser (for indicative purposes only).

A surface condenser is a vessel-type steam-to-water heat exchanger that uses cold water to condense steam. The operation concept of a basic condenser is schematically presented in Fig. 9.8. The exhaust steam from the steam turbine is flowing downwards into the steam chest so that it is exposed to the thin tubes. Cold cooling water, fed from the cooling loop, flows through these tubes, so that the flowing steam condenses on the surface and drops down to the bottom of the vessel exchanging heat with the water tubes. During this process, cooling water absorbs heat from the steam, so that its temperature rises. To cool the water down, it is fed back to the cooling loop where it releases heat. Condensate is collected at the bottom of the condenser vessel and pumped back to the boiler through the hotwell.

In practice, the exhaust steam coming from the steam turbine always contains air which is sucked through various sealing. Hence the vessel from Fig. 9.8 can operate properly only in case a special air removal system is installed. The mix of steam and air flows into the vessel and flows around the heat exchanger tubes. If the air concentration in the mix were to increase, the air pump, or ejector 6, would remove this mix with high air content.

To maximize the condensing efficiency, it is essential for an efficient steam driven engine to have steam condensed at a lower possible pressure. This is delivered through the *condensing unit*, which includes a *condenser*, *condensate pumps*, *ejectors* and external *cooling system*. There are two major objectives of using a condenser in a thermal power plant [10]:

1. To reduce the turbine exhaust pressure and thus increase the output of the turbine and the efficiency of the plant. If the cooling water temperature in a condenser is low enough, it creates a vacuum for the turbine so that the steam flows through the steam turbine much easier. Hence, it is important to use the lowest possible cooling water temperature, which is maintained through the external cooling loop.
2. To recover high quality feedwater in the form of condensate and feed it back to the steam generator without any significant treatment.

A condensing unit also recovers steam from the steam sealing system or other cycle drains, and adds to the cycle high-quality demineralized water from the treatment system in case of losses.

Once-through cooling. Cooling pond

Cooling water can be supplied from different sources. One of the most economically feasible means is the *once-through cooling* which continuously delivers cool water to condenser *1* from the open sources such as a river, big lake or sea, and releases it back.

Cold water from the external source is delivered by the cooling water pump *2* to the condenser, where it absorbs the heat of the steam, and then goes back to be released downstream. In case of reservoirs, the intake and release points should be located far enough to avoid the influence of the released warm water on the cold water intake.

Remark 9.6. The use of once-through cooling system is restricted due to the potential ecological issues caused by thermal pollution and limited number of open water sources in most areas.

One of the alternatives to the once-through cooling system is a *cooling pond* — a natural or constructed source of recirculating water with a relatively low cost of construction. To increase the cooling efficiency, water can be sprayed over the pond and cooled down by the ambient air. The use of a cooling pond may eliminate the negative effect of once-through cooling while requiring less water for cooling needs, however, certain territory should be filled with water to enable enough volume for proper heat rejection. There will also be water loss due to evaporation from the pond, which is the primary cooling mechanism. Makeup water shall be delivered through natural sources such as precipitation and, if necessary, from artificial sources.

Cooling tower

If there is no source of cold water like river or reservoir or its use is restricted by the regulation, a more expensive artificial alternative is available.

A *cooling tower* is a type of heat exchanger to remove the heat from the warm water received from the condenser and feed the cooled water back to the system. The cooling tower reduces the cooling water demand of a thermal power plant. In general, cooling towers use air from the environment to cool down the water, so it can be called as water-to-air heat exchanger.

Cooling towers may either use the evaporation of water to remove process heat and cool the water or rely solely on the ambient air. Hence cooling towers can be of two major types: wet or dry.

Wet cooling tower

Wet cooling towers utilize evaporation process for cooling. Hot water comes from the condenser into the distribution system of the cooling tower, which sprays it over a set of horizontal slats or bars called packing or fill (see Fig. 9.9). The water droplets mix with the ambient air moving through the packing as the water splashes down from one level to another. The temperature difference between the hot water and cold ambient air is large enough, so that water droplets partly evaporate and therefore transfer heat to the dry air. Cold water is collected in a basin at the bottom of the tower and then pumped back to the condenser, while hot and moist air leaves the tower from the top [10].

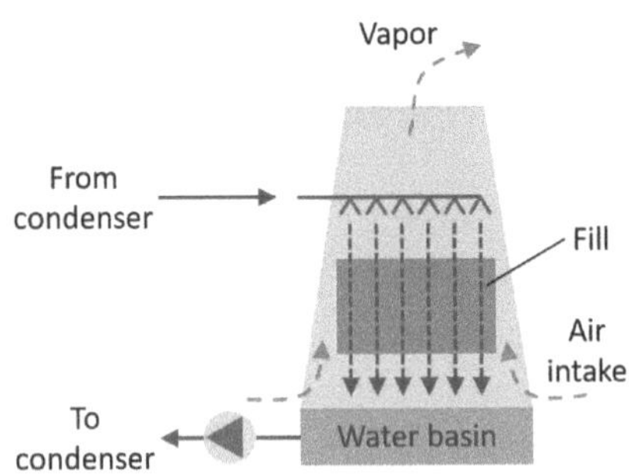

Figure 9.9 ***Schematic of a wet cooling tower.*** Wet cooling tower sprays hot water over the fill and cools it down with the ambient air flow.

When entering the cooling tower, the ambient air is unsaturated (i.e., it contains almost no water vapor and therefore is dry) and as it comes in contact with the water spray, water continues to evaporate, and the moving

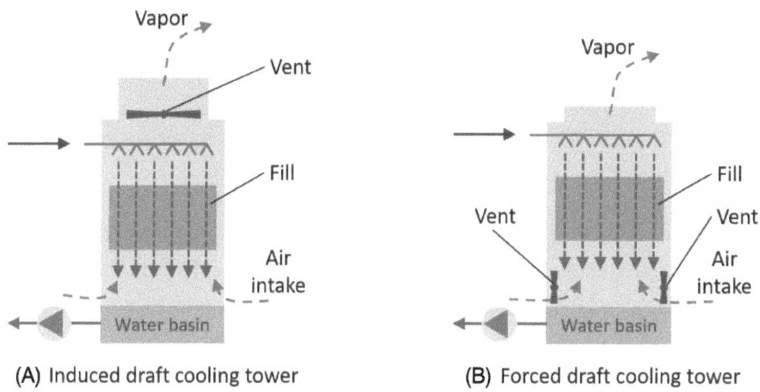

Figure 9.10 ***Mechanical draft cooling towers.*** Simplified schematic diagrams of mechanical draft cooling towers: induced (A) and forced draft (B).

air will absorb the vapor till the air becomes saturated, and carry it away. The humid air continues moving up through the warm water spray, and its temperature rises [10]. Finally, the saturated air goes out of the cooling tower forming a plume out of pure water vapor.

Depending on the way how the ambient air circulates inside the cooling tower, there can be two major types of a wet cooling tower: mechanical and natural draft.

In *mechanical draft cooling towers* the air is moved by power-driven fan motors to force it go through the tower. Here two designs are mostly used:

- In *induced draft towers* the fans are located at the discharge (at the top) and pull the air through the tower as shown in Fig. 9.10(A). Air enters the sides of the tower at low velocity through large openings and passes through the fill, whereas the hot humid air is exhausted to the atmosphere through the ventilator. This maintains the tower at a negative pressure which reduces leakage [10].
- In the *forced draft cooling towers* the blower type fans are mounted on the lower side of the tower (see Fig. 9.10(B)). Though these fans consume less power as they operate in cooler air, such a design has the disadvantages of air distribution problems, leakages and recirculation of the hot and moist air back to the tower [10].

Contrary to mechanical draft cooling towers, where the air circulation is maintained with the help of fans, which consume extra power, natural draft cooling towers can boast a simpler design with no auxiliary power consumption. In *natural draft cooling towers* the airflow occurs due to the natural pressure difference caused by the difference in density between the cold outside air and the hot humid air inside. Simply speaking, the hot air inside the cooling tower tends to go up, while the cold air is sucked through the open bottom part.

Both types of cooling tower — mechanical and natural draft — can be built either out of reinforced concrete or steel.

Remark 9.7. All types of wet cooling tower consume water and release plume that consists of water vapor. This vapor is considered a GHG, as it negatively impacts the climate and therefore may be subject to penalties in some of the countries. However, this is still subject to overall economic trade-off.

Dry cooling tower

A dry cooling tower can be an option where the water resources are limited and where the regulation does not allow any kind of plume.

Since a dry cooling tower is functioning with air cooling technology, there is no water evaporation and hence no need for extra makeup water. Hot water from the condenser circulates through thin tubes which are exposed to ambient air flowing around them. As the air temperature is lower, it absorbs heat from the water and cools it down. To improve the air flow, ventilators can be installed.

Dry cooling towers require more area and have less efficiency. On top, they normally consume auxiliary power and therefore can negatively impact the overall plant efficiency.

Air-cooled condenser

These days, the changing environmental laws and public pressure pushes using completely dry cooling options rather than continuing with once-through or wet cooling towers. In the face of scarce water resources, there are regions where there is not enough water available to satisfy the needs of power generation, agriculture, and people [11].

Compared to the surface condensers, which imply a much smaller footprint, but require a separate cooling loop with a heat exchanger and thus constant water supply, the *air-cooled condenser* (*ACC*) condenses turbine exhaust steam inside the tubes, which are externally cooled by ambient air. As illustrated in Fig. 9.11, steam enters the air-cooled condenser at the top of the heat exchangers through a major steam pipe, flows downward through the heat exchanger tubes, and condenses along its way. Finally, condensate is captured in pipes at the base of the heat exchangers. The condensate

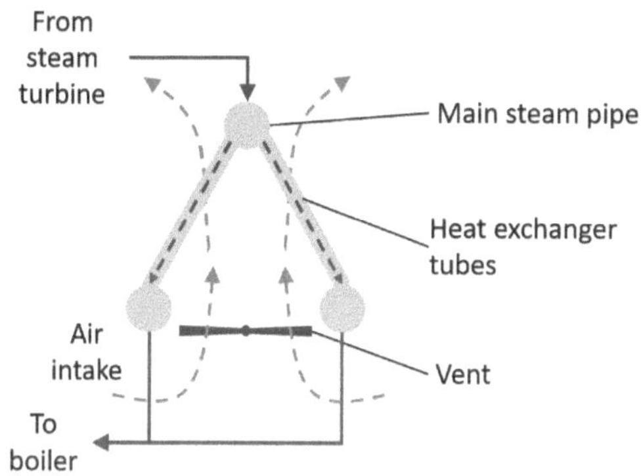

Figure 9.11 ***Air-cooled condenser.*** Simplified schematic of an air-cooled condenser cross-section: air cools down the heat exchangers where steam flows and condenses.

is then returned to the boiler via the condensate pump. To increase efficiency, mechanical fans force ambient air flow over the heat exchanger surfaces.

This condenser operates under vacuum as a conventional surface condenser does. Non-condensible gases, like in a surface condenser, are evacuated in a separate section of the air-cooled condenser via the vacuum pumps or air ejectors.

Remark 9.8. An ACC has a very distinctive "A" shape (with the sides of the letter being the heat exchanger plates) and can be easily recognized on site.

9.6.2.2 Water treatment

Water is the main working medium in a thermal power plant used for running steam turbines and cooling down the equipment. It is usually supplied from an external source and can be contaminated with iron, heavy metals, salts, CO_2, and other substances. Being in direct contact to the equipment, such water would lead to corrosion of material, scale deposits on or inside equipment, foaming during evaporation. This makes the power plant operation inefficient and even dangerous. Moreover, during the operation, there are essential losses of water and steam, which shall be regularly refilled.

Especially for the water and steam cycle in a boiler and steam turbine, water has to be clean enough to ensure proper heat exchange, corrosion protection and production of clean steam. The required quality of the feedwater increases as the steam pressure and temperature increase [12].

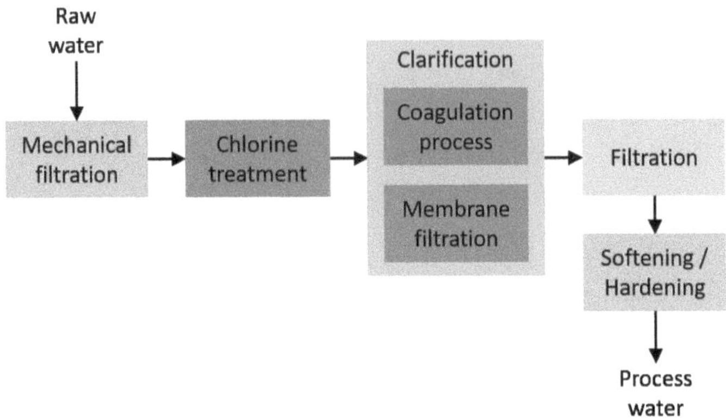

Figure 9.12 ***Water treatment process.*** Simplified process diagram of water treatment: mechanical filtration, chlorine treatment, clarification, filtration and softening or hardening (exact process depends on the requirements and raw water properties).

Water treatment systems can be divided into three major groups:

1. Pretreatment of the raw water that comes from the external source like sea;
2. Demineralization of pretreated water, including feedwater preparation, condensate polishing and deaerating. This is conducted at a water treatment facility;
3. Internal conditioning of water inside the boiler to minimize the impact of scale and deposits in the heat exchanger tubes and drums.

The raw water supplied from an external source shall go through the pretreatment process of feedwater preparation, represented in Fig. 9.12, which can include several stages, for example:

- *Mechanical filtration* of raw water to remove big particles;
- *Chlorine treatment* to remove organic contaminants and protect the system from organic fouling;
- *Clarification*, which removes dissolved contaminants and can be conducted utilizing various technologies, for example:
  - *Coagulation process*, when special chemicals (inorganic salts of aluminum or iron, e.g., $Al_2(SO_4)_3$) are added. Mixed with water, these salts react with the dissolved contaminants and initiate flocculation;
  - *Membrane-base filtration* that does not require any special chemicals and can filter constant flow of water. However, better water pretreatment is required to avoid damages of the membranes;
- *Filtration*, used together with coagulation for removal of the result of chemical reaction — the flocs;

- *Softening* or *hardening* to control water parameters like hardness, alkalinity, and silica. This helps prepare water for further use in the process. Usually the water is treated with lime or a combination of lime and soda ash.

The pretreatment of water, however, does not remove dissolved solids and organic molecules. Therefore, the pretreated water is not yet ready for the boiler due to multiple salts present in its composition. The second stage of water treatment is conducted at a *demineralized water plant* (sometimes called a *DM* or *Demin water plant*). Its process can also have a number of stages:

- *Dechlorination* of water to remove chlorine added at the previous stages to clean the water from organics;
- *Final removal* of salts and other impurities through one of the following processes:
  - *Ion exchange*, where raw water first goes into cation vessel, where dissolved cation salts (Ca^{+2}, Na^{+2}, etc.) are captured and replaced with hydrogen ions leaving the anions as acids. Then the water enters the anion exchange vessel where the anions (SO_4^{-2}, Cl^-, etc.) are removed and replaced with hydroxyl ions leaving purified water. The ion exchange bed must be periodically regenerated with sulfuric acid (cation bed) and sodium hydroxide (anion bed) [12];
  - *Reverse osmosis* (*RO*) is an alternative technology for the removal of the impurities in feedwater. It uses high pressure to force the water through special membranes, which filter out the impurities at the ionic level [12]. Compared to filtration, reverse osmosis provides finer filtration and can clean out inorganic salts and organic molecules.
- *Aeration* or *decarbonization* of water to reduce concentration of carbon dioxide dissolved in water.

Remark 9.9. Water treatment process depends on many parameters: raw water quality, water consumption, type of boiler (its temperature and pressure), and other project limitations. It is therefore important to turn to specialized companies, which would be able to design and propose efficient and cost-effective solution for each particular project.

9.6.3 Electrical

Electrical balance of plant (sometimes called eBOP, or BOPE) incorporates equipment, systems and interconnecting cables to safely and reliably deliver power from the generator to the grid and to supply medium-voltage, low-voltage and DC power to all in-house loads under all service conditions, including normal operation, start-up, and shutdown, and emergency regimes. Major electrical systems, equipment, and protections are designed during the engineering phase of the power plant project and are represented on a special schematic diagram usually called *one-line* or *single-line diagram* (*SLD*). It is a simplified notation used in electrical engineering for representing a three-phase power system: instead of showing each of the three phases with a separate line or terminal, only one line is used, which simplifies the overall diagram and makes it easier to read. This diagram gives information about the circuits and equipment, however, omitting details about exact wiring and operation.

Most of the electrical systems are related to the electrical BOP that serves five major functions:

- Synchronizing generator to the grid frequency and voltage;
- Transferring electricity from the generator terminals to the grid, where electricity is delivered to the consumers. This includes transformations of electricity to the required level of voltage;
- Protecting electrical equipment from the possible damages caused by overcurrent, overvoltage, short circuits, etc.;
- Feeding electricity to all plant systems to ensure their proper functioning;
- Implementing plant control system, data network to ensure safe operation, synchronization, and reaction to the system operator requirements, if necessary.

These functions of the electrical and control systems are implemented through the following equipment:

- *Power evacuation equipment* to supply power from the generator to the grid and to protect the electrical equipment, including:
  - *Busbars* (or *bus bars*), metallic stripes or bars that connect the generator terminals with the major step-up transformer to conduct electric current;
  - *Generator circuit breaker* (*GCB*), which is an electrical switch that is used to protect equipment (generator and step-up transformer) from the short circuit currents, overvoltages and other abnormalities. It is also used for safe maintenance;
  - *Step-up transformer,* to change the voltage of the generated electricity before delivering it to the grid;

- *High voltage switchgear*, including high voltage circuit breaker and control equipment, also used to protect the transformer and electricity distribution equipment from electrical abnormalities;
- Electrical distribution system and line switches, used to safely distribute electricity to electric lines.
- Control and protection equipment, to allow for safe and reliable operation.

- *Auxiliary transformers*, to supply electricity to the plant systems. They transform current and voltage to feed the systems according to their specifications. This is normally carried out through several levels of power supply: the medium voltage system feeds major equipment like big motors, fans, and pumps, while the low voltage is used from other electrical equipment. These systems should always have redundancy to ensure safe and reliable operation;
- *Emergency* and *back-up power supply* through the battery system called *uninterruptible power supply* (*UPS*). This is extremely important for safety-related applications like lube il system, control system, protections, etc.;
- *Protections* include multiple devices, relays, controllers, etc., to ensure safe and reliable operation;
- *Distributed Control system* (*DCS*) of the power plant, which is a digital control system, in which controllers are distributed throughout all major and auxiliary system, and controlled by a central operator station;
- Grounding, lightning arresters and cabling for normal operation of all systems.

Auxiliary power distribution system, which is a major part of the BOP scope, can be considered among the most critical. Its reliability during all regimes of service is crucial to the availability and of the entire power plant unit. Moreover, auxiliary power consumption must be kept to a minimum to increase the plant net efficiency. Therefore, for most of the power plants, electrical BOP is split into several voltage systems with redundancy in supply:

- MV system, including generator, GCB, and step-up transformer with the MV distribution that feeds some of the major equipment like big motors to drive pumps and ventilators (air blowers). These also include step-up and step-down transformers;
- HV system, incorporating an HV substation with HV switchgear and connection to the HV grid;

- LV system, which is used to power all sorts of auxiliary systems, including UPS and some of the smaller motors (for fans, small pumps, etc.);
- DC systems, which power some of the functions of the plant and are used for emergency (lighting, critical pump redundancy, etc.).

In many cases MV and LV systems each can have two separate distribution sections supplying power to the respective auxiliary needs. In case of a failure of one of the sections, the other will be able to redistribute power providing redundancy.

Overall, an auxiliary supply system shall be modular and ensure future extension of the power plant by adding new units (turbines and generators) with their supporting equipment without redeveloping the whole system. This is especially important in case of adding generators with much higher capacities.

9.6.3.1 Circuit breaker

Among most of the electrical BOP components, circuit breakers (or switchgears) represent an expensive and important part of a power plant. They define the overall plant and equipment safety and reliability, and impact significantly the footprint of the plant.

There are two major technologies widely used in the world:

- *Air-insulated switchgear* (*AIS*), also known as a switchyard;
- *Gas-insulated switchgear* (*GIS*). This system has a much smaller footprint but is significantly more expensive and requires indoor installation. The generator circuit breaker is normally of this type, while other breakers (like HV) depend on the plant design and other limitations.

9.6.3.2 Control system

The control system includes the facilities and systems necessary to operate the power plant (including boiler island, turbine island, and BOP). Usually, the operation is carried out from the central control room, providing both automatic and manual control modes. This allows starting the plant in complete automatic mode from the operator station.

The control system also provides asset protection, plant performance optimization, overall reliability and source of data for monitoring purposes. In most cases, a control system has a general platform that comprises several modules linked together. It can include several blocks (arranged in special modules or cabinets) connected to the data highway — a type of local network for control purposes. These cabinets correspond to major and auxiliary equipment and all systems located in a power plant. Basically all control systems can be split into two major groups: process control and rotating machinery control. This network transfers data from instrumentation to the controllers. Unit data highway is also linked to a time server that uses GPS to get the global time signal for synchronization.

Parallel to the unit data highway there is a plant data highway — second network implemented for information exchange purposes. These two networks are interconnected via data gateways.

Major control room accommodates a human–machine interface (HMI) or operator interface, which is connected to both networks and used for operation and control purposes.

References

[1] Dixit U, Hazarika M, Davim J. A brief history of mechanical engineering. Materials forming, machining and tribology. Springer International Publishing. ISBN 9783319429168, 2016.

[2] Hills R. Power from steam: a history of the stationary steam engine. Cambridge University Press. ISBN 9780521458344, 1993.

[3] Leyzerovich A. Steam turbines for modern fossil-fuel power plants. Fairmont Press; 2008.

[4] IEA. Application and development prospects of double-reheat coal-fired power units. IEA Clean Coal Centre 2015.

[5] Reinker JK, Mason PB. Steam turbines for large power applications. GE Power Systems. URL: https://www.ge.com/content/dam/gepower-pgdp/global/en_US/documents/technical/ger/ger-3646d-steam-turbines-large-power-applications.pdf.

[6] Ducker M. The fall of the f-class turbine. Power Engineering; 119.

[7] Siemens. Benson once-through heat recovery steam generators. Siemens AG; 2006.

[8] Breeze P. Coal-fired generation. Elsevier; 2015. 96 pp.

[9] Klempner G, Kerszenbaum I. Handbook of large turbo-generator operation and maintenance. IEEE press series on power engineering. Wiley. ISBN 9781118210406, 2011.

[10] Nag P. Power plant engineering. 3rd ed. Tata McGraw-Hill; 2008.

[11] Wurtz W. Air-cooled condensers eliminate plant water use. URL: http://www.powermag.com/air-cooled-condensers-eliminate-plant-water-use/, 2008.

[12] Niessen WR. Combustion and incineration processes: applications in environmental engineering. 4th ed. CRC Press; 2010. 798 pp.

CHAPTER TEN

Fossil energy economics and project lifecycle

Contents

10.1. Planning and investment decision

The investment decision to build a power plant has to undergo rigorous consideration to answer the major question: will this project be profitable during the anticipated number of years? The idea behind any investment is to earn money or, in other words, to secure proper ROI for an investor, who wants to put money on the table. Power generation projects usually have longer periods of return and are subject to a number of risks, which can certainly be mitigated by various incentives, long-term power sale contracts, or other sale schemes, possibly available on the market.

In order to properly calculate the economics of the project and therefore allocate investment money, in most cases through both debt and equity, there are many critical and industry and market specific questions an investor has to answer. This type of work, usually called a *feasibility study*, is provided by a specialist consulting company that has expertise and resources to perform such a task. The consultant therefore provides a report with the required calculations and reasoning to answer some of the major questions:

Sustainable Power Generation
https://doi.org/10.1016/B978-0-12-817012-0.00021-9

- What is the plant business model (selling base load electricity, operating in a peaking mode, or providing ancillary services)?
- What is the major product to sell (electricity, capacity, emissions allowance, heat)?
- What are the geography and site conditions (is there a potential location, site access, power line, fuel supply, cooling water)?
- What are the required capacity and range of flexibility?
- Which plant design and technology to implement (including the available fuel)?
- What are the economics of the project (CAPEX, OPEX) and ROI?

Preliminary performance calculations and cost estimation are requested from the major equipment manufacturers to define the potential project cost and performances for further calculations. This request, which is sometimes called a *budgetary proposal*, provides a rough estimate of output parameters and pricing of capital costs to estimate the value of the current technology. In general, the up-to-date market data is implemented to do comparison and benchmarking.

There are many value drivers influencing an optimized investment project design of a thermal power plant, including, among other issues:

- Regulatory framework that comprises a tax regime, environmental considerations, emissions limits, local legislation, conditions to obtain permit, and many other sensitive topics;
- Market conditions, which define the electricity price and its evolution, fuel prices, tax regime for electricity suppliers, emissions certificates, competitive landscape, and business models implemented in generation and transmission market;
- Financing scheme based on project financing options, possibilities to borrow money both domestically and internationally, ability to use export funding, etc.;
- People and social aspects, basically personnel costs and availability of skilled workforce for execution of the project and operation of a modern facility. It also covers social acceptance of a project;
- Site conditions, climatic zone, precipitation, availability of water, soil profile, connection to auxiliary needs, and proximity to major transportation routes.

All parameters have to be adequately considered within the feasibility study and therefore require an established process landscape to properly plan and execute the project. While thermal power has been a very established market with a long history, such a framework has been in place for many years. While all of the projects use the same approach for the evaluation, there are certain differences caused by local regulation, codes, standards, financing options, and other requirements. Therefore we consider here a generic approach of a project evaluation, which would give an idea of how to estimate the value of a thermal power plant and make a proper investment decision.

10.1.1 Business model

The choice of a business model depends on the market behavior and regulations. An investor has to understand what is the best way to get money back from the system. This can include multiple options:

- Providing capacity to the market while running a base load power plant with the highest possible capacity factor. Traditionally, base load thermal power plants run about 8000 hours per year for 25–30 years of the plant lifecycle.
- Providing power through the electricity auction (for a certain period of mite defined in the market). This is a semipeak power plant that can provide electricity at a certain period of the day running in a full load operation mode. At other times, instead of shutting the plant down (which is costly and can degrade the equipment), it may run with the lowest possible output and even pay penalties for the excess electricity discharged to the grid. However, the amount of money the plant receives as a payment for semipeak operation is enough to cover all expenses. Such units are normally combined cycle plants with high efficiency and flexibility for partial load and ramp-up timing.
- Operating in a peaking mode for a few dozens of hours a year with an extremely short notice from the grid operator and therefore charging the highest possible price for kilowatt-hour. This requires highly flexible and fast units, usually simple cycle gas turbines.

On top of this, a company can use an efficient modern combined cycle plant with low level of emissions and therefore sell emissions quotas.

All above-mentioned business models require proper understanding of how the market works and how it would work in the coming 20–25 years. The latter is the most important as an investment into a power plant can be considered long-term, and time-related risk mitigation is therefore of vital importance.

10.1.2 Cost of electricity analysis

In order to understand the economics of the project, the investor has to clearly see how long it would take to get the investment money back and what the profit will be. The use of traditional investment analysis tools like NPV is applicable here, however, would require specific calculations related to the cost of electricity.

There are three major approaches to the cost of electricity calculation:

1. Cost of installed capacity, or how much money is paid for the kilowatt of power plant. This refers to the capital expenditures of a power plant and is used to compare capital costs of various technologies;
2. Cost of electricity generation (COE), or how much money is required to produce one kilowatt-hour of electricity with the given power plant. This number requires the above mentioned capital costs, as well as variable cost evaluation, while taking the operating and business model into account;
3. Levelized cost of electricity (LCOE), showing how much money on average is required to generate the kilowatt-hour of electricity over the whole lifecycle of a power plant. Fundamentally, an LCOE is a break-even sales price per kWh needed to justify an investment in a particular power generation facility, which is also used as a convenient tool to compare various power generating technologies.

The investment decision is therefore made on the basis of long-term calculation of the expenses and revenues. The major expenses are therefore the capital costs (distributed for the construction period), operation and maintenance costs (distributed during the whole plant lifecycle), while the revenues come from the sale of electricity.

The cost of installed capacity (capital expenditures or CAPEX) defines the overall project investment (sometimes levelized to kilowatt-hour of production per year) and includes only the project costs without operation. For the indication and preliminary analysis, average industry numbers can be implemented, taking them from any analytical report or a similar project, however, major consultants and owners have a quite accurate number in mind.

Evidently, it is complicated to calculate the exact cost of a power plant without knowing all details and quotes from the manufacturers and construction companies. This is due to many reasons: large amount of equipment supplied by multiple companies (which are subject to their own pricing policies and methodologies), variable labor costs, various regulation in different locations, specific project timeline, volatile oil prices

Table 10.1 CAPEX and OPEX components for various types of thermal power plants.

Technology	Lifetime,[a] years	CAPEX, $/kW	OPEX Fixed, $/kW-year	OPEX Variable, $/MWh
Simple cycle	25	650–800	5.26	29.9
Combined cycle	25	450–1400	6.31	3.67
Combined cycle with CCS	25	3000–3750	18.4	10.0
Pulverized coal-fired	35	1300–3000	23.0	3.71
Pulverized coal-fired with CCS	35	5500–6950	35.2	6.02

Source: Based on [1] and EIC and IIR databases.

[a] Lifetime may vary significantly and can be extended through refurbishment, upgrade, or other life extension procedures. Values given for indicative purposes only.

impacting shipment costs, etc. However, the rough estimation and peer comparison allows for averaged pricing and scenario simulation, providing the minimum and maximum boundaries. In most cases these numbers can be acquired from various market players (OEM and EPC companies) with their budgetary proposals.

Table 10.1 shows average numbers of the capital costs of installed capacity for major types of a thermal power plant. Estimates have been provided in [1] based on the statistics of the industry for reference power plants: 211 MW SCPP, 615 MW CCPP multishaft, and 600 MW single-reheat, respectively.

Remark 10.1. Variable OPEX does not include fuel and other consumables' costs.

It is important to note that CAPEX has a wide range as it depends on many factors: project location, type of gas turbine (e.g., class E, F or H) or coal-fired boiler, overall configuration and scope of power plant, local regulation, and many other factors. In certain cases a power plant CAPEX can be very low, especially when an EPC is an OEM and wants to promote new technology. It is therefore important to build a power plant and have a reference with fired hours in place, while the money is then made out of further unit sales and associated services.

The indicative CAPEX cost structure is represented in Fig. 10.1. The majority of capital costs are related to the expenses for the power island and balance of plant systems, while gas and/or steam turbines occupy up to 20% of costs (for combined cycle). Owner's costs, which can reach up to 17%, include land acquisition, permits for construction and operation,

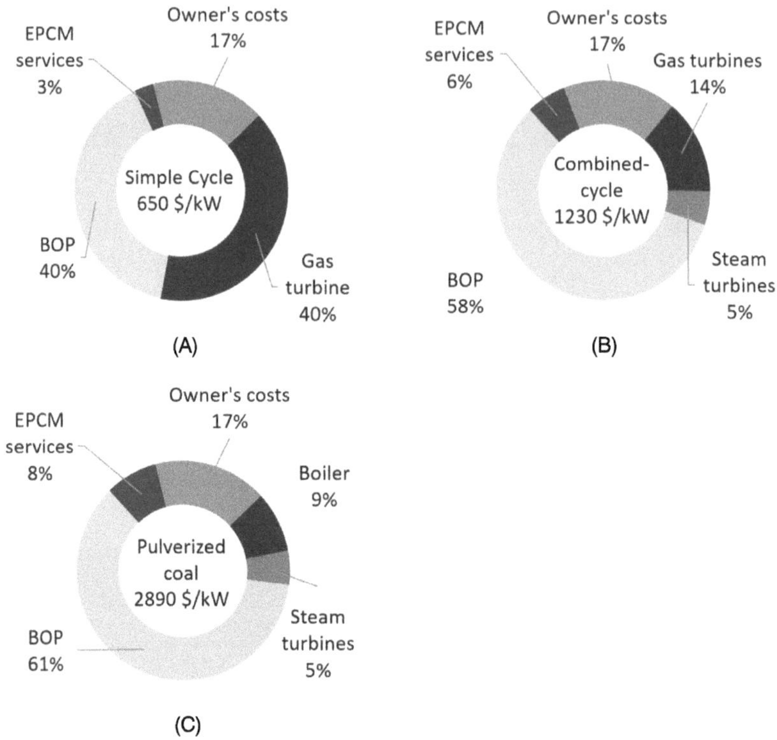

Figure 10.1 ***CAPEX breakdown for thermal power plant.*** Capital cost breakdown for gas thermal power plants: simple cycle (A), combined cycle (B), and pulverized coal-fired steam (C). *Source: Based on [1], EIC and IIR databases and author's estimates.*

connection to the grid, cost of finance, and various administrative costs related, for instance, to fuel supply.

OPEX is spent for operation and maintenance purposes, spare parts, and other related topics.

All these numbers are used to calculate the net present value of the project, which allows evaluating the present cost of the investment and its payback period. The idea is to analyze the expenses and revenues of a thermal power plant according to the construction schedule and operating lifecycle. During the construction period, an investor has to pay to the EPC according to the payment schedule, while the income from the electricity (or heat) sale starts to arrive only after the commissioning of the project.

Basically, an investor has to calculate all cash flows that would take place during the project lifecycle, i.e., for the plant operating years, taking into account risks related to fuel supply costs, electricity sales pricing volatility, financial costs to cover debts, as well as operating expenses and possible failures and outages. Here special attention is paid to servicing plans, especially for gas turbines, as they are more sensitive for maintenance and require some major checks and equipment refurbishments (like combustion systems and blades of the rotors). All this information is plugged into the business model calculations to understand possible risks and try to mitigate them during the planning and negotiation phase.

10.1.3 Site selection

Site selection is an important step in conducting the feasibility study of the project to calculate its profitability and return on investment. Site would significantly impact the schedule of the project and its construction, and partly operation costs. The optimal solution for power plant placement and site arrangement is therefore based on a thorough analysis of economic, social, geographical, and technological, as well as environmental aspects. In most cases each power project belongs to a global power industry development plan or concept within a region or country and would assume connection to the common grid unless operated in island mode.

In some countries there is a power industry strategy for 5–20 years, outlining key direction of demand growth and hence supply. Based on such a conceptual vision, regional and then local approaches are developed for shorter periods of time. This basically defines the amount of capacities required in the coming years with the necessary operating profile (base, semipeak or peak load) and connection to the grid.

The site selection for a thermal power plant is subject to a number of requirements:

- Available land required for construction and potential expansion;
- Geological requirements allowing for construction of a potentially dangerous industrial facility, for instance, proper soil profile to meet the schedule and minimize cost of construction;
- Known seismic conditions, preferably with low activity to minimize technological risks;
- Transportation proximity suitable for delivery of material, especially heavy bulk cargo. In this case, water transport can offer cost advantage;
- Proximity to fuel (coal mine or gas pipeline);
- Availability of water resources;

- Known wind profile to ensure that exhaust gas emissions and possible plume from the cooling tower do not cover residential areas;
- Proximity to existing substation or power line or at least possibility to build a new line at minimum cost.

In any case special attention shall be paid to the ecological requirements and sanitary standards. This is mostly applicable to the coal-fired power plants where solid and gaseous waste treatment is an important topic.

Separately, connection to the existing grid requires a thorough analysis. This is mostly due to the fact that old grids may not have enough capability to transfer power from the newly introduced power plant, especially in case of higher capacities. This study is usually performed either by a grid operator or a separate consultant, who issues a report with the recommendations for the grid upgrade. In the worst case scenario, an investor would have to pay for the new HV transmission lines and substations which will significantly impact the payback period of the project.

10.1.4 Sustainability questions

Recent trends in sustainable development put certain pressure for thermal power plant projects already on the feasibility study phase. This leads to the following questions, which have to be answered in an economic evaluation:

- What is an estimate of the cost impact and efficiency increase if all available technologies to provide "clean" energy are implemented? This would allow making a kind of trade-off between cost of electricity (which shall be low enough to maintain growth of developing countries) and environmental impact.
- What is the amount of CO_2-equivalent emissions that will be released by a power plant? This number would provide a basis for emissions taxes, provisions for future penalties if the regulation becomes more conservative, and allow for better estimation of overall impact;
- What are the project lifecycle and potential upgrades, uprates, and retrofits together with their impact (cost- and performance-wise) to minimize negative effect?

Evidently, fossil fuels and respective power plant projects are vitally important to sustain economic growth now and for the foreseeable future. Further, they must be used to enable the transition to nonfossil alternatives at a pace that will prevent economic disaster. Environmental concerns notwithstanding, fossil fuels and fossil-fired technologies will continue to play a significant, and perhaps dominant, role in the supply of energy for some time to come, due to the current economy [2]:

- Abundance, availability, and proven performance of fossil fuels;
- Existing infrastructure for their production, treatment, transportation, and partly utilization;
- Widespread global use of fossil fuels;
- Continuous development and improvements in thermal power plant technologies towards efficiency increase (gas-fired combined cycle power plants can boast flexibility and extreme levels of efficiency);
- Predictability and therefore straightforward business model of fossil fuels, as well as known performance within various business models: coal-fired for base load with simple and combined cycle being capable of supporting the grid management.

Eventually, a thermal power plant project feasibility study is a complicated investment analysis that has to estimate multiple risks and forecast future (potential) challenges to come up with a reliable business model and approve it for further execution.

10.2. Power plant project execution

When a positive investment decision is taken, the owner is ready to kick-off the power plant project. In an open market, the owner wants to build the most efficient power plant with the highest possible capacity according to the feasibility study, paying as little money as possible. This simple idea, however, cannot be fulfilled due to obvious reasons, therefore the whole project is a tradeoff between various approaches.

Power plant project planning and execution has a number of distinct phases followed by the industry worldwide. These include:
- Bidding stage;
- Detailed engineering and construction stage;
- Commissioning and startup;
- Operation and maintenance.

10.2.1 Bidding process

The feasibility study involved the general definition of a thermal power plant project and the applicable technology, for example, defining the type of a required power plant: simple cycle, combined cycle or coal-fired. Preliminary performance data supplied by the major equipment manufacturers or taken from the market analytics data usually represent some averaged numbers and are not fine-tuned to site specifics. In order to perform a detailed project evaluation and build proper schedule and economics, an investor opens the bidding process with various entry conditions (for example, bidders can be subject to an entry fee). This process is based on certain rules like defined timeline and form and scope of the documents required, with all the guidelines and initial conditions supplied with a document called *request for quote* (RFQ) or *request for proposal* (RFP). Both terms are used in the same way and refer to the documents describing technical and economic requirements to the future power plant an investor would like to build.

The RFQ document is prepared either by the investor (if he/she has the required competence) or by an external consultant specialized in this type of activity (usually in collaboration with the owner). This document is issued in a free format and in most cases includes the following information:

- Contractual, financial, and legal part, usually arranged into *general* and *special terms and conditions*, including but not limited to:
  - Project cost range,
  - Project financial terms (payment, termination, penalties, etc.);
  - Financial data, including major guaranteed parameters;
  - Schedule (delivery, construction, commissioning),
  - Legislation, taxation, termination, delays, force majeure, and other legal consequences of any of the events;
- *Technical requirements*, which usually define:
  - Site information (location, access, conditions);
  - Available resources;
  - Capacity and performance data requirements;
  - High level scope of supply;
  - Regulations and codes and standards;
  - Ecological requirements.

The RFQ document can consist of up to several hundreds of pages, including preliminary drawings, site plans, even some of the detailed design considerations. Approximately half of the information is related to commercial topics, while the rest is covering technical aspects of the project. Theoretically, this type of document does not have any specific details fine-tuned for one or another product in the market, for example, for a single gas turbine. However, this information can be indicated indirectly, for instance, through the required efficiency, number of megawatts per unit, level of emissions, etc.

When the document is issued, the companies that plan to take part in the bidding process to win this project prepare the proposal according to the requirements in one of the forms:

- Free format proposed by the bidder, which can include comments and deviations to the RFQ document if required by regulation (either filled-in table of deviations or redlining of an RFQ);
- Free format of the document with some obligatory forms provided in RFQ.
- Format of the proposal provided in the RFQ and followed by the bidder.

In some countries, the bidders are not allowed to deviate from any of the requirement but can submit a list of clarifications and questions before the final bid and therefore try to justify their deviation. The bidding process may include several stages and may require submitting the following types of proposal:

- Budgetary proposal, when the investor wants to see average pricing for the typical equipment that fulfills basic requirements like performances (for instance, electrical and thermal output). This information can be used to reissue RFP with more details;
- Proposal for complete power plant or for a piece of equipment (e.g., power island or boiler);
- Comments to the RFQ in the form of the list of deviations so that an investor can identify potential pain points before the negotiation process starts;
- Clarifications to the bid.

After submission, the bids are reviewed by the owner side according to set rules and the preferred bidders are selected for further negotiations, clarifications and signature process. The negotiations may take up to several months as both parties have to go through multiple points and come to the mutually agreed solution. Evidently, the investor would try to come up with the most favorable conditions for himself and try to decrease the project financial burden, while the bidder will try to do the same from the opposite side. There can be several negotiations conducted in parallel to choose the best possible bid.

The bidding process can involve multiple stakeholders arranged in a complicated deal structure, and include:

- *Owner*, as a decision maker who is going to build and own a power plant. This is usually a utility company or a power generator (as defined in Section 3.4);

- *Engineering consultant*, typically a company that has the engineering and commercial capability to do the techno-economic analysis and support the owner in preparing the RFQ documents. This consultant is usually one of the established consulting companies well-known in the industry. In certain cases the owner has enough expertise to carry out this task on his own.
- *Bid* or *Legal Consultant*, that is, a company supporting either party during the preparation of the terms and conditions of the contract, and their negotiation. Such a company should have enough expertise in power generation contracts as they include not only legal aspects of the deal but are also closely linked to the technology behind;
- *Engineering, Procurement, and Construction Company*, or an *EPC*, that is, a company, which would do the overall engineering of the plant, procurement of material required, and organize and execute construction process on site till the plant is fully commissioned. This company would act as a project manager. This, however, does not mean that this company is going to carry out all activities on its own, so that usually certain tasks are outsourced to third parties, still keeping the control function.
- *Designer* or *Design Institute*, which is an organization capable of doing detailed design of the power plant. This company is normally hired by an EPC or the owner to support the project. The key criteria for choosing a designing company are its knowledge of local regulation, codes, and standards for design, and reference projects prepared before.
- *Original Equipment Manufacturer*, or an *OEM*, which is basically any manufacturer producing equipment for power plants. However, in most cases these are the suppliers of major equipment like gas and steam turbines, generators, boilers, big pumps, transformers, control system, etc., as such equipment would require a separate complicated contract (or probably a separate bidding process) due to its considerable significance and hence price.
- *Contractors* and *Subcontractors* are various third parties involved in the execution of the project. They would do all kinds of work, including but not limited to design, construction, supply of material, logistics, supply of personnel or tools, etc.

The deal structure of the bid depends on many factors, including the availability of technology on the market, capabilities and competence of the owner, capacity of the project, and so on. In the simplest case, the owner signs the contract with an EPC company on a turnkey basis so that the EPC acts as a prime contractor and supplies design, all material, and construction services. In turn, EPC then has to open its own bid to buy equipment on the market, so that the OEM and other contractors will be called *sub to prime*. In some cases EPC and one of the technology providers make a consortium and act as one single legal entity with the partnership agreement behind, each responsible for his own scope, costs, terms, and conditions. But in front of the owner they behave as one team and represent one project.

In other cases the owner can act as an EPC and does not really open any bid, but acts on a "*build, own, operate*" (BOO) basis. But it still has to open bids and buy major and auxiliary equipment on the market.

When the proposal is signed by both parties — the owner and prime contractor — it has a clear indication of the project scope, price, execution schedule and other deliverables. Within the next several days (or weeks) the investor transfers the first payment to the prime contractor, and this denotes the official project start and is sometimes called *notice to proceed* (NTP).

10.2.2 Project schedule

The speed to market, or the timing between the investment decision and commercial operation date (COD) is becoming more important to the investor, as it drives the whole value game. The overall project schedule is therefore required to be as condensed as possible, with a lot of works being conducted in parallel. While modern power plant equipment is getting larger and more sophisticated, it can require longer lead times for design, manufacturing, delivery, and installation, and therefore has to be fitted to the overall construction schedule. Moreover, as the major equipment is supplied from a few countries, in some areas it can be subject to multiple timing restrictions like customs clearance, transportation windows (e.g., due to frozen rivers), climatic conditions and local routs that require feasibility study upfront. Hence, the overall project schedule has many variables to consider.

The duration of a project from the design till the COD depends on many circumstances but is always influenced by the long lead items: steam and gas turbines, boiler, massive concrete structures (chimney, cooling tower), and the overall size of the plant. In simple cases, when the equipment is typical and highly modular, the schedule can be compressed to 12–18 months; however, for large coal-fired power plants with a specifically designed steam turbine it can easily last 5 years.

The schedule usually comprises several blocks: plant engineering, site and civil works, major equipment for boiler and power island, and mechanical, process, and electrical systems within the balance of plant. As shown in the example, many processes are run in parallel, for example, the site works start with the site preparation, which is conducted alongside with the design. However, civil works for major equipment foundations require at least basic design of this equipment and therefore start only after the load drawings are released.

Also, by the time the equipment is manufactured and shipped to site, the foundations have to be finalized and ready for installation (sometimes even with other requirements like minimum allowable temperature, completed building, etc.). While the shipping option is defined during the bidding phase, its pricing is included into the contract, and the timing of shipment is taken into account within the proposed and signed COD date.

To sum up, proper scheduling requires detailed understanding of the overall design, manufacturing and construction process and is completed only during the project execution phase. A major milestone of the schedule is COD, when the plant has to be connected to the grid to commercially deliver electricity and/or heat to the consumer. If the deadline is missed, the contractor has to pay delay liquidated damages — if this special type of fee defined in the contract.

10.2.3 Engineering, construction, and commissioning

The conceptual design prepared during the bidding phase usually defines major performance characteristics of the power plant and key equipment, but does not specify every detail of the whole plant. When the execution starts, the major designer of the project (either an EPC or a special design company) works on the detailed engineering, including:

- Equipment arrangement and plant detailed layout;
- 3D structural design with foundations, primary and secondary structures, equipment allocation, and critical piping and cable tray routing;
- Civil drawings with major loads, dimensions, and fixings for structures and equipment;
- Process and instrumentation diagrams;
- Other specifications for subcontractors and small equipment manufacturers.

These major specifications and design are the basis for the construction drawings, bills of material for procurement department, and purchase orders for outsourcers' equipment. There is no company on the market that can do all the manufacturing and construction on its own: even for design purposes, big EPCs outsource part of their scope to local engineering

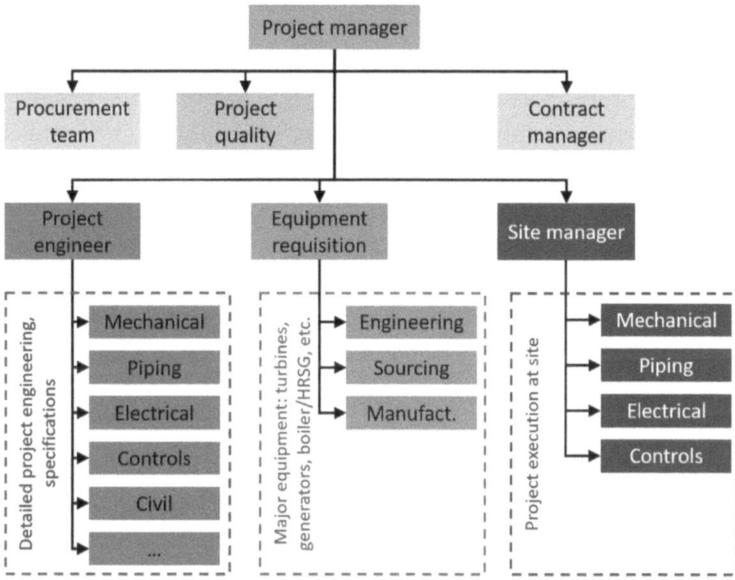

Figure 10.2 ***Simplified project team.*** An organizational chart of the project team is led by a project manager and includes site manager, engineering team, quality manager, and other necessary functions.

companies, for instance, to fit the international design for local codes and standards. Big items, such as cooling towers, are usually designed by the respective companies based on the site limitations and terminal points data provided by the EPC.

Execution of a project is coordinated by a project manager from the major contractor team or, if required, by a special *project management consultant* (PMC) of the owner. This person is responsible for the whole project cycle and has to coordinate all activities so that the project is executed according to the contract terms and conditions, within the budget and following the schedule. Evidently, there is a whole project execution team dealing with various aspects of the power plant, as illustrated schematically in Fig. 10.2. This team involves the following stakeholders:

- Project engineering team, which is responsible for detailed project design and technical specifications of the plant, covering major mechanical and electrical systems, civil and structural works, as well as some of the environmental topics;
- Procurement team, dealing with purchasing and supplying the required material, working on the terms and conditions of this delivery, choosing the right source, etc.;

- Equipment requisition team, which covers topics related to major equipment, namely gas and steam turbines, generators, boilers, or HRSG. This team is dealing with the engineering topics, manufacturing of equipment, proper sourcing and delivery at site, and installation;
- Quality controller, who ensures that the overall project is managed according to the quality standards of the prime contractor. This includes quality of material, scope of works, minimization of waste, optimization of costs and processes, etc.;
- Contract manager, who is covering the contract terms and conditions to ensure the works are executed accordingly. More importantly, this person moderates situations when these terms are breached or there is a risk of such an event;
- Site manager, who is the most important person on site as he manages the whole construction process and has to take major decisions. He is responsible for timely and safe execution, and therefore has a *health and safety manager* in his team.

The construction works are carried out in batches, some of them running in parallel:

- Site allocation with fencing, establishing the guards and access roads, supply of auxiliary power for construction purposes;
- Site clearance, especially important in case of an old construction site, and may include demolishing of old buildings and structures, breaking old concrete, cutting trees, removing excess soil and various waste. Importantly, an ecological investigation has to be performed to ensure there is no chemical or radioactive pollution on site;
- Leveling and metering of the construction area to properly align structures and remove excess soil;
- Excavations of soil and piling for foundations structures according to the civil design;
- Major civil works to complete foundations for main and auxiliary buildings, erect primary steel structures for turbine hall (in case of indoor configuration) and other facilities. These are often completed by a separate civil and structural contractor;
- Erection of large concrete structures like cooling towers or exhaust chimneys (by separate contractors), erection of boilers out of the modules shipped directly to site;
- Installation of primary (turbines and generators) and secondary equipment (pumps, valves, vessels, heat exchangers, etc.);

- Mechanical and process installation;
- Electrical and instrumentation installation.

Each task can be carried out by a specialized subcontractor according to the prime contractor's specifications. These subcontractors, usually smaller local players, take part in the bidding process for each work and then execute their assignments according to the required quality standards and bid contractual requirements. While the site manager coordinates all works, the contract manager and procurement are dealing with the commercial and financial side.

Commissioning of the power plant means that the plant is getting prepared for commercial operation and checked if it has been built according to all contractual requirements. Basically, this process starts with commissioning of each piece of equipment supplied to site and their interconnection, allowing for guarantee tests at the end. Followed by the startup procedure, it is assumed that the conduction of all inspections and quality checks of the built structures and the installed equipment is successfully completed. This is followed by necessary tests of the systems and equipment and confirm that the plant meets the guarantees defined in the contract. The guarantees for a thermal power plant are usually split into two major types: performance and emissions.

10.2.3.1 Performance guarantees

Performance guarantees define the major parameters of the power plant, in a simple case the amount of energy it has to deliver. These can be:

- *Gross*, which do not include any kind of auxiliary consumption by a power plant for its own needs;
- *Net*, which define the final parameters the customer would get.

The owner is always willing to get the net values of the performances to understand how much power he/she will get for sale and hence calculate the income from the sale. This is, however, possible only in case of a turnkey plant when the guarantee provider is responsible for the whole scope and hence can precisely calculate auxiliary consumption of power for the needs of the plant.

Contrary to the turnkey contract, the equipment supply contract guarantees are usually defined on a gross basis so that the EPC, responsible for the plant, would then use these guarantees of the equipment in his own contract to calculate net performances of the whole power plant and provide overall net guarantees to the owner.

There are multiple approaches to define guarantees and specify them within the contract. We name here some of the typical cases:

- Supply of gas turbine. Guarantees would include gross electrical output at the generator bars, heat rate of the gas turbine, and in some cases the flow rate of the exhaust gases, their temperature and maybe energy. The latter three are required for the boiler to calculate all parameters. In some cases start-up and ramp-up can be required by a client to use the machine for semi- and peak load business model;
- Supply of steam turbine. Gross electrical output of the generator and certain parameters of steam in case of extraction (however, subject to the incoming steam parameters only);
- Supply of a combined cycle power island. Gross electrical output of the generator or generators (depending on the configuration), and steam parameters in case of extraction. In certain cases estimation or even guarantee of the auxiliary consumption of the supplied equipment is provided;
- Boiler only. Parameters of steam (pressure, temperature, and flow rate) delivered by this boiler, with the calculations based on the certain fuel quality and conditions;
- Turnkey power plant. Net electrical and thermal output (in case of steam or hot water production), startup time, etc.

Evidently, the values of the guarantees are calculated based on certain assumptions which have to be clearly mentioned within the contract. These may include type and chemical composition of fuel used for the calculation, quality of the water for steam production, test procedures, and parties responsible for carrying out the tests, and many others.

Remark 10.2. The above mentioned guarantees do not cover all possible situations and shall be treated as typical and simplified cases. Performance guarantees is a very complicated topic involving many specialists from both engineering and commercial sides (as they are normally included in a commercial part and provide basis for the penalties calculation), company know-how, optimized configuration of the equipment, and customer requirements.

10.2.3.2 Emissions guarantees

Emissions guarantees are an essential category of the guaranteed parameters stated clearly in the contract. They are also subject to financial penalties unless met within the predefined period of time.

Emissions guarantees are usually split into two major groups. The first defines the maximum (or average) allowable limits for the GHG emissions, particulates in the exhaust air, content of ash and dust, etc. These numbers are ensured through the proper choice of the emissions control equipment or combustion technology.

The second group refers to the noise guarantees and defines the maximum (or average) allowed level of noise around the equipment at working places, as well as around the whole power plant.

Both types of emissions are limited by the local regulation of the country (or region), where the plant is going to be built. Similarly to performance guarantees, emissions guarantees' description is also followed by clear definition of the terms and conditions when, where, and how they should be calculated.

10.2.3.3 Commercial operation date

After the performances and emissions are confirmed to meet the contractual requirements, the plant is prepared for the commercial operation. There may be other guarantees as well, including the time the power plant increases its load (ramp-up time), reliability and availability of the equipment, etc. Only after all of them are confirmed according to the procedures defined by the contract, or at least all the conditions are met, the plant goes through the test run and is finally put into commercial operation. Such a date, usually referred to as *COD* (commercial operation date), is the official start of the plant to generate electricity and deliver it to the consumer provided that all contractual requirements are met.

10.3. Efficiency increase and safe operation

In most cases it pays back to upgrade coal and/or gas-fired power plants: the benefits can include not only increased output in megawatts, but also reduced emissions, more flexible operation and better integration to the grid and hence better collaboration with renewables, and finally big savings over the lifecycle.

State-of-the-art fossil-fired technologies have to operate efficiently and provide flexibility. This is ensured through proper design and construction, as well as maintenance and timely upgrades.

10.3.1 General considerations

Usually an operation and maintenance program is developed for each power plant based on the specific needs of the customer and starts with the power plant developers and owners to identify their primary goals for achieving better results. These services, among other things, can include:

- Daily operation and maintenance;
- Providing maintenance personnel and tailor-made training programs;
- Planned and unplanned services, including spare parts;
- Monitoring and diagnostics of systems and equipment, both on-site and remote;
- Health, safety, and environmental training and programs;
- Fuel and consumables management.

All maintenance programs are aimed at minimizing outage intervals and optimizing schedules, which are influenced by the contractual guarantees (performance, emissions, etc.) to support customers and meet their financial expectations.

The actual performance of a plant compared to its design and "as-commissioned" performance is crucial. As equipment wears, fouls, corrodes, distorts, and leaks, as sensors and instrumentation fail, and as calibrations drift, the plant tends to become less efficient. As well as ensuring integrity, a key requirement of plant maintenance is to maintain peak efficiency. Improved maintenance and component replacement and upgrading can reduce energy losses [3].

In practice, any poorly performing auxiliary equipment or individual components (e.g., fans, pumps, heat exchangers, vent and isolation valves, gearboxes, leaking flanges, and even missing or inadequate insulation) contribute to the overall deterioration of plant performance over time, compounding the effects of deterioration in major components, such as the steam turbine. Significant deterioration can also occur in the steam turbine condenser or cooling-water system, where progressive increases in air ingress and steam and water-side fouling or corrosion can degrade heat transfer. Cooling tower performance is an important consideration in this respect [3].

To attain maximum plant availability and equipment reliability, and to reduce forced outages, operators, together with the equipment and system providers, develop long-term custom maintenance programs. These programs usually focus on the following maintenance approaches [4]:

- *Corrective maintenance* also called "crisis management", focusing on something after it has failed. Corrective maintenance is marked by a high percentage of *unplanned* maintenance activities, high replacement-part inventories, and inefficient utilization of maintenance personnel, but it is sometimes the only choice for certain pieces of equipment;

- *Preventive* or scheduled maintenance, which represents planned maintenance activities to preclude failure. Periodic inspection and maintenance are scheduled at prescribed intervals in an attempt to reduce or eliminate equipment failures. Depending on the intervals set, this can represent a significant increase in inspections and routine maintenance; however, it should also reduce the frequency and seriousness of unplanned machine failures.
- *Predictive maintenance* also called "condition-based" maintenance, replacing the arbitrary time intervals with maintenance scheduled only when the condition of the equipment warrants it. The increase in cost of equipment monitoring should be more than offset by a reduction in unnecessary PM inspections and unplanned machine failures. But it takes time and money to properly perform PdM maintenance, and not every piece of equipment is worth that investment.

In a perfect world, there shall be no corrective maintenance. In reality, the amount of corrective maintenance procedures shall be minimized through a proper combination of predictive and preventive maintenance procedures, and, ideally, high quality of the equipment and construction services, competence of the personnel and absence of force-majeure situations.

Ideally, operation and maintenance program should include continuous training of the personnel. Most of the EPC companies provide the O&M training package together with the construction contract. The cost of this training is normally included in the overall plant (or equipment) contract, while the training is usually a "must have" option. This training is proposed and carried out by the respective OEM for critical equipment and plant engineering and is specially designed for operation and maintenance purposes. The core of this O&M training is focused on the key equipment: gas turbines, steam turbines, combined cycle performance, water and steam cycle, boiler operation, DCS, overall control room for operators, other instrumentation. Training courses usually integrate a combination of classroom training and associated system walk on site. The goal is to provide the understanding of the major equipment, customized to each customer's specific site.

The training is usually conducted along with the commissioning period in order to get plant personnel ready for startup and operation. Depending on the role of an operator or maintenance specialist, training shall be fine-tuned to meet its specific needs. Complementary to the training, the personnel gets the O&M manuals (can be both paper and electronic copies), usually in English or in the local language of the country where the plant is commissioned.

The O&M cost is an important part of a power plant economics along with the capital, financial, and fuel costs. These have to be balanced against lifecycle profitability to finally generate profit as discussed in the beginning of this chapter. These ongoing costs (both fixed and variable) include maintenance, labor costs, site and other equipment management, EHS procedures, and some other tasks. Moreover, this optimization also requires proper planning of the materials and spare parts required for O&M purposes.

10.3.2 Power plant life extension

The decision for a power plant life extension is based on the careful assessment of the condition of all major components and systems. The analysis of all available information on operating history and performance, scheduled outages and emergency shutdowns, as well as its maintenance history play an important part in this. The evaluation of a power plant remaining lifetime is usually conducted by special companies, which are capable of providing lifetime extension strategies based on plant-specific investigations and calculations both for separate components and for the whole plant. These may include material test, analysis of pressurized components such as boiler parts, piping, and turbines, specific measurements of generator parameters, chemical analysis of fluids, etc.

It can be considered as a standard procedure to extend the life of a power plant up to 40 years, and some units have operated for more than 50 years. This is usually achieved by refurbishing major equipment, especially boiler hot gas parts, upgrading the turbines (both gas and steam) to increase their efficiency by minimizing losses, and adding flue gas treatment systems to meet new emission regulations. Life extension is often possible due to the conservative nature of the original plant design and the fact that only a relatively small number of the components are life limited [5].

While for older and smaller power plants, life extension would be disproportionately costly and therefore economically not interesting, the practice or life extension is more commonly applied to larger power plants, mostly coal-fired. Moreover, life extension programs are more feasible and important in countries where the construction of new plant encounters obstacles due to environmental and economic factors [5]. The decision about plant life extension depends on many factors, for instance, local regulations, availability of technology, access to cheaper alternative fuels, etc.

Local regulation against GHG emissions can be among the major reasons of power plant upgrade and its life extension. While installing flue gas treatment equipment to cut emissions of NO_x, CO_2, SO_2, and ash particles, it also involves measures to improve the overall plant efficiency, which would lead to emissions cut-offs as well, e.g., by improving the efficiency of fossil fuel use. It is especially important as coal-fired power plants account for about 38% of power generated in the world [5]. It is estimated that the worldwide average efficiency of coal power stations is below 32%, while new units using supercritical steam conditions can reach headline thermal efficiencies greater than 46%. Improvements in operating efficiencies following a life extension have the potential to give major reductions in greenhouse gases.

Condition assessments and remaining life assessments are prerequisites to a comprehensive plant upgrade [5,3]. This involves visual, material, and non-destructive inspections of the critical parts of the equipment, especially those exposed to high temperatures or extreme conditions and chemicals (e.g., boiler and turbine parts), study of plant operating records, and major revision of the plant design. Depending on the results of these studies, several options are available:

- Changing the operating profile of the plant to minimize the impact on the equipment;
- Upgrading only major equipment parts;
- Retrofitting only emissions treatment systems;
- Converting power plants into other types of fuel, for example, from pulverized coal to natural gas;
- Upgrading all steam systems to increase steam parameters and therefore boost efficiency;
- Shutting down the plant and building a new one (some systems like cooling loop, fuel preparation or water treatment facility can still be reused).

A lifetime extension study may also include considerations for potential improvements of plant operation and its efficiency increase by minor upgrades or retrofit, which are especially important for gas turbines. In these particular cases, gas turbines can be replaced in "flange-to-flange" way, where the old unit is removed and replaced with the new one, while auxiliary systems may remain.

A retrofit project can be a reasonable solution compared to the greenfield construction or a major upgrade due to a number of reasons:

- Lower CAPEX due and reasonable OPEX;
- Limited investment on additional infrastructure (e.g., grid upgrade or extra connection);
- Shorter project cycle;
- Increase of existing capacities;
- Decrease of GHG emissions due to optimized performance of the units and new environmental equipment;
- Much easier project management activities (e.g., easier permit for construction and operation);
- Much easier acquisition of environment permits and persuading the negative public opinion by utilizing existing sites and facilities.

In case major maintenance or upgrade is not economically or technically feasible, other options like shutdown costs and further replacement with a new power plant have to be assessed. While the construction of a new plant is normally more expensive than an upgrade project, there is still a possibility to get certain incentives by building a cleaner plant or replacing the old conventional coal-fired plant with the green option. It important to mention that using new renewable energy technology to compensate the loss of hundreds of megawatts of base or semipeak capacity can still be a big challenge.

10.3.3 Reasons for repowering and upgrades

Base load thermal power plants are usually designed for 25–35 years of service, which corresponds to around 200 000 operating hours. During this period or at the end of its design lifetime, it shall be decided whether the plant continues its normal operation, undergoes lifetime extension, major upgrade, retrofit, or is subject to shutdown. This decision is made upon thorough economic evaluation of operation and maintenance costs, calculation of ROI in case of lifetime extension allowing for 10–20 years of further operation.

During these years of operation, a power plant shall meet the highest possible efficiency, which means careful use of fuel and other consumables (like lube oil, spare parts, etc.), minimized consumption of power for auxiliary needs, and highest possible output to the consumer of electricity and heat. High availability of all equipment stands for shorter outage intervals, which requires proper and timely maintenance procedures and activities supported by skilled workforce.

Therefore, the whole economic performance of a power plant, or, in other words, its profitability, has a strong focus on its operation and maintenance costs, which are driven by the initial design and scope of plant, as well as the established procedures and activities.

Usually older plants are designed and built to accommodate lower steam temperatures up to 500°C compared to USC designs available now. Secondly, major rotating equipment — steam turbines — are usually completely overhauled every 20–25 years. These upgrades can allow power plants increase their output without burning more fuel, i.e., it is all about higher efficiency. Moreover, freshly upgraded plants can boast shorter start-up and ramp-up time to comply with recent requirements of the grid that has been continuously accommodating new renewable energy sources like wind and solar.

The current potential for efficiency gains is tremendous. However, it is mostly driven by economics and legislation: the cost of upgrade and outage of the plant has to be taken finally by customers, which may not be acceptable in developing regions or where economic conditions are not stable enough to plan for more than 3–5 years ahead.

10.4. What is next with fossil power generation?

Coal-fired power plants, also known in the public as power stations, provide a bit less than 40% of global electricity supply. At the same time, these plants account for up to 30% of global carbon dioxide emissions [6], which includes all sources of pollution and is not limited to power generation. Thermal power generation in general, and coal-fired in particular, is likely to remain a key component of the fuel mix for power generation to meet electricity demand, especially in developing countries. To maximize the utility of coal use in power generation, plant efficiency is an important performance parameter. Efficiency improvements have several benefits [6]:

- Prolonging the life of coal reserves and resources by reducing consumption;
- Reducing emissions of carbon dioxide and other pollutants;
- Increasing the power output from a given size of unit; and
- Potentially reducing operating costs.

The implementation of CCS technologies, along with the proper coal pre-treatment, should likely contribute to emissions reduction and efficiency increase, which would still be a tradeoff with cost of electricity.

Worldwide, different scenarios have been proposed on how to meet the fast growing energy needs. However, depending on the jurisdictions and the environmental and economic policies, different paths will emerge. So far only coal can meet these energy demands in the immediate future [7].

Evidently, the future of coal-fired power generation (and partly gas-fired) lies in the hands of world politicians, who need to shape policies for the regulation of carbon emissions and provide a solid climate change framework to encourage carbon-free energy generation [7]. Thermal power generation still possesses a number of obvious advantages, while most of the transition scenarios and public opinion do not really have any viable and fully justified way to replace coal. Most of the claims to shut down coal-fired plants (and gas-fired with them, however, such plants enjoy higher degree of acceptance) and replace them with wind and solar panels do not have real calculations and are simply not possible to bring into life and, on top, maintain growth.

It is important to mention that in most cases thermal energy assumes only coal-fired generation, which is a large, yet not the only, type of conventional thermal power plant that uses combustion of fossil fuels. While coal-fired generation is by far larger in terms of installed capacities, gas-fired power generation, and especially those driven by gas turbines (and steam turbines in combined cycles), offer much lower CO_2-equivalent emissions and much better flexibility. These units are usually seen as a major connection between the current energy system and the sustainable energy system of the future, which will be mostly driven by wind and solar energy sources. The latter have to be backed up with reliable, flexible, and established high-efficiency capacities that may be well represented by combined cycle plants. Moreover, future trend towards distributed power, when multiple local renewable energy source have to be supported by reliable fast power for back-up, shows that the game of small to medium size gas turbines is not yet over.

References

[1] Black & Veatch. Cost and performance data for power generation technologies; 2012.

[2] Tester J, Drake E, Driscoll M, Golay M, Peters W. Sustainable energy: choosing among options. MIT Press. ISBN 9780262304634, 2012.

[3] International Energy Agency (IEA) Coal Industry Advisory Board. Power generation from coal: measuring and reporting efficiency performance and CO_2 emissions. OECD/IEA; 2010.
[4] Peltier R. Competitive maintenance strategies. URL: http://www.powermag.com/competitive-maintenance-strategies/, 2010.
[5] Ambrosini R, ICRCC Centre. Life extension of coal-fired power plants. International Energy Agency Coal Research. ISBN 9789290294221, 2005.
[6] Board ICIA. Power generation from coal: measuring and reporting efficiency performance and co_2 emissions. OECD/IEA; 2010. URL: https://books.google.ch/books?id=cqAEoAEACAAJ.
[7] Letcher TM. Future energy: improved, sustainable and clean options for our planet. Elsevier Science. ISBN 9780080994222, 2014.

PART FOUR

Past, present, and future of sustainable nuclear power

Overview

Civil nuclear technology has emerged from the military projects and proved capable of delivering large quantities of reliable base load electricity when used with consideration and care. It offers potential environmental benefits and becomes even more valued in the face of replacing fossil-burning technologies. Recent developments have shown that new approaches to nuclear technology can extend the potential energy reserves to more than a thousand of years.

However, nuclear energy is still considered dangerous when operated carelessly. Many countries may be unwilling to deploy nuclear power plants and others prefer to shut them down. However, nuclear energy still remains a very promising source of relatively green power and can play a crucial role in sustainable future. In this part we direct our attention to the nuclear basics and currently available technologies, covering a few critical issues: safety, project economics, and decommissioning of old plants.

CHAPTER ELEVEN

Nuclear energy

Contents

11.1. Nuclear energy and fission reaction

The principles of a thermal power plant, described in detail in the previous chapters, are based on the energy chain transformation that starts with the fossil fuels. These fuels can be burnt with a considerable amount of heat released during the chemical reaction. This thermal energy is further used in the steam cycle. Evidently, the chemical energy from the fossil fuels is not the only source of primary energy that can be converted into heat and thus run a similar steam cycle.

The advances in nuclear physics in the 19th and the 20th centuries have brought an understanding that certain minerals can emit radioactivity — another type of energy, which can be utilized in the conversion process. When the possibility of nuclear energy was first recognized in the 1930s and early 1940s, it had the attraction of offering very large amounts of energy from very small amounts of material [1]. Thanks to the development of the nuclear military technology, starting in the 1940s, civil nuclear energy use can already claim more than half of a century history and take an honorary place in the gallery of traditional energy technologies along with thermal and hydro. Theoretically, the two major nuclear possibilities have been considered in detail: fission and fusion. The latter would be inexhaustible for all practical purposes, but developing an effective fusion

Sustainable Power Generation
https://doi.org/10.1016/B978-0-12-817012-0.00023-2

system still remains an uncertain hope [1]. At present, a fission reaction is the only proven and economically feasible alternative for a civil nuclear technology.

Nuclear energy is the internal energy of the nuclear kernels, which is released under certain nuclear transformations. In nature, some heavy elements can emit radioactivity on their own meaning that there are huge amounts of energy stored inside their kernels. This effect is known as *radioactive decay*, which is not that common in nature. Moreover, the decay rate of this reaction is quite slow and of low efficiency. Therefore, it was the discovery of a self-sustained reaction that made it possible to increase the efficiency of energy release and utilize it for practical purposes.

Each element has a number of neutrons and protons. While protons define the nucleus charge, neutrons somehow hold the positive protons within the kernel. However, if the number of neutrons is too high, then the nucleus can be unstable enough and subject to changes if exposed to the impact, and therefore is susceptible to fission. When a neutron hits the nucleus, it becomes unstable and therefore can split into two fission fragments. During this process, a few neutrons are emitted, accompanied with considerable energy release. This means that using only one neutron we can produce enough energy with extra 2–3 neutrons. These extra neutrons would in turn affect other nuclei and sustain the reaction, leading to the *sustained fission reaction*.

This sustained reaction releases considerable number of neutrons and amount of energy in the form of heat in a very short period. If not controlled (through cooling and absorbing extra neutrons), it would lead to a nuclear explosion — a form of *uncontrolled self-sustained nuclear reaction*.

Not all fissionable isotopes can sustain a chain reaction. *Fissile elements* like uranium-233, uranium-235, and plutonium-239 possess this feature and can undergo a self-sustained reaction, others like uranium-238 is fissionable but not fissile: it undergoes induced fission when impacted by a neutron of high kinetic energy. However, the products of this reaction are mostly low energy neutrons and therefore the chain reaction is not sustainable with this isotope.

Conventional nuclear fuel consists of U-235 (approximately 4.5%) and U-238 (which is by far the largest component with 95.5%). This fuel can undergo a self-sustained chain fission reaction, as schematically illustrated in Fig. 11.1:

- When a neutron hits the fissile U-235, it undergoes fission reaction and breaks into two fission products with the release of a large amount

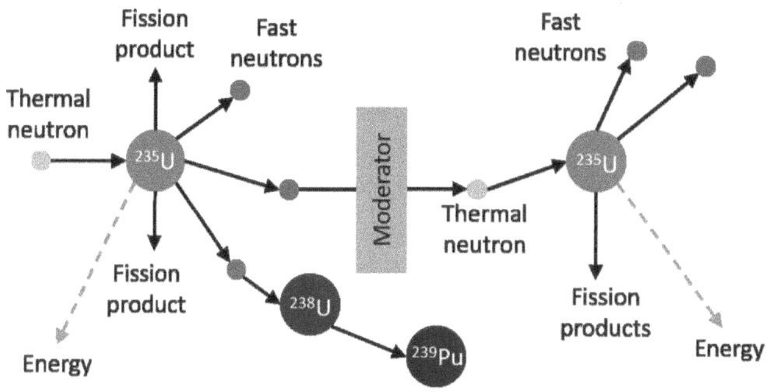

Figure 11.1 ***Schematic of the nuclear fission reaction.*** Nuclear fission chain reaction assumes that a thermal neutron hits the fissile U-235 to produce several fast neutrons and fission products with the release of energy. The moderator serves to slow down the neutrons (make them "thermal") so that they can sustain the reaction.

of energy. Moreover, during this reaction 2–3 neutrons (around 2.5 on average) are released.

- Some of the neutrons can be absorbed by the matter (for example, structures) or leave to the outside.
- Some neutrons will be absorbed by U-238 (which is still part of the fuel) and turn it into plutonium-239, which is also fissile;
- At least one neutron has to be left to be captured by the fissile material to keep the reaction going.

The neutrons released in fission reaction are fast and possess higher energies, between 0.1 and 10 MeV, and are therefore called *fast neutrons. Thermal neutrons*, which possess lower energies, are more efficient in reacting with the nucleus of the fissile matter. Therefore, fast neutrons have to slow down to sustain the reaction. This can be achieved by letting the neutrons diffuse through a substance of low atomic weight, called *moderator*. Neutrons transfer their energy to the nuclei of this matter through a series of elastic collisions, and therefore decrease their energy to thermal values [2].

Plutonium, which is not found in natural ores, can be created from the abundant uranium isotope U-238 through a series of reactions

$$n + {}^{238}\mathrm{U}_{92} \rightarrow {}^{239}\mathrm{U}_{92} \rightarrow {}^{239}\mathrm{Np}_{93} \rightarrow {}^{239}\mathrm{Pu}_{94},$$

as described above. The importance of these reactions is that plutonium-239 easily undergoes fission induced by thermal neutrons. Therefore, both isotopes of uranium can be used in the process of fission induced by slow neutrons: ^{235}U can be used directly and ^{238}U can be converted into fissile ^{239}Pu.

Remark 11.1. Nuclei that are not fissionable by thermal neutrons, but can be transmitted into fissile material by neutron capture like uranium-238, are called *fertile*.

The above described process gives the idea of a *thermal nuclear reactor*, which uses thermal neutrons with lower energies to maintain the chain reaction. An alternative to the thermal reactor strategy is to use primarily fast neutrons to sustain a fission chain reaction. Various fuels and combinations of fuels can provide the required self-sustaining reaction with fast neutrons. For example, highly enriched uranium with more than 20% of U-235 can be successfully used because at these concentrations of ^{235}U, fissions are sufficient to sustain the chain reaction despite the smaller probability of fission with fast neutrons. A more efficient fuel is fissile plutonium or a mixture of uranium and plutonium. Uranium-238, which makes up about 99.3% of natural uranium, is not fissionable by slow neutrons, but can produce energy via fast-neutron fission.

Thus, the physical principles that lie behind the self-sustained chain reaction give a clear split of technologies into two major types: thermal and fast neutron.

11.2. Sustainable nuclear fuel cycle

The *nuclear fuel cycle* is the progression of steps in the utilization of fissile materials, from the initial mining of uranium through the final disposition of the material removed from the reactor. It is called a cycle because in the general case, some of the material taken from a reactor may be used again, or recycled [1].

11.2.1 Types of fuel cycle

Nuclear fuel cycles differ in the nature of the fuel used, it's history in the reactor, and the approach in handling the fuel that is removed from the reactor at the end of the fuel's useful life, usually called *spent fuel* [1].

There are three broad categories of fuel cycle implemented commercially in various countries [1–4]. These cycles have different degree of sustainability and, starting from the least sustainable, are as follows:

1. *Once-through fuel cycle.* Sometimes called an *open cycle* or a *throw-away cycle*, it assumes that spent fuel is treated as waste when it is removed from the reactor. All fission reaction products and the rest of uranium are part of this waste, which is then buried in a waste storage.

2. *Reprocessing fuel cycle.* In the currently employed reprocessing cycle, plutonium and uranium are chemically extracted from the spent fuel. The plutonium is used to make additional fuel, often by mixing it with uranium oxides to produce *mixed-oxides fuel* (*MOX*) for use in thermal reactors. This provides additional energy and changes the composition of waste. At the moment there are developments of the reprocessing fuel cycle to extract both plutonium and minor actinides, and incorporate them in fresh fuel to be used in fast neutron reactors.
3. *Breeding cycle.* For this cycle, the reactor is designed so that there is more fissile material (mostly plutonium-239) in the spent fuel than there was in the fuel put into reactor. Similarly to the reprocessing cycle, plutonium can be extracted from the spent fuel and reused in another reactor. This allows extracting energy not only from the limited ^{235}U but also from the much more abundant uranium-238, hence increasing the energy output from a given amount of uranium. This cycle allows using a much larger resource base even including some old military-related nuclear materials, which would give much better forecast of the nuclear resources to produce large amounts of base load electricity.

A schematic picture of the nuclear fuel cycle is shown in Fig. 11.2, which indicates all three available approaches. Uranium shall be first extracted and preprocessed to deliver the product ready for further processing, usually in the form of uranium oxide. After further enrichment and treatment, it goes into the nuclear fuel production process where the special fuel assemblies are manufactured. These steps of the fuel cycle that precede the introduction of the fuel into the reactor are sometimes referred to as the *front end* of the fuel cycle.

After the fuel elements are loaded in the core of the reactor, they generate nuclear power through the fission reactions. In these processes, the fraction of the ^{235}U isotope is reduced continuously, while radioactive fission products and other isotopes are produced. After having generated energy in the core for the specified period of time (usually 3–4 years), the fuel assemblies are unloaded from the reactor core. This is when the *back end* of the fuel cycle starts.

After some time inside the interim storage on site, the spent fuel is shipped to the intermediate storage facility for spent fuel [5]. After several years in the storage, two options are available for spent fuel:

- Following the once-through cycle, it can undergo the final disposal process to be placed first in long-term storage with further conditioning and transportation into the final waste storage.

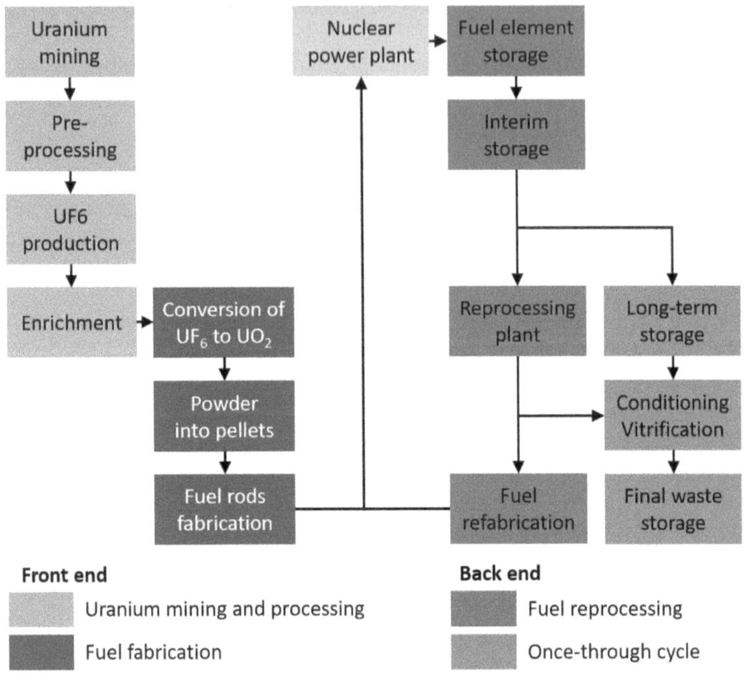

Figure 11.2 ***Simplified nuclear fuel cycle.*** Nuclear fuel cycle starts with the exploration and mining for uranium, followed by its processing, fuel fabrication, its utilization within the nuclear power plant, and further reprocessing and waste disposal, sometimes with fuel reproduction.

- It can go through the special treatment at the reprocessing facility, where uranium and plutonium can be separated from other material. The former can be recycled into the uranium or MOX fuel, while the radioactive waste is further directed into the long-term storage.

Despite major differences of the three types of cycle, all of them involve at least five interconnected stages that constitute a nuclear lifecycle: the front end of the cycle where uranium fuel is mined, processed, enriched, and fabricated; the construction of the plant itself; the operation and maintenance of the facility; the back end of the cycle where spent fuel is conditioned, processed, and stored; and a final stage where plants are decommissioned and abandoned mines returned to their original greenfield state [6].

In the following sections we will discuss major steps of the nuclear fuel cycle in detail.

11.2.2 Exploration for uranium and ore extraction

As discussed in Chapter 2, in certain regions uranium concentration is much higher than elsewhere so that its extraction is much more economically feasible. Such concentrations are usually called uranium ores, which consist of a number of uranium salts and oxides mixed with some wallrock. These ores are usually located in the deposits around one kilometer deep as most of the shallow locations have already been exploited. All known deposits are precisely documented and classified within multiple documents and databases, which provide geological data with possible economic connotation [7].

Two factors influence greatly the exploration and exploitation of ore deposits, namely economic factors and the role of research and technology. Although awareness of the economics has become more acute because profit is fundamental to mining companies, an effective analysis of economic factors is sometimes overlooked [7]. Uranium is considered a commodity and therefore its price is determined by supply and demand as most of other metals, while its effective exploration strategy is determined by a number of factors: economic potential (type, grade, and size) of the deposit and the explorability of the area.

Many methodologies used for a deposit exploration, such as geological mapping, remote sensing, geophysics, and geochemistry [7], allow for proper estimates. Once completed, the decision of the exploitation type is made. There are several major approaches to uranium extraction:

- *Open pit development* method is used where a deposit is shallow. This type of mining method includes the use of bulldozers to strip and develop the deposit, bulldozers and shovel loaders for loading, and dump trucks to remove the ore from the open pit. This type of mine usually requires large openings on the surface, larger than the size of the deposit itself. This leads to the fact that the amount of material that must be removed to access the ore deposit may be huge. After completion of mining operations, the pit is then backfilled with overburden, which would require moving even bigger amount of material and performing reclamation work on restored surface. These days such mines are limited to a few locations.
- *Deep underground mines* are used in case of clearly detectable mineral veins and imply the same logic as any other ore mining. This option is commonly used to extract higher grades of ores to cover higher costs of deep mining. Deep mines are more dangerous, for instance, because of various explosive or radioactive gases.

- *In-situ recovery* (ISR) or *in-situ leaching* (ISL) means that the process is carried out at the premises of the deposit underground. When a uranium ore deposit is detected, companies drill a borehole straight to the deposit with a number of boreholes around. Then, the water or a special acid is pumped through the central well where it leaches out uranium and goes back to the surface through the side boreholes. This does not impact the environment as there is no sagging and disturbance of the ground surface, off-balance and waste ore dumps, while the acid is quite harmless at those depths. This borehole method of uranium deposit development entirely prevents any sources of dust and decreases emission of radioactive substances into the air compared to conventional methods. Thus, the ISL method is the most environmentally friendly among mining methods.

The radiation hazard of open pits for the personnel is usually lower compared to underground mining. However, both types of ore extraction contaminate surface and ground waters and hence cause problems for the ecosystem around the mine. Moreover, upon completion of the mining process, the pits have to be backfilled with waste rock with further restoration of the soil layer and wildlife. Overall, the mining process is subject to strict regulation and environmental protection measures, especially before ground breaking during the planning phase. Unlike conventional mining methods, where the uranium mineral and host rock are excavated together and the uranium is recovered on the surface, ISR technique removes the uranium while leaving the host rock in place [7].

Therefore, in-situ leaching is by far more environment-friendly process which has been widely implemented in the mining industry in general. Importantly, the choice of a mining technology is a tradeoff between regulation, efficiency of production and overall economics, while the general sustainability topic is largely left aside.

However, uranium mining (as any type of mining) contributes to the overall GHG emissions, the amount of which depends on the quality of the uranium ore used in the nuclear fuel cycle.

11.2.3 Processing and enrichment

Both open pit and deep mining deliver ore in bulk pieces. At conventional mines, the extracted bulk ore goes through a mill where it is crushed and then ground to the consistency of sand [5]. These fine particles are then separated from the empty filler material of the ore, for instance, on the radioactivity level. Sometimes ore can undergo the roasting process to remove water or other waste additives. The obtained concentrate is sometimes ground in water to produce a slurry of fine ore particles. This slurry is then leached with the use of sulphuric or nitric acid or soda ash to dissolve the uranium oxides. The additive chemical depends of the composition of the concentrated ore and is subject to economic feasibility of the process. After leaching, the solids and liquids are separated, and, as a result, concentrated salts of uranium are obtained, which go through the uranium refining process. The end product is the uranium oxides, which are mildly radioactive.

In most cases, these steps are carried out at the enrichment facility located near the mine to minimize the bulk material transportation.

If the uranium is extracted through the ISL process, the uranium solutions retrieved from the wells contain a number of additives, which are removed through the ion exchange process and further extraction or scale build-up. After heating, the uranium is transferred in the same form of the uranium oxide, usually U_3O_8, which is called *yellow cake* because of its color.

For the application as a reactor fuel, these natural uranium concentrates must then be enriched, so that the uranium concentration has to be raised up to around 5%. Therefore, the uranium oxides are transported from the processing plant to a special facility, where they are converted into gaseous uranium hexafluoride UF_6, for the necessary enrichment steps in isotopes separation and enrichment plant. The process separates this gaseous substance into two streams using the centrifuge process several times. This would allow splitting the lighter uranium-235 and heavier uranium-238 into two volumes with higher concentration of each of the isotope.

11.2.4 Nuclear fuel fabrication

Fabrication of nuclear fuel assemblies for a nuclear power plant is a step in the front end of the fuel cycle when nuclear fuel is transformed from a commodity into a highly engineered product, with designs that are tailored to specific needs [7].

The enriched UF_6 is transported to the fuel fabrication plant where it is converted into the uranium dioxide UO_2 powder — the material for the fuel pellets. The powdered uranium dioxide is mixed with an additive to ensure that the pellet is properly pressed. After being pressed into small pellets, they go through the heating process exposed to very high temperatures for several hours. As a result, a pellet becomes a hard ceramic material.

During each stage, strict quality control procedures are applied to ensure each pellet (a few grams in weight) complies to the specifications.

The assembly line then inserts these pellets into the metal tubes manufactured out of zirconium or a similar alloy. These tubes, usually referred to as *fuel rods*, are about 4 m long each. These rods are the very elements, which would release heat inside the reactor. They are set together into the structures called *fuel assemblies* — the final fuel product that is delivered to the nuclear power plant and installed into the reactor core. Depending on the technology and country of origin, these fuel assemblies can be of different cross-section shape, for example, rectangular or hexagonal.

The number of fuel rods depends on the requirement of the fuel assembly, the design of which is specific to a given reactor type and a specific power plant fuel management strategy (which considers cycle length, capacity factor, existing fuel in the reactor core, etc.) [7].

11.2.5 Fuel utilization and handling of spent fuel

In a nuclear reactor, the fuel and other materials are arranged in a suitable way to produce a self-sustained chain reaction, where on average just one of the neutrons released by each fission goes on to cause a further fission. At that point, the reactor is said to have reached *criticality* or is said to be *critical*. In this situation, the energy released per unit time is maintained at a constant level and one speaks of a *controlled chain reaction*. The minimum amount of fissionable material for a given set of conditions needed to maintain a chain reaction is called a *critical mass* [2].

In case the fraction of neutrons producing new fissions is lower than one, the chain reaction is not self-sustainable and is called *subcritical*. It may die away without any appropriate corrections. On the contrary, if the average number of the neutrons giving rise to new fissions is greater than one, the reactor becomes *supercritical* and the energy release per unit time would increase. If left uncontrolled, it could lead to partial or total meltdown of the fissile mass [2]. However, the reactor cannot explode on its own as the concentration of the fissile material is still very low.

In the latter case, a moderator is used to slow down the speed of occasionally fast neutrons. The control of the self-sustained reaction is based on the control of the number of neutrons exposed to the heavy atoms. This is carried out with the *absorber* elements — special matter used for absorbing extra neutrons to maintain the *neutron multiplication factor* or simply the difference between the numbers of neutrons generated at each stage and those being lost (either released outside the boundaries of the core through the reflector of the system or absorbed without new release). In *fast breeder reactors*, however, there is no moderator as the principle is based exactly on the use of high-energy neutrons.

Evidently, nuclear fuel burnup and reactivity control are important points in the core design of a nuclear reactor. The fuel burnup analysis generally evaluates the time-dependent core power distribution and reactivity, which are necessary to know the proper thermal, hydraulic, and fuel design of the reactor. The core design for reactivity control predicts reactivity changes during the operation and therefore determines its optimal control methods, e.g., the insertion and withdrawal of neutron absorbers (generally referred to as control rods), implementing a burnable poison or a chemical shim, depending on the technology [8].

Periodically, a portion of the fuel in the reactor is removed and replaced by fresh fuel. Typically, an average sample of fuel remains in the reactor for three years, and approximately one-third of the fuel is removed each year, with a shutdown time for refueling and maintenance of up to two months [1]. The trend now is to extend the interval between refueling and decrease the maintenance time to increase the plant economic efficiency as every shutdown results in loss of energy sales.

The spent fuel assemblies removed from the reactor core after their utilization are very hot and still very radioactive due to the fission products. Therefore the spent fuel is stored in special pools under water, which provides both cooling and shielding against radiation. It is important to highlight that these basins are not equipped with any pump or water mixing equipment as the whole process is carried out due to natural circulation. Hot water naturally circulates upward, while the cold water tends to go down to cool down the spent fuel rods.

After around 10 years under water, spent fuel can be transferred into the interim storage facility. This can be either wet storage, where spent fuel is kept in water pools, or dry storage, where it is kept in concrete casks that provide air-cooling, shielding, and physical protection [2,6].

11.2.6 Fuel reprocessing and radioactive waste disposal

In a spent fuel reprocessing plant, fission products are separated from the remaining uranium and from neutron-bred plutonium. When spent fuel elements enter a reprocessing plant, they are first cut open in a special room by remotely operated robotized arms. The contents are dropped in nitric acid solution, which forms various chemical salt solutions with uranium, plutonium, and lower-mass fission products. Uranium and plutonium are then removed via special chemical reactions to be further used within nuclear fuel [3].

The remaining salt solutions containing fission products are treated to extract isotopes used in medicine and other important applications. What is finally left has low commercial value and is therefore treated as waste. Being heated and vitrified into solid glass-like material, it is contained and sent to the disposal process.

Radioactive wastes are defined as spent radioactive materials and materials that are contaminated with radioactive materials and bound to be disposed without any further use [9]. In some countries, which do not conduct reprocessing and follow the once-through cycle, spent fuel has no further use and is therefore considered as waste.

There are several types of solid radioactive waste that is generated during the fuel cycle [9]:

- High-level radioactive waste from reprocessing facilities. These are highly radioactive actinides that come from the separation from uranium and plutonium;
- Medium and low-level radioactive waste, which can come from various sources:
  - MOX fuel fabrication facilities;
  - Uranium waste from the enrichment and fuel fabrication process;
  - Nuclear power plant waste in the form of spent control rods, reactor core parts, etc., as well as filters, pieces of concrete and metal from the upgrade or decommissioning projects.

The radioactive waste treatment and disposal process, which has been implemented nowadays, includes the following steps [9]:

1. Nuclear waste comes from the normal operation and maintenance of the nuclear power plants and other facilities, including decommissioning;
2. The waste is pretreated to remove decontaminated gases and liquids, separate non-radioactive waste and waste below clearance limit (which can be disposed with traditional methods);
3. Radioactive wastes undergo treatment and further conditioning to convert them into forms that are safe to handle, transport, store, and dispose of. Wastes are solidified, sealed into steel containers, and additionally packaged for transportation;
4. Waste forms are then transported to disposal sites and placed in a specially designed facility, usually deep underground. They have to be built

in stable rock formations, which are not exposed to earthquakes, eliminating the potential risks to people and the environment.

5. At a disposal site, when a specific volume of waste forms has been placed, that section of the site is backfilled. This process, usually called *burial of waste forms*, continues till all sections of the facility are filled, then the site is closed;
6. Closure of the nuclear waste repository does not mean that all activities are stopped. Control and monitoring from authorities and operator of site continue for a specified period of time.

There were various approaches to store the containers. Initially they were placed into special water basins with water acting as a naturally convecting coolant and shield. Placement of the radioactive waste containers on a basaltic deep ocean floor was also considered. However, more easily accessed land-based underground storage in abandoned mines or similar excavations is favored now [3].

Engineers see no difficulty with securely storing radioactive wastes underground in deep mines in salt-bed formations or similar excavations. Salt-beds are known to be geologically stable for millions of years [3]. Permanent geological repositories must provide protection against every plausible scenario in which radionuclides might reach the biosphere or expose humans to dangerous levels of radiation. These risks include groundwater seeping into the repository, corrosion of waste containers, leaching of radionuclides, and migration of contaminated groundwater towards areas where it might be used as drinking water or for agriculture [6].

On top of this, the overall radioactive waste management encompasses policies to define the safety levels of radioactivity related to waste. Some types of waste may require attention and ongoing safety assessment for the period of a few tens of thousands of years. Therefore technological efforts for radioactive waste management should offer quantitative and comprehensive basis for social decision-making [9].

11.3. Applications of nuclear energy

It is not much of exaggeration to say that the birth of civil nuclear energy originated from the military developments which were pursued during the World War II. While the initial investigations started in Germany, the Allies took over and came up with the first nuclear bomb in 1945, which had been developed and built in the US. However, this US monopoly for nuclear weapons lasted until the early 1950s when the technology was obtained by the USSR and other countries, leading to the appearance of a new type of political instrument — nuclear deterrence. Further use of nuclear power spread over the naval propulsion, mostly for submarines to ensure longer autonomy, as well as into intercontinental ballistic missiles which were supposed to deliver nuclear explosives over the oceans.

While there had been many attempts to extend nuclear engines to aircrafts (mostly for strategic bombers to further extend their range and autonomy), spacecrafts, and vessels, only the latter succeeded and became rather common type of propulsion for naval ships or large icebreakers.

Evidently, this initially military technology was sought to be applied in everyday life. Therefore the history of civil nuclear energy dates back to the middle of the 20th century: several reactors for research and weapon plutonium production purposes appeared in the US starting 1943, while the first nuclear power plant built solely for civil purposes was the AM-1 Nuclear Power Plant in Obninsk (Soviet Union), launched in 1954. Its electric capacity was around 5 MW. Two years later, the first purely commercial nuclear power plant, at Calder Hall in Sellafield (England), started operation in 1956 opening a new era of purely commercial nuclear power which began to grow in the late 1960s [10].

Presently, apart from electricity generation and weapons, nuclear energy is used in multiple applications, from medical devices to food production and research.

11.3.1 Medical, food, and agricultural applications of radioisotopes

According to the World Nuclear Association [11], some isotopes are used in medicine. Isotopes are variants of a given chemical element that have nuclei with the same number of protons, but different numbers of neutrons. Some isotopes are referred to as "stable" as they are unchanging over time. Others are "unstable" or radioactive since their nuclei change over time through the loss of alpha and beta particles. The attributes of naturally decaying atoms, known as *radioisotopes*, give such atoms several applications across many aspects of modern day life.

The applications of radioisotopes differ and include, for example [11]:

- Nuclear medicine uses radioisotopes to provide diagnostic information about the functioning of a person's specific organs, or to treat them.

- Radiotherapy can be used to treat some medical conditions, especially cancer, using radiation to weaken or destroy particular targeted cells.
- Sterilization of medical equipment can be carried out with radioisotopes.

Medical radioisotopes are produced in special research nuclear reactors.

In the food industry, isotopes and radiation are used to expose foodstuff to gamma rays to kill bacteria that can cause food-borne disease, and to increase shelf-life. It has the same benefits as when food is heated, refrigerated, frozen, or treated with chemicals, but does not change the temperature or leave residues [11].

Isotopes can be also used in agriculture:

- Radioisotopes are added into fertilizers to properly estimate the rate of consumption by the plant and therefore provide better dosage of the fertilizer into the soil.
- Radiation is used to sterilize insects and prevent them from having offsprings and spreading over large agricultural areas.
- Ionizing radiation is sometimes used to induce mutations to come up with new genetic lines.

11.3.2 Radioisotopes in industry

Industrial applications of radioisotopes include, for instance [11]:

- Industrial tracers, when radioisotopes are used by manufacturers as tracers to monitor fluid flow and filtration, detect leaks, and gauge engine wear and corrosion of process equipment. Small concentrations of short-lived isotopes can be detected whilst no residues remain in the environment. By adding small amounts of radioactive substances to materials used in various processes, it is possible to study the mixing and flow rates of a wide range of materials, including liquids, powders, and gases and to locate leaks.
- Industrial gamma radiography to inspect metal parts and their internal structure, quality of production, etc. It works more or less in the same way as medical X-rays or security scanners.
- Use of radioisotopes as gauges for measuring thicknesses of sheet materials without physical contact.
- Carbon dating which is mostly used in geology, anthropology, archaeology, etc., to measure the abundance of naturally occurring isotopes and hence determine the age of material.

The same ideas of isotope tracers are implemented in environmental studies to detect and analyze pollutants, assessment of groundwater resources, in production of special detectors (e.g., for smoke), and so on.

All these applications have been due to the research of nuclear reactions and development of nuclear reactors. While the electricity and heat generation is the most obvious application of nuclear energy, it still has a lot to offer.

References

[1] Bodansky D. Nuclear energy: principles, practices, and prospects. New York: Springer. ISBN 9780387269313, 2007.
[2] De Sanctis E, Monti S, Ripani M. Energy from nuclear fission: an introduction. Undergraduate lecture notes in physics. Springer. ISBN 9783319306513, 2016.
[3] Eerkens JW. The nuclear imperative. A critical look at the approaching energy crisis. Topics in safety, risk, reliability and quality. Springer; 2006.
[4] Hore-Lacy I. Nuclear energy in the 21st century: World Nuclear University Press. Elsevier Science. ISBN 9780080497532, 2010.
[5] Kessler G. Sustainable and safe nuclear fission energy: technology and safety of fast and thermal nuclear reactors. Power systems. Berlin, Heidelberg: Springer. ISBN 9783642119903, 2012.
[6] Sovacool B. Valuing the greenhouse gas emissions from nuclear power: a critical survey. Energy Policy 2008;22:2940–53.
[7] Hore-Lacy I. Uranium for nuclear power: resources, mining and transformation to fuel. Elsevier Science. ISBN 9780081003336, 2016.
[8] Oka Y, Kiguchi T. Nuclear reactor design. An advanced course in nuclear engineering. Springer Japan. ISBN 9784431548980, 2014.
[9] Nagasaki S, Nakayama S. Radioactive waste engineering and management. An advanced course in nuclear engineering. Springer Japan. ISBN 9784431554172, 2015.
[10] Tester J, Drake E, Driscoll M, Golay M, Peters W. Sustainable energy: choosing among options. MIT Press. ISBN 9780262304634, 2012.
[11] World Nuclear Association. URL: www.world-nuclear.org.

CHAPTER TWELVE

Modern nuclear power plant

Contents

12.1. General arrangement and major components

As discussed in Chapters 3 and 11, the concept of a nuclear power plant is based on three energy transformations: the energy of the nuclear fuel into heat, heat into mechanical energy, and lastly mechanical energy into electricity. The heat from the nuclear reactor is used to evaporate water and produce steam, which is admitted to the steam turbine. Similarly as in a conventional steam power plant, steam is the working medium in a nuclear plant and therefore the same idea of running a turbine and generating electricity is used.

Importantly, nuclear power plants can be also considered as thermal as they use heat to evaporate water. However, instead of direct burning of fuel, nuclear fuel undergoes the fission chain reaction. Still, in some countries the term "burning" is utilized with respect to nuclear fuel spending within the reactor, which is mostly due to the similarities of the steam cycle in both types of power plant.

Sustainable Power Generation
https://doi.org/10.1016/B978-0-12-817012-0.00024-4

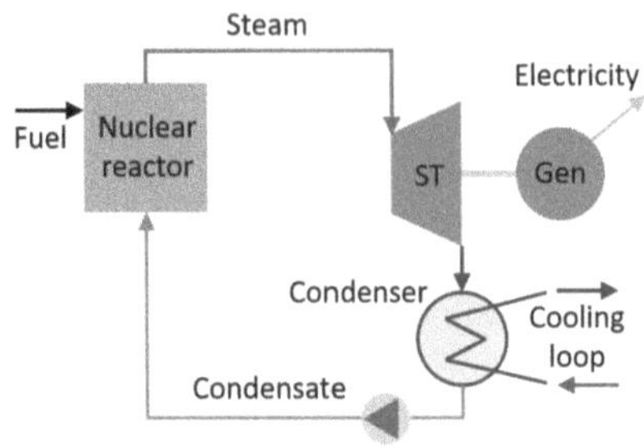

Figure 12.1 ***Simplified nuclear power plant process diagram.*** Schematic process diagram of a simplified nuclear power plant: reactor acts as a steam generator, which is fed to the steam turbine. Cold water from the condenser is pumped back to the boiler.

12.1.1 Heat balance diagrams

Consider first a simplified steam cycle of a typical nuclear power plant illustrated in Fig. 12.1. In this case, water is fed to the nuclear reactor area where it evaporates exposed to the heat released during the fission reaction. The steam is then admitted to the steam turbine, where it expands and transfers its energy into the mechanical energy of the rotating shaft of the turbine, coupled to the electrical generator. Exhaust steam is then turned back into water inside the condenser and is fed back to the cycle by the main circulating pump in the form of condensate.

A major part of the nuclear cycle is similar to the steam cycle of a conventional thermal power plant. The main difference is in the way the steam is generated, which clearly impacts the properties of this steam and therefore the overall cycle. Nevertheless, the approaches used to increase the efficiency of the steam cycle for a thermal power plant, described in Chapter 7, can be applied for this cycle as well.

This means that, in order to increase efficiency, we can implement a two-pressure steam turbine as shown in Fig. 12.2. This allows increasing the efficiency, however, may lead to certain issues. The steam from the nuclear reactor is of much lower quality compared to that generated in a coal-fired boiler, therefore when it passes through the HP section of the steam turbine, its temperature and pressure may go down significantly, enough to further decrease its quality. This means that this steam will be saturated, that is, represented by a mix of steam and water. In order to protect the steam turbine from corrosion and at the same time maintain its efficiency, water has to be separated from the steam. This is carried out inside special devices called *steam separators*: they use HP steam from the reactor steam

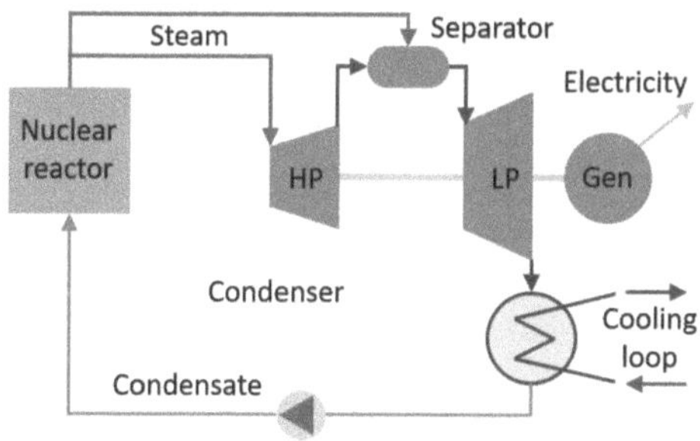

Figure 12.2 ***Simplified diagram of a nuclear plant with steam separator.*** Schematic process diagram of a simplified nuclear power plant with steam separator and two-pressure steam turbine. Steam separator conditions exhaust steam from HP section to admit it further to LP section.

generator in their heat exchangers to condition this saturated steam and reheat it.

Similarly to the optimized heat balance of a traditional thermal power plant, additional steam extraction can be implemented to preheat the feed-water and therefore increase the cycle efficiency of a nuclear power plant. A set of heat exchangers, as shown in Fig. 12.3, are employed similarly to a coal-fired power plant, however, the steam turbine that is used in a nuclear facility differs in size and structure from the one used in coal-fired power plants.

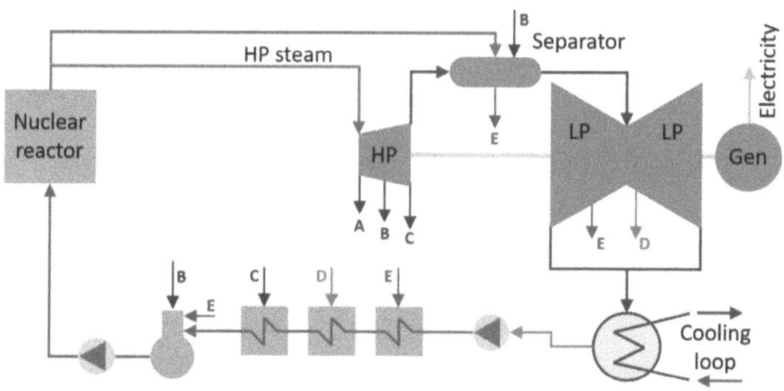

Figure 12.3 ***Simplified process diagram of an optimized nuclear power plant.*** Schematic process diagram of a simplified nuclear power plant with the optimized heat balance.

12.1.2 Equipment areas

The heat balance diagram of a nuclear power plant, represented in Fig. 12.3, can be split into two major parts: nuclear reactor area, which also supplies steam, and the rest of the cycle. Such a split, though looking simplistic, has been mostly driven by safety levels implemented in each of these areas. Evidently, the nuclear reactor area and some other equipment and systems around it are subject to much higher safety requirements and therefore are organized in a different way. The rest of the equipment, namely the overall steam turbine cycle, is similar to the conventional steam turbine driven power plant.

Therefore, a nuclear power plant comprises two major equipment areas:

1. *Nuclear island*, which is the heart of the nuclear power plant. It is formed by containment building, auxiliary building and fuel handling area. The arrangement of a nuclear island is specific for each nuclear power plant project and depends mostly on the type and configuration of the nuclear reactor. However, for all nuclear island configurations, it houses the *nuclear steam supply system* (NSSS) with all necessary safety systems.
2. *Turbine* or *conventional island*, which accommodates the steam turbine with its generator, condenser, heat evacuation equipment, and other balance of plant systems.

Fig. 12.4 illustrates major equipment areas of a nuclear power plant with key components. While the nuclear island of a power plant is unique, the conventional island is mostly similar to the traditional coal-fired power plant and incorporates the following key components:

- Steam turbine, which is usually a two-pressure machine of high capacity to convert the energy of the massive steam flow. The major peculiarity of these units is their size and number of low pressure sections to convert as much energy of the low quality steam as possible.

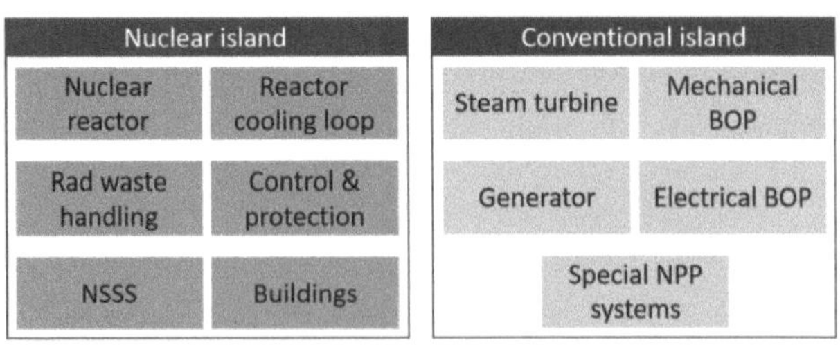

Figure 12.4 ***Equipment areas of a nuclear power plant.*** Nuclear power plant can be schematically split into two major areas: nuclear island and conventional island.

- Electrical generator;
- Heat exchangers, e.g., condenser and moisture separator reheater. The latter uses steam extraction flows to dry the major steam, which comes from the HP section and is guided to the LP section;
- Balance of plant, including mechanical (lube oil, water treatment, compressed air, cooling, etc.) and electrical systems (electricity evacuation, power supply, instrumentation and control of the equipment).

All equipment is located inside the building called the turbine hall. This building shall protect the equipment from the environment (like temperature variations, wind, precipitation) and provide stable and favorable conditions for operation and maintenance. On top, a nuclear site has a number of auxiliary buildings for administration, safety regime, security, fire fighting brigade, warehouse, etc.

12.2. Nuclear island as the core element

The nuclear island of a nuclear power plant is basically responsible for converting water into steam of the required parameters, which is further used in the steam cycle. Comparing it to the conventional power plant, the nuclear island is similar to the boiler island of a coal-fired power plant.

According to the definition from [1], the *nuclear island* is a part of a nuclear power plant which incorporates all equipment, systems, installation and control, and other relevant hardware installed within the reactor and reactor auxiliary buildings. The nuclear island is considered to be the most critical area of a nuclear power plant. This is mostly due to the potential danger of the technology implemented in the reactor — nuclear fission reaction.

The major and essential component of a nuclear island is a nuclear reactor, which can be named the "heart" of a nuclear power plant. By definition, a *nuclear reactor* is a device which allows for controlled sustained nuclear fission chain reaction. One of the most important points here is that the reactor has to provide full control of the energy release process and therefore control of the power plant operation.

Apart from a reactor, the nuclear island also incorporates other important systems:

- Radioactive waste handling systems to treat solid, liquid and gaseous forms of low level waste. In some cases, spent fuel storage ponds are associated with the nuclear island; however, we keep them in a balance of plant scope due to their lower impact on safety;

- NSSS, which depends on the type of the reactor;
- Reactor cooling loop designed for a specific reactor configuration;
- Control equipment to maintain and control the fission reaction;
- Buildings and protections.

The exact configuration of the nuclear island is always subject to the type of the reactor. The latter defines the cooling medium and a number of loops, type of NSSS and its configuration, safety and emergency systems. Moreover, overall nuclear containment building size and plan are fully subject to the reactor type.

12.2.1 Classification of reactors

Nuclear reactors are usually classified according either to the purpose or type of the nuclear fuel, operating mode, and arrangement of the reactor [2, 3].

The first nuclear reactors were built to produce ^{239}Pu for nuclear weapons. Subsequently, reactors have been used for many other purposes, of which electricity generation is now, by far, the most prominent. Further uses have been to propel ships (naval vessels, submarines and ice breakers), produce radioisotopes, and, to a limited extent, supply heat. Many additional reactors have been built for teaching or research, much of the latter involving the study of the properties of materials under neutron bombardment and the intrinsic properties of neutrons and other subatomic particles [4].

Therefore, in terms of *purpose*, currently operated stationary nuclear reactors are subdivided into two broad categories:

- *Power reactors*, which are mainly used in nuclear power plants to generate steam used in steam turbines to produce electricity. The same technology can be also used for running desalination plants to produce fresh water, provide hot water for district heating purposes, and conversion of fertile material into fissile. Power plant reactors are normally up to several hundreds of megawatts.
- *Research reactors*, which are usually operated at laboratories, research institutions and universities in many countries. They usually generate neutrons for various purposes, for instance, to produce isotopes for medical devices, testing materials and conducting research. In general, these reactors are usually of low power up to several megawatts.

Often their applications can be combined, for instance, both for electricity generation and production of plutonium-239 for military purposes.

In terms of *operating mode*, depending on the speed of neutrons at the time they are absorbed by the fissile material, the reactors can be classified into three types:

- *Thermal reactors*, in which neutrons are slowed down to thermal energies, and most of the fission reactions are with low-energy thermal neutrons. The speed of neutrons is decreased by the special matter called *moderator*;
- *Fast reactors*, in which most of the fission reactions take place with fast neutrons with energies more than 100 keV. These reactors do not use any moderator, and in these reactors a mix of fuels can be used;
- *Intermediate reactors*, which occupy the place between fast and thermal: here neutrons are slowed down to the energies in the middle range.

Based on the *geometry of the fuel element and the moderator arrangement*, the reactors can be of two major types:

- *Homogeneous reactors*, which have fuel homogeneously dispersed in the moderator. The free neutrons therefore enter the homogeneous medium like slurry. This design provides better heat transfer, while the reactor fuel can be added or reprocessed during operation;
- *Heterogeneous reactors*, where the fuel elements are in the matrix of the moderator so that the free neutrons have to face a heterogeneous medium. This system is considered to be the easier to design and construct.

The nuclear reactor technology has been under development since the first commercial operation in the early 1950s, and this development is usually represented in many broad categories called *generations* of reactors. Each generation of reactors has implemented a number of technological innovations and numerous safety features compared to the previous generation. At the moment, it is commonly accepted to distinguish between four generations:

I Early prototype reactors designed and built from the 1950s to late 1960s. These were reactors of various design with low capacities usually up to 50 MW_e.

II Commercial power reactors of the unit capacities of hundreds of megawatts up to 1.2 GW_e, which have many representatives in the currently operated fleet. These are various reactors like LWR, BWR, CANDU, and RBMK. Such reactors have been designed for 30–40 years of operation with the possible life extension based on extensive studies and operating experience worldwide. These reactors, covering the period from the 1970s to 1990s, have been upgraded to maintain required safety regulation.

III An evolution of LWR technologies with extended lifetime and improved performance. These reactors, being designed and built in many countries (China, France, Finland, Russia, the USA, and others) within the period of the late 1990s till 2010s, had implemented advanced safety features.

IV Future generation of innovative nuclear reactor technologies, which shall implement up to 60 years of life, simplified maintenance, modularization, adding special equipment and systems to mitigate severe accidents, and standardization for pre-licensing. Moreover, these reactors shall employ enhanced safety system (introduced already at the design stage) and improved security to stop or at least minimize proliferation of nuclear material, which can potentially be used for weapons.

Research and development in the direction of new generations of reactors have come up with a number of innovative approaches, which are still in the design and feasibility study phases. One of the approaches is to develop fast breeder reactors that would allow closing the nuclear cycle and decreasing the amount of extremely radioactive waste. The commercial operation of the IVth generation reactors is likely to happen in the 2040s, though even currently available types of reactor can serve as a basis and research platform for next developments.

12.2.2 Types of reactor and nuclear steam supply system

As of December 2018, there were 454 nuclear reactors in operation (i.e., connected to the grid), in 30 countries, with an installed electrical capacity of more than 400 GW_e [5]. At the same time, there were 54 nuclear reactors under construction with the planned capacity of around 55 GW_e. Table 12.1 lists the countries where there are nuclear power reactors in operation, under construction and long-term shutdown, as of December 2018. The table also gives the relevant installed capacity of the nuclear power plants.

12.2.2.1 Types of reactor

Although a variety of reactor configurations and types have been tested on an experimental basis since the beginning of a commercial nuclear era in the 1950s, essentially there are six different types of reactor and therefore

Table 12.1 Nuclear power plant units in operation, under construction and in long-term shutdown, as of December 2018.

Country	Reactors in operation		Reactors under construction	
	Units	MW_e	*Units*	MW_e
USA	98	99 333	2	2234
France	58	63 130	1	1630
China	46	42 800	11	10 982
Japan	42	39 752	2	2653
Russia	37	28 264	6	4573
Republic of Korea	24	22 494	5	6700
India	22	6255	7	4824
Canada	19	13 554		
Ukraine	15	13 107	2	2070
United Kingdom	15	8918		
Sweden	8	8612		
Germany	7	9515		
Belgium	7	5918		
Spain	7	7121		
Czech Republic	6	3930		
Switzerland	5	3333		
Finland	4	2769	1	1600
Hungary	4	1889		
Pakistan	5	1318	2	2028
Slovakia	4	1814	2	880
Taiwan	4	3844	2	2600
Argentina	3	1633	1	25
Brazil	2	1884	1	1340
Bulgaria	2	1926		
Mexico	2	1552		
Romania	2	1300		
South Africa	2	1860		
Armenia	1	375		
Iran	1	915		
Netherlands	1	482		
Slovenia	1	688		
United Arab Emirates			4	5380
Belarus			2	2220
Bangladesh			2	2160
Turkey			1	1114
Total	454	400 285	54	55 013

Source: [5].

configurations of nuclear island which are now employed for commercial operation:

- Light-water cooled reactors:
  - Boiling light water reactor (BWR),
  - Light water cooled, graphite-moderated reactor (LWGR),
  - Pressurized light water reactor (PWR),
- Pressurized heavy water reactor (PHWR),
- Gas-cooled reactor (GCR),
- Fast breeder reactor (FBR).

The majority of the commercially implemented and operated reactor technologies in the world are light-water cooled reactors as illustrated in Fig. 12.5. Historically, the choice of the particular plant design has been

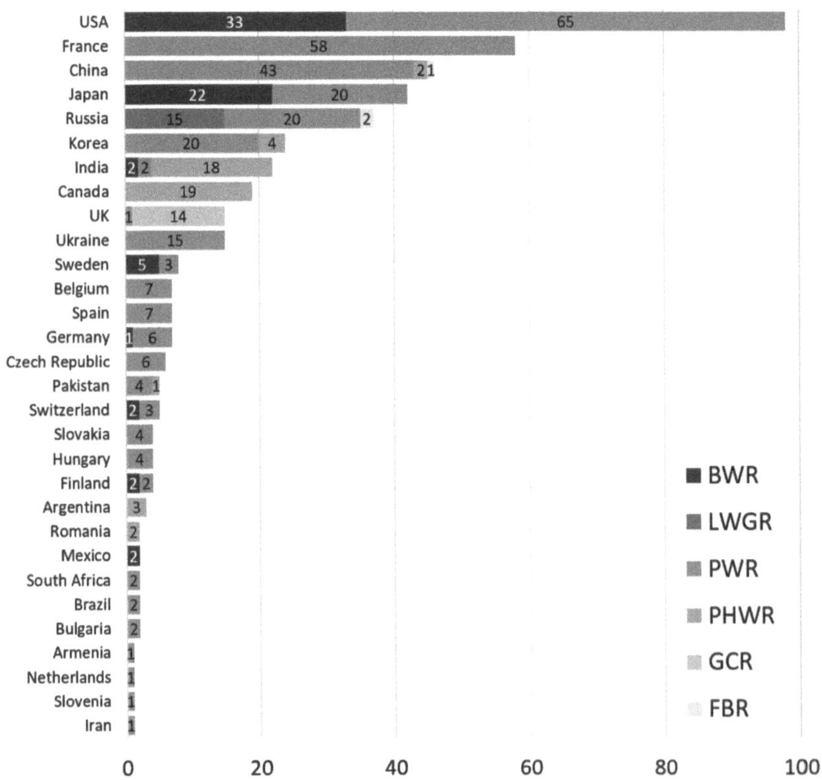

Figure 12.5 ***Commercially operational reactors as of December 2018.*** Commercially operational reactors per type and per country: the absolute majority of reactors are of BWR and PWR types. *Source: based on data from [5].*

influenced by many factors, including availability of fuel, technology, local competence in design, construction, and operation, as well as the previous experience and political influence of the technology owners.

Except for three breeder reactors, all other reactors are of thermal type. Therefore, we consider the major components of a thermal reactor as the mostly widespread technology. The main distinction between different reactor types lies in the difference in the choices of fuel, moderator, and coolant [4].

The moderator is an essential part of a thermal nuclear reactor, as it is required to slow down neutrons to thermal energies. There are limited options for moderators now implemented in commercial reactors: light water, heavy water, or graphite. Any of these can be used with enriched uranium. However, with natural uranium, it is not possible to achieve a chain reaction with a light water moderator, but it is practical to use heavy water or graphite, both of which have high moderating properties [4].

The main function of the *coolant* is to transfer thermal energy from the nuclear fuel to the steam turbine, either directly or through intermediate steps. During the plant operation, the cooling loop is an inevitable part of the energy conversion chain. However, in a nuclear reactor, cooling has a special additional importance, because radioactive decay causes continuous heat production even after the reactor is shut down and electricity generation has stopped. It is therefore essential to maintain cooling to avoid melting of the reactor core [4].

The coolant used in thermal reactors can be either liquid or gas. For thermal reactors the most common coolants are light water, heavy water, and carbon dioxide. The type of coolant is commonly used to designate the type of the reactor. In contrast, fast reactors usually use liquid metals as coolants, for instance, sodium.

Remark 12.1. In PWR, BWR or PHWR designs, coolant can also serve as a moderator.

Table 12.2 illustrates the major differences in nuclear reactor designs. In further sections we will consider each technology separately, together with the overall steam cycle, as well as advantages and disadvantages of each scheme.

12.2.2.2 Pressurized water reactors (PWR)

A *pressurized water reactor* (*PWR*) is a light water cooled and moderated nuclear reactor. It uses slightly enrich uranium or MOX as a fuel. This

Table 12.2 Comparison of various types of nuclear reactors in operation.

Feature	PWR	BWR	PHWR	LWGR	GCR	FBR
Fuel	LEU MOX	LEU MOX	Natural uranium LEU	LEU	Natural uranium LEU	Plutonium mixed with natural or depleted uranium
Moderator material	Light-water	Light-water	Heavy-water	Graphite	Graphite	None
Neutron energy	Thermal	Thermal	Thermal	Thermal	Thermal	Fast
Reactor coolant	Light-water	Light-water	Heavy-water	Light-water	Carbon dioxide	Liquid metal
Number in operation	290	78	49	15	14	3
Capacity, WM_e	272 935	75 320	24 634	10 219	7720	1369
Main countries	USA, France, Japan, Russia, China	USA, Japan, Sweden	Canada	Russia	UK	Russia

Source: aggregated from [3,5] and public data of OEM companies.

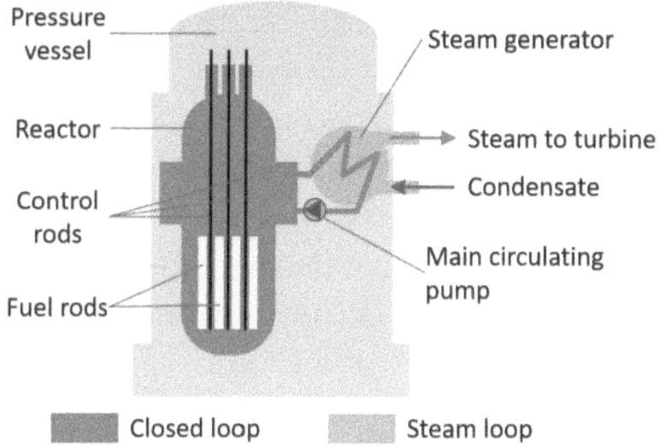

Figure 12.6 ***Schematic diagram of a PWR.*** Pressurized water reactor employs a closed loop of pressurized water that delivers heat to the steam generator and steam cycle.

reactor, as schematically shown in Fig. 12.6, consists of the reactor core, located inside the steel pressure vessel. The vertically arranged fuel rods contain fuel pellets of uranium dioxide.

The operating process in this reactor is quite simple. The pressure vessel contains vertically arranged fuel elements and control rods, which are admitted from the top to control the reaction. Light water is fed by the main circulating pump through the pressure vessel, where it is exposed to the heat released from the fuel elements due to the fission reaction. This water absorbs heat and therefore serves both as a coolant and as a moderator to slow down the neutrons. The average temperature of the coolant is slightly higher than 300°C with the pressure kept at the values around 15 MPa [2, 3], which prevents it from boiling inside the vessel.

Hot pressurized water flows to the special heat exchanger called steam generator (a part of an NSSS). Within this water-to-water heat exchanger, the coolant transfers heat to the second loop and, being cooled down, is pumped back to the reactor. The first loop is therefore completely closed.

The second loop uses demineralized water like in a conventional power plant. This water, when passing through the steam generator, absorbs heat from the coolant and therefore evaporates. Steam is then admitted to the steam turbine, which runs the generator. After being cooled down in a condenser and turned into water, it is pumped back to the steam generator through the feedwater preheater, which increases the overall efficiency.

The obvious advantage [3,4,6] of the PWR design is that a leak of radioactive nuclides in the core would not transfer any radioactive contaminants to the turbine and the condenser, as both loops are separated. Indeed, only the water in the primary loop can become radioactive as it flows through the core. To enhance the safety of the system, the primary loop, including the steam generator, is confined within the containment building usually made of reinforced concrete. This means that any leak from the primary loop will be also confined inside the containment building, which provides the required level of safety.

Due to its simplicity and reliable design, these reactors became by far the most widespread technology, occupying around 65% of market share and being constructed and operated in almost all countries, which possess commercial nuclear energy.

12.2.2.3 Boiling water reactor (BWR)

A *boiling water reactor* (*BWR*) is the second most widespread technology with around 18% of share. Similarly to PWR, it uses the same type of fuel and light water as a coolant and moderator. The major difference of a BWR technology is that there is only one loop: the same water is used as a coolant, moderator and working medium in the steam turbine cycle.

BWR consists of a reactor, which accommodates the fuel rods. Light water is pumped through the vessel and absorbs heat, which is released in the reactor core during the nuclear fission reaction. While water is kept under low pressure of around 7 MPa, it would boil in the core at about 290°C [3] so that is why it is called a boiling water reactor. The steam is then fed directly to the steam turbine, so that the coolant acts as a working fluid as well (see Fig. 12.7).

Steam then flows through the steam turbine and is cooled down in the condenser. The water is then pumped back to the cycle by major circulating pump. This way, in the BWR design, the primary loop goes outside of the containment building. Therefore a leak in that part of the loop may lead

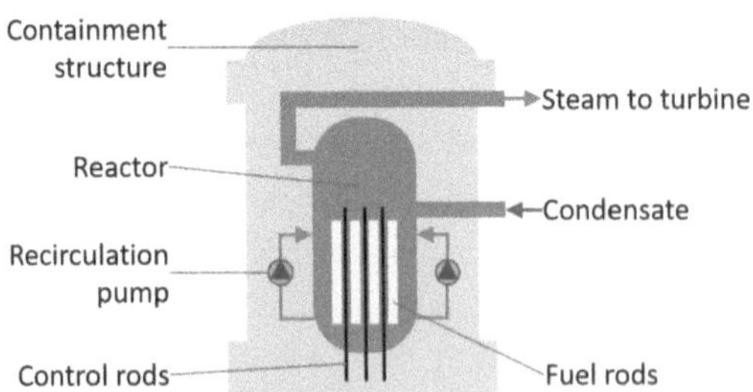

Figure 12.7 ***Schematic diagram of a BWR.*** Boiling water reactor boils light water that is fed to the reactor. The evaporated steam is then fed directly to the steam turbine.

to a spread of radioactive water that would not be as strongly confined as within the containment building. This means a greater concern about possible contamination of the environment around the reactor.

On the other hand, the presence of steam in the upper part of the reactor vessel (which serves as a large steam generator), has worse properties to moderate neutrons. Therefore, in case of overheating, there will more steam and therefore less capacity to slow down neutrons, and they will be less efficient in causing new fissions. This means that the presence of steam is a natural safety feature of this design [3].

Due to technological reasons, the control rods are located at the bottom of the reactor, while at the top there is special equipment to separate water from steam and to send steam to the cycle. Because to this fact, safety control rods have to be pushed up instead of being able to fall down in case of emergency.

12.2.2.4 Light-water graphite-moderated reactor (LWGR)

A *light-water graphite-moderated reactor* (LWGR) belongs to a class of reactors, whose most known representative is the RBMK, Russian abbreviation for the High Power Channel-type Reactor [3]. This class of reactors occupies a little bit more than 3% of the world operating fleet and is present only in Russia.

The LWGR is a boiling water reactor, which uses light-water for cooling and graphite as moderator. Due to its dual purpose, basically for electricity generation and for plutonium production, it has a very specific design as shown in Fig. 12.8. The tubular fuel elements are located inside the pressure tubes, where the water flows. These concentric structures

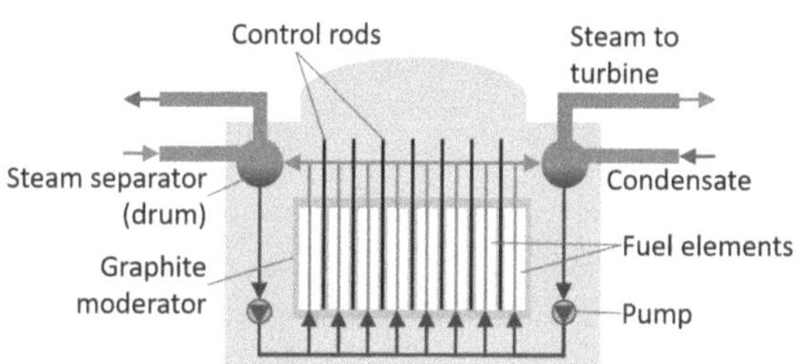

Figure 12.8 ***Schematic diagram of an LWGR.*** Light-water graphite-moderated boiling water reactor (of RBMK type) uses light water as a coolant and working medium: it flows through the fuel elements and absorbs heat. Steam separators then supply steam to the steam turbine cycle.

(basically water pipes with fuel rods located inside) are arranged within the graphite moderator blocks whose total mass is very high.

While the water flows through the pressure tubes, it absorbs heat released by fuel rods and therefore evaporates. This water and steam mix is headed to the steam separators. While the water is fed back to the pressure tubes for evaporation, the steam is headed to the turbine. At the outlet of the steam drums, the average temperature and pressure are 280°C and 6.38 MPa, respectively, with a net efficiency in the conversion to electricity of around 31% [3].

The advantage of RBMK reactors is that they can work with a low fuel enrichment level and offer the possibility to replace fuel tubes during reactor's operation (up to 5 replacements per day). However, graphite reaches very high temperatures and, if the cooling water is lost, the neutron multiplication coefficient increases. Both these factors make earlier RBMK reactors unsafe, as it was dramatically shown in Chernobyl accident [3]. A number of significant design changes and improvements have been implemented after that disastrous accident to address safety problems, so that there are still 15 of these reactors in commercial operation.

12.2.2.5 Pressurized heavy-water reactor (PHWR)

An alternative design of a thermal nuclear reactor, which uses natural uranium and does not require any enrichment process, is a *pressurized heavy-water reactor*. The use of heavy water, which contains deuterium (a hydrogen isotope) as a more efficient moderator, ensures that less neutrons are lost in the moderation process with respect to light water.

The principal representative of this type of reactor is the Canadian-built CANDU reactor (Canadian deuterium uranium reactor), and more recently, a new design of a PHWR has been developed in India [3]. This technology occupies around 10% of commercial reactors' market share.

A schematic diagram of a CANDU reactor is shown in Fig. 12.9. Since heavy water contains deuterium, which is twice as heavy as hydrogen, the geometry of the PHWR is significantly different from other technologies. The CANDU design is based on the pressure tubes, immersed in a heavy water tank called calandria, which acts as a moderator. The pressure tubes are arranged horizontally and contain fuel elements, loaded inside. Heavy water flows at a high pressure of about 10 MPa at around 290°C [3] through the pressure tubes and absorbs heat from the fuel elements, so that it serves as a coolant. The high pressure prevents heavy water from boiling. The coolant is then fed to the steam generator, where it transfers heat to the

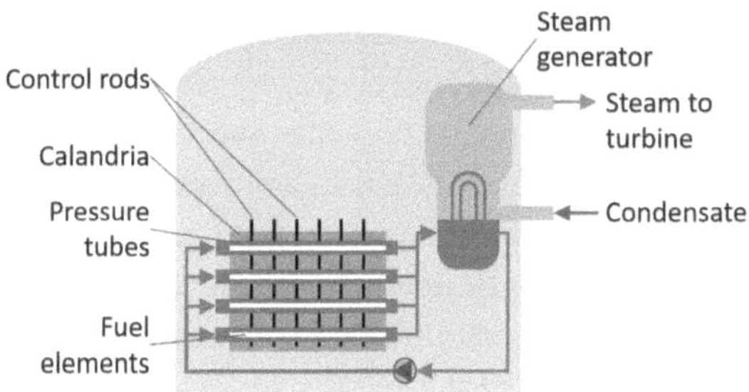

Figure 12.9 ***Schematic diagram of a PHWR.*** Within a PHWR the pressure tubes go through the heavy water moderator tank (calandria). As the tubes contain fuel elements, the water circulates through them and absorbs heat, which is then transferred in a steam generator to evaporate water.

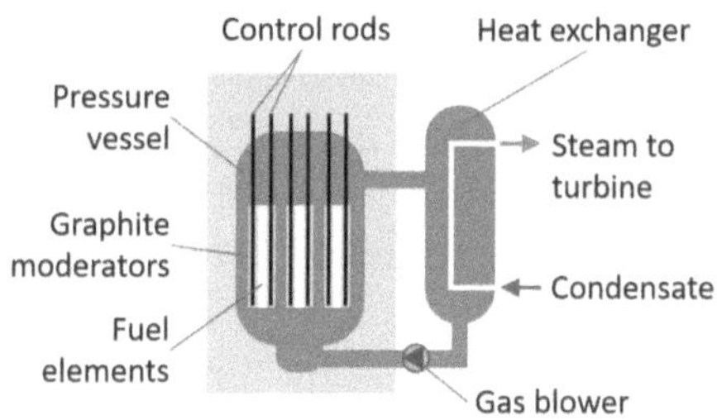

Figure 12.10 ***Schematic diagram of a GCR.*** Gas-cooled reactor uses gas for inner cycle to bring thermal energy from the fuel elements and transfer it within the heat exchanger to evaporate water.

second loop. While the coolant temperature is not very high, the steam is about 260°C at 4.7 MPa, which does not allow increasing the overall electrical efficiency above 28% [3].

The absence of a big pressure vessel leads to lower capital costs of this reactor, and further developments are aimed at avoiding heavy water at least for cooling purposes (though using slightly enriched uranium). This is especially important as heavy water production is quite expensive and outweighs the costs of fuel enrichment.

12.2.2.6 Gas-cooled reactor (GCR)

Gas-cooled reactors use graphite as a neutron moderator and carbon dioxide gas as the coolant. With the 3% of market share, all of them are installed in the UK. These reactors use natural or slightly enriched uranium as a fuel.

As illustrated in Fig. 12.10, the carbon dioxide circulates through the core, absorbing the heat from the fuel elements and reaching 650°C. Then, it flows to the heat exchangers, which are located outside of the reactor concrete pressure vessel. These are the gas-to-water heat exchangers that use the once-through principle to boil the water flowing through it. The water is then used for the conventional steam cycle.

In this design, boron control rods are used to penetrate the moderator and control the reaction. Besides, there can be a secondary shutdown system involving nitrogen injection into the coolant.

In the second generation of the GCR, the steam generators are located inside the concrete pressure vessel, which requires a much bigger structure and therefore higher capital costs.

12.2.2.7 Fast breeder reactor (FBR)

All thermal reactors described above have two major disadvantages [3]:

1. Low efficiency in fuel usage: only 1–3% of the overall uranium fuel element (including both U-235 and U-238);
2. Usually low steam parameters (except for GCR) used in the steam turbine cycle caused by the requirement to use water as coolant in the primary cycle.

Because only around 0.7% of natural uranium is directly fissionable U-235, and 99.3% is the less fissionable U-238 isotope, large amounts of uranium would be wasted unless the U-238 is converted to fissionable plutonium-239 (which does not occur in nature) by the breeding reaction with fast neutrons. This reaction takes place in all power reactors that use 3% enriched uranium but with low efficiency as the technology is still aiming at thermal neutrons and uses moderators to slow down the fast ones.

The breeder technology can still be an extremely important topic in the face of depleting uranium resources. Converting U-238 into fissile Pu-239 can increase the amount of available fissile material with the factor of 60, so that uranium reserves can provide nuclear energy for 1800 to 3000 years if breeder reactors are utilized [6]. These fast neutron reactors represent a design different from thermal ones: they have no moderator and fast neutrons produced in fission reactions are not slowed down to lower energy intentionally. Accordingly, the components and materials used in these reactor types — thermal and breeder — differ significantly [3].

There are only three breeder reactors currently in operation, however, only two of them, both located in Russia, have commercial application and considerable capacity. The third one in China, though considered in operation, is small and is used for development purposes only. Recently, Russia commissioned and put into operation a new breeder reactor unit

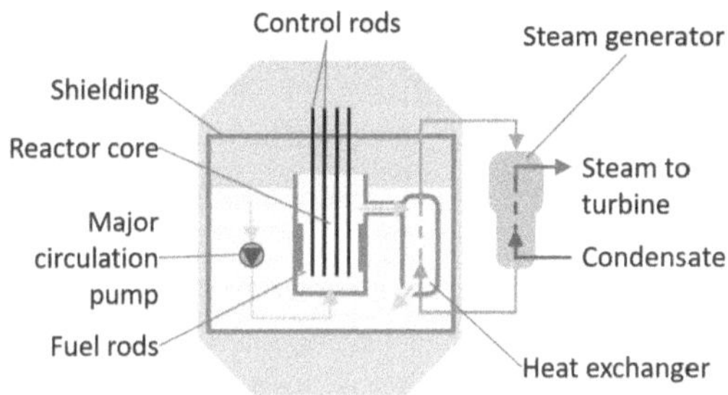

Figure 12.11 ***Schematic diagram of an FBR.*** A liquid-metal fast-breeder reactor employs three loops: the first one transfers heat from the reactor core to the heat exchanger, the second brings this heat to the steam generator, and the third supplies steam to a steam turbine.

named BN-800 (from Russian meaning, fast with sodium cooling) with a net capacity of 789 MW_e, which is a further development of a BN-600 reactor operated in the same plant.

The design implemented in the Russian BN-600 is schematically illustrated in Fig. 12.11. This design accommodates three loops:

1. The primary loop uses liquid sodium as a coolant. Major circulation pumps push coolant through to absorb heat from the reactor core, which contains fuel elements, and bring it to the heat exchangers. The sodium is heated up to around 550°C.
2. The intermediate loop also uses liquid sodium, which transfers heat from the primary loop to the third, or power generation conventional loop. Within this loop, the sodium is heated up to approximately 520°C, which allows boiling water and producing steam in a steam generator.
3. The third loop is therefore a conventional one to supply steam to the steam turbine.

The idea to have three loops is due to the fact that sodium gets particularly radioactive within the primary loop. In air, sodium can burn, while a direct contact with water would produce flammable hydrogen and may lead to an explosion.

The largest advantage of a fast breeder reactor is that it makes much better use of the basic uranium fuel, as it can operate with MOX fuel. Moreover, since a fast breeder reactor breeds new fuel, there are subsequent savings in fuel costs as the spent fuel can be reprocessed to recover usable plutonium. In a thermal reactor, the limited amount of uranium-235 content in the fuel is utilized as fissile material and the abundant uranium-

238 is largely unused and is therefore disposed as waste, while it would be possible to use it in a fast reactor [3].

With the concept of breeding in an FBR, the abundant U-238 can be converted into fissile plutonium-239 and hence a better fuel utilization is possible in fast reactors. On top of this, fast reactors are capable of burning long-lived actinides (the waste from thermal reactors), thereby reducing the problem of long-term storage of radioactive wastes [3]. This is due to the fact that all long-lived actinides can be fissioned by fast neutrons and therefore leave waste, which is significantly easier to treat and handle.

12.3. Conventional island technology and balance of plant systems

12.3.1 Overview

The major function of a conventional island of a nuclear power plant is to transfer the energy of the pressurized steam into the electricity. Secondly, it has to supply the water back to the nuclear island with the pretreatment, if required by the cycle.

The name *conventional island* comes from the similarity between the steam cycles of nuclear and thermal power plants. Both use steam in a steam turbine, condense it back to water and pump to the steam generator to close the cycle. Therefore, a conventional island, sometimes called a *turbine island*, is a part of a nuclear power plant that incorporates the following components:

- Steam turbine;
- Generator;
- Auxiliary systems supplied with major equipment;
- Mechanical balance of plant systems, including reheater and moisture separator, feedwater heaters, condenser, and other heat rejection equipment;
- Electrical balance of plant, which employs power evacuation system, auxiliary power supply (low and medium voltage), emergency power supply, instrumentation and control, etc.;
- Special nuclear power plant services and systems, including local fire brigade post, medical checkpoint, security and defense systems, etc. This is linked to the special level of security that has to be maintained at a nuclear site.

All major equipment is located inside the building called turbine hall. This building shall protect the equipment from the environment (like temperature variations, wind, precipitation) and provide stable and favorable conditions for operation and maintenance.

12.3.2 Steam turbines and generators

A power island of a nuclear power plant employs the shaftline based on a steam turbine coupled to an electric generator. Though been similar to large-scale coal-fired conventional power plants, the steam generated in a nuclear reactor may require certain changes in a steam turbine design.

12.3.2.1 Steam turbine

The implementation of a steam turbine in a nuclear power plant dates back to the 1950s when the first commercial reactors were put into operation. From the very beginning of the nuclear era, steam turbine was considered as a major machine to drive the electrical generator. Similar to a conventional fossil-fired power plant, nuclear power plant was supposed to operate on steam and therefore required special steam turbines for that. The major challenge here was, and still is, to design such a machine that would be able to admit steam of lower parameters and efficiently convert its energy into the mechanical energy of the rotating shaft.

The thermal megawatts or MW_{th} is the nominal thermal power output of a nuclear power plant. It depends on the design of the reactor and relates to the quantity and quality of steam the reactor can produce. The major difference in steam parameters compared to a conventional power plant is that the pressure of the steam is always subcritical (HP steam pressure usually around 6–7 MPa for thermal reactors, compare to the values in Table 8.1), as it is linked to the parameters of the coolant within the reactor cooling loops. While its temperature directly depends on the type of the reactor, it may lead to very saturated steam generated in the NSSS, which results in too high moisture content that would require additional treatment throughout the cycle.

All this requires quite high steam flow to maintain high electrical output of hundreds of megawatts. Therefore, nuclear power plants employ one large steam turbine per reactor. The type of the steam turbine, its configuration and steam parameters depend on reactor type and overall plant specifics, and are usually designed and built for a specific project.

Steam turbines for nuclear power plants can be subdivided into two major groups based on the speed of the rotating shaft:

- Fast or full-speed steam turbines revolve at a "traditional" speed of 3000 or 3600 rpm (50 or 60 Hz, respectively) to run the synchronous

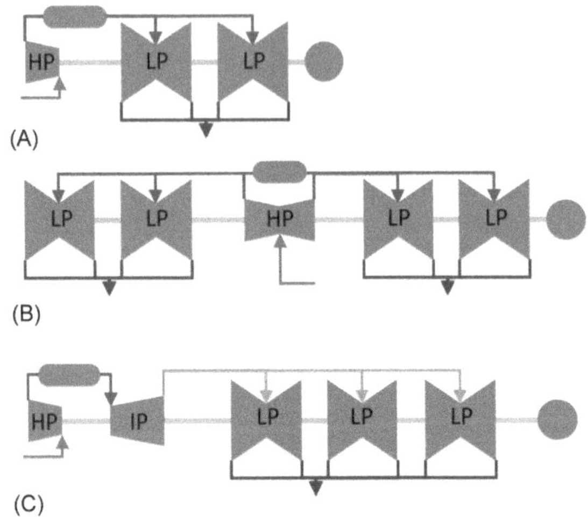

Figure 12.12 ***Simplified steam turbine configurations for nuclear power plants.*** Typical configurations of steam turbines for nuclear power plants: two pressure configurations with single (A) and double (B) flow HP sections; three-pressure steam turbine (C).

generator. They use the design experience and lessons learned from conventional steam turbines. However, their size increases significantly with the increase of the output. On top, dynamic loads and stresses tend to grow dramatically with the massive steam flow.

- Slow or half-speed steam turbines have their speed twice lower than the fast ones: 1500 or 1600 rpm, depending on the market frequency. For lower capacity units, these machines are by far bigger and heavier compared to their fast colleagues, however, with the increase of capacity this difference in size tends to go down. Moreover, slower units can have fewer LP sections due to longer LSB.

There are several major designs of steam turbine readily available for the nuclear power plants and various types of reactor. They are of reheat type with several pressures, as illustrated in Fig. 12.12:

- Double-pressure, when the steam first goes through the HP section. The exhaust steam is then reheated and separated from moisture content and further admitted to the LP sections of the steam turbine. From the low pressure cylinders, it is exhausted to the condenser (see (A) and (B)).
- Three-pressure units employ a more efficient scheme. Similarly to a double-pressure configuration, steam first flows through the HP section. After being reheated, it is admitted to the IP section, where it

transfers part of its energy to the shaft and is further guided to the LP sections (refer to (C)).

The number of sections of each pressure depends on the overall capacity of the unit. In practice, similar to a conventional steam turbine (as it uses the same principles), the steam flows have to be organized in a way to minimize the impact on the shaftline and the support bearings. This leads usually to an axial double-counterflow configuration of cylinders, at least for LP sections. In a three-pressure design, HP and IP sections have only one flow each, but with different directions along the shaftline, which compensates the impact of pressurized steam to the overall structure.

All of the modern steam turbines for thermal reactors employ a reheater and moisture separator to condition steam after it comes from the HP section.

All steam turbines, both for the nuclear and thermal power plants, have the same components as they are designed according to the same underlying principles. For instance, a typical two-pressure steam turbine for a nuclear power plant with the capacity up to 2000 MW_e may features a double-flow HP cylinder and up to three LP cylinders, which provides a broad range of capacities.

12.3.2.2 Generator

Uniformly, the overall concept of the generator is the same as for those implemented in a traditional thermal power plant, which is discussed in Section 9.5. However, the particular design of the generator depends on the capacity of the steam turbine and its speed. While for the 50 or 60 Hz units the number of poles of the generator is equal to two, half-speed generators would have four poles.

A typical generator for a nuclear steam turbine consists of the frame, which accommodates the stator of the generator. The stator windings of this particular model are water-cooled as they heat up considerably during operation. There are two bearings installed on the stator to hold the rotor with windings, which is cooled by hydrogen.

However, the cooling system of the generator depends on the manufacturer, and can be either water, air or hydrogen based. The latter is among the most efficient but would require additional hydrogen supply system at the power plant. For maintenance purposes, an additional nitrogen system is required to purge the generator before doing any activities to ensure that there is no explosive gas inside the system.

The exciter is located at the end of the generator and provides a low voltage in the winding of the rotor. When the rotor starts turning, it would induce the voltage in the stator and therefore produce alternating current as the winding revolves. To maintain the excitation voltage within the required limits, a modern generator is always equipped with the special automatic voltage regulator (AVR).

In order to maintain safe operation, bearings shall be supplied with lubricating and hydraulic oil. Due to friction, its temperature goes up significantly, meaning that the oil has to continuously circulate through the cooling device. This is ensured by a redundant oil pump system.

12.3.2.3 Accessories and auxiliary systems

All steam turbines and their generators, no matter for which type of power plant they are designed and used, have a number of typical support systems necessary for normal and safe operation. These include mechanical, fluids and electrical systems and may include (for details refer to Section 9.2):

- Hydraulic and lubricating oil systems, which supply oil from the oil reservoir to the bearings of the steam turbine and generator;
- Shaft sealing system to prevent steam leakages within the steam turbine cylinders;
- Cooling system to evacuate heat from cooling medium (air, water or hydrogen), as well as from the lubricating oil. On top of this, special air blowers can be installed at the generator terminal side to cool down the electrical bus bars that evacuate electricity from the generator;
- Steam turbine and generator enclosures and acoustic treatment to safeguard personnel from noise and heat;
- Various electrical, instrumentation and control systems used for operation and control purposes.

A steam turbine is usually supplied with the necessary stop and control valve, which are designed according to the steam cycle. These include HP, IP (if applicable) and LP stop and control valves.

12.3.3 Balance of plant systems and their importance

Balance of plant systems are an essential part of a nuclear power plant. They include all mechanical and electrical equipment, systems, and interconnecting cables and pipes that support operation of nuclear and conventional islands and unite the whole plant into one system.

12.3.3.1 Mechanical BOP

Mechanical balance of plant is aimed at proper operation of fluid and mechanical systems around nuclear and conventional islands, interconnecting them through various piping (like gas, water, or steam) and providing support functions. These systems may include, but are not limited to:

- Heat rejection systems to condense steam, turn it back into water and supply to the NSSS, including all necessary pumps, preheaters, deaerators, etc.;
- Moisture separators and reheaters to remove excess water and superheat the steam, thus avoiding erosion corrosion and possible erosion in the intermediate (if any) and low pressure sections of the steam turbine;
- Water treatment system for the conventional cycle. In case of BWR, this water is treated within the nuclear island because of its radioactivity. For other types of reactor, where water and steam are not exposed to radiation, this system supplies demineralized water as in a traditional thermal power plant. Similar to conventional thermal power plants, water has to be properly cleaned and demineralized to avoid sediments and clogging inside the water–steam cycle;
- Oil treatment system to provide clean oil to the rotating equipment for lubrication and hydraulic purposes;
- Compressed air system to power smaller valves;
- Drainage to remove excess water and non-radioactive waste to the drainage system for further waste treatment and disposal;
- *Emergency diesel generator* used to supply emergency power. There can be one or two diesel generators to produce electricity. These machines are run on liquid fuel (fuel oil), which is stored on site in a tank. The capacity of emergency fuel supply shall be enough for several days of operation, depending on the local regulation;
- Fire protection system for the whole power plant. It includes both active (water, mist or compressed CO_2 disposal) and passive (special coatings, penetration closings, etc., to stop smoke and fire or to control temperature increase) fire protection;
- Heating, ventilation and air conditioning, usually excluding nuclear island since it has its own system;
- Civil and structural works that incorporate all foundations, buildings, structures and other site facilities like fencing, gates, camp, office buildings, etc.;

- Hydrogen and nitrogen systems for the generator cooling and maintenance purposes (in case of hydrogen-cooled unit);
- Spent fuel storage ponds;
- Safety services, including medical, radioactive control, and fire fighting brigades;
- Security systems to ensure complete security of site and surroundings against any possible breach from outside.

Let us consider some of these systems in detail.

Heat rejection system

The design of a heat rejection system has a major impact on the overall efficiency of a nuclear power plant. This system includes equipment with interconnection and necessary monitoring and control to optimize the overall plant heat balance and therefore efficiency of the steam cycle.

This system normally includes the following components:

- Condenser, which is located below the LP exhaust of the steam turbine. It admits steam and turns it back to water. Condensers for a nuclear power plant have the same design as for conventional ones, which is described in detail in Chapter 9. However, it is unlikely to use an air-cooled condenser in a nuclear power plant due to its lower efficiency and massive steam flows. Therefore traditionally a steam-to-water type heat exchanger is implemented;
- Cooling system, which supplies cold water to the condenser to maintain vacuum and evacuates heat released during condensing process;
- Condensate pump, feedwater heaters, feedwater pump, deaerator, etc., to optimize the thermal conditions of the water within the cycle in order to increase its efficiency and minimize losses.

The major question that has to be investigated and answered during the feasibility study is the type of a cooling loop. Due to high amounts of steam and size of the unit, the cooling loop has to provide a sufficient temperature drop to the steam at the exhaust of the steam turbine. Theoretically, this can be managed through various systems, which are implemented in a traditional thermal plant, however, nuclear facilities require much higher cooling water flows. Depending on the site characteristics, cooling is usually conducted by one of the following means (some of them are schematically illustrated in Fig. 12.13):

1. Water only, in a once-through arrangement, when large amounts of water are freely available from the nearby river or sea. This type of cooling is very efficient, but may cause heating polluting of the en-

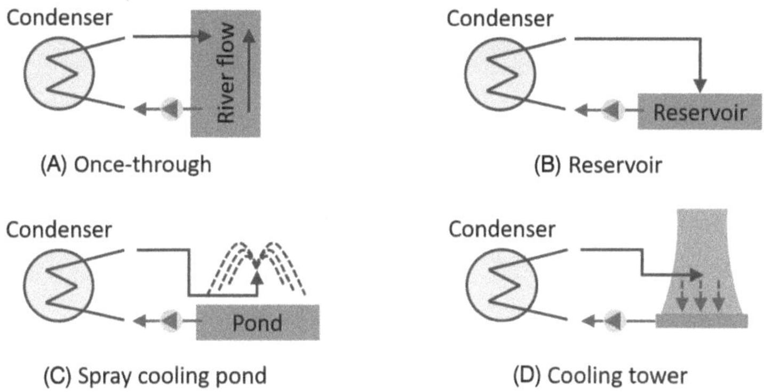

Figure 12.13 ***Types of cooling system used in nuclear power plants.*** Schematic diagrams of various cooling systems which can be implemented in a nuclear power plant: (A) once-through cooling, (B) reservoir, (C) spray cooling pond, and (D) cooling tower.

vironment or, in the unlikely event of accident, may easily spread the contaminated cooling water;

2. Water from the artificial reservoir or cooling pond. This is as efficient as once-through cooling, however, requires additional space, which shall be inside the nuclear area for safety reasons;
3. Spray cooling pond, where spray coolers are implemented over the smaller artificial pond to cool down the water with the ambient air. It requires much less space than reservoir and has very high efficiency. However, it still requires a lot of space;
4. Water and ambient air through a cooling tower of either type (mechanical or preferably natural draft). These coolers are more expensive to build, but have a much smaller footprint.

Spent fuel storage ponds

Spent fuel from the reactor still releases heat and radiation due to radioactive decay and cannot be recycled or disposed immediately. It is therefore required that spent fuel is stored in a special storage pond on site for several years. These ponds contain special racks where the fuel elements are stored under water. This water serves two purposes:

- It reduces the radiation level considerably and hence serves as a shield;
- Due to natural circulation, it evacuates heat from the elements stored deep. This circulation usually does not require any auxiliary power, however, special equipment is normally utilized for redundancy purposes.

Water treatment is usually carried out within the radioactive waste treatment system. Backup water sources are required in case of emergency refilling.

This pool is located in a special enclosed building, which is usually connected to the containment building to allow for storing unloaded fuel in case of reactor maintenance. Fuel handling operations can be controlled either manually or from the control panel.

12.3.3.2 Electrical BOP

Electrical balance of plant systems of a nuclear power plant incorporate equipment, systems and interconnecting cables to safely and reliably deliver power from the generator to the grid and to supply high, medium and low voltage power to all in-house loads under all service conditions, including normal operation, start-up and shutdown, and especially in emergency regimes.

For any power plant, these systems serve the following purposes:

- Synchronizing generator to the grid frequency and voltage;
- Transferring electricity from the generator terminals to the grid;
- Protecting electrical equipment from the possible damages;
- Feeding electricity to all plant systems to ensure their proper functioning;
- Implementing plant instrumentation and controlling system and data network to ensure safe operation, synchronization and reaction to the system operator requirements, if necessary.

These functions are performed with the help of several systems implemented in a power plant with a certain degree of redundancy to ensure the highest possible level of safety. These systems include, among others:

- Power evacuation;
- Auxiliary power supply;
- Electrical power network (wiring, cabling, switchgears, etc.);
- Instrumentation and control system with control networks.

Power transfer system

The major power transfer system is used to deliver electricity from the generator terminals to the grid. It has the same structure as for a conventional steam turbine driven power plant and normally incorporates the following equipment:

- Generator bus bars to carry electricity from generator terminals;
- Generator circuit breaker, which is used to protect equipment (either generator or transformer downstream) from overvoltage and overcurrent;
- Main step-up transformer to increase the voltage to HV level of hundreds of kilovolts before it goes to the grid;
- Substation or switchyard to distribute electricity to the grid and to isolate the plant from the external grid, if required.

This power transfer system can operate in both directions: in normal operation regime it delivers power to the grid, while in emergency it can be used to power the plant systems from the grid as an emergency source of electricity.

Auxiliary power supply

There are three auxiliary power distribution systems within a nuclear power plant:

- High voltage AC (for instance, 10 or 13.8 kV) supplies power to the critical rotating equipment of high capacity: feedwater pumps, reactor coolant pumps, circulating water pumps, condensate pumps, etc.;
- Medium voltage AC (e.g., several hundreds of volts) is used for moderate capacities and emergency supply like auxiliary and emergency feedwater pumps, HVAC system, various motor operated valves, and other equipment;
- Low voltage AC (for example, 110 or 220 V) powers the rest of the systems, including various small pumps, lighting, and so on;
- Emergency DC supply system is activated in case of loss of main power (generator shutdown). It employs a number of batteries which are charged during normal operation and, if the voltage goes down, switch on automatically to support major safety-related functions of a nuclear plant;
- Emergency AC power from the diesel generator is the second option which is activated straight after the batteries due to some natural time gap between possible loss of power and diesel startup and ramping time.

Contrary to conventional power plants, a nuclear power plant has to have a much higher rate of safety and redundancy to ensure continuous operation of all critical systems, especially cooling of the reactor. Therefore, electrical systems of a nuclear power plant have more sources of power supply.

Instrumentation and control system

The *instrumentation and control* (I&C) system architecture, together with plant operations personnel, serves as the "central nerve system" of a nuclear power plant [5]. It consists of various elements [5]:

- Sensors, interfacing with the physical processes within a plant and continuously taking measurements of plant variables such as neutron flux, temperature, pressure and flow;
- Control, regulation and safety systems that process measurement data to manage plant operation, optimize plant performance and keep the plant in a safe operating envelope;
- Communication systems for data and information transfer through wires, fiber optics, wireless networks or digital data protocols;
- Human–machine interfaces (HMI) to provide information and allow interaction with plant operating personnel;
- Surveillance and diagnostic systems that monitor sensor signals for abnormalities;
- Actuators (like valves and motors) operated by the control and safety systems to adjust the plant's physical processes;
- Status indicators of actuators (for example, whether valves are open or closed, and whether motors are on or off) providing signals for automatic and manual control.

This system senses major parameters (like temperature, pressure, acceleration, velocity, etc.) of the equipment, monitors performance of the plant and accumulates and integrates information, which is fed to the control system. The automated control system makes necessary adjustments to the plant operation according to the predefined algorithms.

More importantly, the automated control system responds to any abnormal performance and possible failures to maintain safe, reliable, and efficient operation. All information is represented on HMI workstation screens to ensure that the personnel of a nuclear power plant are controlling the situation.

Modern digital instrumentation and control systems have a number of redundant control and protection systems covering all equipment in a nuclear power plant: reactor, feedwater, cooling, steam, turbines, etc. The control information comes from the monitoring devices to the process control stations, analyzed and delivered to the operator workstation through the plant data highways, while the control commands are transmitted in the same way back to equipment. Same type of information comes from the protection devices, however, through a separate data network due to process criticality and to ensure redundancy.

On top of this, operators can make manual tripping commands if required by the operating procedures.

12.4. Nuclear power plant safety

Ensuring that a reactor can be operated safely is one of the most important goals of reactor designers and operators. In order to minimize the potential hazard to the public from the radioactive material contained within an installation, a number of principles and provisions have been developed and incorporated into the design and operation of nuclear power plants. Collectively, these principles are summarized in the golden rule of reactor safety, which can be stated as [3]: there is a minimum risk to the public and the environment from the reactor operation, provided that at all times:

- The reactor power is controlled;
- The heat generated in the core is removed;
- The radioactivity is contained.

The basic underlying approach towards the nuclear island safety is ensuring the highest possible standards in overall plant design, construction, reliability of components, and quality of materials. Major concerns about possible accidents and the lessons learned have led to particularly intense efforts to achieve high standards [4]. This has been implemented on a continuous basis, and from generation to generation nuclear reactors, they have been subject to continuously increasing safety standards and regulations.

This has resulted in a number of approaches and safety measures, which have been implemented in all types of nuclear power plant, both in operation or under development and construction. These measures include active and passive safety, redundancy, and defense-in-depth [4].

12.4.1 Nuclear accidents and their consequences

According to IAEA [7], there are seven levels on events classified according to their consequences (see Fig. 12.14):

1. Anomaly is the lowest type of an incident that has no impact to people and the environment, but may lead to slight overexposure of a member of the public in excess of statutory limits and cause minor problems with safety components. However, defense-in-depth remains.

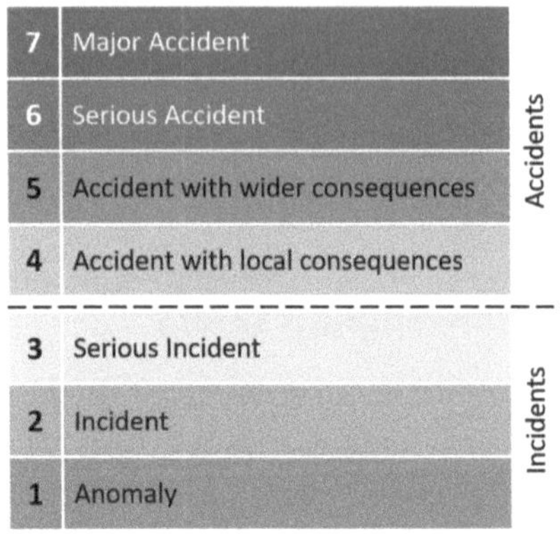

Figure 12.14 ***The international nuclear event scale adopted by the IAEA.*** The events are classified according to the consequences and the impact. *Source: [7].*

2. Incident with exposure of a member of the public or the worker in excess of the statutory limits. This may lead to significant contamination in some areas of a plant.
3. Serious incident with exposure in excess of ten times the statutory annual limit for workers with non-lethal effects. This might cause severe contamination in the areas not expected by design.
4. Accident with local consequences with minor release of radioactive material with at least one death from radiation. This usually does not require implementation of countermeasures except for food control. Examples include Saint-Laurent (1969, France), Tokai-Mura (1999, Japan).
5. Accident with wider consequences that causes limited release of radioactive material (causing several deaths) and possible implementation of some planned measures. This type of accident took place at Three Mile Island (1979, USA) with fuel melting, or Windscale (1957, UK).
6. Serious accident that causes significant release of radioactive materials and therefore possible implementation of planned countermeasures. For instance, one can name Kyshtym accident in a military reprocessing plant (1957, USSR).
7. Major accident leading to a major release of radioactive material, health and environmental effects. It requires implementation of extended countermeasures. Examples include Chernobyl (1986, USSR) with fuel meltdown and fire, and Fukuhsima Daiichi (2011, Japan) with fuel damage, radiation release, and massive evacuations.

The truth is that perfect safety does not exist. Despite the adoption of rigorous safety measures and regulations, nuclear plants, like any other industrial activity, cannot be entirely risk-free. Potential sources of problems include human errors and external events having a greater impact than anticipated. Accidents may happen and, in fact, a number of them have occurred. Analyses of the causes of their occurrence have provided valuable knowledge and lessons learned, leading to progressive improvements in safety [3].

12.4.2 Active and passive safety

There a distinction between *active* and *passive safety* implemented in a nuclear power plant in general, and in a nuclear island in particular [4]. An active safety system depends on the proper operation of reactor equipment and overall systems, which are used in emergency (e.g., cooling, control rod introduction, etc.). Passive safety features are usually readily available due to the overall plant design and physical principles of the matter and technologies implemented in a plant. These systems do not require any specific action of the operator, automation, or even any of the mechanical, electrical or pneumatic drive (which in the unlikely event may fail).

Both systems can be well combined with each other. For example, the gravity-driven fall of the control rods into the reactor core has to be initiated by the control system or operator (active safety), however, does not require any further action to physically move the rods into the core as it will be done purely by physical principles.

Passive safety, also known as *inherent safety*, is implemented during the conceptual design phase. In the extreme case of an accident, when the reactor is left completely unattended for some time, with coolant and auxiliary power supply being cut, the passive safety system shall be capable to turn itself on to shutdown and gradually cool down the reactor core with no damage. Unfortunately, such an ultimate level of safety is difficult to achieve, though there have been certain developments in this field, and the Generation IV is expected to be of very high passive safety standards.

12.4.3 Redundancy

Modern design and manufacturing techniques have improved dramatically during the last decades. The reliability factors of various pieces of equipment and systems allow for multiple hours of continuous operation, if planned maintenance procedures are applied accordingly. However, despite multiple levels of quality checks, sophisticated monitoring and control devices, and special software for predicting the changes in operation, state of material and therefore possible failures, there is still a probability that something may go wrong. This is an inevitable and therefore unavoidable process linked to natural tear and wear, structural changes in the matter, material fatigue and failure, or extreme and unpredictable environmental conditions. This can be minimized to a certain extend, but cannot be 100% avoided.

However, the likelihood of any sort of accident caused by a failure can be significantly reduced to a minimum by *redundancy*. This term means that a device or a system has a kind of backup: if the major one fails, the other would immediately continue its operation without causing an accident and therefore give time for maintenance and repairs.

Redundancy can be implemented in a number of ways:

- Use of identical units of the same type and functionality. This is mostly applicable to motors, pumps, or other critical equipment which perform safety-related functions. It does not mean that the redundancy has to be 100% operational, but at least ensure that one failure would not lead to simultaneous problems of all components.
- Parallel diverse systems can be implemented to perform the same functionality. For example, these can be different cooling methods for emergency, various power supplies for pumps (from the auxiliary power, external grid, emergency diesel generators and UPS system), or multiple communication and control channels wired through multiple routs.

Overall, redundancy is a very common approach used in almost all areas related to reliable and safe operation. However, in a nuclear power plant, redundancy is extremely important and considered as a requirement for safety reasons.

12.4.4 Defense-in-depth

The *defense-in-depth* concept assumes multiple safety barriers to ensure several levels of independent levels of protection. The principle of defense-in-depth is seen in considering the barriers that prevent or minimize exposure due to the release of radioactivity from a reactor. It usually consists of several barriers [4]:

1. Nuclear fuel pellets have ceramic-like properties and can retain most of the radionuclides, although some gaseous fission products may escape;
2. The special zircalloy cladding of the fuel assemblies (rods) has extended corrosion resistance and traps most or all of the gaseous fission products that may escape from fuel pellets. The design is specifically done

to allow for the gases stay inside the fuel rod without increasing the pressure.

3. The pressure vessel, manufactured out of special steel (with wall thickness around 200 mm), together with the cooling loop, retain fission products that escape from the fuel assemblies due to theoretically possible defects in the rods or in case of an accident or overheating.
4. Nuclear reactor containment building has a number of its own safety systems and is designed to retain all radionuclides that escape through the cooling system or, in a case of a very severe accident, from the pressure vessel. It normally consists of the outer (reinforced concrete) and inner walls to maintain containment of the inner volume. It is designed to withstand earthquakes, floods, hurricanes, direct impact of a commercial aircraft, and all other possible natural disasters.

All these barriers are usually designed and then tested to the extend of the most unlikely event. In case the reactor containment building and all other levels of safety systems fail, and there is a significant release of radioactivity, then the population has to be evacuated and rescue measures need to be performed.

References

[1] Davies G, editor. Glossary of nuclear terms. Burges Salmon; 2014.
[2] Gupta M. Power plant engineering. PHI Learning. ISBN 9788120346123, 2012.
[3] De Sanctis E, Monti S, Ripani M. Energy from nuclear fission: an introduction. Undergraduate lecture notes in physics. Springer. ISBN 9783319306513, 2016.
[4] Bodansky D. Nuclear energy: principles, practices, and prospects. New York: Springer. ISBN 9780387269313, 2007.
[5] International Atomic Energy Agency (IAEA). Power reactor information system (PRIS). https://www.iaea.org/PRIS, 2017.
[6] Eerkens JW. The nuclear imperative. A critical look at the approaching energy crisis. Topics in safety, risk, reliability and quality. Springer; 2006.
[7] International Atomic Energy Association (IAEA). The international nuclear and radiological event scale user's manual. Vienna: IAEA; 2013.

CHAPTER THIRTEEN

Development of sustainable nuclear power plant project

Contents

13.1. Nuclear power plant project justification

Nuclear power plant construction can be considered among most difficult and challenging infrastructure projects due to its complexity, specific regulation, siting limitations, high upfront investment and, more importantly, social acceptance. It normally takes long to conduct a proper project justification and its further development.

13.1.1 Feasibility study and project key drivers

General perceptions of the nuclear energy, among both the public and policy makers, have undergone several dramatic shifts in the past decades. As nuclear energy emerged in 1945 from scientific obscurity and military secrecy, it began to be talked of in speculative terms as an eventual power source. As experience was gained in reactor construction, it would become much more economical; and because reactor would emit virtually no pollutants during operation, it would be relatively clean, especially in contrast to fossil-fired technology. More generally, nuclear energy was favored by an image of it as a source of abundant, clean energy, by the possible technological imperative to move ahead because it was possible to do so, and by

https://doi.org/10.1016/B978-0-12-817012-0.00025-6

the ongoing recognition of an eventual limit to fossil fuels that started to appear on the horizon [1].

Contrary to other traditional power plants, the future of nuclear power does not fully depend on pure economic prospects. Economics in this case implies the operation of nuclear power plants, but also the facilities going with them to supply nuclear fuel and dispose of nuclear waste [2]. The economic advantage here lies in its relatively low fuel costs, which means that the low volatility of the uranium price will only have moderate effects on the overall electricity generating costs.

The nuclear power plant business model implies stable base load generation, however, can be well extended to other applications like desalination and production of fresh water or district heating — both opportunities, however, have not yet been commercially implemented.

Thus, because of strong regulatory frameworks, a nuclear power plant project goes beyond common economic evaluation, and the social and political concerns play a major role in a decision-making process. Therefore the overall process is by far more complicated and involves multiple stakeholders over much longer investment decision and project cycle. This is even more impacted by the considerably higher capital expenditures of the project compared to thermal power plants, where private companies are capable of doing it without significant efforts. Nuclear power generation is much more regulated and impacted by the governments, therefore private companies have to undergo thorough feasibility study processes and comply with regulations before making an investment decision and start looking for money.

13.1.1.1 Feasibility study

A feasibility study covers both the preparation work and the scope of the feasibility report itself. An investigation of the energy conditions in a country at the pre-project stage and the planning of the energy sector development needs to be established after conducting an analysis of the structure of the national energy market and of the supply and demand for energy in the country [3].

The decision that nuclear energy would be part of the power industry of a country is taken on the government level. In those countries, where nuclear segment has developed for decades, the legal framework and the requirements are already on the high level and therefore would require less effort of approving new capacities or even technologies. However, in any country the decision of building a new or extending an old nuclear station requires close consideration of various aspects: technology, degree of infrastructure development, available resources, social-economic aspects, legal framework, safety, fuel supply, and many others.

The country's nuclear energy act, its laws and decrees should cover all matters related to the national energy policy. The government should formally commit to the use of nuclear energy for peaceful purposes, adhere to international conventions and treaties, enforce environmental protection, codify occupational health and worker safety policies, and establish emergency preparedness and management policies and the role of various government levels, stakeholders, and the public, and their interrelationships in enacting these policies. Regulation should also cover foreign investments, vendor and contractor relationships and expected ethical behavior, intellectual property rights, preservation of national traditions, cultural values, promotion of economic benefits, and technological development, control of nuclear material, fuel cycle issues, and the national insurance coverage policy [3].

Taking into account the overall complexity of the industry in general and nuclear projects and with long lead times, a nuclear power plant project lifecycle would last for decades and literally through several generations of people. This poses a serious question of a thorough feasibility study of a nuclear technology deployment.

The feasibility study for a nuclear power plant project is the process to determine whether the nuclear technology deployment is advantageous to the country, and identify the most suitable characteristics and conditions for a possible plant [3]. This study shall involve all possible stakeholders directly or indirectly impacted by the new technology. These include, for example, government bodies, NGOs of various kinds, industrial and residential consumers, residents of the country or at least of the surrounding region. In certain cases, an international panel discussion is required in case the power plant is planned near the border or for cross-border electricity supply.

13.1.1.2 Electrical system analysis

A national or even regional cross-boundary electrical system analysis is usually performed to understand the capacity and needs of the demand of the electricity market, as well as the capacity of the grid to accommodate these additional capacities.

An electrical system analysis assumes a number of activities and includes:

- Current electricity demand and its forecast for the whole lifecycle of a nuclear power plant;

- Supply characteristic of the generated electricity;
- Electricity market structure and its organization with the potential influence with new plant;
- Base scenario of the overall electricity system development;
- Nuclear power plant project impact on the grid and the possibility to transfer full capacity with the current grid assets.

This study can be carried out together with other planning activities of new capacity introduction compared with the regional electricity demand. This analysis is normally based on historical data, and can include:

- Thorough study of the electricity consumption scheme;
- Load demand, e.g., base, semi-peak, and especially peak;
- Duration of each demand and its nature (seasonal, weekly or daily);
- Efficiency study of the transmission and distribution system, understanding current and possible future losses and capacity bottlenecks;
- Historical supply of electricity, including fuel and electricity pricing, as well as various flows of energy imports and exports.

Finally, electricity demand projections for the nuclear power plant project lifecycle time are proposed. There are multiple approaches, including statistics-based extrapolation or consumer analysis, however, the result shall give a clear understanding of the market energy demand. This understanding is crucial to define the scope of the future power plant, namely electrical capacity and the capability of the generator to respond to the grid requirements.

The analysis should be conducted with relevant and consistent macroeconomic and microeconomic data, so that electricity demand projections can be reliable and consistent with demographic, economic, and industrial development projections [3].

Remark 13.1. While the same type of analysis is conducted for any type of a power plant, it is by far more important for nuclear projects. This is due to much higher investment costs and rather low flexibility of the plant to match market demand, if needed. While thermal plants require less money to build, and big hydro facilities can operate with certain flexibility (to allow for better market fit within semi- and peaking mode), nuclear energy is subject to much higher risks.

The result of the analysis shall clearly highlight the advantages of an addition of nuclear capacities within the market, for example, to cover growing base load electricity demand, growing emissions charges to be mitigated with low carbon technology, energy independence of the region from foreign power imports, etc.

13.1.1.3 System capacity

Naturally following the electricity system analysis and demand forecast, the feasibility study shall answer the question about the nuclear power plant capacity and overall system integration, yet not going into technical details.

The size of the system is the overall capacity of a nuclear power plant that can consist of several units, so that the size of a nuclear unit in this context refers to the maximum electrical power output in megawatts, usually denoted by MW_e, that it can deliver to the grid.

There has been a steady increase in the unit size, partly due to economies of scale, so the designs of nuclear units that are currently available from international nuclear plant vendors are large. This may introduce additional requirements if the grid system is relatively small [3]:

- The need to control large and rapid changes in frequency, voltage, and power flow that will occur after a trip of the nuclear unit or if a fault in the transmission system disconnects the nuclear unit (primary reserve);
- The need to have sufficient generation to meet electricity demand during periods that the nuclear unit is shut down, whether for planned maintenance or following a fault or unplanned trip (secondary and further reserves).

In order to integrate the unit into the energy system and ensure power evacuation from the generator terminals to the grid, suitable HV lines are required supported by appropriate substation. The number of HV power lines is driven by the capacity of the plant and number of blocks, however, the overall grid connection is defined after the site location is confirmed, as it directly drives the length and therefore cost of line construction.

13.1.1.4 Siting

Siting is the term used to describe the process to select where a nuclear installation shall be built and whether the decided location is suitable for it. Siting is one of the most important decisions during the feasibility study of a nuclear energy project.

Nuclear site selection is subject to a number of restrictions and requirements coming from both technology and regulation. The legal framework of site selection comes from the regulators represented by the government of the accommodating country, international bodies like IAEA, and other organizations that may have an impact. These guidelines define the safety requirements to the design and construction to ensure that the deployment would not put people and the environment in danger.

Along with the legal and economic requirements, social aspects of nuclear power play a dominant role. Recent events have shown that the public acceptance (or denial) can bring a destructive influence to the nuclear industry, not only through the demonstrations and objection of the selected site, but also the complete ban of the new construction and even possible closure of the existing plants. With regards to new build plants, the public is usually invited to the open discussions and is free to propose restrictive measures that would influence the overall site selection procedure, surroundings involved in the footprint, and the safety of the project. For instance, the government can propose a number of possible locations for public acceptance and make the process of information exchange as transparent as possible.

Site selection is among the major steps to be conducted during the economic evaluation of the nuclear power plant project. At the very beginning of the feasibility study, large areas of the country planning to introduce or expand nuclear power are investigated to determine potential nuclear sites.

This step usually involves activities to check all necessary parameters of sites available theoretically, so that by the final decision there are a few site possibilities left. These possible sites are then studied in deeper details to more precisely estimate the capital costs at each location. At this point, the most suitable areas are identified and candidate sites are selected [3]. In general, the least expensive site is favorable from the economic point of view.

When the shortlist of several preferred sites is prepared, the site characterization process starts and the focus is shifted to the most promising site in order to demonstrate its final acceptability and suitability, and to derive the design basis for site related safety aspects [3]. Site requirements are usually driven by the available technology and resources. These include, but are not limited to, the following points [4]:

- The site has to provide enough space to build the complete nuclear power plant of the required capacity with all necessary infrastructure, network connection and building and construction camp.
- The proximity to the nearest villages, towns, cities or any other inhabited places from site has to comply with local (if any) and applicable international regulation. In most cases there shall be a sanitary protected area around site to ensure the required safety level. The possibility of allocating the safety zone around the power plant means that a nuclear site must be located in a remote place of about 15–30 km away

from densely populated areas to minimize potential radiation or other hazards.

- Prevailing wind direction shall be preferably in the opposite side to the nearest densely populated areas to minimize the possible risk of contamination with radioactive clouds and precipitation.
- Adequate water supply for multiple needs of the plant. The major consumer of water is the cooling system which usually requires twice more cooling water than the conventional steam power plant. One of the best solutions will be to have a big water reservoir available both for cooling of the equipment, spent fuel and for emergency needs. Obviously, a nuclear power plant requires massive cooling towers, cooling ponds or once-through cooling loop.
- Availability of land with the required soil profile and the least possible seismic impact to ensure stable and safe construction and operation. As the major buildings are erected out of the highly reinforced concrete, while the major equipment is manufactured of special steel, the soil bed shall withstand much higher loads compared to other types power plants. On top of this, the nuclear power plant design is subject to much stricter seismic requirements for all structures, especially those of a nuclear island and cooling loop.
- The whole site shall be carefully studied against all possible natural disasters in their worst case. For instance, strongest possible earthquakes, hurricanes, tsunamis, floods, etc., shall be taken into consideration to ensure that both site location and the overall design comply with the safety standards. The site shall not be subject to any seasonal flooding and has to be located higher than the highest possible water level within the region. It shall not be located downstream of hydropower plants to minimize the risks of the dam failure. Ground water levels shall be below the lowest possible construction point of the nuclear power plant to ensure stability of all structures.
- Proximity to major roads and other transportation paths allowing for the delivery of oversize and heavy material. In the best case, these shall be either by railroad, which is free of tunnels and bridges to allow for oversize transportation, or by water. The latter would require additional unloading facilities capable of accepting the vessel with bulk goods. These roads shall be also used for radioactive waste disposal.
- Network connection with the overall community has to be maintained with 100% at any time. This is important in the potential cases of emergency to make sure the information flow is stable at any circumstances.

- Availability of auxiliary power for the period of construction and proper connection to the grid. A related substation should be capable of accepting and transmitting high stable outputs of the power plant operating in a base load mode.
- Availability of an emergency fuel for auxiliary systems like diesel generators to run the cooling loop.

Moreover, there are many safety related human induced events that also have to be considered [3]. These include potential airplane crashes and other transportation hazards, fires and explosions, chemical and radioactive hazards, terrorist attacks, proliferation of nuclear weapons, and so on.

Siting factors and criteria are important to ensure that radiological doses from normal operation and postulated accidents are acceptably low and that natural phenomena and potential human induced hazards are appropriately accounted for in the design of the plant [3].

Finally, social acceptance by the public of the preferred site location can be the limiting factor of the whole project. Modern nuclear development activities try to involve public as early in the study as possible to make sure that people understand the benefits and the safety measures that would be implemented at site.

13.1.1.5 Preliminary site layout

Preliminary site layout is an essential part of a feasibility study, as it allows estimating the following project-related aspects:

- Constructibility of the power plant facilities and spacing between the buildings and structures;
- Efficient and safe layout for personnel to ensure safe working conditions and emergency evacuation ways;
- Site access roads, transportation limitations and further development, if required;
- Potential facilities and houses for relocation;
- Proximity to cooling water source and the ways of water delivery and discharge.

Overall, the site layout has to support the decision of the site selection and define the design basis for the power plant.

13.1.1.6 Technology and fuel cycle assessment

Based on the electric system analysis, defined system and unit sizing, and selected site, appropriate nuclear technology shall be defined. These activities are a logical conclusion of the previous studies, and therefore are aimed at defining a clear project requirements' envelope and high level scope.

This part of feasibility study normally includes [3]:

- Reactor technology assessment to identify preferred reactor type to be implemented in the power plant according to the required capacity. Ideally, it shall be done through evaluation of financial and technical value against competition, however, there is always certain (usually large) element of politics behind;
- Nuclear safety and performance assessment to define project-specific boundary requirements;
- Detailed project risks assessment, for instance, dealing with the schedule, technology maturity, quality assurance, pre-qualification of contractors, etc.;
- Fuel cycle considerations, especially focusing here on fuel supply, interim storage and radioactive waste treatment.

13.1.1.7 Environmental impact assessment

Environmental protection should receive serious attention when a nuclear program is contemplated, ensuring no harm comes to humans or the environment at the site, within the country, in the region or globally. This should be a high priority goal during all phases, namely, design, construction and normal operation, and under accident conditions and during decommissioning activities [3].

The ecological impact estimation should assess not only possible direct damage of the worst possible case of an accident, but also the damages to the environment during construction (site clearance, excavations, roads, etc.), as well as direct operational influence (like cooling water consumption or heat pollution). Its results shall be well implemented within the cost modeling, including the calculation of lifecycle CO_2 emissions.

13.1.2 Nuclear power plant financing mechanisms

Nearly all nuclear power plants operating today were financed and built in regulated utility markets. Thus, they were guaranteed both future customers and high enough electricity prices to ensure a profitable rate of return. Under these conditions, cost overruns and project delays were covered by higher electricity prices and were ultimately paid for by customers. In addition, in many countries much of the financing for nuclear power plants was provided by governments or with government backing or government guarantees of some kind [5].

According to the analysis conducted by IAEA [5], the economics of nuclear power depends on national or even local conditions, including the costs of capital, labor, and materials, the regulatory environment, and the availability and costs of alternative generating technologies. Moreover, political opinions on nuclear power are often divided and strongly held.

For large multi-billion projects like a nuclear power plant, financing by a combination of debt and equity is generally required, as 100% equity or financing from one source is highly unlikely (except for the government). The split between equity and debt depends on the nature and financial position of the project sponsors, on the local conditions where the plant is to be built, and on the viability, structure, and evolution of the electricity sector in which the plant will operate [5].

There are several options for nuclear power plant project funding:

- Initial financing with some government funding for preliminary project assessment and feasibility studies, regulatory considerations, etc.;
- Capital market instruments like securities, stocks, or bonds;
- Directly allocated development funds from various international aid organizations, development banks, export credit agencies, etc., especially in developing countries;
- Infrastructure funding from institutional investors like pension funds, insurance companies, large investment funds, etc. In this case, part of the ownership of the plant can be transferred to the investor.

Evidently, nuclear power cannot be introduced in a country without some government support [5] as it is first a political decision. Governments not only have the power to establish favorable financing conditions, but also organize the legal framework and regulatory environment for further nuclear licensing, construction, and operation. Governments can therefore have the ownership of the nuclear sector (partial or full), equity participation, risk sharing, etc.

On top, governments can support debt financing by guaranteeing certain conditions for future operator (like electricity sales price), while utilities still have to be profitable in the end. This means that any nuclear power plant project has to go through a specific analysis to prove that the economics is within the competitive range compared to other power generation sources. The financial model shall consider all components of the cost incurred during the lifecycle of the nuclear power plant. These elements shall include:

- Licensing costs for design, construction, and operation (can be also allocated as owner's costs, see below);

- Capital costs, or CAPEX, for EPC services, preliminary and detailed design, construction, equipment manufacturing, installation and commissioning;
- Owner's costs, including land acquisition, water supply, project management and other associated activities;
- Plant operating and maintenance costs with the budget for future life extension;
- External costs dealing with possible risks;
- Decommissioning costs, which have been ignored for many years.

Capital costs can represent up to 60% of the total nuclear power plant project costs and may vary from country to country due to multiple reasons: variations in labor and material costs, economies of scale, local or imported materials and technology, managerial differences and regulatory aspects, transportation, etc.

Proper business modeling of the project should ensure stable and reliable cash flow for the plant planned operation and the shortest payback period. This is among the most important prerequisites for successful project financing. It is also important to consider exchange rate risk that can be a major consideration for nuclear power plant construction, especially in countries that import nuclear technology or reactors and foreign experts. Hence, careful contracting and strong project management are the key elements in commercial risk allocation and mitigation, even for governments [5].

Overall project risk mitigation is usually the responsibility of the project owner and, in most cases, the government. The latter plays a dominant role in establish the regulatory environment for risk management. These risks can include political uncertainties, mostly dealing with the future of the project, stability of the situation in the country, unrest or civil war. Moreover, the country in general has to ensure non-proliferation of nuclear weapons, comply with IAEA safety requirements, etc.

Having said that, the overall project costs for a nuclear power plant can differ dramatically from country to country. According to [6], the average CAPEX is around 3500–5600 $/kW of installed electrical capacity, with the LCOE being in the range of 90–150 $/kWh, as shown in Table 13.1. On the high end usually are projects in developed countries like France or Finland with the reactors that are currently under construction. On the low end are projects in the Middle East or China. However, there is a high rate of uncertainty on many nuclear power plant projects, mostly due to the very

Table 13.1 Typical nuclear power plant costs.

CAPEX, $/kW	OPEX, %/year of CAPEX	Capacity factor, %	LCOE, $/kWh
3500–6520	1.5–1.9	85–92	0.09–0.15

Source: Based on [6] and EIC and IIR databases analysis.

long planning and construction cycle relative to other generation options, so that nuclear projects can be subject to significant cost overruns [6].

Additionally to capital and operation and maintenance costs, it is required to consider nuclear fuel cycle costs and decommissioning and waste management costs.

The fuel cycle cost estimation depends on two major parameters:

- Type of nuclear fuel, since these days only two options are commercially available: uranium oxide or mixed oxide;
- Type of fuel cycle, once-through or treatment and recycling (refer to Section 11.2).

Some studies [3] suggest that fuel cycle costs may be around 20% higher than the once through estimate of 8.28 $/MWh, if reprocessing is adopted.

Since few NPPs have been fully decommissioned, the exact costs of implementing this last phase in the life of a plant can only be predicted. Estimates vary from 10% to an indefinite multiplier of the construction costs. Data are often not published owing to contractual arrangements and property rights [3].

13.1.3 Schedule and lifetime estimation

Nuclear project schedule is usually complicated to estimate and hence meet due to many challenges and issues, including, but not limited to:

- Technical risks like design delays and changes, regulatory approvals, contractual performance of vendors and equipment manufacturers, supply chain issues, quality control, transportation issues (e.g., climate-limited windows), and skilled labor for critical works;
- Economic risks caused by financing delays;
- Social and political risks that may be caused by general public attitude, political turmoils, regulatory issues, etc.

Feasibility studies, including the environmental impact assessment, siting and government decision on the construction till the moment when a construction permit is obtained may take several years and is subject to many factors. During this time, many of the bidding processes are held to understand the equipment suppliers.

Construction, installation, and commissioning works depend on the proposals and commitments from suppliers and contractors and also take up to 10 years till the commercial operation date. Construction schedules of nuclear power plants, from the first placement of structural concrete to grid connections, have ranged from less than 5 to more than 12 years [7].

13.2. Nuclear power plant lifecycle management

The main criteria for continued plant operation are that the plants must remain safe to the satisfaction of the regulators but also economically viable to meet the requirements of the utilities and stakeholders [8]. Safe, reliable, and economic nuclear power plants typically exhibit careful, conservative operation and rigorous, well-planned maintenance activities to minimize risks to workers, the public and the environment [9].

The operator is responsible for the plant's safety and therefore carries out regular assessments of essential structures, systems, and components, while independent regulatory bodies conduct safety inspections to enforce action, if required. The IAEA's safety standards, developed on the basis of international consensus, address all aspects of safety in the operation of nuclear power plants and regulatory activities. They include requirements for the management of safety and the organizational and technical aspects of safety during the life cycle of nuclear power plants. They include implementing guides for all major types of facilities [9].

Operational safety mostly refers to the performance of the plant personnel and the management of plant safety at the facility [8] and includes, among others:

- Individual protection of the personnel;
- Monitoring of radioactivity discharge both on and off-site;
- Waste management safety measures;
- Spent fuel management and treatment before its transportation to recycling factory or storage.

The personnel therefore undergoes a continuous training process followed by examination, practice on simulators, and overall knowledge and competence buildup programs.

While the designed life of a nuclear power plant is usually up to 40 years, it is likely that with the appropriate investment and refurbishment, the lives of nuclear power plants may extend up to 60 years and beyond [8]. Since many of the present day reactors are approaching the end of their initial designed life, as shown in Fig. 13.1, there are considerable efforts undertaken

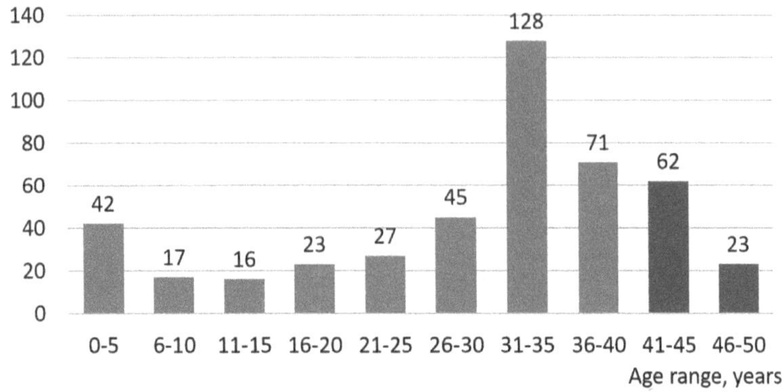

Figure 13.1 ***Age structure of the nuclear reactors in operation.*** Many of the present day reactors are approaching the end of their initial designed life with the peaking 128 units within the range of 31–35 years of life, and 85 of them have already reached the 40 years limit. *Source: Based on data from [9] as of December 2018.*

to extend their operational life by various technological means: retrofitting, upgrades of systems, changes in operational and maintenance practices and approaches, implementation of new safety features, etc. Without life extension, significant investment would be required to shutdown and replace old nuclear capacities by new ones. This is the situation in many countries in Europe, the USA, and Russia.

There are two major points of going into nuclear plant life extension:

- The economics of extending the life of existing plants, compared with the capital costs of building a new plant, is very favorable. The construction of a new plant, even of a different type (e.g., coal or gas-fired) of a similar capacity usually turns out to be more expensive than life extension of a currently operating facility. It is also much easier to get a license for extension (either site or unit life) rather than for a new nuclear construction at a different site.
- Nuclear decommissioning, which would logically be the next step after shutdown, is by far more complicated and cost-intensive activity, as discussed further in detail in Section 13.3.

However, the decision on the life extension depends on a range of issues to be thoroughly considered by the owner and regulator [8]:

- Technical feasibility to evaluate the effect of aging. It requires the guaranteed performance of major components and systems for the extended life period. While aging process reduces equipment operational boundaries and margins (e.g., pressure vessel, steam generators, cooling loops, piping, containment structures), it requires careful assessment and established procedures to set new operational limits.

- Plant safety for intended period of operation depends on many factors and current safety procedures and requirements. While the initial design (and therefore safety) is based on the experience and the obtained knowledge, findings of the maintenance inspections and results of tests, it can be still complicated to make forecasts for the next 20 years of further operation and therefore ensure complete safety.
- Regulatory framework to extend the operational license, as it fully relies on the decision-making process and rules of the regulatory body. While in some countries, like in the UK, a regulator can easily grant permit, others may still imply stricter rules and complicate the process, mostly driven by high rate of uncertainty.
- Public acceptance and political influence can as well be a limiting factor not only for the life extension, but for complete nuclear industry.
- Economic considerations to justify the investment.

Normally, the economic evaluation of the life extension project follows the same approach as for any other upgrade or rehabilitation, though many more factors have to be considered next to those mention before. According to IAEA [9] and [8], lifetime extension costs are much lower compared to those of a new construction, and can be around 200–1000 $/kW.

13.3. Nuclear decommissioning

Nuclear decommissioning is a long, expensive and complex process with a multidisciplinary nature [10,11]. According to the IAEA definition [12], decommissioning includes administrative and technical actions taken to allow for the removal of some or all of the regulatory controls from a facility (except for a repository which is closed and not decommissioned).

However, the scope definition of "nuclear decommissioning" is not internationally agreed [11], for instance:

- Administrative and technical actions taken to allow for the removal of some or all of the regulatory controls from a facility and to restore the site to new use [10];
- All clean-up of radioactivity and progressive dismantling of the plant. For practical purposes it includes defueling and removal of coolant. It concludes with license termination after decontamination is verified and waste removed [13].

- The process of safely closing a nuclear power plant (or other facility where nuclear materials are handled) to retire it from service after its useful life has ended. Decommissioning begins after the nuclear fuel, coolant, and radioactive waste are removed [14].
- Process of decontaminating the facility to reduce residual radioactivity, dismantling the structures, removing contaminated material and finally releasing the facility for other uses [15].

This process shall be considered at early stages of a nuclear project development as overall process of nuclear decommissioning may take up to 50 years (and even longer) depending on residual radioactivity, and fully depends on the design of the nuclear reactor and other systems. Therefore, it requires extensive financing. These are usually the owner and the government that are responsible for nuclear decommissioning and the overall legal and safety framework around this process.

This stage of a nuclear power plant lifecycle can be described as post operation activities leading to decommissioning of the plant and management of the nuclear waste within the frame of the country's long-term waste management program [8]. Many nuclear facilities will undergo decommissioning in the coming years as has been illustrated in the age structure depicted in Fig. 13.1. And the facilities include not only the nuclear power plants themselves, but also the whole value chain that consists of the uranium enrichment plants, fuel fabrication and reprocessing facilities, waste management, specialized transportation infrastructure, etc. This means that we can expect major investments till 2050 globally in decommissioning activities. These activities can be extremely expensive from a finance perspective, since most of the nuclear facilities that require decommissioning were not designed with this in mind and their financing modeling did not take these costs into account [8].

13.3.1 Nuclear decommissioning methodologies

There are three major internationally accepted stages, originally defined by IAEA [15]. These stages may be separated by extended periods of care and maintenance with the appropriate security and radiological surveillance, or they may follow directly one after the another in a continuous, systematic, and progressive manner [16]:

1. Reactors are completely defueled and the fuel usually shipped away from the reactor. All heat transport fluids and readily removable contaminated materials are removed as well. For the facilities outside the nuclear island, all radioactive sources and readily removable equipment

are removed. All contaminated material is either stored safely or recycled, and the access to the contaminated areas is controlled by physical barriers and administrative procedures.

2. Contaminated areas are decontaminated to the extent appropriate. Remaining areas with unacceptable residual radioactivity levels (for example, reactor core structure) are sealed to prevent unauthorized access. All contaminated parts that can be easily dismantled are removed or transferred off-site or into special plant areas that are to be sealed. Some parts of the plant can undergo refurbishment and can be used for other purposes.
3. All materials, equipment, and structures in which radioactivity levels exist above prescribed limits are removed to an approved storage or disposal site. Therefore the site and remaining equipment can be released for other purposes without any radiological protection restrictions. No further surveillance, inspection, or tests are required.

This allows the facility to undergo decontamination within a few years or even months after the shutdown, however, the overall project activities can take up to 25 years. This strategy targets major structures, equipment and parts be decontaminated and dismantled to ensure unrestricted access to site, or with restrictions imposed by regulator. In the literature it is referred to as an *immediate dismantling* strategy. Immediate dismantling is the preferred decommissioning strategy [15].

In some cases the immediate dismantling is not possible due to various reasons, so that other strategies are implemented [15]:

- *Deferred dismantling* is the strategy in which the final dismantling of the facility is delayed and the facility is placed into long term storage where it is maintained in a safe condition. This approach assumes that the facility is defueled first and is therefore made relatively safe for later dismantling. The deferred period can be from 10 years with almost no upper time limit, while the site will be maintained until radioactivity level is reduced to lower levels or preserved to ensure required safety. The deferred dismantling option is often used at multifacility sites when one or more of the facilities are shut down while others continue to operate. This is especially true of facilities that share some common systems.
- *Entombment* is a strategy when the nuclear power plant is defueled after shutdown and enclosed within the concrete building until the radioactivity decays to a level permitting unrestricted or partly restricted access by a regulator. This strategy is suitable only for special cases such as re-

mote or small reactors or in case of emergencies after the accident. This type of decommissioning would require long-term maintenance and monitoring. The fact that radioactive material will remain on the site means that the facility will eventually become designated as a near surface waste disposal site and criteria for such a facility will need to be met.

All these approaches have found their place in the latest recommendations of IAEA which defines the phases of a nuclear power plant lifecycle [16]:

- Operational phase to prepare for future decommissioning via data collection, proper maintenance and budget allocation;
- Shut-down transition phase, which includes defueling, post-operational clear-out (POCO) to take out readily removable contaminated material and equipment;
- Preparation for safe enclosure (either safestor or entombment) if such approach is chosen, which is followed by the safe enclosure period; and
- Final dismantling.

According to the IAEA [15], an important consideration when selecting the decommissioning strategy is the desired end state of the facility following completion of the decommissioning project. The preferred end state would be to achieve a situation whereby the site is released for unrestricted use as a green or brown field, or a situation whereby it may be released for restricted use. Other end states could involve partial release of a site or release of a site under restricted conditions to control its future use. The foreseen end states for the three different strategies are shown in Fig. 13.2.

13.3.2 Decommissioning strategy

Until recently, decommissioning did not enjoy a high profile for operators, decision-makers and other stakeholders in a number of countries. In the early days of the nuclear era, the focus was on design, construction and operation of a nuclear facility. Decommissioning was seen as a remote event [10]. It was perceived as a future concern with the resources to be somehow available at the time they are required. The immediate result of this approach was inadequate or no planning for decommissioning.

Evidently, the first step towards a successful decommissioning project is to develop an early strategy that is both acceptable to the regulator, and achievable in terms of program and costs [10]. Compliance to regulations and consistence with the national policy on the management of radioactive waste [17] are among the most critical factors of the decommissioning

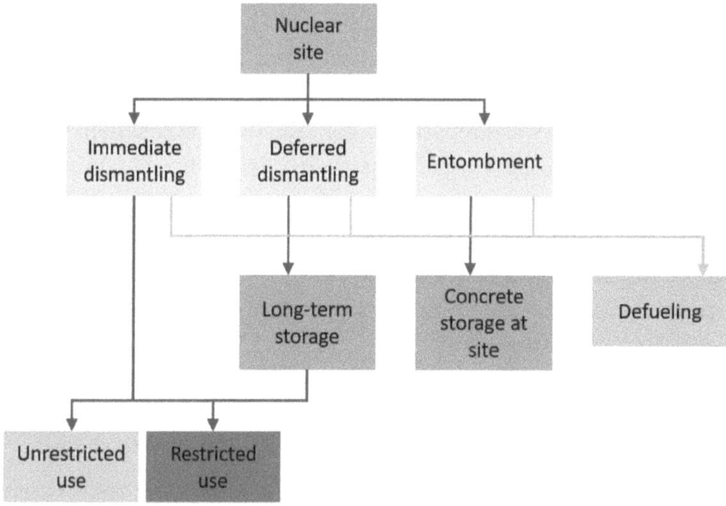

Figure 13.2 ***Major decommissioning strategies and their results.*** Possible end states for the use of areas in the three decommissioning strategies. *Source: Based on [15].*

strategy. In many countries, atomic energy legislation was framed primarily to cover siting, design, construction, commissioning and operation of nuclear facilities and there were no or very limited requirements regarding nuclear facility decommissioning [10]. However, as the old facilities are approaching the age of shutdown, local atomic energy regulators are issuing specific documents, licensing procedures, and laws, which still delays the overall process.

Evidently, the preferred decommissioning strategy shall be immediate dismantling as the most direct way towards a greenfield site with the unrestricted use — the ultimate goal the public would prefer. However, it is the responsibility of the license owner to identify a proper decommissioning strategy. The licensee shall demonstrate that, under the strategy selected, the facility will be maintained in a safe configuration at all times and will reach the specified decommissioning end state, and that no undue burdens will be imposed on future generations [17].

The licensee shall prepare a proper decommissioning plan and shall maintain it throughout the lifetime of the facility, in accordance with the requirements of the regulatory body, in order to show that decommissioning can be accomplished safely to meet the defined end state [17].

The general decommissioning strategy can incorporate several stages:

1. Planning the decommissioning starts when the plant is still in operation. It normally involves the cost analysis of the project and the sources of funding;
2. Post operation is the preparation for the shutdown when the maintenance activities are carried out to get ready for decommissioning keeping the plant safe. The fuel is still in power plant, and the power plant is still operating;
3. Characterization is the step of checking the radioactive components, structures, and systems to create a complete inventory and ensure the content of the radioactive waste is defined along with the associated risks. At this stage costs and timeline are fine-tuned and agreed;
4. Defueling of the plant when the fuel is taken out from the plant and transported to the recycling facility;
5. Decontamination is the activity to reduce radioactivity from the structures. Various components like pipes, floors, walls, etc., are released to reduce the amount of radioactive waste and its further exposure to the personnel. This is normally done through mechanical or chemical means;
6. Physical dismantling as a real demolition activity. It can be also conducted by remotely operated machines and tools to minimize radioactivity exposure;
7. Safe storage assumes that the plant is left dormant for residual radioactive decay. This, of course, includes routine maintenance and safety construction;
8. Demolition of remaining buildings;
9. Land refurbishment to make a greenfield.

All these stages are subject to a number of factors that generally affect the overall strategy and its execution:

- Safety factors cover the potential for harm to people's health, safety, and security, or the environment, from loss of control of the hazards at the facility [10];
- Financial factors as decommissioning is very rarely seen as an investment but rather a net cost activity;
- Technology factors may include both the design of the facility, its operational history and current state, timeline for radioactive decay, and mostly the availability of technical solutions for efficient decommissioning;
- People factors may be twofold: a project may mobilize new jobs and therefore be supported by the public, but can also be negatively ob-

jected by regulation, NGOs and the public due to potential safety concerns.

The problem of the cost estimation of the decommissioning has not been solved yet for many reasons. Primarily, this is due to the fact that the data on decommissioning is limited to a few projects that have undergone or are undergoing this nowadays. The early cost estimates for decommissioning were limited and based on then-available experience, which was very little. Similar experience from conventional power plants is applicable only to conventional islands for some of the plants, so that it cannot contribute to a realistic cost estimate.

In theory it looks quite obvious: owners and licensees of nuclear power plants are generally responsible for developing cost estimates of decommissioning, and a good understanding of these costs is fundamental for the development of estimates based on realistic decommissioning plans. Transparent cost estimates also provide a basis for accumulating the necessary funds with the aim of ensuring that these are available when needed to cover the actual cost of decommissioning activities [18].

The cost of decommissioning is greatly influenced by several factors or drivers, which must be carefully managed to avoid escalation and overruns. Some useful qualitative recommendations may be provided for each of these drivers [18]:

- Decommissioning policy and strategy needs to be defined as soon as possible, ideally when building a plant. This will allow for sharing of responsibilities between the diverse actors and for a streamlining of the process of cost estimation and its revision;
- Roles and duties of the diverse actors, and the regulatory framework;
- Planning and preparation phase prior to decommissioning, and site characterization. Costs of decommissioning will be influenced by the nature and level of radioactivity of the materials being handled;
- Management of spent fuel and operational waste;
- Dismantling operations and related waste management;
- Prospects for waste (final) disposal, including spent fuel;
- Final stage of decommissioning, de-licensing, site restoration and reuse;
- Manpower management, contractors. The cost of manpower would appear to be the main contributor to decommissioning costs, whether for preparation activities, project management, implementation of decommissioning activities, or waste management and surveillance;
- Risks management, uncertainties, and contingencies.

Moreover, the cost of managing spent nuclear fuel following the defuelling of the reactor, in particular interim storage of the fuel, is not always included in the cost of decommissioning, but is often treated separately. This is even worse in the case for the final disposal of fuel or related waste, which is a major source of costs in waste management, particularly for high-level waste [18].

Unfortunately, it is almost impossible to give an indication of decommissioning cost per, lets say, MW_e of installed capacity or initial investment. The values may vary greatly and start from a few hundreds of dollars or euros per kW_e of installed capacity without any upper limit, which means that the overall project costs end up in hundreds of millions or even billions of dollars.

Having said that, the nuclear power plant in general may have a number of great advantages: lower GHG emissions, large capacity, competitive cost of electricity. However, many economic evaluations of the project do not take decommissioning costs into account, which has to be included in the energy price to be collected by the time the facility becomes old to undergo decommissioning.

13.3.3 Concluding remarks

According to [10], it is generally agreed by technical experts that old nuclear facilities can be decommissioned safely, cost-effectively, and with no adverse impact to the humankind or the environment. All the techniques and approaches are quite straightforward and well understood — in theory. However, as with any other type of industrial and complicated project, it is the matter of experience which is essential for reducing cost and risks. Hence, proper planning, development of complete strategy, detailed and careful preparation, allocation of right people, delegation of responsibilities, mobilization of knowledge, and the right available technologies are essential to ensure safety and quality.

Selection of the optimal decommissioning strategy is complex. Decisions on the best or optimal strategy for decommissioning facilities using radioactive material may well go far beyond purely radiological protection considerations, therefore, these decisions require multiple criteria to be taken into account. The topic of nuclear decommissioning has become an emerging area not only for energy managers or ecologists, but also for sustainability scientists: it poses the question of the existence of an economically viable energy option, which can prove to be safe and cost-effective on all stages of its lifecycle, including decommissioning, which is equally critical for a nuclear plant.

13.4. Motivation for sustainable nuclear power generation

The role and compatibility of nuclear power with sustainable development objectives cannot be assessed in isolation, but only in comparison with existing alternatives on a level playing field. The final choice of the energy mix in any given country will be a sovereign decision reflecting the country's situation and needs [19].

Nuclear power generation has a number of significant advantages that can become crucial on the way towards sustainable energy mix. These include, for example, the following [19]:

- Uranium resources are available and abundant. The currently prevalent technology is mostly based on a once-through cycle and hence uses a small percentage of the energy content of the nuclear fuel. The use of fast reactors (FRs) operated in a closed fuel cycle could increase uranium resource efficiency by at least a factor of 60, extending the longevity of conventional resources to nearly 6000 years at current rates of consumption.
- The spent fuel produced by nuclear reactors can also be regarded as a resource and not solely as waste. It can be reprocessed and further used as MOX fuel which is currently used in fewer than 10% of present reactors.
- Lifecycle CO_2 emissions of nuclear power are far less than those from thermal power and are mostly associated with construction (steel and cement production). The estimates show less than 15 g of CO_2 equivalent emissions per kWh, which can be a good approach to a decarbonized power system.
- Despite multiple speculation, nuclear power has a much lower impact on human health. Natural exposure to ionizing radiation, such as terrestrial or cosmic radiation, is several orders of magnitude higher than artificial, human-made radiation, especially the radiation stemming from the nuclear fuel cycle. Radioactive waste disposal emits insignificant doses.
- Nuclear energy in general is important for the economy as it creates highly skilled jobs.

Among the concerns and potential issues on the way towards sustainable energy systems that implement nuclear technology one can name the following:

- The size and complexity of nuclear power plant projects and their long lead times are usually underestimated and therefore delayed, causing extended deadlines and runs over the budgets.
- Nuclear waste and its treatment is still an unsolved issue. Even though the waste related to spent and recycled fuel is relatively small in volume, waste associated with decommissioning may be crucial. Moreover, waste management involves multiple future generations which shall be also backed with financing, skilled labor and responsible management to ensure the highest possible level of safety.
- Direct comparison of COE may put nuclear among the most affordable and cheapest base load energy sources. These, however, do not include the future decommissioning costs which are still a big challenge.
- Nuclear power plants are still thermal in their nature and require large amount of water, mostly for the heat dissipation system. Using once-through cooling is efficient but induces thermal pollution, and the use of hybrid systems with cooling towers and ponds still requires water, though in lesser quantities.
- Plant siting is usually an issue, especially in highly populated countries as it requires larger safety areas. Still, compared to coal-fired plants with associated mine areas, or solar PV plants of the same capacities, nuclear plants can be much more advantageous.
- Nuclear technology, unfortunately, is not only related to peaceful electricity generation. Maintaining and deploying nuclear capacity worldwide requires the most stringent measures on safety and against proliferation of nuclear material and weapons. This has to be ensured by countries and international bodies like the International Atomic Energy Agency.

Overall, the attitude towards nuclear energy differs dramatically throughout the world. Risks associated with dread and the unknown, such as the perceived lack of control and the potential of catastrophic events with fatal consequences, often drive public opinion [19].

Nuclear industry is unfortunately going through the information vacuum and therefore is not fully and, more importantly, properly covered in the mass media while the specialists isolate themselves and prefer to keep distance. This disconnect between the nuclear specialists and the public is tremendous and is continuously increasing, or at least not decreasing. It does not mean that nuclear would become a single source of power in the coming future, or that it would be challenged massively by the society. It is all about the global perception of people and their desire to critically address the challenges of modern world, so that the picture of the cooling tower at a nuclear site would not bring the negative thoughts of "dirty smoke" into their minds [20].

References

[1] Bodansky D. Nuclear energy: principles, practices, and prospects. New York: Springer. ISBN 9780387269313, 2007.

[2] Kessler G. Sustainable and safe nuclear fission energy: technology and safety of fast and thermal nuclear reactors. Power systems. Berlin, Heidelberg: Springer. ISBN 9783642119903, 2012.

[3] International Atomic Energy Agency (IAEA). Preparation of a feasibility study for new nuclear power projects. IAEA nuclear energy series, no. NG-T-3.3. Vienna: IAEA. ISBN 978-92-0-145610-6, 2014.

[4] Dubrovskiy V. Construction of nuclear power plants. Moscow: Energoatomizdat; 1987 [in Russian].

[5] International Atomic Energy Agency (IAEA). Financing of new nuclear power plants. IAEA nuclear energy series, no. NG-T-4.2. Vienna: IAEA. ISBN 978-92-0-107208-5, 2008.

[6] World Energy Council. World energy perspective. Cost of energy technologies. London: World Energy Council. ISBN 9780946121304, 2013.

[7] International Atomic Energy Agency (IAEA). Project management in nuclear power plant construction: guidelines and experience. Vienna: IAEA; 2012.

[8] Lillington J. The future of nuclear power. Elsevier Science. ISBN 9780080532240, 2004.

[9] International Atomic Energy Agency (IAEA). Power reactor information system (PRIS). URL: https://www.iaea.org/PRIS, 2017.

[10] Laraia M, editor. Nuclear decommissioning: planning, execution and international experience. Woodhead publishing series in energy. Elsevier Science. ISBN 9780857095336, 2012.

[11] Invernizzi D, Locatelli G, Brookes N. How benchmarking can support the selection, planning and delivery of nuclear decommissioning projects. Progress in Nuclear Energy 2017;99:155–64.

[12] International Atomic Energy Agency (IAEA). Glossary. IAEA official website. URL: https://www.iaea.org/ns/tutorials/regcontrol/intro/glossaryd.htm#D, 2016.

[13] World Nuclear Association. URL: www.world-nuclear.org.

[14] United States Nuclear Regulatory Commission (US NRC). Decommissioning. NRC official website. URL: https://www.nrc.gov/reading-rm/basic-ref/glossary/decommissioning.html.

[15] International Atomic Energy Agency (IAEA). Decommissioning strategies for facilities using radioactive material. IAEA safety report series, vol. 50. Vienna: IAEA; 2007.

[16] Bayliss C, Langley K. Nuclear decommissioning, waste management, and environmental site remediation. Elsevier Science. ISBN 9780080537788, 2003.

[17] International Atomic Energy Agency (IAEA). Decommissioning of facilities. Vienna: IAEA; 2014.
[18] Nuclear Energy Agency, OECD. Costs of decommissioning nuclear power plants. OECD; 2016. 260 pp. URL: https://www.oecd-ilibrary.org/content/publication/9789264255555-en.
[19] International Atomic Energy Agency (IAEA). Nuclear power and sustainable development. Vienna: IAEA; 2016.
[20] Belyakov N. Revisiting nuclear power: public acceptance and informational isolation. URL: https://www.linkedin.com/pulse/revisiting-nuclear-power-public-acceptance-isolation-nikolay, 2016.

PART FIVE

Sustainable hydropower

Overview

Hydropower is the largest renewable source of energy that has been utilized on a large scale. For centuries the humankind has realized the benefits of the flowing water and has used it for mechanical work, and, lately, for electricity generation.

Hydroelectric power plants boast very low levels of associated GHG emissions and provide flexible capacity that can deliver both stable load and grid management. This part is devoted to sustainable hydropower and covers the concept of the plant and the underlying technologies. Special attention is paid to the sustainability topics covering both positive and negative attributes.

CHAPTER FOURTEEN

Traditional hydropower plant technology

Contents

14.1. Concept of sustainable hydroenergy utilization

Water covers more than 70% of the Earth with around 96% of it concentrated in surface waters like oceans, seas, rivers, and lakes, excluding surface and floating ice. Water carries a large amount of kinetic and potential energy which circulates through the water cycle powered by the Sun, so that hydropower can be considered as a derivative from solar energy. It is therefore logical that power can be obtained from this abundant resource, which belongs to the renewable primary energy sources.

Capturing water energy and converting it into electricity can be technologically carried out in multiple ways so that there are various concepts of hydropower plants. These include hydroelectric power plants that use the natural flow of water to spin turbines and power electric generators,

Sustainable Power Generation
https://doi.org/10.1016/B978-0-12-817012-0.00027-X

tidal plants to convert the energy of water moving during the tides, wave plants to capture energy of ocean and sea waves, or even plants to capture underwater streams captured by turbines similar to those for wind. Apart from the conventional hydroelectric power plants, which have been historically widely utilized and technically advanced, other types are still in the development or early implementation phase without mass commercial operation, which shall be topic of Chapter 21. However, they promise to contribute to the world generating capacity mix in the coming future as the water energy seems abundant and generates almost no greenhouse gas emissions.

Having said that, in this chapter we shall focus on the traditional hydroelectric power generation, also called hydropower plants. Historically, it has been used since many centuries to provide mechanical power and run mills and blacksmith facilities. Similarly to steam power development, it was the invention of the alternator that made it possible to use hydropower for electricity generation purposes.

14.1.1 Hydroenergy cycle and conversion

As explained in detail in Chapter 2, exposed to the Sun radiation, water evaporates from the surface reservoirs such as rivers, lakes, seas and oceans to the atmosphere and circulates there along the air flows. After being condensed mostly in the upper (colder) layers of the atmosphere, water is delivered back to the surface in the form of precipitation (like rain or snow) and refills the water reservoirs. Moreover, exposed to the gravity, water flows to the rivers and then downstream to the seas and oceans, where the evaporation process continues. The water cycle does not change the thermal balance but only redistributes energy which makes it one of the cleanest natural sources of energy.

Water has enough mechanical energy and gradually loses potential energy along its flow. A certain part of it is transformed into kinetic energy of the moving mass, while the majority is wasted due to the friction with surface, in between the particles, and for mechanical changes of the surface where it flows, to be finally dissipated in the form of heat back to the environment. However, due to the continuous water cycle, mechanical energy of the water, also known as *hydroenergy*, is continuously renewed.

As the energy of the water flow is dissipated along its length, its efficient usage is therefore quite complicated. To calculate the capacity of the water flow between two points A and B, we assume that the run of the river has constant flow rate Q (m^3/s) and constant cross-section of the way without any in- and outflows along the way from A to B. The flow rate basically defines the amount of water that flows through a certain point in one second.

The capacity of the water flowing from the height H_{AB},

$$N_{AB} = \rho g H_{AB} Q, \tag{14.1}$$

defines its capability to perform work and is measured in watts W (refer to Chapter 1 for details). Here ρ is the density of the water (measured in kg/m^3), while $g = 9.81\ m/s^2$ is the gravitational acceleration.

Looking at (14.1), in order to increase the capacity Q of the water flow, we can influence only the values of the head H_{AB} and the flow rate Q as the density and the gravity acceleration are given values. Therefore building special artificial engineered structures, the *head of the river* can be concentrated in one place to get a higher water head H to increase the capacity of the water flow. Increasing the head will allow for higher flow rate Q as there will be more water available. Installing special equipment like hydroturbines and generators captures this energy and delivers bigger certain capacity. This energy of the water is first converted into mechanical energy of the rotating equipment which powers the electric generator to produce electric power and deliver it to the consumer.

14.1.2 Hydroenergy potential

The idea of capturing the energy of water underlines the whole model of the hydropower industry. Having the above logic in mind, an investor faces a number of questions before starting a hydropower plant project. These are usually linked to the following points:

1. What is the right place to capture the maximum hydropotential, i.e., the highest water head H and the biggest water flow Q to maximize the capacity?
2. How to utilize this capacity to ensure the minimum possible energy losses?
3. What are the consequences of the hydropower plant deployment in terms of ecological and social, as well as financial, impact on the project cash flow?

The first major question can be partly answered by a hydroresource specialist to understand the potential of the hydroresource available in the region. All water resources of the Earth are estimated at the level of 1.5 billion cubic meters, which comprise oceans, seas, rivers, lakes, glaciers, and steam in the atmosphere, for instance, in the form of clouds. The absolute majority, however, is the ocean and sea water corresponding to about 90% of all water resources and is therefore not available for traditional hydropower industry. Water shall also be considered as one of the most important resources needed for a human to maintain his/her well-being. Its usage can be broadly split into two major groups:

- Direct water consumption, when water is directly consumed in multiple activities and processes. These include water for drinking, washing, cooling and heating, agriculture, etc. To close the water cycle, it shall go through the filtration and waste treatment process before being released back to the environment.
- Water usage, when water is used but not consumed, so that it may transfer some of its energy during the process. It includes water transport, fishing, hydropower, and some other activities.

The amount of water available for usage is huge and possesses a big amount of energy, which can be potentially captured and transferred into electricity. The estimation of the hydropower potential is usually carried out in several stages by calculating:

- *Theoretical potential*, which defines the overall available energy of the hydroresources;
- *Technical potential*, which is the share of the theoretical energy potential available for capture at the current level of technology;
- *Economic potential*, which is the part of the technical potential, economically feasible compared to other sources of energy.

The technical potential for hydropower is usually estimated at around 15 000 TWh/yr, or about 35% of a theoretical potential derived from the total annual runoff of precipitation [1]. This technical potential would require a global capacity of 3750 GW at 4000 full load hours. The percentage of undeveloped technical potential is highest in Africa (92%), followed by Asia (80%), Australia/Oceania (80%), and Latin America (74%). Even in the most industrialized parts of the world, the undeveloped potential remains significant, at 61% in North America and 47% in Europe. This significant share is due to many reasons: technological difficulty of capturing water energy, cost of production, leading to loss of competition to other sources of energy, lack of domestic technology and expertise, political issues, etc.

Globally, only around 20% of hydropower potential has been developed. On a regional scale, this situation is quite heterogeneous and differs from region to region [1]:

- The African continent has the largest proportion of untapped hydropower potential, with only 8% currently developed. Regional instabilities and political issues, as well as the absence of necessary infrastructure, remain among the main challenges in developing the hydroenergy segment. However, recent developments, especially in small hydro, have shown the ongoing way forward.
- Central and South America have about a half of electricity generated from hydropower. The available but unexploited hydropower potential is approximately 540 GW, distributed among almost all countries of the region. Therefore most of the countries in the region have been actively developing hydropower.
- North America has always been active in hydropower. The US Department of Energy aims to double the hydropower capacity mostly through the upgrades of the existing plants, while Canada is entering into a new and significant period of hydropower development. Already producing about 60% of its electricity from hydro, Canada still estimates that there remains a twice bigger unused potential, which is distributed across the country.
- Asian hydropower development has been mostly driven by massive Chinese investments into massive hydropower development projects. On top of electricity production, they are aimed at flood control measures and irrigation activities. India, facing the energy poverty, is still on the way of massive development. The rest of the South-East Asia continues to face power shortages and therefore tries to tap the used potential which is still very high. The overall undeveloped potential of Asia sums up to 80%.
- At present, only about half the technically feasible potential for hydropower in Europe (excluding Eurasia and Russia) has been developed. Most of the easily accessible and economically attractive hydropower resources have already been utilized.
- Russia and Eurasia possess enough water resources with the hydro share of around 20% on average. The major issue is that most of the capacities are old and therefore would require upgrades or decommissioning. The former may well increase the overall installed capacity level or utilization factor.

Hydropower projects can bring significant benefits to developing and emerging economies, providing essential services such as electricity generation, freshwater management and climate adaptation [2]. According to the "Key Trends in Hydropower" report presented by IHA [2] at the end

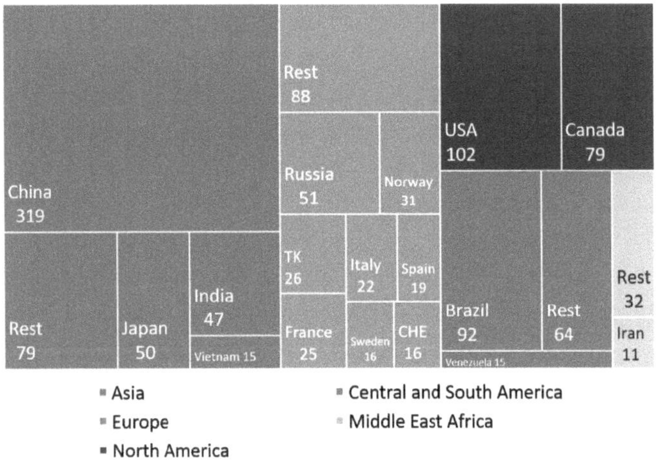

Figure 14.1 ***Total installed hydropower capacity per region, GW.*** Asia accommodates the largest installed capacity of hydropower, with China becoming the largest player at the end of 2015. *Source: Based on [2] and IIR database.*

of 2015, the world's total installed hydropower capacity reached 1211 GW, including 145 GW of pumped storage as illustrated in Fig. 14.1.

Following the recent trend, China continued to dominate the market for new development and total installed capacity with adding 19.4 GW of hydropower plant capacities, including 1.2 GW of pumped storage in 2015. Other countries leading in new deployments include Turkey (2.3 GW), Brazil (2 GW), India (1.9 GW), Iran (1 GW), and Vietnam (1 GW). The overall hydropower growth is driven not only by general electricity demand, but also by the need for reliable, clean, and affordable energy. Looking forward, significant undeveloped potential remains across all regions, particularly in Asia, Africa, and Latin America. Combined with the strong electricity demand in the developing countries, hydro would remain one of the growth drivers within these regions.

In the future, however, with a changing climate, the hydropower potential could change due to effects of climate change on the water resources that hydropower depends on [3]:

- Changes in river flow volume, variability, and seasonality;
- Extreme flow events like floods or droughts;
- Changes in sedimental loads which could affect reservoir storage capacity.

14.2. Types and configurations of hydropower plant

Hydroelectric power plants differ significantly in size, major equipment and overall facility arrangement, type and capacity of the units, water head, etc. Naturally, the most favorable water head H is quite impossible to achieve without considerable changes of the water flow and the landscape. This is why such power plants are extremely site-specific and tailor-made to fit into local geographical, hydrological and economic conditions.

Hydroelectric plants are classified commonly by their hydraulic characteristics, i.e., by the amount of water flowing through the turbines. In general, there is an established classification of hydropower plants into three functional categories:

1. Run-of-river and diversion;
2. Reservoir or storage;
3. Pumped storage.

14.2.1 Run-of-river and diversion

Run-of-river schemes utilize the natural stream flow of a river as illustrated in Fig. 14.2. In most cases a weir or a small dam can enhance the continuity of the flow [1] and guide the water to the power block without any water storage being provided.

Generally, this type of hydropower plant would be feasible only on such streams, which maintain a certain flow of such magnitude that makes it possible to generate electricity throughout the year. This is especially important for the dry season. Since the flow usually varies throughout the year, the plant has to be able to manage the variations and not be shut down during low seasons. At peak monsoon seasons, when the levels of

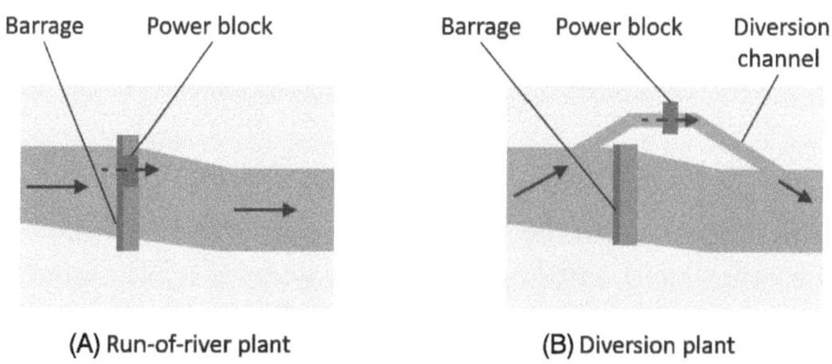

Figure 14.2 ***Run-of-river and diversion hydroelectric power plants.*** Simplified drawings of the run-of-river (A) and diversion (B) hydroelectric power plants.

precipitation or snow-melt waters are high, a small dam is built to take care of the variation of flow in the river downstream. In this case it is worth studying the economic feasibility of installing more units for peak seasons.

Moreover, during off-peak hours of electricity demand, for example, at night, some of the units may be stopped and the water conserved in the storage area upstream, which is again released during peak hours for power generation.

Similarly to the run-of-river, the *diversion* concept assumes that water is channeled from a river, lake or dammed reservoir to a remote powerhouse, where the equipment is located [1]. Diversion is mostly used for the mountain conditions with rather low water flow and big river bed slope. In the beginning of the area (Fig. 14.2) water is guided from the natural river bed to the canal or penstock to a hydrosite. A water head is created due to the difference between the natural river bed and artificial slopes.

The use of diversion concept on big rivers with massive water flow and medium to low water heads is not reasonable economically as it would require massive construction of long and big penstocks or canals. Such plants are mostly built on mountainous rivers with suitable conditions. The same concept, however, can find place in case of a water head between two water sources (e.g., two rivers or lakes) located nearby at different heights. In this regard, water is guided from the upper water source to the lower via a canal with the generating equipment installed in it.

14.2.2 Reservoir and storage

The *reservoir* or *storage* concept assumes that the water flow passes a dam at a hydrosite, which collects water and therefore increases the water head due to an artificial reservoir for the wide site span. The *headwater level* H_h is the level of water in the reservoir at the dam, while the lower water level H_t at the dam where water is discharged is called the *tail water level*. The difference between these levels

$$H = H_h - H_t$$

stands for the *water head* of the hydro power plant (see Fig. 14.3).

On the other hand, creating the dam provides a stable water flow and hence power output, ignoring the seasonal changes of the water supply from the catchment area. This water storage would compensate for scarce water during the off-season, as well as provide water supply for other needs (e.g., irrigation, normal consumption). In order to minimize costs and

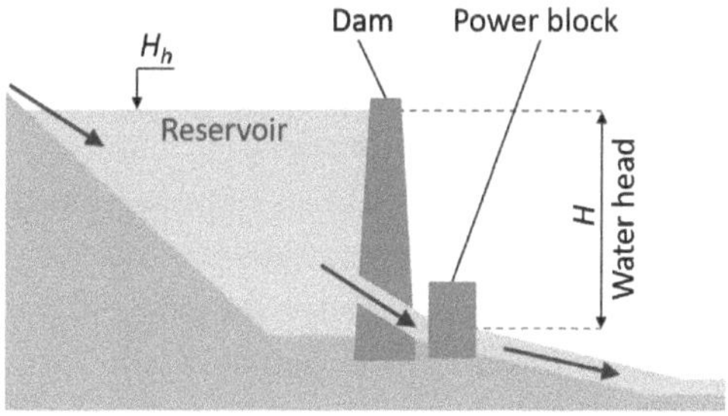

Figure 14.3 ***Schematic of the reservoir hydroelectric power plant.*** Water head is created by the construction of a dam: it collects the water in a reservoir to increase the level difference between the upper and the lower points.

impact to the environment, the dam can be partly natural, especially in mountainous areas.

The essential equipment is usually located near or within the dam itself.

It should be noted that it is practically almost impossible to build a massive hydroplant that would utilize the whole water head from the upstream level of a river to the downstream. However, in order to benefit from most of the potential, the river can be divided into several parts along its flow, each having its own dam and hence own water head. This system is also known as the *cascade* of hydropower stations.

Remark 14.1. Both storage and run-of-river hydropower plants can have diversion schemes. For example, in case of a mountainous river with the difficult road bed profile, the dam can increase the overall water flow and therefore water head, while the diversion approach will further increase the head and simplify the construction of the plant: the penstocks would guide the water to the turbines located somewhere nearby the dam in a separate place (for instance, in a space within a mountain). However, such a power plant would still be usually called a reservoir one.

14.2.3 Pumped storage

The *pumped storage* concept assumes that there are two reservoirs with different heights so that there is a water head in-between them. These reservoirs are connected with the penstocks and special hydroturbines, which are able to operate either as ordinary hydroturbines or as water pumps.

The idea of this type of a plant is that it can operate as natural battery to store power. When the demand of electricity in the grid is low, the units would operate as pumps and transport water from the lower reservoir to the upper. During the periods of high electricity demand, the water is flowing naturally from the upper to the lower reservoir and it therefore powers the units that operate like normal hydroturbines.

Hydropower's storage capacity and fast response characteristics are especially valuable to meet sudden fluctuations in electricity demand and to match supply from less flexible electricity sources and variable renewable sources [1]. In case of low overall power demand, other sources can be used to power the pumps and fill the reservoir, while at peak times or loss of capacity a pumped storage hydroplant can be utilized as an emergency.

Remark 14.2. Pumped storage plants are among a few real possibilities which allow enough power storage capacity to meet commercial needs. Despite certain complexity of their construction and potential difficulty of choosing the right site location, pumped storage power plants are of great importance and might be considered as a good contributor into green power. At the moment we do not see a mass boom of their construction (due to some objective reasons), however, these plants are quite popular in many countries.

14.2.4 Head height classification

Hydroelectric power plants are sometimes classified according to the head of water causing the turbines to rotate. This information is important to understand the type of a turbine to be used to generate electricity.

Despite various countries and companies having their own classification, we can roughly split it into three major heights of water head [4]:

- *Low head* of water at the turbine is assumed less than 30 m (this number varies in the literature). Typically, a dam in this type is small and in some installations can be only a few meters in height. The available head largely determines the type of turbine to use. A combination of low head and high flow will usually favor an axial flow propeller type of reaction turbine, such as Kaplan with pitch-controlled blades.
- *Medium head* will typically have between 30 and 300 m of head at the turbine. This type of hydropower plant will generally be situated in a mountainous region and be equipped with Francis (reaction) type of turbine or an impulse turbine of the cross-flow type.
- *High head* usually exceeds 300 m, so that a station is considered to be high head. Impulse turbines such as the Pelton or Turgo wheels are often used for these installations, although Francis turbines are more usual at the larger power levels.

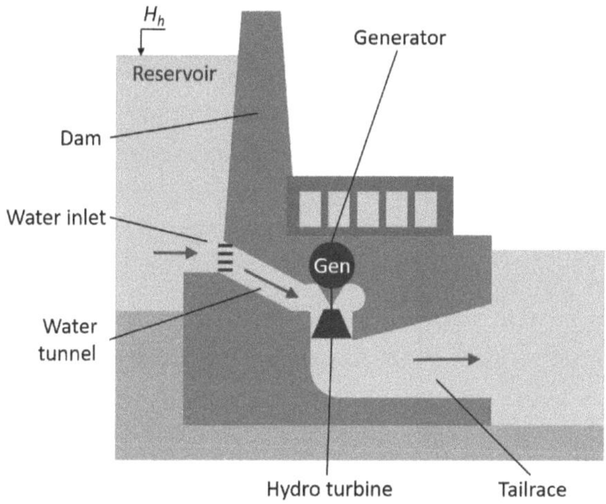

Figure 14.4 ***Schematic illustration of hydroelectric power plant equipment.*** Schematic configuration of a reservoir hydropower plant with major equipment areas and components: waterway structures, power block, and tail race.

14.3. Modern hydropower plant

A typical hydropower plant consists of several major parts represented schematically in Fig. 14.4. These parts follow the energy transformation chain carried out in a hydroelectric power plant and include:

- *Waterway civil structures* for water storage and/or diversion, like a dam or a barrage. Water from the reservoir or river flows into the water inlet and is guided through the water tunnel to the hydro turbines.
- *Power block* comprises a hydroturbine, which rotates the rotor of the hydrogenerator so that current is produced. This current goes along the bus bars to the main step-up transformer and then to the substation. Finally, high voltage electricity is delivered through the grid to the final consumer.
- *Tail race* conveys water from the hydroturbines through a set of channels or discharging conduits back to the river downstream of the power plant.

Although the above components are common for all hydropower development schemes, the general arrangements for high and medium head hydropower plants are similar. The low head power plants, which are usually of run-of-power type schemes, often have a slightly different arrangement.

The run-of-river hydropower plants mostly have their turbine hall structure integrated into the dam so that they form one structure. This is feasible due to lower water heads and therefore lower water forces against the structure. There can be one single dam accommodating water flow control (spill-over or gates), major generating equipment and other structures. In case of specific site or climatic conditions, there can be two adjacent or even separated dams.

For the rivers with navigation, various dam locks are built, either adjacent to the dam or within a separate diversion channel. For the large rivers, where the dam has to close a much wider span of the river bed, there can be several dams with the integrated spill over and lock for navigation. In practice, the lock (if any) is the first structure to be erected to ensure continuous navigation during the major construction period. The lock is usually protected from the main water flow with the civil structure to ensure safe navigation, especially for smaller watercraft.

In case of higher water heads, which leads from the run-of-river to a small reservoir type of hydropower plant, it is worth implementing the turbine hall into the main dam structure to hold the hydrostatic pressure. This is mostly applicable in the areas with more complicated site conditions, caused, for instance, by mountainous landscapes, higher river banks, significantly complicated geological conditions combined with lower water flow rates. In such cases, the power house is located near the main dam within an integrated civil structure.

Another type of hydropower plant arrangement can be implemented in case the geological conditions do not allow locating the turbine hall straight at the dam. While the dam serves its main purpose to collect water and form the reservoir, the water is diverted with the special water tunnel to the turbine hall located nearby. Depending on the configuration, the power block can be located either above the ground or underground. The latter requires massive excavations inside the neighboring rocks, while the extracted material is used to erect the rock-filled dam.

The above described examples of the hydropower plant configurations outline only typical cases and do not cover all possible solutions. Each hydropower plant site is unique and requires special and original engineering of all civil and engineered structures. However, all hydroelectric power plant components can be schematically grouped into three major areas: waterway civil structures, power block, and tail race.

14.4. Civil structures and waterways

Waterway structures are the largest part of a hydroelectric power plant. It is the system of engineered and civil structures to guide the water from the natural source (river and/or reservoir) to the hydroturbines of the power plant. These structures include the following major components:

- *Main dam* or *barrage*, which is used to increase the water head to designed level;
- *Overflow spillway* or *gate* to control the level of the water and prevent the hydropower plant and surrounding areas from the floods and other potential accidents;
- *Water inlet*, where the water is taken into the *water tunnel* or *penstock*, guiding water further to the turbine. The inlet is located at the dam or barrage. The tunnel can be located both inside the body of the dam or in the neighboring rocks. The latter is called the *diverter water channel*;
- *Side dams*, which are used to close the gaps around the main dam and prevent the water flow around. Altogether they can be part of one continuous dam or can be situated separately;
- *Dam lock* used for raising and lowering the watercraft;
- Special *sideways for fish* to allow them going upstream and downstream avoiding the hydroturbines.

All of these civil structures are supported by many engineered systems, which ensure their proper operation.

While any hydropower plant has a dam or barrage to guide the water to the water inlet further through the penstock or tunnel, not every plant incorporates a water lock or other civil structures. Similarly, some plants do have overflow spillways, while others employ special channels to guide excess water from the upstream of the plant. The final scope of waterway civil and engineered structures depends on the site layout and general design of a plant.

14.4.1 Dam

Generally speaking, a *dam* or *barrage* is a barrier that impounds water. It guides the water and creates a reservoir with the required water head and controls the amount of water downstream. The dam is a major part of the hydropower plant water way structure. Its shape and material depend on the project location, geological conditions, and project design and requirements.

14.4.1.1 Rock-fill and earth-fill dams

One of the historically established ways to create a dam is to use natural material like pieces of rock and soil to close the river and increase water head. These dams are called *rock-filled* or *earth-filled* and use local material produced out of the excavations on site and from the surrounding rocks during construction of the tunnels (if any). This type of dam consists of the following structural parts:

- Dam body as the major part, which withstands the overall water pressure and ensures dam's stability and durability;
- Insulation to maintain its water resistance;
- Drainage to collect water, which goes through the insulation, and release it from the dam structure.

There are multiple designs of rock-filled dam while the choice depends on the specific site and project requirements. The material used in such a dam can be rock, soil, or a mix of them.

The rock-fill dams, sometimes also called *embankment dams*, are designed to withstand the water pressure only by utilizing the weight of the dam body material. However, in case of water spill-over, the earth-filled or rock-filled dam can be exposed to quick erosion, therefore it requires diversion water channels or spillways to control the water head. Moreover, this type of dam usually assumes that the power block is located nearby the dam and uses special waterways (tubes or channels) to guide water from the upstream. This is mostly due to the fact that the dam body is not capable of accommodating heavy equipment inside its rock-filled structure.

14.4.1.2 Concrete dams

Reinforced concrete provides an alternative to the dam construction material. There are three major engineering approaches to build dams out of concrete:

- *Gravity dam* is designed to hold back the horizontal water pressure by utilizing dam's own weight only. It has the same design as a rock-filled embankment dam, but can tolerate minor overtopping flows much easier. Such dams can be out of common or reinforced concrete, the latter with the hollow design due to better performance of the material. Most of the gravity dams are straight, however, can have a curved dam axis, too.
- *Arch dam* is a concrete dam with the dam axis curved. The arch dam design ensures that the water presses against the arch, compressing and strengthening the structure as it pushes into its foundation or sides. This

type of dam is most suitable for narrow gorges with steep side walls of stable rock to support the structure. While these reinforced concrete structures are much thinner, they require less construction material.

- *Buttress* or *hollow dam* is a dam with a solid, water-tight upstream side that is supported at intervals on the downstream side by a series of buttresses or supports. A buttress dam is a good choice in wide valleys where solid rock is rare. Contrary to arc and gravitational dams, their stability is ensured only through the massive buttresses.

The choice of dam is made at the design stage according to the geography and economics of the project. While concrete dams can have a longer lifetime, they require local concrete production on site to minimize material and transportation costs.

14.4.2 Spillway and overflow channel

Intentional release of water from the upstream reservoir is confined to water releases through special water paths called *spillways* or *overflow channels*. They maintain the normal water level in the reservoir and pass excess water downstream the dam without erosion (which is especially important for rock-fill dams). This allows controlling the water level and overcoming floods caused by precipitation, melting of snow or other additional water flows upstream.

There are two major types of spillway:

- *Uncontrolled*, when the water flows automatically upon reaching a certain level upstream the dam and starting to flow over the crest of the spillway.
- *Controlled*, when special gates or other mechanical structures are used to regulate the flow of water and therefore its level upstream. When the water reaches a certain level, the gates open and release water. There are usually several gates operated by motors or pulled up and down by special cranes.

Technically, there are several opportunities to construct spillways and guide excess water downstream the hydropower plant:

- Open channels over or near the dam. This is mostly erected out of concrete and used on concrete dams to prevent erosion;
- Closed channels or tunnels inside the dam body or rock nearby;
- Bell-mouth or siphon spillway, when the water can enter around the perimeter of the inlet and go down the channel.

There can also be an emergency spillway that functions during the emergency conditions to prevent overtopping and damages of the structures. It can be in the form of an excavated channel in earth or rock near the dam to discharge water further downstream to protect the dam structures. Its size is designed to manage the "design flood" — the largest possible magnitude of flood in the site area.

14.4.3 Water intake, penstock, and tailrace

The *water intake* structure is used to get water from the reservoir and deliver it through a special channel called *penstock* to the hydroturbines. Water intake is usually equipped with a gate system used for water flow control and maintenance purposes. When operated in polluted water conditions, the whole system has to be properly and frequently cleaned and serviced, and the gate allows closing the system and isolating it from incoming water. The water intake is also protected with the special *trash rack* to collect trash and debris, and prevent the hydroturbines and valves from damages.

Penstocks can be made out of concrete inside the dam body or as separate channels or large tubes that guide water to the turbines, depending on the configuration of the hydropower plant.

After the water flow passes through the turbine and transfers part of its energy to the unit, it is released through a *draft tube* to the downstream part of the waterway called *tailrace* or *water discharge.*

14.4.4 Other waterway structures

There are other civil and engineered structures included into the waterway solution:

- *Fish ways* and *fish ladders* are used to ensure that fish can migrate up- and downstream without being damaged by the running hydroturbines. Despite modern development in fish-safe technology, the traditional solution still remains valid: during the breeding period fish usually go upstream to the small rivers, and dams can damage the whole species habitat. The fish ladder allows for safe and quite easy navigation alongside the dam, but still requires an effort to pass it.
- *Lock* is built to allow watercraft go along the river up and downstream, when the dam creates a water level difference. There are multiple designs of this type of engineering structure, however, the easiest one operates like a big basin closed with two gates on the sides. Assume a vessel is coming from the upstream of the lock. The water level within the closed basin is aligned to that where the vessel stands, and the gates on the upstream side open. The vessel enters the basin, the gates close,

and the water level goes down to the required level. As soon as the level is reached, the downstream gates open and release the watercraft.

14.5. Energy conversion equipment within power block

The *power block* is the heart of a hydropower plant where the energy of the water flow is transformed into the electrical energy. The power block contains major and auxiliary electromechanical equipment to provide reliable and safe operation of the plant as well as production of electricity. It normally consists of:

- *Hydroturbine* with *generator* and *inlet valve block*;
- Turbine hall, where the equipment is located. The turbine hall can be situated in a building above the ground or underground depending on the overall configuration of the plant;
- Mechanical and electrical BOP including the related substation, control room, and allowing for normal operation of the major equipment;
- Switchyard and HV transmission line to transfer electricity to the grid.

Remark 14.3. In some cases, penstock, turbine chamber and the draft tube are considered as a part of a hydroturbine and valve block, however, we assume here that they belong to civil structures.

14.5.1 Efficient hydroturbine

A hydroturbine is a rotating machine that transform the energy of the water flow into the energy of the rotating shaft. Simply put, it consists of a wheel-type structure mounted on the shaft: impacted by water flow, the wheel starts to rotate. Water can be guided to the turbine in two ways:

- Through the nozzles, passing through which it hits the buckets of the turbine and rotates the wheel. Such turbines, also called *active* or *impulse*, use only kinetic energy of the water (basically, the faster the flow, the higher the energy) that strikes the buckets to cause rotation;
- Through special inlet guide vanes that direct water to the turbine wheel. While the inlet pressure before the turbine is higher (due to the water head pressure), water has both potential energy (the weight of the water) and kinetic energy (speed of flow) and it therefore pushes the water through the wheel. A special form of the buckets transfers pressure to the wheel and therefore pushes it to rotate. Such machines are called *reaction turbines*.

Table 14.1 Comparison of various types of water turbines used in hydropower plants.

Type	Bulb	Kaplan	Francis	Pelton	Pump
Unit capacity, up to MW	80	300	1000	400	500
Head, m	0.5–30	30–90	up to 700	up to 1800	1300
Efficiency, %	92	95	96	95	94
Plant type	Run-of-river with low heads	Small reservoir and run-of-river	Reservoir	Big reservoir	Pumped storage

The choice of a proper turbine technology depends on the characteristics and the configuration of the hydropower plant, mostly the available water head and water flow. The final type of turbine can be chosen only after all studies of site conditions and operational limits are completed. While these parameters may differ dramatically from site to site, there are various hydroturbines available on the market to meet the specific site conditions.

There are four major types of hydroturbine usually installed in large utility scale hydropower plants, with the brief comparison represented in Table 14.1:

- *Bulb* or *pit*, which possess an important feature of the horizontal positioning inside the water way shaft, resulting in a reduced size compared to a vertical shaft arrangement. The generator is located in a watertight housing called the "bulb". These turbines are mostly employed in run-of-river hydropower plants with lower water heads up to 30 m. Their unit capacity is less than a hundred megawatts.
- *Kaplan* type hydroturbine, which was invented by an Austrian professor V. Kaplan in the beginning of the 20th century. The Kaplan turbine is a propeller type turbine with a number of adjustable blades. This reaction turbine is usually made of stainless steel, utilizing special shape of the blades to optimize its performance. Water is directed through a special camera that has a "snail" shape to guide the flow uniformly to all blades.
- *Francis* turbine, which was invented by J.B. FRANCIS in Lowell, Massachusetts, USA, in the mid-19th century. The Francis turbine is a reaction turbine that is employed for water heads up to 700 m. It has a number of buckets mounted on the wheel and coupled from the outside. Being the most common water turbine today, it can deliver capacity of more than 1000 MW per unit.
- *Pelton wheel*, which is an impulse type turbine invented by an American engineer L. PELTON in the 19th century and has proven to be among

the most efficient hydroturbines. The water is directed with high speed through nozzles against the buckets, arranged around the drive wheel. This design is the most efficient for very high heads and stable water flows, allowing for several hundreds of megawatts of unit capacity.

- *Pump* turbine, which can be used in either way: as a hydroturbine or as a water pump to drive water in the opposite direction. Implemented in pumped storage hydropower plants, it can be of various configurations and types.

To control the water flow to the turbine, special valves are installed. They are normally used for safety, maintenance, and shut-off, as well as for flow and pressure regulation. Moreover, some turbines employ special regulated inlet guide vanes to manage the flow and direct it properly to the wheel. This is normally done to control the output of the unit, which is required to maintain the grid frequency, deliver extra capacity or follow the instructions of the grid operator. While water flow regulation can be done rather rapidly, most of the hydroturbines therefore are considered flexible in terms of electricity dispatch.

Hydroturbines are installed in turbine buildings on special bearings that enable rotation. As they are coupled to the generator, the whole systems is installed as a single unit. Some of the wheels can have a horizontal (like Pelton wheel or bulb turbine) or vertical axis (like Francis or Kaplan) that impacts the whole plant design.

All hydroturbines are exposed to water. Even though this water is usually fresh, due to variable water conditions and quality, turbines have to withstand certain chemical impact as water can carry large amounts of sediments. If the sediment contains hard minerals such as quartz, guide vanes, runners and other steel parts are subject to high abrasion, rapidly reducing efficiency. New solutions include coating steel surfaces with a hard-wearing ceramic top layer [5]. Moreover, modern research is focusing on creating fish-friendly turbines that would allow fish to pass through the unit without injury.

14.5.2 Hydrogenerator

A hydroturbine is coupled to a *hydrogenerator* (also called alternator) via a single shaft with the common support and bearing system. This structure, including both the turbine and generator, is sometimes called a *hydroelectric unit*.

Depending on the hydropower plant characteristics, the turbine used and the required speed, the following generator configurations are available:

- Vertical shaft generators, which are coupled to vertical shaft hydroturbines of any size with outputs up to 840 MVA;
- Horizontal shaft generators for Pelton type hydro turbines, which are installed vertically;
- Motor-generators, which are used in pumped storage plants to generate electrical energy and to drive turbines when they operate in pump mode;
- Pit or bulb type generators are used for smaller, low-head applications and are coupled via step-up gear to the horizontal turbine shaft. Pit generators propose robust design and cope reliably with the high runaway speeds required for hydropower plants.

Normally hydroturbines have much lower speeds compared to gas and steam turbines, therefore the generators have to be capable of generating the current of the required frequency of 50 or 60 Hz. This therefore requires more poles to generate current of the same frequency (see Section 9.5 for details).

Evidently, hydrogenerators also have a stator and rotor with their own windings and therefore require an excitation system as any other generator installed in a traditional thermal or nuclear power plant. The major difference, however, is in the size of the generator due to usually much larger capacity and lower rotation speed, which results in the way generator is mounted on the foundation, usually with vertically oriented shaft line, through special bearings.

A *cooling system* is an essential system to dissipate heat from the operating hydrogenerator. This heat is released within the stator and rotor windings. Depending on the capacity, hydrogenerators are usually air-cooled, water-cooled, or with mixed air and water cooling:

- In air-cooled generators, air is blown through the structure and absorbs heat. Then it goes to the air-to-water heat exchangers, where it is cooled back, and flows again to the generator. The cycle is normally closed as it requires filtered and cold air.
- With water cooling, distilled water is circulating within the cooling and absorbs heat. It is then pumped back to the water-to-water heat exchanger where it is cooled down and pumped back through the system to close the loop.
- In case of a mixed cooling system, water is used to cool down the stator windings, while the rest is cooled with air.

In case of turbine shutdown for maintenance or emergency, the water valve stops the water flow to the turbine, and the generator is disconnected from the grid. The machine starts to spin in the idle run regime, which may last for a long time due to the high momentum of the rotor connected to the turbine, especially for large units. The situation may even worsen in case of valve malfunction and even low water flows directed to the turbine. Therefore to stop the rotor, a special breaking system is used: the break pads are activated and hold the special breaking ring located on the shaftline usually when the speed of the rotor goes below 30% of the nominal.

This is extremely important for safety reasons when the maintenance personnel is conducting some works down the turbine.

14.5.3 Balance of plant systems

Similarly to other types of power plant, balance of plant systems provide support for the major equipment, namely hydroturbine and generator, to ensure its operation at all service regimes. These systems include mechanical and electrical equipment implemented at any hydropower plant, usually within or nearby the power block.

Mechanical balance of plant systems cover the functions dealing with fluids, mechanical operation and maintenance, and may include:

- Cooling water system, to cool down all other systems that release heat. These are, first of all, hydrogenerator, motors, pumps, and other equipment. Cooling water can be used either directly or through heat exchangers;
- Dewatering and drainage systems, which evacuate excess water from the civil structures like dam. While concrete is the most common material for construction and has low permeability for water, water can still get inside through gates, tunnels, microscopic cracks, etc.;
- Compressed air system, to supply clean compressed air to the switches, instrumentation, and other needs;
- Emergency diesel generators, which are turned on in case of power loss;
- Overhead cranes located inside the turbine hall building, which are required for installation and maintenance purposes. While the parts of the turbines and generators are heavy, they are lifted and removed by heavy lifting equipment during planned maintenance and inspection procedures or unexpected repairs;
- Fire protection, both active and passive, which is implemented to safeguard equipment and prevent it from potential fire damage. Contrary to thermal power plants, where there is a potentially explosive environment (mostly due to fuel), a hydropower plant is still a hazardous

site with HV electricity, and short-circuits or damages of the electrical equipment can cause fire. Moreover, high temperature liquids (lube oil) or parts of material can lead to fire as well, unless properly treated;
- Heating, ventilation, and air conditioning, to optimize the working conditions for personnel and equipment, especially in electrical rooms;
- Oil treatment system, to clean and cool down lubricating and hydraulic oil, and to supply it to the rotating equipment.

Electrical BOP systems ensure electricity supply and its distribution within the plant systems, as well as implementing control functions. They normally consist of:
- Power evacuation equipment, including:
  - Bus bars, which connect generator terminals with the main transformer;
  - Main step-up transformer, to increase the voltage of the electricity;
  - Circuit breakers and HV switchgear (either air or gas insulated), to deliver power to the high voltage transmission line;
- Auxiliary power supply:
  - Auxiliary transformer, to supply power to MV and/or LV auxiliary system;
  - MV and/or LV power supply system to feed the hydropower plant systems. It is usually split into two parts. The first and the most critical feeds the hydroturbines and generators (cooling, lubricating oil pumps, water supply, excitation, fire protection, lighting), while the second feeds all the rest like ventilation, compressed air, cranes, drainage, etc.;
  - Emergency power supply, to ensure safe operation of the most critical equipment. Several batteries are located in the special protected compartment to ensure reliable supply of power for critical systems for a period of time required by local regulation;
- Control system and protections, to ensure safe operation with the highest possible efficiency and compliance to the grid requirements;
- Grounding, lighting protection and interconnecting cabling.

References

[1] International Energy Agency (IEA). Technology roadmap. Hydro. France: OECD / IEA; 2012.

[2] IHA. https://www.hydropower.org/2017-key-trends-in-hydropower.

[3] Letcher TM. Future energy: improved, sustainable and clean options for our planet. Elsevier Science. ISBN 9780080994222, 2014.

[4] Munoz-Hernandez G, Mansoor S, Jones D. Modelling and controlling hydropower plants. Advances in industrial control. London: Springer; 2012.
[5] IFC. Hydroelectric power. A guide for developers and investors. International Finance Corporation; 2015.

CHAPTER FIFTEEN

Hydropower project lifecycle

Contents

15.1. Introductory

Historically, hydropower development has had many drivers, depending on economic and social circumstances in various regions of the world [1]. A hydropower project, from the initial investment decision to its commercial operation, has always been subject to a number of careful considerations and, due to a considerable scale and risks, required the application of the internationally recognized environmental and social standards.

Hydropower plants are unique compared to other conventional electricity generating options — thermal or nuclear. Hydroelectric plants are always custom-designed site-specific projects. They require large capital investment, offering quite low operating costs and long operating lifespans of 40–50 years that can often be extended to 100 years through retrofit or upgrade projects [2]. Therefore a hydropower plant project results in very competitive electricity costs and hence favorable return rates. Another major difference is the financial modeling of a hydroelectric power plant: while other traditional facilities take fuel costs largely into account and, therefore, depend on its volatility, hydroelectric power does not have such a cost component at all.

Hydropower project development is usually a challenging activity with multiple risks and uncertainties associated with the hydrology and geology of a potential site, licensing and permitting process, construction in a re-

Sustainable Power Generation
https://doi.org/10.1016/B978-0-12-817012-0.00028-1

mote location with difficult transportation conditions, environmental and social risks, as well as financial risks caused by long payback periods and uncertain electricity tariffs.

Having the above logic in mind, an investor usually faces a number of challenges before starting a hydropower plant project. These questions are usually linked to the following points:

1. What is the right place to capture the maximum hydropotential?
2. How to utilize this capacity to ensure the minimum possible cost of kilowatt-hour production?
3. What are the consequences of the hydropower plant deployment in terms of ecological and social, as well as financial, impact on the project cash flow?

It is therefore extremely important to ensure comprehensive and thorough risk assessment and proper project development and execution. Thus, a typical hydropower plant project development process can be generally split into three major blocks:

- Opportunity feasibility study, which covers major conceptual and planning topics. During this stage, the overall business opportunity is estimated and an investment decision is made;
- Implementation of the project, which includes initiation of the project financing, choice of the main contractor, detailed engineering, procurement of material and equipment, and construction of the plant. This stage ends up with the commissioning activities, when a new power plant is put into commercial operation;
- Operation and maintenance activities, when the plant generates revenue.

15.2. Feasibility study and economic drivers

One of the most complicated and critical stages of the project development takes place during the opportunity feasibility study phase since the majority of the investigation, planning, and design studies are completed in this phase. Moreover, it is the result of this phase that gives the green light for an investment decision and kicks-off the implementation.

When a company thinks about a new investment into a hydropower generating facility, it has first to answer the question whether this investment brings money back and when. This is extremely important to understand for hydropower plant projects, as they require massive upfront investments and complicated execution before money starts to flow in from the sale of electricity. Hence the first point to consider is whether the market demand for electricity is (and will be) favorable enough to absorb hundreds of megawatts at a reasonable price.

If the answer is positive, the overall planning process starts with the assessment of available potential locations and plat configurations that can accommodate a hydroelectric power plant within the region. This assessment usually covers a wide range of topics and requires professionals from many disciplines to take part in this activity. Preliminary assessment can reveal several most promising locations with the rough understanding of the load profile and installed capacity, as well as the basic idea of technology.

At this stage, engineering, geological and hydrological studies are carried out to collect all possible information and identify several options for a potential site location and configuration of a power plant. This allows making a rough economic evaluation and doing preliminary calculations of the project economics, especially investment budget.

When certain locations and configurations match the requirements of the investor (for instance, in terms of available capacity, proximity to existing consumers, etc.), more detailed study is undertaken for the chosen sites. This pre-feasibility study also follows a staged process [3]:

- Engineering studies using available information to assess existing technologies and the feasibility of hydroelectric development at particular site. This normally requires multiple site visits, preliminary layout preparations, hydrology estimations, geological assessment and high-level technical proposal for the plant;
- Preliminary calculation of the project budget for each site and the cost of electricity estimation in terms of annual cost per kilowatt-hour;
- Assessment of the permit needs for all site locations;
- Analysis of potential social and environmental risks and the ways to mitigate them.

The conclusion of this stage is a recommendation for a preferred option. The successful identification and evaluation of the alternatives requires experienced and knowledgeable specialists in hydropower project development. In case an investor does not have these competences, it is advisable to hire a special consulting company that can perform such studies based on the previous experience: planning, engineering and construction. The latter is especially important to define the potential schedule and cost estimation.

Table 15.1 Typical hydropower plant costs.

Technology	CAPEX, $/kW	OPEX, %/year of CAPEX	Capacity factor, %	LCOE, $/kWh
Large hydro	940–8400	2–2.5	25 to 90	0.02–0.19
Small hydro	1300–8000	1–4	20 to 95	0.02–0.27
Upgrade	500–1000	1–6		0.01–0.05

Source: Based on [4,2], EIC and IIR databases analysis and calculations.

Finally, the feasibility study phase of the project is carried out to do the analysis of the preferred option and further minimize risks. This usually includes:

- Technical and financial evaluation of the preferred option;
- Final assessment of the budget and financing options;
- Detailed understanding of the permit issue process, social and environmental risks.

The overall opportunity feasibility study has to come with a clear conclusion about the investment decision, type of technology, site location, timeline of the project and full economic evaluation. While the preparation of the final report can be executed by a third party (engineering consulting company), the decision has to be done by the developer. Therefore in each of these stages, the alternatives have to be studied to the depth necessary for reaching a conclusion on their suitability for the stated purpose. If any of the alternatives fails to meet the feasibility study requirements, it has to be eliminated. This funneling approach allows minimizing potential risks, saving money and time during execution of the project and therefore increasing the ROI as much as possible.

According to [2,4] and EIC[1] and IIR[2] databases, a typical CAPEX for a large hydropower plant depends on country and site, and can be within a wide range of 940–8400 $ per kilowatt of installed capacity (see Table 15.1 for details) with the average around 2800 $/kW. It is important to say that there is no major correlation between the EPC company and OEM involved in the project and the cost of installed kilowatts: in most cases, cost is driven by local regulation, availability of land, permit issue processes, and overall construction practices. Moreover, the economies of scale usually play one of the key roles, leading to lower cost per kilowatt of installed capacity for larger power plants of the comparable type.

[1] The EIC datastream: https://eicdatastream.the-eic.com/.

[2] Industrial Info Resources: http://www.industrialinfo.com/.

Remark 15.1. While the operating and maintenance costs are about 2% per year of the CAPEX, there is no charge for fuel (which is a major cost for a thermal power plant), so that the variable part of the cost is much more stable as it is not influenced by the volatility of fuel prices.

The capacity factor of a hydro power plant is another variable to be considered, and can be as low as 25% depending on the mode of operation (peaking or base). This leads to an average LCOE pricing of roughly 0.02–0.19 $ per kilowatt-hour [4,2], which is a break-even point for an investor: the sales price of electricity shall be above this number to make a positive return.

Knowing CAPEX and OPEX for a future project allows an investor to calculate and analyze the net present value of the project and consider various financing options to estimate the ROI and payback period.

Another point that becomes inevitably important these days is the lifecycle carbon emissions cost, or basically, the overall carbon emissions associated with the planning, construction and operation of a hydropower plant. Despite zero carbon emissions from the plant operation, it still has certain level of emissions associated with it, mostly due to production of concrete, delivery of material to a remote site location, and possible impact on the vegetation damaged by the reservoir.

15.3. Design and construction

When the feasibility study is completed and the investment decision is made and approved, the investor (normally a financial company and/or an operator) has to choose the developer for the project — usually an EPC company (or several companies in consortium) that would be capable of preparing a detailed design and then building a hydropower plant. The common procedure here is to start the bidding process, when the investor issues a request for proposal and allows companies to propose their solutions and pricing. This is mostly done with the help of a technology and engineering consultant — a specialized company that helps the investor specify in detail what is wanted, often already during the feasibility study phase. There are only a few huge and world known players that are acting as owners, partially as investors (with the governmental funding and loans), and project designers and managers simultaneously.

15.3.1 Bidding process

During this bidding or contracting process, plant construction contractors are selected based on the following selection principles, depending on a procurement strategy [2]:

- EPC or prime contractor to execute the project on a turnkey basis. It can be a preferred solution to have a "one stop shop" for the investor: a company or a joint venture that acts as a single company bearing all responsibility for the whole project. It wold be responsible for overall engineering, procurement of material, and construction activities on site until commercial operation. An EPC would outsource some of the work to its subcontractors, but would still be a single contact point to the owner. However, the prime contractor will bear all risks and therefore can increase the project price.
- Separate contracts for various parts of the plant project, for example, for detailed design, procurement of major equipment, waterway construction, etc. While the owner assumes all responsibility here, it can be a demanding activity that requires experience and expertise. The investor either outsources project management activities to a consultant, or drives the project with his own project management team.

Traditionally, large hydropower plant projects are executed on a turnkey basis through an EPC contract. In many cases, there is a joint venture between a construction company and major equipment manufacturer (hydroturbine and generator), which wins the project as it can propose much more favorable conditions, proper references and better schedule.

After the primary contractor (or a project manager) is awarded, the project must achieve financial closure to ensure allocation of the budget (normally through debt due to large investment values). There are two major options for financing a hydroproject [2]:

- *Corporate finance*, when a lending institution offers financing to the investor that would to the developer to implement the hydroproject and assume responsibility for debt servicing (interest and capital repayments). The lender focuses on the company's ability to service the interest and repay or refinance the debt.
- *Project finance*, which focuses on the project to generate cash after its execution. Under this scheme, project assets may serve as collateral to reduce lender risk.

15.3.2 Design and construction

The construction process of a hydropower plant consists of the following major steps, which may well overlap:

1. Detailed design of a project, to issue main construction drawings and specifications for procurement of major, and secondary equipment and systems;
2. Site preparation activities;
3. Civil works, which start in parallel with detailed design with site preparation and infrastructure development (camp, roads, power supply, etc.). Then, the major civil works include erection of dams, penstocks and waterways, and supplementary structures, as well as the power house and foundations for equipment. The latter require details from the manufacturers about the foundation loads and dimensions of the equipment;
4. Major equipment installation is done when the pedestals are erected and equipment is delivered to the site. As most of the big hydropower plants are located in remote areas, delivery of equipment can become a challenge especially for bulk items such as turbines, generators, and transformers. Special road surveys, precise scheduling (taking into account climate conditions) and expertise of specialized logistics companies plays an important role and may cause significant project delays;
5. Electrical and mechanical works, which are conducted around the major equipment in order to erect supplementary equipment also called balance of plant. These are auxiliary pumps, motors, electrical systems, instrumentation, control room, etc.;
6. Commissioning period, which involves all necessary test procedures to check the construction works, equipment and connections to the grid. At this time the guaranteed parameters of the power plant (i.e., power output into the grid) should be met;
7. Finally, commercial operation, which starts after all the test procedures are successfully accomplished and the generated electricity delivered to the grid.

After the financial closure and EPC contract signature, the general contractor will start a detailed design, which is the basis for plant construction. Conducted by an EPC company, the detailed design is aimed at releasing proper drawings for construction and erection of foundations, structures and buildings, manufacturing and installing major and secondary equipment, etc. In most cases part of the work is outsourced to other companies that specialize in certain works: civil, structural, electrical, mechanical, equipment, etc. When ready, the EPC company places orders for long lead items like hydroturbines, generators, valve blocks, and defines the specifications for other equipment.

Site works of a hydropower project comprise several major steps:

- Site preparation, i.e., clearing the site for the temporary camp of the workers. For a massive hydropower project the amount of workers employed at a site can be up to several thousand people at peak times. This would require specially designed and constructed camps with all necessary services (electricity and water supply, drainage and sewage, transportation, and other recreational facilities) available for the overall lifecycle of the construction and commissioning.
- Site supply chains, which should be established along with the camp construction. These shall provide power and utility supply for the period of construction, as well as ensure delivery of material and equipment. As most of big hydroprojects are located in remote areas, transportation opportunities can be limited and nontrivial: separate special roads are usually built to ensure heavy material delivery. On top, some sites can be equipped with newly-built concrete factories to allow for production on the premises.
- Excavations and soil works, usually conducted in batches to ensure the water source (especially downstream) and the nature are damaged as little as possible.

In mountainous areas, various approaches are used, usually explosion or special tunnel cutting machinery.

Depending on the design, construction process may differ. However, it is normally done in several steps:

- Construction of temporary waterways to guide water around the construction site;
- Erection of dams, including main and side dams (if applicable);
- Construction of the turbine hall structures (either in a separate building or inside the rock);
- Erection of major water intakes and penstocks;
- Finishing of the concrete structures, gates, spillways, etc.

After major civil works are completed, so that all foundations are properly erected to accommodate major and auxiliary equipment, mechanical and electrical systems are installed. Their delivery to the site is subject to transportation limitations, so that it is important to provide good transport accessibility for the site in advance.

The installed equipment is then interconnected with pipes for various fluids: water (cooling, drainage, fire protection), compressed air, lubricating oil, hydrogen and nitrogen (if required), etc. Finally, electrical interconnections like electrical and control cables are routed through the hydrosite.

15.3.3 Commissioning

The commissioning phase is usually performed by an EPC with the help of the requisition engineers of the equipment manufacturers or by a separately hired specialist company as this procedure requires multiple areas of expertise for all systems to check what has been built and to ensure required level of quality.

Before the commissioning procedures, all required information, such as quality certificates, test procedures, test results from equipment installers, and other related documentation, are collected by the responsible commissioning engineer. While the major tests normally include performance guarantees, the hydropower plant has to fulfill several other requirements [2]:

- Dry tests: after installation, equipment is tested in dry conditions to verify basic functionality;
- Wet and load tests: with water collected in the reservoir (which may take some time after the dry tests);
- Performance testing: verifying turbine hydraulic efficiency and generator electrical efficiency, and the overall plant output according to the numbers and conditions specified in the contract;
- Trial operation and reliability run: this requires up to 30 days to ensure all systems are functioning properly.

Remark 15.2. All these procedures are usually performed according to the contract conditions and are subject to financial penalties. Depending on the contract, there can be various conditions and guarantees specified during the negotiation process before the EPC contract signature.

At the end of the commissioning phase, the responsible party provides detailed test reports with all results and all the operation and maintenance manuals and documentation, including "as built" drawings. The final certificate issued by the owner declares project acceptance and is usually linked to the release of the EPC from its obligations covered by the final payment [2].

Table 15.2 Average life expectancy for hydropower plant parts and systems.

Part or system	Years
Waterway steel structures (penstock, gates, trash racks, etc.)	60–80
Waterway concrete structures	70–100
Major mechanical equipment, including hydro turbine, generator, valve block	40–50
Mechanical BOP systems (pumps, cooling and compressed air systems, HVAC, etc.)	30
Transformers, HV and MV switchgear, and related equipment	30–40
Instrumentation, control, and protection systems	20
Rest of electrical BOP (LV AC and DC power supply, emergency diesel generator)	30
External transmission lines	70

Source: Aggregated from [2,1].

15.4. Lifecycle management

The average life expectancy of hydroelectric power plant systems and components under normal conditions of operation and maintenance are summarized in Table 15.2. These values may significantly differ from plant to plant as they depend on site specific conditions (water quality and chemical composition, climatic conditions), plant operating profile, and operation and maintenance procedures. While the given average numbers can give an idea about the frequency and scope of major rehabilitation and upgrade works, they have still to be assessed and planned according to the real conditions.

15.4.1 Operation and maintenance activities and costs

Operation and maintenance activities are designed to ensure safety, availability and reliability of a hydropower plant, so that it can generate electricity for sale according to the planned mode. Since the O&M costs represent the largest share of the operating expenses, proper planning is important already during the design and construction phases.

The future O&M staff usually receives proper training from the EPC and especially from the OEM before the plant is put into commercial operation, so that people can be involved in commissioning activities and testing. All equipment and systems are supported by an O&M manual provided by the manufacturer and/or EPC. In many cases, OEM representatives can be involved directly into O&M activities or provide remote control and monitoring solutions — all activities covered by a long-term servicing agreements.

The number of personnel involved in the O&M activity depends on the project size, complexity, and local regulations of a country, and can be from several local operators supported by a crew of O&M staff from the owner (or other contractor) up to several dozens of people for a larger site in a less developed area or country.

Similar to other power plants, the hydroelectric power plant O&M costs can be divided into two major categories [2]:

- Electrical and mechanical equipment costs. The recommended annual budget for electromechanical equipment maintenance is 2.0–2.5% of initial investment costs, with 40% going for yearly maintenance and the remaining 60% accumulated in a reserve fund for future rehabilitation and upgrades.
- Civil works costs are limited to around 0.4–0.6% of initial investment costs and are used for underground works, reservoir and site maintenance, etc.

In most cases, standard O&M procedures do not include major upgrade projects.

15.4.2 Hydroelectric power plant upgrades

There is a general belief that hydropower is a mature and established technology with substantial capital investment in place, it cannot be markedly influenced by modern technology [5]. However, despite its long-lasting history and the status of one of the oldest technologies, there are still multiple opportunities for its upgrade, retrofitting, and optimization. These mostly target such sustainability issues as the problem of fish migration, changes of the hydroregimes, decrease of oxidation, etc.

Major upgrades of a hydroelectric power plant are considered as a major investment project that targets overall plant economic (through technological) efficiency. While the concrete structures of the waterway have a much longer lifecycle, the upgrade project usually focuses on major equipment: replacement of turbines and/or generators, substation or switchyard (e.g., switch from AIS to GIS), and balance of plant systems (like motors, compressors, batteries, instrumentation, protections, and DCS). Still, in many cases, upgrades also include civil works, especially for rock-fill and earth-fill dams.

As it incorporates many works, which can be compared to almost a complete hydroelectric power plant construction, it is sometimes called an *upgrade program* that consists of several separate projects. This approach has two reasons:

1. It allows avoiding large investment at a time and provides lower financial burden;
2. It gives an opportunity to minimize the amount of work at site with optimal outage time.

Upgrade programs require significant investments in R&D operations, including CFD modeling of the water ways, modern turbine design to increase its efficiency and decrease impact, generator design optimization and overall process improvements. Secondly, future development may focus on the design and development of the technologies suitable for much lower water heads that may significantly broaden the hydropower resources that are available for use without major changes in water regimes. These activities have been undertaken both by major manufacturers and research institutions, and we may expect novel commercial applications available in the nearest future.

15.5. Hydropower as part of sustainable energy system

Hydroelectric power generation is currently the world's largest installed renewable conversion technology to produce electricity. Being a historically establish and mature, hydropower has many advantages compared to other sources of electrical energy generation, especially to thermal power plants: it is renewable, clean, largely carbon-free, and well suited for integration into the grid, often supporting the stability of a grid [6]. Hydroelectricity is ideal for providing peaking power and smoothing intermittent power sources like wind and solar. When it is in spinning reserve mode, it can get connected and provide electric power within 15–30 seconds [7]. Depending on the country, hydropower can be used either mostly for base load generation or in peaking regimes.

Hydropower's flexibility of operation can be one of the enablers for a sustainable electricity generation system by:

- Providing stable and large-scale base load capacity in areas with high water resources potential;

- Acting as a grid balancing technology to accommodate variable energy sources like solar and wind;
- Enabling energy storage by collecting potential energy of water in the reservoirs, or using excess energy from variable source within pumped storages;
- Using excess energy to produce hydrogen as one of the possible future energy carriers.

But hydropower can also have negative impacts, especially on the local level, where dams, weirs, diversions and changes in river flow may disturb ecosystems and create problems [6]. Some of the potential issues discussed internationally are presented in the following example case.

With the need for more power for rapid development and to reduce GHG emissions, in 2002 China broke the ground for another hydropower plant known as Xiaowan Dam. Located on the Mekong river, which basically feeds China, Myanmar, Laos and Thailand, this concrete arch-type dam is 292 m high, which makes it among the highest in the world.

The commercial operation started in 2010. The Chinese claimed that opening another dam would help people living downstream to cope with floods and droughts by a controlled release of water. Still many countries were concerned about the consequences. As reported by [8], the change in water quality and flow could disrupt fish migration patterns, increase the need to use more artificial fertilizers because nutrient-rich sediment would deposit behind the barrage instead of flooding into people's fields, or people may face the possibility of more droughts and greater river bank erosion, while others will have to rethink their own plans to build hydroelectric plants downstream because the volume and speed of water downstream will be affected.

However, in 2017 it was confirmed that while the Mekong communities were concerned about the adverse effects of Chinese dams on such issues as sediments and fisheries, these dams had not reduced the flows downstream during the dry season [9]. The example shows that the truth is, as always, somewhere in the middle, and a good and sustainable in theory project is driven by pure economics and therefore turns out to be unsustainable.

Regarding CO_2 equivalent or other GHG emissions, the hydropower technology does not emit carbon dioxide, nitrogen oxides, hydrocarbons, or particulates during its operation directly. Still, this is an underestimation since, especially in the construction phase, CO_2 emissions are very significant, as in many other technologies. They are mostly associated with production and consumption of concrete (which is especially high for large reservoir-type plants with concrete dams), manufacturing and delivery of equipment, and so on. However, taking into account the long life of hydropower plants, which may be up to 80 years, the impact of CO_2 emissions per kWh produced is considered among the smallest of all renewable technologies [6].

The majority of lifecycle GHG emission estimates for hydropower lie between about 4 and 14 g of CO_2 equivalent per kWh of generated power, but under certain scenarios there is a risk of much large quantities of GHG emissions [10]. Moreover, the reservoir may be responsible for the release of methane, a greenhouse gas far more potent than CO_2. This can be due to changes in the river ecosystems due to reservoir construction, chemical composition, etc.

Hence, even though hydroelectric power plant construction can improve the local economy and provide flexible and reliable source of electricity, there are still concerns regarding its sustainability in the long run. One of the aspects to be considered is the effect that climate change can have on hydropower production. These studies are especially important given that hydropower systems have lifetimes that are extended over time and span several generations and, consequently, long-term climate variations can have an important impact on the projects currently being developed [6].

References

[1] International Energy Agency (IEA). Technology roadmap. Hydro. France: OECD / IEA; 2012.

[2] IFC. Hydroelectric power. A guide for developers and investors. International Finance Corporation; 2015.

[3] Keeyask generation project. Project description supporting volume; 2012.

[4] Andrews J, Jelley N, Jelley N. Energy science: principles, technologies, and impacts. Energy science: principles, technologies, and impacts. Oxford: OUP. ISBN 9780199592371, 2013.

[5] Tester J, Drake E, Driscoll M, Golay M, Peters W. Sustainable energy: choosing among options. MIT Press. ISBN 9780262304634, 2012.

[6] Guerrero-Lemus R, Martínez-Duart J. Renewable energies and CO_2: cost analysis, environmental impacts and technological trends – 2012 edition. Lecture notes in energy. Springer. ISBN 9781447143857, 2013.

[7] Jacobson MZ. Review of solutions to global warming, air pollution, and energy security. Energy & Environmental Science 2009;2:148–73. https://doi.org/10.1039/B809990C.

[8] Dam the consequences. The Guardian 2007. URL: https://www.theguardian.com/world/2007/apr/06/outlook.development.

[9] Mekong River Commission for Sustainable Development. The effects of Chinese dams on water flows in the lower Mekong basin. URL: http://www.mrcmekong.org/news-and-events/news/the-effects-of-chinese-dams-on-water-flows-in-the-lower-mekong-basin/, 2017.
[10] Letcher TM. Future energy: improved, sustainable and clean options for our planet. Elsevier Science. ISBN 9780080994222, 2014.

PART SIX

Emerging sustainable energy systems

Overview

The largest portion of electricity is retrieved from traditional primary energy sources: coal, natural gas, oil, nuclear fuel, and hydro energy. While these sources can serve as a bridge towards a pure sustainable technology, they shall be diversified.

Recent trends show several emerging technologies that stem from other energy sources: wind, solar radiation, biomass, waste, geothermal energy, or ocean movements. While they still occupy a small portion within the electricity generation mix, their growth rate is amazingly high, so that they are gradually increasing their share. In this part we cover such emerging technologies and their development.

CHAPTER SIXTEEN

Wind energy

Contents

16.1. Wind resources and installed capacity

Wind energy is one of the oldest sources of energy employed by the humankind, comparable only to the use of animal force and biomass [1]. Since the dawn of the human civilization, wind energy has been captured by sails to run the ships or spin wind mills to pump water or air for blacksmiths, or run grinders to mill cereals.

The 20th century has brought a new application of wind mills, that is, for electricity generation. By the middle of the century, millions of windmills were in use worldwide, many of them deployed to produce mechanical power or as decentralized electricity supply for remote locations. At those days, no-one has ever though of wind mills as a replacement of traditional utility scale power plants or as a remedy against global warming, which has not been the case discussed by the public. Moreover, when the electricity grid reached most of the remote areas, the use of windmills for electricity production decreased dramatically until the middle of the 20th century, when wind power generation started to be considered as an economically feasible and promising technology.

Whatever the application of the wind mills, their performance is always strongly dependent on the quality of the local wind resources [2], or in simple words, the speed, flow direction, and stability at various heights.

Sustainable Power Generation
https://doi.org/10.1016/B978-0-12-817012-0.00030-X

Therefore, a wind park strongly depends on the wind resources and their potential.

Wind energy, in its essence, can be well considered as a form of "solar" energy. As discussed in detail in Chapter 2, it is the solar radiation that heats up the atmosphere and creates zones of different pressure that basically initiate air to move, which is called winds. On average, the ratio of total wind power to incident solar power is on the order of 2%, reflecting a balance between input and dissipation by turbulence and drag on surfaces. The general mechanism of the wind systems is affected by the Earth's terrain which would influence the direction of the surface winds and their speed at lower altitudes. Therefore only a small fraction that is close enough to the Earth's surface could be practically accessible, and only selected locations have winds that are sufficiently strong and steady to be attractive for exploitation [2].

However, even these potential wind resources for power generation purposes are immense, but still are subject to several constraints on use that limit near-term exploitation to roughly 20% of electric grid capacity [2]:

- Winds vary in speed, and hence in incident energy flux, during the day and from season to season, and would not follow the demand pattern;
- Its nondispatchable nature limits the portion of wind power in global power generation mix. Therefore it requires provisions for primary and secondary reserves and grid stability;
- Wind energy is available according to the natural pattern and, in case of excess, energy should be stored, however, there are limited opportunities for that;
- The best wind fields may be located in remote areas and not in a reasonable proximity to the consumers. This may require expensive transmission lines (subject to power losses). Moreover, suitable wind sites are limited and the best have already been used.

Still, wind energy generating technologies have become and will continue to be an important technology in a global sustainable electricity generation mix.

Globally, around 1700 TW of wind energy are available over the world's land together with the ocean surfaces at the heights of about 100 m if all wind at all speeds were used to power wind turbines. This reference height is in the range of the hub height of modern wind turbines. However, the wind power in locations over land and near shore where the wind speed is 7 m/s or faster (the speed necessary for cost-competitive wind energy) is around 72–170 TW [3–5], but only slightly over half of this power is in locations that could practically be developed. These are large regions of the USA, Canada, Northern Europe, the Gobi and Sahara deserts, much of the Australian desert areas, and parts of South Africa and Southern South America, and South Africa [5]. In the USA, for example, wind from the Great Plains and offshore the East coast could supply all US energy needs. Other windy offshore regions include the North Sea, the West Coast of the USA, the East Coast of Asia, and some others.

16.1.1 Wind quality and speed

Wind power quantifies the amount of wind energy flowing through an area per unit time, or the flux of wind energy through an area. *Wind energy* by definition is the energy content of air flow due to its motion, which is called kinetic energy and is a function of fluid's mass and velocity, given by

$$E_k = \frac{mv^2}{2}, \tag{16.1}$$

where m is the mass, and v is the speed (m/s), which is assumed constant.

Wind power is the rate of kinetic energy. The power contained in the wind flowing through the area S per unit time is

$$P = \frac{dE_k}{dt} = \frac{1}{2}\frac{dm}{dt}v^2 = \frac{1}{2}\rho v S v^2 = \frac{1}{2}\rho S v^3,$$

measured in watts, where ρ (kg/m^3) is the density of the air. Therefore, the wind power is proportional to the cube of wind speed, density of the air, and the area $S = \pi r^2$ that is swept by the wind turbine blades of length r.

However, the wind turbine cannot extract the total power of the flow, therefore turbine's real power P_t will be lower than the total power P, and their ratio

$$C_p = \frac{P_t}{P} < 1 \tag{16.2}$$

is called the *power coefficient* or *turbine efficiency*. Hence, turbine power capture is given by

$$P_t = \frac{1}{2}\rho S v^3 C_p.$$

This capacity is limited by a theoretical *Betz limit* [6], meaning that power coefficient C_p cannot be greater than 16/27, which means that a conventional wind turbine efficiency cannot beat a theoretical limit of 59.3%.

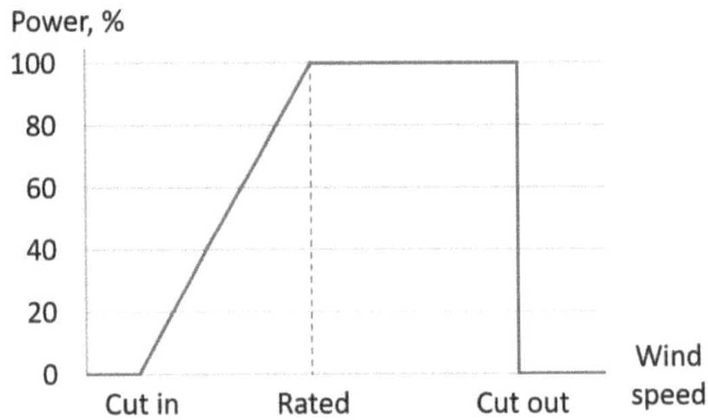

Figure 16.1 ***Wind turbine power curve depending on the wind speed.*** Wind turbines starts its operation at cut-in wind speed and increases its power gradually to 100% as the wind speed grows. Wind turbine operates at full load till the cut-off wind speed (curve for indicative purposes only).

The quality of wind resources is variable and localized so that accurate surveys of long-term sites are a prerequisite to deployment. Wind speed varies with time and height, and is also affected by the landscape, vegetation (trees), structures, or other wind turbines that alter the wind field. In general, there is much less fluctuation in airflow at greater heights.

Therefore, the theoretical wind turbine power is a function of variable wind speed and can either be calculated on a mean wind speed basis, or using the exact wind speed distribution based on real observations and measurements. Technologically, a wind turbine has to target wind speeds that lie in the limits between:

- *Cut-in speed* around 3.0–4.5 m/s, which is the lower wind speed limit that allows operating the wind turbine economically;
- *Cut-out wind speed* of around 30 m/s that requires a wind turbine to stop due to unsafe operation.

As illustrated in Fig. 16.1, when a turbine starts to operate at cut-in speed, it increases the power output as the wind speed grows up to a rated level. For the wind speeds in the range between rated and cut-off speeds, the power output of the turbine is maximum.

While the wind speed is not constant and usually follows the statistical Weibull distribution, it is important during planning to consider this distribution and the operating limits of the wind turbine to maximize the power and continuity of operation. It may result in the fact that a turbine operates less than theoretically possible with strong and constant wind, which is characterized by the *capacity factor* — the ratio of actual total yearly power of the turbine to its rated power. Table 16.1 shows a few examples of total

Table 16.1 Average sizes and capacity factors of onshore and offshore wind turbines and wind parks.

	Onshore	Offshore
Capacity, MW		
— per unit	up to 3	up to 7
— per wind park	up to 200	up to 1000
Capacity factor		
— to date	23%	40%
— new installations	34%	49%

Source: Based on EIC and IIR energy databases and public data of OEM companies.

capacity factors of currently installed turbines onshore and offshore and some of the new installations that, due to advanced technologies, can provide better wind capacity utilization.

16.1.2 Installed capacity

The global wind power installed capacity has increased by an average of 23% per year from 2004 till 2014, driven by the onshore technology, which accounts for more than 96% of the total installed capacity [7]. Despite this deployment, wind still accounts for no more than 6.4% of the total installed capacity globally, supplying around 3.6% of electricity [7].

Wind power is playing a greater role in power supply in a growing number of countries. In 2016, wind energy covered an estimated 10.4% of the EU demand and equal or higher shares in at least 11 EU member states, as well as in Uruguay and Costa Rica. At least 24 countries around the world met 5% or more of their annual electricity demand with wind power. In the United States, utility-scale wind power represented over 5.5% of total electricity generation and accounted for more than 15% of generation in nine US states, including Iowa (with the impressive 36.6%). Two German lands had enough wind capacity at year's end to meet over 86% of their electricity needs, and four had enough capacity to meet over 60% of their needs. Thus, globally, installed wind power capacity by the end of 2016 was enough to meet an estimated 4% of total electricity consumption [8].

The year 2016 has shown about 55 GW of wind power additions, resulting in 487 GW of total installed wind capacities. This was mostly due to Chinese developments with more than 23 GW, which became the largest wind capacity contributor. The second largest contributor was the US with more than 8 GW, followed by Germany (around 5 GW) and India (additional 3.6 GW).

16.2. Wind power plants

Wind power plants, also known as *wind farms* or *wind parks*, consist of wind turbines and associated balance of plant, or BOP. Balance of plant is basically everything around wind turbines, and includes civil works, electrical equipment and rarely transmission line (if it is part of the project scope).

16.2.1 Wind turbine and its components

Historically there have been several approaches to design a wind turbine — a machine that converts the power of the wind into electricity. Generally, they can be split into two broad categories:

- Horizontal-axis wind turbines or HAWT, when the axis of rotation is parallel to the ground. Depending on the location against the air flow, these turbines can be split into upwind (when the turbine faces the flow) and downwind, respectively. These turbines can have a different number of blades (usually 3), possess various speed and torque values.
- Vertical-axis wind turbines or VAWT have the vertically-oriented rotational axis and can generally be lift or drag force driven.

However, modern economically feasible wind turbines are mostly of HAWT type with 3 blades as illustrated in Fig. 16.2. A typical wind turbine of HAWT configuration has the following major components:

- Foundation, to accommodate turbine, which can be made out of concrete for onshore units and as a steel structure for offshore implementation;
- Tower, either out of sections of rolled steel or concrete. The height of the tower can range from 40 to more than 100 meters to capture the wind of the highest quality;
- Nacelle, which accommodates major systems of the turbine, namely the drive train;
- Main frame of the nacelle, to mount it on top of the tower, which also accommodates the yaw system — a special drive to rotate the nacelle to follow the changing wind direction;
- Rotor, which comprises blades and a hub. The blades may well exceed 100 m in length and have a special variable (along length) wing-shape defined by air dynamics. They are manufactured in specially designed

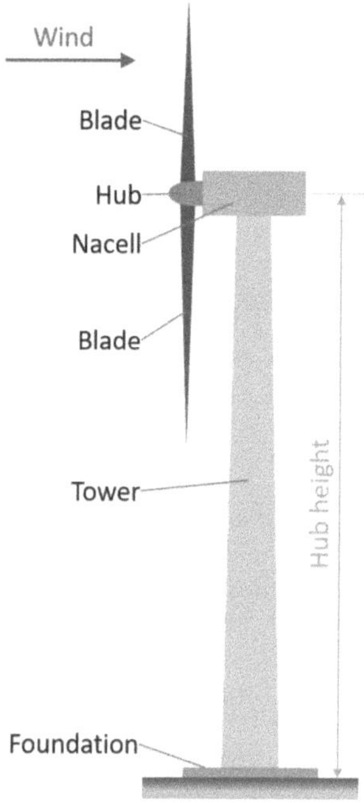

Figure 16.2 ***Wind turbine and its key components.*** Key components of a wind turbine: foundation, tower, nacelle, and rotor with blades.

moulds from composite materials like mix of glass fiber and epoxy resins. The blades are held in their position by the hub made out of cast steel. Pitch control gearboxes installed in the hub serve the purpose of setting wind rotor blades at the best angle to the wind.

The current trend within the industry goes towards increasing the turbine rated power and therefore the diameter of the rotor. The historical development of the rotor span, as shown in Fig. 16.3, gives an idea of a rapid technological progress from small diameters of 15 meters in the early 1980s up to more than 160 m with the capacity of 9.5 MW offered these days. From 1980 to 2020, the estimated average annual growth rate may well exceed 7%. Recently, GE Renewable Energy has announced an offshore wind turbine Haliade-X that boasts 12 MW capacity with 220 meter rotor and hence becomes the largest wind turbine to date.[1]

[1] https://www.genewsroom.com/press-releases/ge-announces-haliade-x-worlds-most-powerful-offshore-wind-turbine-284260.

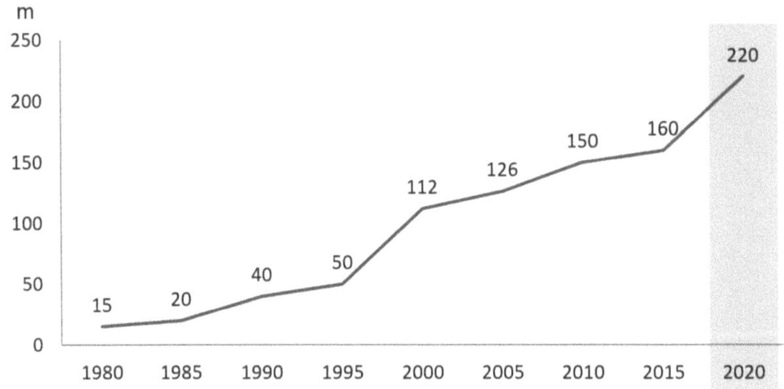

Figure 16.3 ***Evolution of the rotor diameter for wind turbines.*** The diameter of the wind turbine rotor has increased more than 10 times since 1980, and expected to grow further till 2020 and beyond. *Source: Based on [2,9,10] and author's estimates.*

Such large rotating machines have to be properly installed for safe and reliable operation. While the onshore turbine's foundation is a relatively simple civil structure, fixing a turbine offshore to the seabed becomes a challenging task. For various water depths, there have been several types of foundations implemented commercially (see Fig. 16.4):

- Mono pile foundations, usually made out of steel, that are pushed into the seabed for about 30 meters with a hammer machine;
- Gravity foundations, manufactured out of concrete and filled with sand, stones or gravel to be further fixed at the depths down to 30 m;
- Tripod steel structure used to fix at the seabed with three smaller piles;
- Jacket foundation, which is a steel structure similar to a lattice tower that is fixed at the seabed. This type of fixation can be used down to 60 m of water.

For deeper waters it is not economically feasible or even not possible to install the complete wind turbine at the seabed. In this case floating foundations have been developed and tested, some of them are already in operation, for instance, tension leg or semi-submerged platforms [11]. The deployment of floating wind turbines can dramatically increase the quality of energy that can be captured by the wind parks, as the wind profile over the ocean is usually better than over the continents.

The *drive train* of a typical wind turbine consists of the following components and systems:

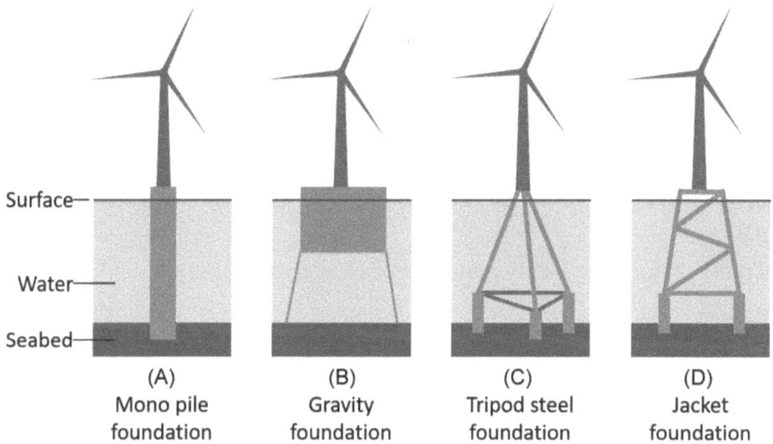

Figure 16.4 ***Major types of offshore wind turbine foundations.*** There are several major types of offshore wind turbine foundations: (A) mono pile, (B) gravity type, (C) tripod steel structure, and (D) jacket.

- Main shaft that transfers the rotational force of the rotor with blades to the gearbox. The rotor and the shaft are supported by several bearings to withstand forces and minimize friction losses;
- Gearbox, which increases the slow speed of the rotor to the speed required by the generator;
- Electrical generator (usually a permanent magnet alternator), to convert mechanical energy of the rotating shaft into electricity;
- Brake system, to stop rotation of the whole drive train, if needed.

Most of the wind turbines generate alternating current that cannot be fed directly to the grid as it is not synchronized and varies in frequency and voltage. Sometimes it is called "wild power". Hence, wind parks require additional electric equipment provided within the balance of plant.

16.2.2 Balance of plant

Balance of plant of a wind farm comprises everything around wind turbines and is usually split into civil and electrical scope. While civil scope of an onshore wind mill is relatively simple and incorporates foundations, roads, drainage system, etc., the electrical BOP part for both on- and offshore wind parks is more complicated and may include:

- *Electrical substation*, to convert AC produced by generator into AC consumed by the grid. This can be done through passing AC through a bridge rectifier with further conditioning using a variety of tech-

nologies. Finally, this DC can be either used directly (if needed) or converted into AC by inverter;
- *Switchgears*, to protect equipment and disconnect it from the grid;
- *Instrumentation and control system*, together with the connection to a control room, to collect performance and maintenance data;
- Interconnecting cables, to evacuate power from the turbine to the substation;
- Transmission line or its part, to connect to the grid, if included in the scope.

16.3. Wind farm project lifecycle

Despite a possible opinion that wind turbines look much simpler and wind parks look smaller than conventional power plants, its development and project management can be as complicated as for traditional technologies. A wind farm development project may be subject to a thorough and comprehensive economic evaluation thanks to the developing nature of the technology, however, recent projects indicate that wind has become competitive to traditional energy without any incentives. Still, wind projects have to be considered with the same attention as any other complicated energy projects.

16.3.1 Feasibility study

There are several major requirements that should be met for a wind farm development project:
- Availability of sufficient wind resources, to maintain certain level of electricity production;
- Access to the area (land or offshore) and its availability for construction, including topographic and geological assessment;
- General acceptance of the project by the community and the public;
- Ability to gain permits from authorities regarding environmental impact;
- Access to the market through the available transmission line and substation, or an opportunity to build such a line, and further acceptance of the power by the grid;
- Market for the electricity generated from the future plant, i.e., guarantee that electricity will be sold at predictable and acceptable price like power purchase agreement;

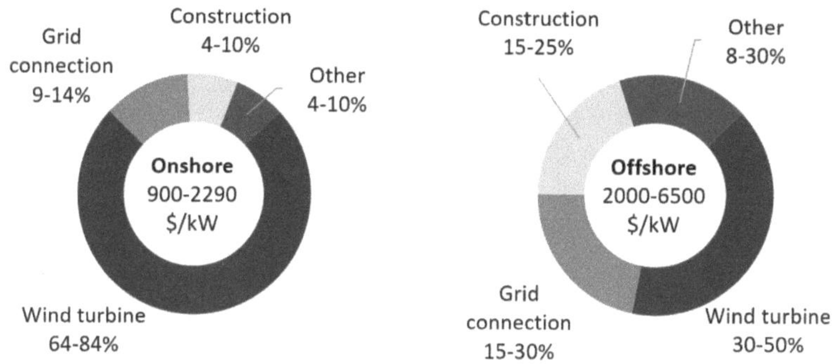

Figure 16.5 ***Wind power cost breakdown structure.*** Wind cost breakdown structure for onshore (left) and offshore (right). *Source: Based on [12], EIC and IIR database and OEM data analysis.*

Table 16.2 Typical wind park costs.

Technology	CAPEX, $/kW	OPEX, $/kWh	LCOE, $/kWh
Onshore	900–2290	0.005–0.025	0.06–0.12
Offshore	2000–6500	0.027–0.054	0.10–0.21

Source: Based on [12,13] and EIC and IIR databases analysis.

- Access to the capital to finance the project with reasonable interest rate, to ensure market ROI.

A wind power project starts with the assessment of the annual wind regime in the location selected for the project and calculation of its economics. Typically offshore turbines are at least 20% more expensive per unit of installed capacity, and prices of the foundation are at least 2.5 times higher for the same size of turbine. It is evident that water depth, distance from the shoreline and project size influence the final investment cost for specific offshore wind farms [9]. This cost is very site-specific and currently concentrated in a small number of markets [13], therefore the CAPEX of the project has a wider range throughout the world (see Fig. 16.5).

For wind power plants, the O&M costs are usually lower compared to conventional power generating technologies. Therefore, wind farms can be considered capital intensive compared to conventional ones, for which fuel costs play a dominant role [9]. Typical annual values of O&M costs for wind farms are represented in Table 16.2. However, this statement remains untrue for offshore wind parks where the O&M costs are largely driven by the costs of the maintenance vessel and its schedule. Still, with the development of new parks and the increased level of experience, these costs are already falling down.

The trend towards using wind turbines on taller towers with larger rotor diameters for a given generator capacity is working towards generating more electricity for the same installed power. Therefore, the LCOE from wind generation has continued to decline and for onshore wind plants it was around 0.06 USD per kWh in 2017, and offshore wind had the average LCOE at the level of 0.14 USD per kWh [8,14]. It is important to highlight that both technologies (and onshore especially) are continuing to establish their positions within the same cost range as fossil-fired conventional power.

The LCOE has been going down mostly due to technology improvements, competitive procurement, and a large base of experienced, internationally active project developers [14]. There were record low bids in tenders in Chile, India, Mexico, and Morocco, and prices fell rapidly in some offshore tenders in Europe. Onshore wind was the most cost-effective option for new grid-based power during 2016 in many markets, including Brazil, Canada, Chile, Mexico, Morocco, South Africa, Turkey, and parts of Australia, China, Europe, and the United States [8].

In addition, as the amount of wind capacities and its share of total generation have increased, so have grid-related challenges, which include lack of transmission infrastructure, delays in grid connection, lack of public acceptance, and curtailment where regulations and current management systems make it difficult to integrate large amounts of variable renewables [8].

While the LCOE for wind technology is steadily decreasing, which makes wind much more competitive than a decade ago, wind project assessment has to be done carefully based on real market data and should not depend solely on historical values.

16.3.2 Wind park project execution

The phases of a wind farm project execution are the same as for other energy projects and include the following stages:

1. Bidding phase, when the utility, owner or developer opens a bid to get quotes for equipment or a complete turnkey plant based on the RFQ documents prepared based on the results of the feasibility study. It finishes normally with the contract signature;
2. Detailed engineering phase, when the project is developed by an EPC or developer to fit the equipment to real ambient conditions of the site;

3. Procurement phase, which deals with the BOP and civil works contracting and allocating related subcontractors for various works;
4. Construction phase, when the actual site construction are carried out by respective contractors and supervised by EPC, developer and/or owner.

Remark 16.1. Even though these phases seem separate, they can still overlap: procurement of equipment and construction works like site mobilization and excavations may start before detailed engineering is fully executed.

As wind farms can be generally split into two major equipment areas — wind turbines and BOP — several options for contract structure are possible:

- Equipment supply only, which includes supply of wind turbines, related control system, and construction supervision and commissioning services. This type of contract is signed with the turbine manufacturer;
- Equipment and installation, which include the first option, together with the transportation and installation services and equipment (e.g., heavy lifting equipment and cranes);
- Turnkey, which adds BOP (electrical equipment and related civil works) on top. This type of contract means that the owner finally gets a complete wind park that is commissioned and ready for operation.

In many cases BOP scope is erected by a separate company different from wind turbine manufacturer, or, in some cases, two different contractors responsible for civil works and electrical systems, respectively.

16.3.3 Repowering and decommissioning

Wind turbines, as any other complicated rotating machines, require proper maintenance during operation. Due to wind's unstable nature, complicated monitoring and control systems, as well as condition monitoring and data analytics, are being installed to optimize their performance and lifecycle. Cutting costs by preventing failure instead of running wind turbines until they break lies at the heart of operations and maintenance [15]. The principles of condition monitoring are not new, but it is the interpretation and analysis of the measured data, with increasingly sophisticated computational analysis, that is now coming to the fore [15].

According to the industry estimates [16], as wind turbines are designed with lifespans around 20–25 years, roughly a fifth of the world's installed wind turbines may be halfway through their lives. One consequence of this technology's aging is reduced performance. As wind turbines age, their reliability drops, spare parts may well become scarce which increases the O&M costs. Moreover, their capacity factor drops down, so that a wind turbine loses approximately 1.6% of its output per year [16]. This is especially important since in those countries, which have been deploying onshore wind turbines on a large scale, most of the good wind resource sites have already been occupied. Therefore, there is a question of increasing the efficiency and output of the existing wind parks as they reach the end of their lifecycle.

The initial idea was to upgrade or to *repower* old wind turbines by changing some of its equipment. Reliability can be generally improved with measures such as retrofitting modern control systems. There is also the option of *partial repowering* [17], but this approach has serious limits. Installing a 3 MW nacelle on top of a 250 kW tower is "not recommended", to say the least. The increase in blade size will often cause rapid wear in gearboxes and other rotating machinery and bearings as the early infrastructure's parameters are exceeded [16].

Therefore old wind turbines are decommissioned — dismantled and recycled, and the site can be used again for new installations. This approach, sometimes also referred to as *decommissioning for repowering*. This full repowering shall address a number of important questions:

- Should the site area be kept within the existing limits or be extended?
- Should larger units be installed, increasing the overall capacity, or the capacity and unit size be kept on the same level?

This basically leads to the complete business modeling of a project with the major difference that the wind profile has already been evaluated in every detail along the existence of the previous site, which may well support the optimized design of a new site.

While onshore wind parks have been in place for a longer period of time, their repowering has been pretty well established. Despite the ongoing discussions about the onshore activity, no offshore wind turbines have been repowered as per the end of 2016, and only a handful of offshore developers are considering doing this [16]. Offshore wind farms are relatively new, and therefore owners or operators have been mainly concerned with improving installation techniques and achieving operational efficiency. Despite the potentially large costs, logistical difficulties and environmental impacts, decommissioning has been given little attention to date. This is an important matter because decommissioning is a significant part of any project, and should be considered from the very beginning, from the design stage, as if not done then, impacts can become more severe and costs can be higher than expected [18].

For offshore wind parks, similar to onshore, two approaches are considered: partial or full repowering (decommissioning for repowering). This decision depends on factors such as the site, size of the project, regulations applicable, power price and operating costs, extent of reusable infrastructures, and the expected profitability of the repowered project compared to the actual market [19].

As decommissioning is a relatively new topic, there is not much legislation on it. Therefore a variety of options exist and will vary depending on the country. They can range from complete removal of the different elements and their reusage or recycling where possible, or appropriate disposal, or leaving them on site after operation is ceased [18].

Hence, the experience in offshore wind decommissioning is very limited to date. Therefore most of the published sources vary in cost and time estimates for such projects. While estimated timing of decommissioning projects varies from a few days to more than a hundred per turbine (based on data from [18]), these numbers are not a reliable reference as some of them do not reveal all details (for instance, whether the onshore project management and procurement timing is included).

The estimated cost for decommissioning an offshore wind farm also varies in extreme range from around 2–3% of the total capital cost up to 60–70%. This means that the developer needs to save money during the lifecycle of the wind farm in order to be able to pay its end of life [18]. Evidently, as these are the initial estimates with new technologies, the costs are most likely to go down dramatically once the experience is gained.

16.4. Sustainability attributes

Wind energy is a freely available renewable energy source, which has been in the focus of humankind for centuries. Evidently, capturing this free energy (which is initiated by the solar radiation and hence is unlimited on the humankind time scale) is one of the best solutions for sustainable development. In this regard, wind power generation has enjoyed enough investment into research, development, construction, and the special regulations and economic incentives. This has resulted in an ongoing growth of the installed capacity, increase of efficiency and size, and development of new technologies like floating wind turbines to capture more wind resources of higher quality.

Evidently, wind power generation has a number of valuable advantages [6], for instance:

- The amount of lifetime GHG emissions associated with wind turbine technology is limited to the production, transportation, and erection of the wind farms, and is the lowest compared to other technologies. According to the studies [20], wind has among the lowest lifecycle CO_2 equivalent emissions at the level of 2.8–7.4 g per kWh of generated electricity. Wind energy is also free from particulates or sulfur dioxide, which are considered a major problem of coal-fired power plants.
- Wind energy is fully sustainable as it is kind of "solar" energy: whenever the sun shines and therefore the wind blows, energy can be harnessed and sent to the grid.
- Wind turbines have a small footprint and thus can be combined with other activity like farming and growing crops. They can be located in remote locations or mountainous regions where winds are strong but the location does not allow any other activity.
- Wind resources are distributed more evenly throughout the countries compared to fossil fuels, hence they can contribute to national energy security on a higher level.
- Compared to any type of thermal energy or biomass, a wind power plant consumes very limited water resources, roughly 170–320 l per kWh of generated electricity [6]. This is especially important in the face of growing fresh water scarcity and water conservation problem.

As for any other technology, there are still some limitations that may impact the deployment of wind power generation. These are, for instance:

- The intermittency of wind does not allow the wind farms to generate stable base load electricity. Therefore, integration of wind parks requires extra capacities to compensate for losses of wind and shutdowns through primary and secondary reserves. For the moment, these are done by gas turbine driven power plants (usually simple cycle, however, modern combined cycle can also propose fast ramp up and flexibility rates) or hydropower.
- Good sites are usually in remote locations, while the easily accessible sites have already been used for wind generation.
- Shortage of rare earth element, neodymium (Nd), needed to manufacture permanent turbine magnets. Wind turbines containing such magnets manufactured out of neodymium, iron, and boron (NdFeB)

have demonstrated a series of advantages in terms of efficiency, weight, dimension and maintenance, which are key requirements in wind applications, especially for offshore [21]. These permanent magnets, which are much stronger than traditional iron magnets, are required in the generator to produce electricity at slow rotation speed. At the moment, the supply of neodymium has been dominated by China with the US now reopening their mines [6]. The direct substitution of rare earth materials like Nd in permanent magnets or the development of new materials with a similar functionality to the NdFeB magnet is still at the research stage. However, progress has recently been made on magnet manufacturing techniques and wind generator design that allow the reduction of the amount of rare earths used in a wind turbine [21].

It is almost generally accepted that the utilization of renewable energies is, besides energy saving, the best means to reduce pollution and decelerate climate change, and wind energy has proved its efficiency and positive contribution to combat climate change. On the other hand, there are still some issues regarding the environmental impact in a broad sense [22]:

- Noise emissions due to the aerodynamics of the wind mill, especially of the rotating blades. Besides noise in the audible frequencies so-called infra-noise has also been the subject of concern [22].
- Oscillation shadow due to the rotating blades causes optical disturbance for the residents. Depending on the local conditions, minimum distances may be required by the regulation.
- Impacts of wind energy development on wildlife can be direct (collision fatality) or indirect (functional habitat loss or barriers to movement). The mitigation strategies may include careful site consideration, repowering of sites (replacing smaller old units with larger ones as the latter decrease the mortality rate by more than 50%), curtailing operations at high-risk periods for birds, or using special sound devices or visual approaches (applying paints on blades) to warn birds [23].
- Visual appearance and disturbance of the landscape, especially in tourist popular areas.

Still, wind energy is a growing contributor to sustainable power generation system. It has also shown its credibility within hybrid installations, particularly with solar PV or hydroenergy. Used on industrial scale, wind energy has been supported by the developed regulation, and recent projects have demonstrated its economic advantage over conventional competitors.

References

[1] Wagner H, Mathur J. Introduction to wind energy systems: basics, technology and operation. Green energy and technology. Berlin, Heidelberg: Springer. ISBN 9783642329760, 2012.
[2] Tester J, Drake E, Driscoll M, Golay M, Peters W. Sustainable energy: choosing among options. MIT Press. ISBN 9780262304634, 2012.
[3] Archer CL, Jacobson MZ. Evaluation of global wind power. Journal of Geophysical Research 2005;110:D12110. https://doi.org/10.1029/2004JD005462.
[4] Lu X, McElroy MB, Kiviluoma J. Global potential for wind-generated electricity. Proceedings of the National Academy of Sciences 2009;106(27):10933–8. https://doi.org/10.1073/pnas.0904101106.
[5] Jacobson MZ, Delucchi MA. Providing all global energy with wind, water, and solar power, part i: technologies, energy resources, quantities and areas of infrastructure, and materials. Energy Policy 2011:1154–69. https://doi.org/10.1016/j.enpol.2010.11.040.
[6] Letcher T. Wind energy engineering: a handbook for onshore and offshore wind turbines. Elsevier Science. ISBN 9780128094297, 2017.
[7] URL: http://www.energy-transition-institute.com/Insights/Wind.html.
[8] REN21. Renewables 2017 global status report. REN21 Secretariat. ISBN 978-3-9818107-6-9, 2017.
[9] Guerrero-Lemus R, Martínez-Duart J. Renewable energies and CO2: cost analysis, environmental impacts and technological trends – 2012 edition. Lecture notes in energy. Springer. ISBN 9781447143857, 2013.
[10] Letcher TM. Future energy: improved, sustainable and clean options for our planet. Elsevier Science. ISBN 9780080994222, 2014.
[11] Dvorak P. Foundations that float. Windpower Engineering & Development 2017. URL: https://www.windpowerengineering.com/offshore-wind/foundations-that-float/.
[12] IEA-ETSAP (International Energy Agency – Energy Technology Systems Analysis Programme), IRENA (International Renewable Energy Agency). Wind power. Technology brief. IEA-ETSAP, IRENA; 2016. URL: https://www.irena.org/-/media/Files/IRENA/Agency/Publication/2016/IRENA-ETSAP_Tech_Brief_Wind_Power_E07.pdf.
[13] International Energy Agency Wind (IEA Wind). 2015 annual report. IEA Wind. ISBN 9780990507536, 2016.
[14] IRENA (International Renewable Energy Agency). Renewable power generation costs in 2017. Abu Dhabi: IRENA; 2017.
[15] Milborrow D. Breaking down the cost of wind turbine maintenance. Wind Power Monthly 2010. URL: https://www.windpowermonthly.com/article/1010136/breaking-down-cost-wind-turbine-maintenance.
[16] Buchsbaum L, Patel S. Wind turbine repowering is on the horizon. Power. URL: http://www.powermag.com/wind-turbine-repowering-horizon/.
[17] Lantz E, Leventhal M, Baring-Gould I. Wind power project repowering: financial feasibility, decision drivers, and supply chain effects. NREL; 2013.
[18] Topham E, McMillan D. Sustainable decommissioning of an offshore wind farm. Renewable Energy 2017;102:470–80.
[19] Dodd J. Decommissioning: should they stay or should they go?. WindPower Monthly 2015. URL: http://www.windpowermonthly.com/article/1349270/decommissioning-stay-go.

[20] Jacobson MZ. Review of solutions to global warming, air pollution, and energy security. Energy & Environmental Science 2009;2:148–73. https://doi.org/10.1039/B809990C.

[21] Pavel CC, Lacal-Arántegui R, Marmier A, Schüler D, Tzimas E, Buchert M, et al. Substitution strategies for reducing the use of rare earths in wind turbines. Resource Policy 2017;52:349–57.

[22] Stiebler M. Wind energy systems for electric power generation. Green energy and technology. Springer-Verlag. ISBN 9783540687627, 2008.

[23] Arnett EB, May RF. Mitigating wind energy impacts on wildlife: approaches for multiple taxa. Human–Wildlife Interactions 2016;1(10):28–41.

CHAPTER SEVENTEEN

Solar energy

Contents

17.1. Solar energy potential and conversion

The amount of the Sun energy that reaches the Earth and can potentially be captured is significant and ranges from as low as 15 750 EJ to 49 837 EJ [1] (which is around 4375–13 843 PWh per year) as discussed in Chapter 2. The average amount of solar energy received at the border of the Earth's atmosphere is around 342 W m^{-2}, of which about 30% is scattered or reflected back to space, leaving around 70% or approximately 239 W m^{-2} available for harvesting and capture [2]. Solar energy is by far the largest energy resource on the Earth that is readily available on the surface.

Unfortunately, this potential is not entirely exploitable. Solar is the intermittent source of energy as it is subject to the rotation of the Earth (so that the Sun light hits the Earth only during the day period) and its rotation around the Sun, providing seasonality of the solar energy.

The promising fact is that solar energy is the intrinsic source of other forms of renewable energy like wind, bioenergy, ocean, and is the initiator for major cycles and fossil fuels. Directly, solar energy has been used by humankind since centuries for heating and cooking purposes. The idea to convert solar energy into other forms, and especially into electricity, has been of vital importance among the scientists and engineers, and today we

Sustainable Power Generation
https://doi.org/10.1016/B978-0-12-817012-0.00031-1

have two major options for solar energy use based on the conversion chains discussed in Chapter 1. These are [3,2]:

1. Passive solar technologies involve the accumulation of solar energy without transforming thermal or light energy into any other form. This is mostly used, for instance, for collecting, storing, and distributing solar energy for heating purposes.
2. Active solar technologies collect solar radiant energy and use special equipment to convert it into other forms of energy, e.g., heat or electricity. These technologies can be further grouped into two major categories:
   a. Solar thermal technology that collects and concentrates solar energy by special devices and further converts it into electricity through other forms, and
   b. Photovoltaic technology that enables the direct conversion of solar energy using semiconductor devices.

On the industrial scale, both active solar technology options — photovoltaic and solar thermal — are implemented. The intense research efforts of energy scientists with regard to solar options have helped to yield an improved efficiency of photovoltaic technology [2], which enabled increasing the speed of solar photovoltaic deployment for industrial electricity generation scale. Similarly, solar thermal technologies have also been economically feasible for large electricity generation. This is achieved through the concentrating solar power technology — an approach that allows collecting solar radiation and using its energy to convert liquid into steam and employ steam turbine cycle for electricity generation.

Fig. 17.1 shows the development of the installed capacities of photovoltaic (PV) and thermal solar (CSP) technologies for the last decade. For the last five years, Asia eclipsed all other markets for PV technology, accounting for about two-thirds of global additions. The top markets — China, the United States, Japan, India, and the United Kingdom — accounted for about 85% of PV additions in 2016. For the cumulative PV capacity, the top countries included China, Japan, Germany, and the United States [4]. While China continued to dominate both the use and manufacturing of solar PV, emerging markets on all continents have begun to contribute significantly to global growth. By the end of 2016, every continent had installed at least 1 GW of PV capacities, at least 24 countries had 1 GW or more of PV capacity, and at least 114 countries had more than 10 MW [4].

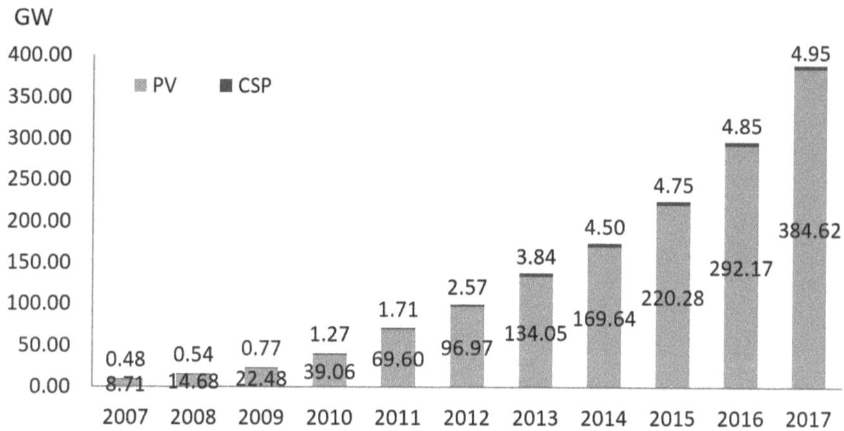

Figure 17.1 ***Solar energy installed capacity.*** Solar energy installed capacity growth shows that PV technology dominates the market due to its massive use both in private and industrial electricity generating applications. Solar thermal CSP technology is much smaller but is utility scale only. *Source: Based on data from [5].*

Contrary to PV, solar thermal technology has been showing moderate growth during the last decade. In 2016 South Africa led the market in new additions, becoming the second developing country to do so after Morocco in 2015 [4]. CSP market growth continued to be driven outside of the traditional markets of Spain and the United States, which have been among the top drivers of CSP installations and still have the largest share.

The two most economically and technically feasible solar energy options — photovoltaic and solar thermal — are the viable option for sustainable future energy mix. Though the installed capacities seem large, their estimated share in the global electricity generation mix varies from less than 1% to around 1.5%, depending on the data source. Still, as solar energy is the abundant source of power, these technologies prove to be an important contribution to the generation mix and will definitely express growth and technological advancement in the coming years.

Further on, we consider these technologies in detail and discuss their economics, benefits, and potential drawbacks from the sustainability point of view.

17.2. Concentrating solar power

On a larger scale, solar thermal power plants employ technology to concentrate the energy of the Sun as a heat source. The heat is then used

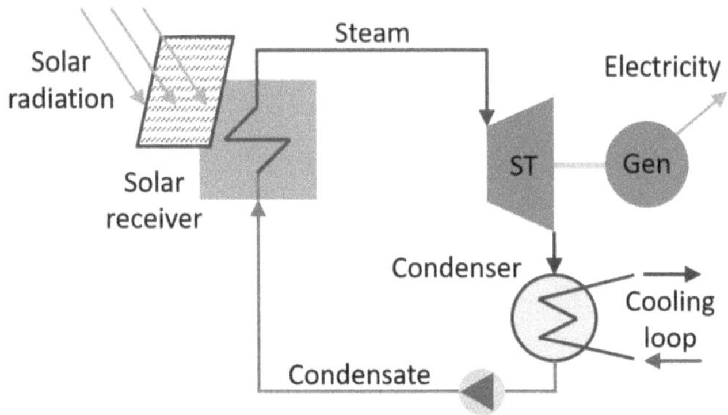

Figure 17.2 ***Simplified process diagram of a solar thermal power plant.*** Solar radiation is concentrated on a solar receiver (heat exchanger) that evaporates water. Steam is fed to the steam turbine that spins the electric generator.

to heat a fluid, which then directly or indirectly runs a turbine and an electrical generator. Concentrating the Sun's rays allows for the fluid to reach working temperatures high enough to ensure fair efficiency in turning the heat into electricity, while limiting heat losses in the receiver [6].

17.2.1 Plant configuration

Consider a simplified process diagram of a solar thermal power plant that represents the generic approach to this technology and is schematically illustrated in Fig. 17.2. The solar radiation is concentrated by some technological means on a special type of heat exchanger called *solar receiver* that transfers its thermal energy to the fluid which flows through it. In the simplest case it can be water or a special liquid. This water evaporates, and the high pressure steam is fed to the steam turbine, where it expands and transforms its energy into the mechanical energy of the spinning shaft as in a conventional steam power plant. Being coupled to the generator, the rotational mechanical energy is converted into electricity.

Cold steam that has flown through the steam turbine is then turned back to water within a condenser, and the cold condensate is pumped to the heat exchanger within the receiver to close the cycle. The process is similar to the conventional steam power plant, where solar energy receiver and heat exchanger act as a boiler.

Depending on the amount of energy that can be concentrated, i.e., the temperature of the fluid that can be achieved by concentrating solar energy, the process may be optimized to increase its efficiency. Fig. 17.3 illustrates

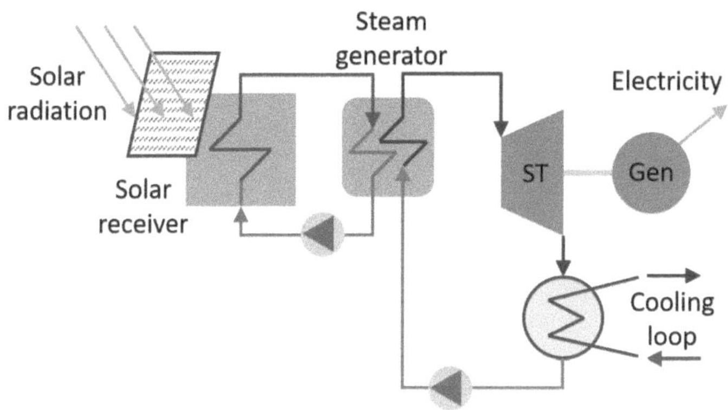

Figure 17.3 ***Simplified process diagram of a solar thermal power plant with two cycles.*** Solar radiation is concentrated on a solar receiver that heats up special fluid circulating in the first circle and bringing energy to the steam generator – a fluid-to-water heat exchanger, where water evaporates. Steam is fed to the steam turbine and, after flowing through, is condensed back into water and pumped to close the cycle.

the schematic of the electricity generation with solar thermal technology using two separate cycles:

1. The working fluid, for example, a molten salt mixture, absorbs heat from the concentrated solar energy within the heat exchanger at the receiver. Then, this hot pressurized liquid is fed to a water-to-liquid heat exchanger and, after releasing its heat, is pumped back to the solar receiver.
2. Water within the second cycle flows through the heat exchanger, absorbs heat from the working fluid, and evaporates. Steam is then fed to the steam turbine, where it transfers its energy to the spinning shaft, and then condenses to be pumped back to the steam generator.

In order to further optimize the cycle and balance the variations in solar energy due to clouds or even the day and night change, thermal storage is implemented into the cycle. Fig. 17.4 illustrates the process where the working fluid of the first cycle is pumped through the thermal absorber or hot silo where it is stored for several hours at temperatures ranging from 500 to 600°C. After flowing through the steam generator and giving its energy to boil water, the fluid is pumped back through the cold silo to be further reheated in the solar receiver.

Therefore, a thermal solar power plant can be generally split into several equipment areas:

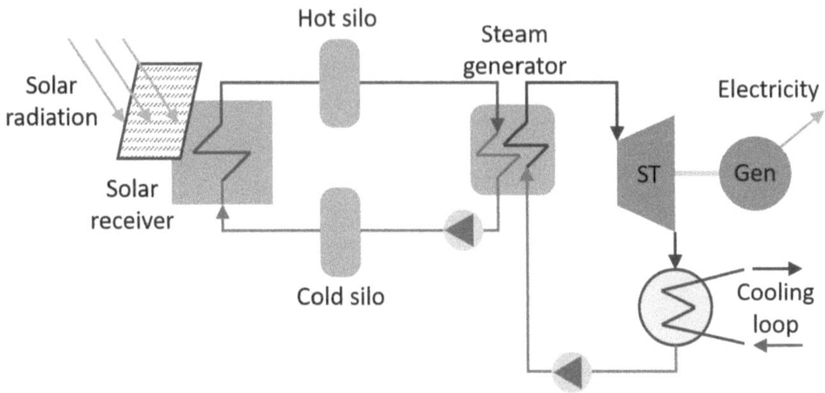

Figure 17.4 ***Simplified process diagram of a solar thermal power plant with two cycles and thermal storage.*** Solar radiation is concentrated on a solar receiver that heats up special fluid circulating in the first circle and bringing energy via the thermal storage to the steam generator – a fluid-to-water heat exchanger, where water evaporates. Steam is fed to the steam turbine and, after flowing through, is condensed back into water and pumped to close the cycle.

1. Solar field of the solar energy receivers, heat exchanger and supporting systems;
2. Thermal storage that is incorporated into the fluid cycle;
3. Power island with the steam generator, steam turbine, electrical generator, and necessary systems; and
4. Balance of plant, both mechanical and electrical.

Remark 17.1. Depending on the employed technology and the type of solar receivers, a solar field may incorporate a boiler.

17.2.2 Solar receivers

There are three major approaches for the solar field to collect and guide solar radiation to the receiver heat exchanger: parabolic trough, solar tower and parabolic dishes.

17.2.2.1 Parabolic trough

Parabolic trough is a set of concave mirrors that concentrate solar rays on the receiver tube that is located in the focus. These troughs can track the Sun around one axis, typically oriented north–south to ensure the highest possible efficiency. The fluid flows through this tube and absorbs heat from the concentrated solar energy.

Similar to a parabolic trough is a linear Fresnel system. These collectors resemble parabolic troughs but use long flat Fresnel mirrors. This technology is much cheaper to install but has lower efficiency. Examples are limited to a few plants.

Synthetic oil or molten salt mixture is usually used to circulate through the heat exchanger tubes and absorb heat concentrated by the mirrors. This fluid usually gets heated up to 400–600°C. The tubes then deliver this oil to the process.

17.2.2.2 Central receiver or power tower

Central receiver, which is sometimes called a power tower, is a point-focus solar energy collector that can provide much higher temperatures compared to parabolic trough or linear Fresnel. This technology uses a circular array of mirrors called heliostats which track the Sun, reflect its radiation and concentrate it onto a fixed receiver at the top of a tower that stands in the middle of the field. This allows reaching temperatures up to 1000°C and even higher, which substantially increases the efficiency of the thermal cycle as it allows generating superheated steam.

The working fluid in this case can be water with a single cycle design, molten salt or other liquids with a double loop design. Moreover, due to such high temperatures, pressurized gas or even air can be utilized in a single-loop cycle and be fed to a gas turbine directly replacing natural gas [7].

17.2.2.3 Parabolic dish

The third major category of the solar thermal technology receiver is the parabolic dish. Here, solar radiation is concentrated by a parabolic concentrator, which has a single energy absorber at the focal point. Heat is absorbed and either directly applied to a heat engine generator that is mounted near the focal point [3], or transported by the working fluid to a separate power house.

The reflector is installed on a two-dimensional tracking system to follow the Sun and maximize the amount of energy absorbed by the system. Because of this high concentration of energy in a single point, the temperatures of the absorber can reach 600–1500°C [3]. Due to this, the heated working fluid (which ranges from water to gas) is then used to generate electricity in a small piston engine or microturbine attached to the receiver [7].

Table 17.1 Summary of the characteristics and costs of major solar thermal technologies for electricity generation.

	Parabolic trough	Solar tower	Parabolic dish
Size, MW	10–320	10–200	5–25
Operating temperature, °C	up to 600	up to 1000	up to 1400
Annual capacity factor, %	23–50	20–77	25
Commercial status	Commercial	Commercial	Prototype tests
TIV, MUSD per MW[a]	2–6.5	2.5–6.5	4–12

Source: Based on [3,6,7] and EIC and IIR databases.

[a] Indicative numbers, as the TIV depends on site location, design, local regulation and incentives.

This technology can be ideal for remote locations or distributed applications as the capacities of solar dish systems are in the 5–50 kW_e range. For many years, there has been significant design and operational experience developed with dish engine systems, due to advances in dish structures, high-temperature receivers, use of hydrogen as the working fluid, as well as some experiments with liquid metals and improvements in Stirling engines [7]. This should bring the technology to commercial use, though it is still expensive in terms of capital and operating costs per kW of generated electricity.

The solar thermal portfolio for commercial use has a considerable diversity in terms of scale, ranging from small units of a few kilowatts of electric capacity up to 100 + MW_e utility-scale facilities. Table 17.1 represents a summary of the key attributes of these technologies.

17.2.3 Thermal storage

Solar thermal technology was initially highly sensitive to the operating conditions and could be interrupted or at least significantly impacted, for instance, by the clouds. This was one of the reasons to integrate a thermal energy storage into a system. Until recently, its capacity could provide up to one hour of full-load storage to maintain operation during the cloudy periods or assist startup and shutdown, which ultimately supported the grid.

Modern thermal storage systems can provide up to 15 hours of thermal storage using molten salt, which allows operating a solar thermal power plant on a 24/7 basis. The first remarkable example was the Gemasolar CSP plant with a tower receiver built in Spain. The facility was commissioned in May 2011 and had the capacity of 19 MW_e. Being a remarkable example, it clearly showed the potential of the thermal energy storage system which can support a solar thermal plant during cloudy or evening peak times and even provide electricity on demand.

There are three main types of thermal energy storage system (further discussion regarding energy storage is conducted in Chapter 23):

- *Sensible (single-phase) heat storage*, which uses a liquid or solid storage medium, such as water, molten salt (nitrates, carbonates, chlorides), or solids (sand, ceramic, graphite, concrete, rocks), heated or cooled to store energy. Generally, the first choice for the CSP plant project with a thermal energy storage system is molten salts, which can withstand extremely high temperatures to maintain high levels of cycle efficiency. They can maintain up to 10–15 hours of energy storage. This is the least expensive option, but it requires large volumes of the medium due to their low energy density.
- *Latent heat storage*, which relies on the storage medium changing states, for instance, from solid to liquid, that use latent heat to store energy. These materials include molten salts, metallic alloys, etc. While this technology is not commonly used, recent developments have shown promise in some applications. Being expensive, this type of storage boasts a much smaller footprint due to high energy density of the medium.
- *Thermo-chemical storage*, which uses chemical reactions to store energy. It converts energy and stores it in chemical bonds through decomposition or synthesis reactions. Even though it offers the highest energy density, it is still costly: the cost of the equipment is still very high compared to other technologies and the medium itself, even though the latter is also very expensive.

According to [5] nearly all CSP power plants of parabolic trough or tower type employ a thermal energy storage system, mostly of the sensible heat storage type. Trough plants are now designed for 6–7.5 hours of storage on average, which is enough to allow flexible operation.

17.2.4 Power island and balance of plant

Power island of a utility-scale CPS power plant that uses steam turbine technology and has a lot in common with the conventional thermal power plant as it employs the same steam turbine technology. Therefore, it consists of the following major equipment:

- Steam turbine, which can be of the same type as for small to medium size conventional power plant, when the steam conditions are comparable to those in a conventional plant. The number of pressures and final

turbine configuration depend on the overall plant design and respective heat balance;
- Electrical generator, which is usually supplied together with the steam turbine;
- Necessary auxiliary systems, including lubricating and hydraulic oil, cooling, instrumentation and control, steam safety and control valves, etc.

In the case of other than steam cycle, different engines and rotating machines to convert energy of the high pressure medium can be used. These may include gas turbines or engines.

The balance of plant systems depend on the overall plant scope, but in general include systems which are typical for most of the thermal power plants, including conventional ones:
- Mechanical systems usually incorporate:
  - Heat transfer system to supply cooling medium to the condenser and other necessary equipment. Various types of cooling tower can be implemented, however, as the solar thermal plants are mostly located in arid deserts or semiarid regions, water resources are scarce, which dictates the choice of the air-cooled condenser technology;
  - Steam generator (evaporator) with the required deaerator, water treatment, valves, pumps, and piping for the water–steam cycle;
  - Multiple support systems (oil, cooling, compressed air, fire protection, etc.), to ensure safe and reliable operation.
- Electrical systems to support local plant needs and evacuate power to the grid. These involve:
  - Power evacuation system with the generator bars, main transformer and substation or switchyard to safely connect to the grid;
  - HV, MV, and LV power systems of the plant to provide electricity for local needs, including DC systems and batteries;
  - Instrumentation and control systems to optimize performance and, especially here, control the solar field.

17.2.5 Project cost analysis

Solar thermal project development follows the same process that starts with the feasibility study, where resource assessment and siting play the major role. A solar field has a large footprint and the plant has to be located in the region with near to ideal solar energy conditions — very low cloud cover and atmospheric moisture levels and high year-round insolation levels. Therefore, siting is crucial for a CSP plant.

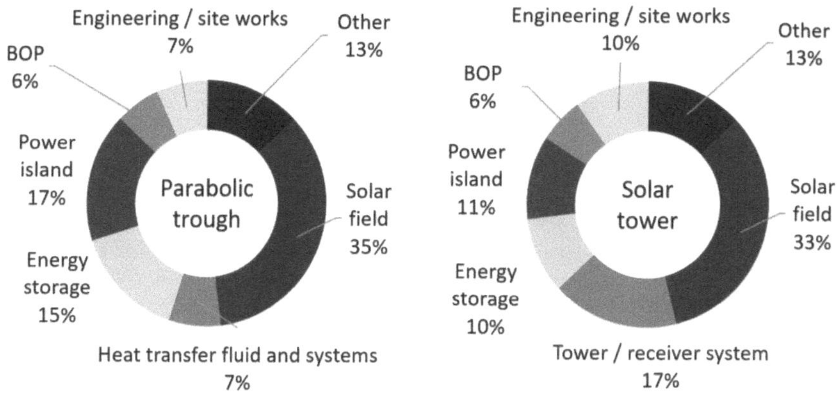

Figure 17.5 ***Solar thermal power plant project capital cost structure.*** Capital cost structure of the two major solar thermal commercial technologies: parabolic trough and solar tower. *Source: Based on [5], EIC and IIR databases and calculations.*

Secondly, the choice of technology depends on the business model: thus, for utility-scale plants the choice of the storage capacity would drive the sales model, i.e., if the plant is capable to generate electricity at evening peak times or support the grid should the need arise.

Moreover, a CSP plant can be considered as an add-on to an existing traditional power plant, for instance, a combined cycle one. Such hybrid plants can accommodate a solar thermal cycle to expand steam generation and increase efficiency and capacity of the combined cycle plant, or use gas turbines for peaking or night generation. Such hybrid plants can contribute to decreasing GHG emissions by partially utilizing solar energy instead of burning fossil fuels.

All the economics of the solar thermal plants has to be calculated on a project basis, with the average numbers of the investment per MW_e of installed capacity ranging largely (see Table 17.1 for details). The approximate capital cost structure of a solar thermal project is represented in Fig. 17.5 for two major commercial technologies: parabolic trough and solar tower.

As seen from the data in Fig. 17.5, the solar field accounts for more than 30–45% for both technologies. It usually includes the support structure, receivers, mirrors, heat transfer system and fluid. The thermal storage cost depends on the technology but can be considered as equal for both technologies of the same capacity and may be roughly estimated at the level of 9–20% on average, which is dominated by the cost of medium (about 5–11%) [8]. The O&M costs of a solar thermal plant are relatively low compared to conventional coal-fired power plants. A typical 50 MW

Table 17.2 Typical thermal solar power plant costs.

Technology	CAPEX, $/kW	OPEX, $/kWh	LCOE, $/kWh
Parabolic trough	2500–13 200[a]	0.02–0.03	0.14–0.35[b]
Solar tower	2500–13 200[a]	0.03–0.04	0.14–0.35[b]

Source: Based on [8,9] and EIC and IIR databases analysis.

[a] The installed cost depends largely on the capacity of the thermal storage: without a storage, the cost may be 2550–11 200, and adding 4–8 hours of storage drives it up to 6050–13 200 USD/kW.

[b] Auction prices show much lower values around 0.06 $/kWh, but they have to be treated with caution due to auction prices being dependent on a set of obligations and terms in the contract that can be very market and project specific.

parabolic trough plant would require around 40 O&M employees for operation, maintenance and solar field cleaning (see Table 17.2).

Solar thermal power plants with thermal energy storage are becoming a viable option with full flexibility and quite competitive cost model. Recent bids have shown a stable decline in LCOE, so that we can expect it would continue to occupy its niche in the solar radiation rich countries.

17.3. Solar photovoltaic system

The second commercially available active solar technology to capture radiation and convert it directly into electricity is the photovoltaic, or PV, technology. PV devices employ the photoelectric effect in semiconductor materials which were discussed in detail in Chapter 1. This technology was developed in the 1960s due to the rapid deployment of the space programs in Russia and the United States, where solar panels were mounted on the spacecrafts. These panels were made out of crystalline silicon and cost a fortune. As the research and development of new semiconductor materials continued, the prices of the semiconductors have dropped significantly, so that solar panels became an essential part of many gadgets and applications. In the late 1970s, the first solar panels for power generation emerged, and during the next few decades their capacity has grown dramatically, as the prices dropped from about 250 000 USD/kW_e down to 2500 USD/kW_e and even further, that is, more than a hundred times [3,10]. But the costs continued to go down and down, mostly because of the development of technology, economies of scale and the introduction of new manufactures from multiple countries, especially from China. Nowadays the selling price is targeting 300–400 USD/kW_e and is likely to continue its decrease [9].

As the costs fell down, the global PV market has boomed in the last decade, so that the cumulative global installed PV capacity grew from around 6 GW at the end of 2006 to roughly 291 GW at the end of 2016 with the average annual growth rate of 28%, and the trend is continuing.

17.3.1 Types of solar PV systems

Solar PV systems that are used for power generation applications consist of several cells mounted on some surface, with electrical connections and connecting equipment. The grid-connected systems employ inverters to transform DC into AC and connect to the grid. The balance of systems (or the balance of plant) includes this electrical equipment, wiring, control and monitoring devices, and structural components for installing panels. The panels can be permanently fixed or may utilize special tracking system to follow the Sun.

PV systems are attractive thanks to their relative simplicity [3]: there are no moving parts, no thermodynamic cycles or flowing hazardous medium. They are noise-free, require limited maintenance (except for washing the panels in some environments), and are scalable due to their modularity.

The number of cells, the overall system capacity and its connection to the grid depend on a type of energy system [7]:

- Industrial and utility-scale power plants of more than 1 MW produce considerable quantities of electricity and their capacities are usually of megawatts. The solar panels of such facilities are mainly mounted on frames on the ground and require a large footprint. However, they can also be installed on large industrial buildings (e.g., warehouses, transportation terminals) to optimize spacing and cover local energy-intensive customers.
- Residential (with less than 10 kW) and commercial (with around 10–100 kW) systems are usually smaller PV systems connected to the grid, which cover the needs of residential and business customers. Usually they are connected to the local network and sell electricity to the market, while the customers get energy from the grid and not from the system itself. This allows having uninterrupted supply of energy if the system is unavailable (during cloudy conditions, for instance). This system is ideal for use in areas of unreliable power supply or if companies have free spacing to install such systems also for sustainability purposes.
- Stand-alone, off-grid PV systems have no connection to the local grid and operate in an island mode. They usually have to employ batteries to balance the output changes and outages so that power can be still used at night. Some of the typical off-grid applications include:

- Off-grid systems for rural electrification that bring electricity to remote rural areas or developing countries. They can range from small home systems which cover a household's basic electricity needs, up to larger solar isolated mini-grids which provide enough power for several homes, a community or cover the needs of a small business.
- Off-grid industrial applications that are deployed in remote areas to power smaller industrial consumers (e.g., mobile phone network nodes, traffic signals, navigational aids, lighting, signs, small utility plants, etc.). These can be either full PV solutions of hybrid systems, especially for critical infrastructure applications.

Most of the PV systems can be ground-mounted, but due to the size utility-scale facilities are not normally located on rooftops or integrated into buildings, however, the trend is likely to change.

In many cases solar PV systems become the second source of power within a hybrid system. These facilities are powered by the Sun when it is available and by other source of energy during the night and extended cloudy periods. Solar PV can be combined with wind parks, biomass plants or diesel generators to ensure a consistent supply of energy.

17.3.2 Major technologies

The key performance metric of any PV technology is its net efficiency of converting sunlight into electricity [3]. In many cases the reported values are obtained in laboratories where conditions are properly controlled and therefore are near to perfect. In field operation these values are usually lower for single modules and go further down on a scale of a complete system. The efficiency also depends on the durability of the modules and the operating conditions. Thus, in arid desert areas with high probability of sand storms the maintenance costs will be extreme, as continuous cleaning of the panels is needed.

Therefore, the major driver of the solar PV development is the increase of efficiency, decrease of cost and maintenance needs that ultimately results in lower cost of electricity, which is the driver for the customer. Multiple approaches have been proposed, tested and deployed for commercial use.

Solar cells are usually named after the semiconducting material they are made of. These materials must have certain light absorption characteristics to be used for PV cells. Generally, photovoltaic technologies can be divided into three major groups (as shown in Fig. 17.6):

1. Wafer-based, also called the first generation of PV cells, which incorporate traditional technologies:

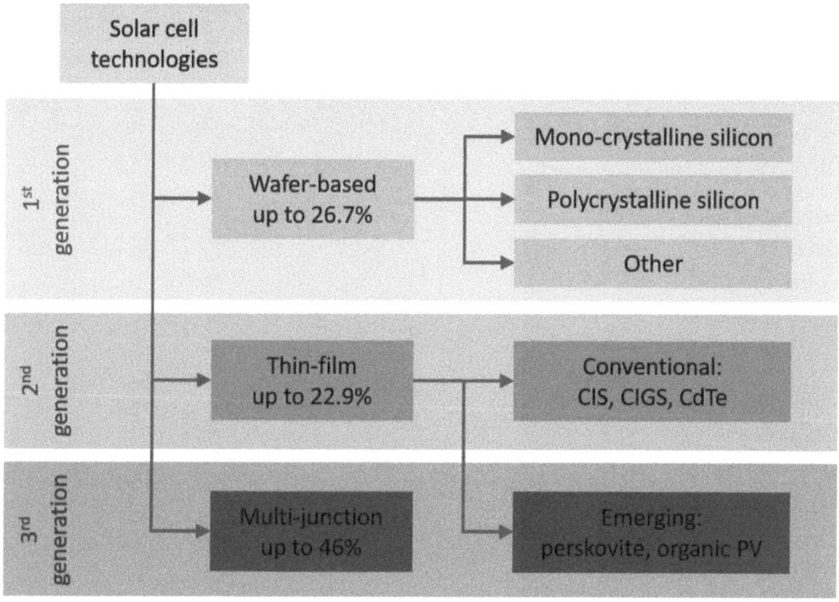

Figure 17.6 ***Photovoltaic technologies.*** There are several generations of photovoltaic technologies: wafer-based (1st generation), conventional thin-film (2nd generation), and emerging thin-film and multi-junction (3rd generation).

- Usually silica-based system, both monocrystalline and polycrystalline, and
- Some other technologies like gallium-arsenide (GaAs) which did not gain large commercial spread;

2. Thin-film cells, which normally absorb solar radiation more efficiently that silica-based, allowing the use of extremely thin film structures. These technologies can also be split into:
   - Conventional thin-film technologies, also sometimes called the second generation of solar cell technology, which includes amorphous silicon, cadmium telluride (CdTe), copper indium diselenide (CIS) and its further development copper indium gallium selenide (CIGS);
   - Emerging thin-film technologies, or the 3rd generation of solar cells, like perovskite or organic photovoltaic (OPV);
3. Emerging multi-junction solar cells that use several p–n junctions to increase efficiency. These technologies can also be considered as the third generation and emerging.

Table 17.3 Solar cell average efficiency values for lab cells and modules as of 2017.

Classification	Cell efficiency, %	Module efficiency[a], %
1st generation: wafer-based		
Monocrystalline silicon	26.1	24.4
Polycrystalline silicon	22.3	19.9
2nd generation: thin-film		
CIGS	22.9	19.2
CdTe	22.1	18.6
Amorphous silicon	13.6	
3rd generation: emerging technologies		
Multi-junction	46.0	
Perovskite	23.3	11.6
Organic PV	12.6	

Source: Based on [11,12].

[a] Module efficiency is on average 2% lower than the cell efficiency, numbers for indicative purposes only.

17.3.2.1 Wafer-based systems

The first commercially available solar cells were made from crystalline silicon, or c-Si — a pure form of silicon. The cells were made from thin slices or wafers cut from a single crystal of silicon or from the block of crystals. The former, called monocrystalline silicon cells, were highly efficient, but their manufacturing process was rather slow and costly. Another option was to manufacture cells from a block of silicon crystals, which is called polycrystalline or multicrystalline. They have lower efficiency but their cost can be several times lower than that of monocrystalline. Therefore most commercial PV applications use polycrystalline silicon systems that are produced as thin wafers cut from cast ingots or drawn as a thin ribbon from molten silicon [3].

The efficiency of the best commercial c-Si modules now exceeds 26% in laboratory conditions and manufacturers are already targeting higher values, together with significant cost reductions. The module efficiency is a few percent lower, but is already above 24%, as represented in Table 17.3.

17.3.2.2 Thin-film

There have been other types of silica-based cell, for instance, amorphous silicon. The idea dates back to the 1970s, when it was proposed to deposit a thin film of noncrystalline, amorphous silicon directly on an inexpensive support material [3]. The concept, designated as a-Si, showed excellent absorption characteristics but turned out have lower efficiency of around 12.5% which could become even worse during operation.

Of course, silicon is not the only material that can be used for solar cells. Scientists have found other materials that can provide similar optical properties to silicon but lower energy intensity and potentially lower cost of the final product. Among others, copper indium diselenide ($CuInSe_2$ or CIS) layers have shown higher efficiencies than a-Si devices. This was the creation of the second generation of solar cells — the thin-film technology.

A thin-film solar cell is made out of one or several extremely thin layers, called thin film, of photovoltaic semiconducting material deposited onto a substrate base out of glass, stainless steel or even flexible plastic. The materials used were CIS and its further development copper indium gallium selenide (CIGS), or cadmium telluride (CdTe). The laboratory experiments have shown that the new thin film technology has evolved in the way that its efficiency is approaching (and sometimes exceeding) polycrystalline silicon cells, as shown in Table 17.3.

However, developing CIS cells on a large scale raises some concerns: the process is extremely toxic, and the minerals used for production may be limited in some regions and lead to supply constraints.

17.3.2.3 Emerging technologies

The efficiency of the traditional single-junction photovoltaic solar cell to convert solar energy into electricity is limited only to approximately 31–33% even in an ideal case. This limit, known as Shockley–Queisser (S–Q) limit [13] is due to multiple physical constraints, but the search to exceed the limits and propose new and more efficient solar cell technologies has never stopped. New materials, optimized design and production technology have driven the costs down, and completely novel and innovative approaches resulted in new solar cells, for example:

- *Multi-junction cells*, which are solar cells with several p–n junctions made out of different semiconducting materials. Each p–n junction will absorb certain wavelength and produce electric current in response. The use of several materials and p–n junctions allows absorbing a much broader range of wavelengths, which considerably improves the overall efficiency of the cell. Recent developments in two-, three-, and even four-junction solar cells have demonstrated amazing efficiency of around 46%, however, this has been achieved in laboratories and with the special concentrators to collect the solar radiation on the panel.

Their upfront cost still remains high, and it increases dramatically with a number of junctions per cell. Still, the research has shown that such levels of efficiency are achievable, so that we can expect that further development would drive the cost down.

- *Perovskite cells*, which are named after the perovskite-structured compound that has the same crystal structure as calcium titanium oxide. This compound serves as the light-harvesting layer of the cell. Perovskite materials, e.g., methylammonium lead halides, are quite cheap to produce and quite easy to manufacture. These materials possess valuable properties required for PV applications, for instance, they can absorb a broader spectrum. Overall, they are now considered as a very promising material for modern solar cells. The major issues that hold the perovskite technology back are the cost of the high-efficiency material and the lifespan of cheaper panels. Moreover, scaling up the cells also drives the efficiency down, so that the values of 23.3% look very promising but require further development.
- *Organic PV cells*, a rapidly emerging solar cell technology, consisting of one or several photoactive semiconducting layers sandwiched between two electrodes. Organic solar cells incorporate carbon into their photoactive layer to offer several advantages over conventional or "inorganic" cells. Carbon-based organic cells can be potentially cheaply manufactured in thin rolls, so that they can be incorporated into various designs (buildings or devices) and can have various colors or even be transparent to fit almost any design. However, the efficiency of the currently available laboratory specimen reaches 12.6%, which is far below the wafer-based monocrystalline or polycrystalline silicon technologies. Still, the research continues in the direction of increasing efficiency, and recent results show the levels of around 14% or above [14]. However, in order to reach the levels of efficiencies for cell exceeding 15% and module exceeding 10%, have large areas, fabricated through roll-to-roll solution processes, and exhibit device life times greater than 10 years will, however, require interdisciplinary efforts by both scientists and engineers [15].

17.3.3 Project cost analysis

Thanks to an incredible growth rate and the fact that the bulk of PV production is situated in Asia, with China being the ultimate leader, the competition in this industry continues to be strong. This has been driv-

Table 17.4 Typical solar PV power plant costs.

Technology	CAPEX, $/kW	OPEX[a], $/kW	LCOE, $/kWh
Solar PV	1000–3500	10–18	0.07–0.31[b]

Source: Based on [9] and EIC and IIR databases analysis.

[a] The cost is calculated for USD/kW of installed capacity. Land-lease costs vary from region to region and can be very significant, and the given costs are average values for the US.

[b] For utility-scale systems.

ing the PV costs down and, taking the continuous efficiency growth into account, it favorably impacts the overall project economics.

According to [9], module prices have dipped as low as 400 USD/kW of capacity for good quality products, and continue to go down. However, for utility-scale solar PV projects the balance-of-system costs continue to play an increasing role, and the overall project capital costs differ for various regions. The average CAPEX for large scale PV facilities ranges from as low as 1000 USD/kW up to 3500 USD/kW, with the trend towards the cost decrease (see Table 17.4 for details).

The O&M costs for utility scale plants are reported to be around 10–18 USD/kW of installed capacity per year, with the average capacity factors of around 18% [9].

The LCOE development has shown PV's growing competitiveness compared to other power generating systems. The utility-scale facilities now claim LCOE as low as 0.07–0.31 USD per kWh of generated electricity, however, it depends on many factors. PV has a large footprint and is subject to variable solar radiation, and there may be a tradeoff between its lifecycle and efficiency. Despite some issues, solar PV has shown economic feasibility and is increasing its share.

17.4. Sustainability attributes

Solar energy offers a broad range of energy generating technologies with capacities varying from as small as a few watts up to hundreds of megawatts of a utility-scale facility. Utilizing solar radiation, which is abundant and renewable, is an important contribution towards sustainable development.

Concentrating solar power is a utility-scale thermal solar electricity generating technology that has a number of sustainability advantages:

- The established thermal cycle provides high efficiency, reliability, and flexibility;

- It is an almost GHG-free energy technology with limited lifecycle GHG emissions of around 14–32 g CO_2 equivalent per kWh, so that each square meter of the CSP concentrator surface is enough to avoid the annual production of 0.25 up to 0.4 t of CO_2 [16];
- The use of CSP technology within a hybrid power plant reduces the use of fossil fuels.

However, the CSP technology is limited to a few locations and has a relatively large footprint (parabolic trough or tower technologies) which may not always be available or may impact local agriculture. Moreover, these plants still require water, either for the water–steam cycle or for cleaning the mirrors.

Solar photovoltaic technology also proposes a number of great advantages:

- It does not generate any type of solid, liquid or gaseous byproducts during its operation when producing electricity;
- PV systems do not have noise pollution;
- Solar PV panels can be installed on multiple locations, including rooftops, facades, ground, which gives the best freedom of location;
- Its modularity and range of capacity can fulfill multiple applications.

Even though solar PV can be perceived as a truly sustainable solution, it still has a number of important drawbacks which are very rarely communicated:

- Solar PV technology is not emissions-free. The estimated lifecycle CO_2 equivalent emissions are significant at the level of 30–80 g/kWh. This is mostly due to PV panel production, which involves some toxic or even explosive materials, corrosive liquids, GHG emissions during the manufacturing process, operation, maintenance, and disposal [16].
- Recycling the material in decommissioned PV modules is generally economically viable, mainly for large-scale applications. However, it requires special procedures, established processes, and regulations, which are not yet in place.
- Even though PV plants do not consume any water for operation, in certain locations water is required for cleaning, resulting in nonnegligible quantities.
- Utility-scale PV fields have a large footprint, which cannot be completely avoided by putting cells on rooftops.
- Finally, the PV technology requires expensive batteries to store electricity, which is still a challenge.

The overall trend towards cheaper solar technologies and their increased efficiency, especially notable for PV systems, would decrease the need of incentives and make them more competitive in the near future. The benefits of these technologies have already been realized, as demonstrated by the ongoing growth of installed capacity. It is not much of exaggeration to say that solar energy capturing technologies are among the most important contributors to sustainable power generation development, and definitely not in the long run. The evident source of primary energy — solar radiation — is up there, and it is the task for researchers, engineers, manufacturers, policy makers and the whole society to capture it and minimize the impact of this move.

References

[1] United Nations Development Programme (UNDP), World Energy Council. World energy assessment: energy and the challenge of sustainability. New York: UNDP; 2000.

[2] Kabir E, Kumar P, Kumar S, Adelodun A, Kim K. Solar energy: potential and future prospects. Renewable and Sustainable Energy Reviews 2018;82:894–900. https://doi.org/10.1016/j.rser.2017.09.094.

[3] Tester J, Drake E, Driscoll M, Golay M, Peters W. Sustainable energy: choosing among options. MIT Press. ISBN 9780262304634, 2012.

[4] REN21. Renewables 2017 global status report. Paris: REN21 Secretariat. ISBN 978-3-9818107-6-9, 2017.

[5] Data retrieved from: https://www.irena.org/solar.

[6] IEA O. Technology roadmap. Solar thermal electricity. Paris: International Energy Agency; 2014.

[7] Teske S, European Renewable Energy Council, Greenpeace International. Energy [r]evolution. A sustainable world energy outlook 2015. Greenpeace International; 2015.

[8] IEA-ETSAP (International Energy Agency – Energy Technology Systems Analysis Programme), IRENA International Renewable Energy Agency. Concentrating solar power. Technology brief. IEA-ETSAP, IRENA; 2013. URL: https://www.irena.org/Publications.

[9] IRENA (International Renewable Energy Agency). Renewable power generation costs in 2017. Abu Dhabi: IRENA; 2017.

[10] Deambi S, Institute E. Solar PV power: a global perspective. Energy and Resources Institute. ISBN 9788179933893, 2011.

[11] Green MA, Hishikawa Y, Dunlop ED, Levi DH, Hohl-Ebinger J, Ho-Baillie AW. Solar cell efficiency tables (version 51). Progress in Photovoltaics: Research and Applications 2017;26(1):3–12. https://doi.org/10.1002/pip.2978.

[12] NREL. Best research-cell efficiencies. URL: https://www.nrel.gov/pv/assets/pdfs/pv-efficiencies-07-17-2018.pdf.

[13] Shockley W, Queisser H. Detailed balance limit of efficiency of p–n junction solar cells. Journal of Applied Physics 1961;32(3):510–9. https://doi.org/10.1063/1.1736034.

[14] Che X, Li Y, Qu Y, Forrest S. High fabrication yield organic tandem photovoltaics combining vacuum- and solution-processed subcells with 15% efficiency. Nature Energy 2018;3(5). https://doi.org/10.1038/s41560-018-0134-z.

[15] Su YW, Lan SC, Wei KH. Organic photovoltaics review. Materials Today 2012;15:554–62. https://doi.org/10.1016/S1369-7021(13)70013-0.

[16] Renewable energy sources and climate change mitigation. Special Report of the Intergovernmental Panel on Climate Change. Cambridge, UK, New York, NY: Cambridge University Press; 2011.

CHAPTER EIGHTEEN

Energy from municipal solid waste

Contents

18.1. Waste and its management strategies

Since the beginning of their history, humans have generated solid waste and disposed of it in a variety of ways, such as in makeshift waste dumps or incinerations [1]. The industrial revolution has dramatically increased the production of goods and their consumption, which in turn resulted in growing waste discarded by the society, catalyzed by the ongoing urbanization. It therefore became necessary to provide landfills and incinerators for disposing waste, supported by special transportation, logistics and management services.

The amount of waste has increased tremendously during the last years and consequently became an issue. Thanks to new materials like polyethylene or other plastics, glass, and resins, their complete natural breakdown may take too long and damage the environment. In response, some of the most advanced countries started to develop various measures to reduce waste by optimizing production and consumption patters, developing new natural or recyclable materials, packaging, etc. New approaches to sort waste, recycle materials, burn and recover heat from the waste decreased the amount of landfill, however, the scale is still too small.

Almost any human activity generates waste, but its nature and composition differs according to the source. The municipal solid waste usually

Sustainable Power Generation
https://doi.org/10.1016/B978-0-12-817012-0.00032-3

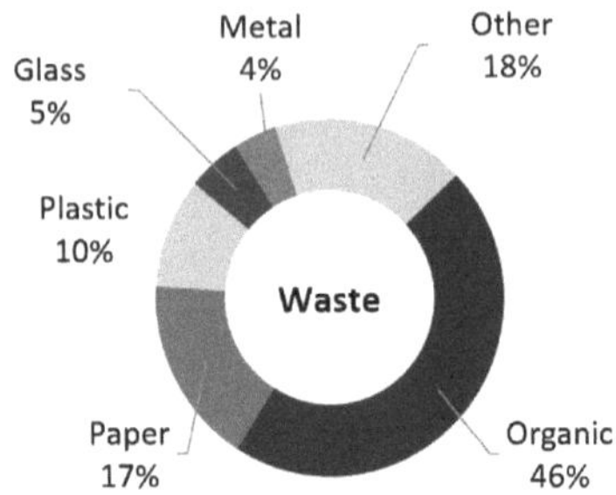

Figure 18.1 ***Municipal solid waste composition worldwide.*** Almost half of the waste comes from organics, with paper being second with around 17%, plastics have 10%, and glass amounts to 5%. *Source: aggregated from [2,3].*

incorporates residential, commercial, and municipal services waste, as well as some industrial waste if it is treated by the municipality [2,3]. Construction and demolition waste is treated separately in many cases as it allows for proper waste management and disposal. While process waste, as well as medical and agricultural, usually has specific ways of treatment due to its unique nature, municipal solid waste is a growing complicated compound mix and therefore needs different treatment.

According to the studies [3], the average municipal solid waste composition worldwide is dominated by organic waste, followed by paper and plastics (see Fig. 18.1). However, this composition differs for various countries and location, e.g., urban and rural areas, and depends on the average income rate of the country.

It should be noted that even at the European Union level there is no clear and established definition of the municipal waste, therefore the practices of its treatment in the EU, as well throughout the globe, differ. Unfortunately, the availability and quality of annual data are major problems for the waste sector. Solid waste and wastewater data are lacking for many countries, data quality is variable, definitions are not uniform, and inter-annual variability is often not well quantified [5]. Around the world, waste generation rates are rising. In 2012, the worlds' cities generated 1.3 billion tonnes of solid waste per year, amounting to a footprint of 1.2 kilograms per person per day on average [6].

The current annual municipal solid waste generation is estimated at the level of 1.3–1.9 billion tonnes and is expected to increase to approximately 2.2 billion tons per year by 2025 [2,3]. Importantly, solid waste yearly generation rates range from less than 0.1 tons per capita in low income countries to more than 0.8 tons per capita in high-income industrialized countries. Even though labor costs are lower in developing countries, waste management can constitute a larger percentage of municipal income because of higher equipment and fuel costs [5]. Available data shows the average split of MSW generation per region with the OECD countries being responsible for 44% of total world waste generation [3].

Waste disposal data are the most difficult to collect as many countries do not do this on the national level. Furthermore, in cases where data is available, the methodology of how disposal is calculated and the definitions used for each of the categories is often either not known or not consistent [3]. It has been estimated [1,3,5,7] that the recoded post-recycling solid waste in 2017 already amounts to over 1.4 billion tons per year, of which more than 70% are placed in landfills and dumpsites, 19% is recycled or recovered and 11–15% is led to energy recovery facilities. Moreover, only 20% of the landfilled solid waste is disposed of in sanitary landfills that reduce aqueous and gaseous emissions into the environment [1,7]. And this situation is also due to the fact that the number of people that lacks access even to the most elementary waste management services is estimated to at least 3.5 billion which may well increase to around 5.6 billion by 2050 if the businesses continue without any significant change in waste management practices [7].

The waste management sector follows a generally accepted hierarchy that responds to financial, environmental, social and management considerations and also encourages minimization of GHG emissions. In the order of preference, this hierarchy includes [3,8–10]:

- *Reduction of waste* initiatives, seeking to reduce the quantity of waste at generation points by redesigning products or changing patterns of production and consumption. A reduction in waste generation has a two-fold benefit: reduction of waste and material during production and elimination of the emissions associated with the avoided waste management activities.
- *Reuse, recycling and materials recovery*, to reduce quantities of disposed waste and return the materials to the economy. A value-based conception of waste, along with appropriate collection infrastructure, could prevent the loss of valuable waste and increase the timely reuse and recycling of used products [10].
- *Incineration* of waste with further energy recovery, which can reduce the volume of disposed waste by up to 90%, however, only for waste streams with very high amounts of packaging materials, paper, cardboard, plastics and horticultural waste. Recovering the energy value

embedded in waste prior to final disposal is considered preferable to direct landfilling — assuming pollution control requirements and costs are adequately addressed. Typically, incineration without energy recovery (or non-autogenic combustion, the need to regularly add fuel) is not a preferred option due to costs and pollution.

- *Landfilling*, which is a common final disposal for waste and ideally should be engineered and operated to protect the environment and public health. Landfill gas (LFG), produced from the anaerobic decomposition of organic matter, can be recovered and the methane (which is approx. 50% of LFG) burned at high temperatures with or without energy recovery to reduce GHG emissions. Proper landfilling is often lacking, especially in developing countries.
- *Controlled dumping*, which is the least preferred option, when the waste at least undergoes registration and controlled placement/compaction. Controlled dumps usually have surface water monitoring, but no engineering measures and no landfill gas management systems.

A wide range of mature technologies and established approaches are available nowadays to mitigate GHG emissions from waste. They include:

- Landfilling with landfill gas recovery (for instance, for its combustion) to reduce the release of methane emissions directly to the atmosphere;
- Post-consumer recycling to mitigate waste generation;
- Composting of selected waste fractions to minimize GHG emissions generation;
- Processes that further reduce GHG generation compared to landfilling. They are based on thermal treatment, including incineration and industrial co-combustion, mechanical-biological treatment with further landfilling of residuals, and anaerobic digestion.

Therefore, the mitigation of GHG emissions from waste relies on multiple technologies whose application depends on local, regional and national drivers for both waste management and GHG mitigation [5].

While post-consumer waste is a relatively small contributor to global greenhouse gas emissions (< 5%) with total emissions of approximately 1300 million tons of CO_2-equivalent emissions in 2005. The largest source is landfill methane (CH_4), followed by wastewater and nitrous oxide (N_2O). In addition, minor emissions of carbon dioxide CO_2 result from incineration of waste containing fossil carbon C (from plastics or synthetic textiles) [5].

Existing waste-management approaches can provide effective mitigation of GHG emissions and provide public health, environmental protection, and sustainable development co-benefits. Collectively, these technologies can:

- Directly reduce GHG emissions through landfill gas recovery, improved landfill practices, engineered wastewater management,
- Avoid significant GHG generation through controlled composting of organic waste or state-of-the-art incineration. The latter can benefit from additional energy recovery.

Today, an estimated 250 million tons of MSW are worldwide thermally treated annually to produce electricity and heat. This relevant global industry is usually called waste-to-energy (WTE) and is rapidly growing: an estimated 250 WTE facilities were built during the first 15 years of the 21st century, mostly in Europe and East Asia [1]. The great majority of these plants are based on the *grate combustion technology* of received or post-recycled MSW, and produce both electricity and heat. While there are some plants that implement gasification technology, the trend is towards very large facilities based on the principle of combustion on a moving bed [1]. Although various WTE technologies such as waste incineration, anaerobic digestion, gasification, and some others exist, mass incineration is still the most common [4].

Remark 18.1. In some literature, waste-to-energy is classified as part of biomass energy. In this book we consider biomass energy in a separate Chapter 19.

Municipal solid waste is not the only type of waste that can be potentially utilized in utility-scale power generation. Many industrial processes have a by-product gas called *synthetic gas*, or *syngas*, which is a mixture of hydrogen, carbon monoxide, and other gases, e.g., carbon dioxide or nitrogen. Some gases may have much higher percentages of nonflammable gases, but can still be burnt, for instance, blast furnace gas, which is not directly attributed to syngas. Considered in some processes as waste products, they are either burnt in a flare or used as a fuel.

18.2. Waste-to-energy concept

The dominant technology for energy recovery from municipal solid waste is based on a direct combustion of waste, when the heat is used to generate steam, which is fed into a steam turbine. This technology is similar to conventional steam power plant that burns solid fuels. However, the amount of energy recovered from the MSW combustion can vary significantly with the characteristics of MSW fed into the boiler (its composition, mass flow rate, heating value), combustion technology, configuration and features of the boiler, and the characteristics of the thermodynamic cycle [4].

18.2.1 Traditional steam cycle

A WTE steam cycle is relatively simple and similar to a smaller size conventional thermal power plant driven by a steam turbine (see Fig. 18.2). A boiler burns fuel, represented by MSW, with the presence of oxygen (i.e., air). Its chemical energy of the combustion reaction is transferred to water that flows through the heat exchanger tubes inside so that it heats up and therefore evaporates. Further heating leads to the increased pressure of steam, which is then fed to the steam turbine, where it expands and transfers its energy to the turbine's rotor. A steam turbine spins an electric generator that delivers electricity to the consumer.

Cold steam from the steam turbine goes to the heat dissipation system, either in the form of condenser or district heating heat exchanger, where it gives its energy away to cooling or heating water, respectively. Finally, cold condensate is pumped back to the boiler.

Similarly to a conventional steam power plant, WTE power plant cycle can be optimized to minimize its energy losses and therefore increase efficiency. As discussed in Chapter 7, major heat losses of a thermal power plant occur inside the condenser when the heat of the exhaust steam is

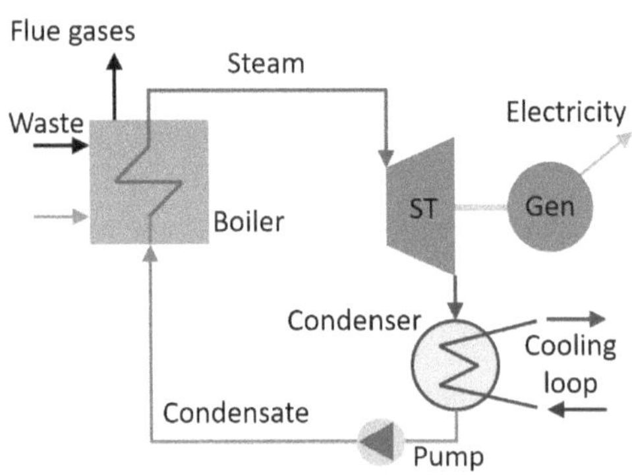

Figure 18.2 ***Simplified process diagram of a WTE plant.*** Similarly to a conventional thermal power plant, boiler burns waste to evaporate water, and the steam is used to spin the steam turbine and associated generator.

taken away by cold water, leading to around 50% of energy waste. One of the easiest ways to recover this energy is to utilize this WTE plant for a cogeneration application, usually for local district heating needs. In this case the condenser can be used as a heat exchanger that transfers heat from the cold steam, released from the steam turbine, to the district heating fluid.

Certain part of energy, however, cannot be transferred to the water inside the heat exchangers due to technological reasons and hence this energy "leaves" the boiler through the exhaust stack into the atmosphere. The amount of heat wasted to the atmosphere can be around 5–15% of the initial chemical energy of MSW and depends on the technology, fuel composition and plant operating conditions.

Similar to a traditional coal-fired steam power plant, WTE plant's thermodynamic cycle can also be optimized in many ways, including proper heat recovery throughout the cycle:

- In a cogeneration application, steam may be extracted directly from the steam turbine and fed to the end user or to district heating heat exchangers. This may decrease the amount of steam that finally goes to condenser and reduce cooling water consumption.
- Cold condensate that is pumped from condenser can first flow through the special heat exchanger called a preheater that uses the temperature of hot gases downstream the boiler. These gases still possess enough energy after they have heated and evaporated water.
- Air required to burn waste is taken from the outside environment, and may be cold enough, which would require extra chemical energy to maintain reaction. Therefore, this air may also be preheated by the exhaust gases before entering the incinerator.

The above mentioned ways are among the few that are always considered by designers of any type of a thermal power plant. As a WTE plant is usually smaller and located in highly populated areas, especially in developed countries, its design and operating parameters must be as competitive as possible to drive the costs down. This discussion will be conducted in detail in Section 18.4.

18.2.2 Combustion of syngas in gas turbines or gas engines

While the combustion of solid waste is usually carried out in special boilers, which generate steam to implement the steam turbine cycle, the use of syngas is slightly different. Of course, syngas can also be burnt in boilers to produce steam, however, it is much more favorable to feed syngas to a gas turbine. While this approach seems obvious, it may lead to some issues regarding the combustion process: traditional gas turbines are designed to burn natural gas which contains more than 95% of methane and the process is optimized for lower emissions. Syngas has a very different composition which may not be easy to burn in such a combustion system, so that either it is mixed with natural gas for a more favorable composition, or the combustion chamber is optimized for this fuel.

The process of a gas turbine based thermal power plant that burns syngas is the same as for traditional natural gas fired one (refer to Chapter 7) and may differ only in some special equipment for fuel treatment, control system and combustion chambers. Such cycles are implemented in industry where syngas (as a waste byproduct of the process) can be further utilized to generate electricity for the local needs. Moreover, using a gas turbine allows implementing a combined cycle process, that is, increasing efficiency beyond 40% of a steam thermal plant. Moreover, gas turbines offer higher flexibility of the process.

If the amount of syngas is limited, the project does not require large capacities, or the composition does not allow using a gas turbine, another technology is applicable, namely, one can use reciprocating gas engines. Their capacity varies from a few kilowatts to a dozen of megawatts, and they can be combined as modules to rapidly increase the capacity of the whole facility.

18.2.3 Use of landfill gas

Landfill gas that is the result of waste storage in landfills and contains approximately 50% of methane with roughly 50% of CO_2, and is contaminated with a small amount of other pollutants. In the worst case, it is freely released to the atmosphere from the landfills which are not properly managed, and causes significant damage due to its strong GHG nature.

In order to prevent this, landfill gas is usually flared or, in simple words, wasted by a direct burning process. Though this combustion process has lower negative impact compared to its free release to the atmosphere and much lower GHG emissions levels compared to conventional coal combustion, it is still wasteful. In an ideal case, this gas should be gathered and fed to a power plant equipped with either reciprocating engines or gas turbines with the combustion chamber designed for specific fuel composition. Such power plants are similar to traditional simple cycle power plants driven by a heavy duty gas turbine, however, they are smaller in size and are usually built in modules. Moreover, they require special attention to the fuel composition. While gas engines are less sensitive to such fuels, traditional gas turbines may need fine-tuning of its fuel gas module, combustion chamber and control system to ensure reliable and efficient operation.

Remark 18.2. Most commonly built WTE power plants burn municipal solid waste as base fuel and employ steam cycle based on boilers and steam turbines.

18.3. Waste-to-energy plant configuration

The process of generating electricity in a conventional mass-burn conventional waste-to-energy power plant has the following major stages:

1. Waste is delivered by garbage trucks into a large receiving pit for pretreatment and further delivery to combustion chamber within the incinerator;
2. The incinerator or a special boiler is designed to efficiently use the chemical energy of the burning waste to evaporate water and supply superheated steam of certain parameters;
3. The high-pressure steam turns the blades of a turbine generator to produce electricity;
4. An air pollution control system removes various pollutants from the combustion gases before they are released through an exhaust stack to the atmosphere;
5. Ash and other waste products are collected from the boiler and the air pollution control system.

Based on this process, the overall conventional WTE plant configuration can be arranged into the following major parts and equipment areas, as schematically represented in Fig. 18.3:

- Boiler island, which comprises municipal solid waste or fuel management system, incineration system with a steam boiler, and emissions control and treatment equipment;

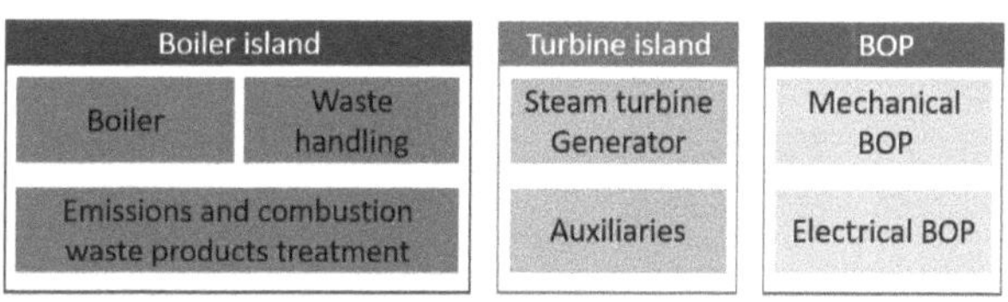

Figure 18.3 ***Equipment areas of a typical utility-scale waste-to-energy power plant.*** A typical WTE plant implies three major blocks: boiler island, turbine island, and balance of plant systems (mechanical and electrical).

- Power island with steam turbine, generator and necessary subsystems, supplied with them, and
- Mechanical and electrical balance of plant systems.

Remark 18.3. A WTE plant that operates on syngas as a product of waste gasification would have configuration and equipment areas similar to:

- Simple cycle or combined cycle power plant, if this plant is based on gas turbine technology, either alone or in combination with the heat recovery equipment and steam turbine;
- Gas engine based modular plant (usually container-based);
- Conventional thermal gas-fired boiler and steam turbine, if syngas is burned within a traditional boiler.

Evidently, a hybrid WTE plant would have a different configuration, which will be similar to a CCPP or a combination of a CCPP and a conventional WTE plant, depending on the design. While these plants are still limited, we will focus only on the conventional technology, yet paying specific attention to the sustainability questions which are similar to any type of WTE technology.

18.3.1 Boiler island

Similarly to a conventional coal-fired power plant, the boiler island of a waste-to-energy plant unites those systems related to fuel (i.e., waste) handling, its pretreatment and combustion to generate steam, as shown in Fig. 18.4.

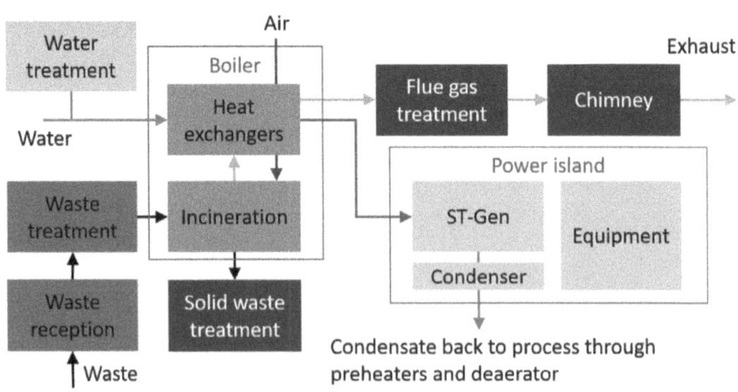

Figure 18.4 ***Simplified process diagram of a WTE plant.*** Schematic diagram of the process including boiler island configuration: waste reception and treatment, incineration in a boiler, emissions and process waste treatment system.

18.3.1.1 Waste reception and treatment

The waste is usually delivered to a power plant by special trucks and discharged into the reception bunker, which can hold enough waste for several days of operation. Ideally, this bunker should be closed and kept under lower pressure to avoid discharge of gaseous emissions from the waste inside and prevent the neighboring community from a harmful impact. This bunker has to be frequently cleaned thoroughly to avoid deep contamination according to strict environmental, health and safety procedures imposed by regulation.

In many cases waste is kept in a bunker for a few days for fermentation which should reduce its humidity and increase its heating value.

The waste in the bunker is mixed and delivered by overhead cranes with special grabs to the charging hoppers, which are a part of the waste feeding system. This system delivers waste to the combustion chamber of the boiler, either continuously or in batches, depending on the type of an incinerator implemented in a plant. Sometimes waste may undergo a milling process to ensure that it has relatively fine structure for combustion.

Finally, the pretreated waste is fed to the incinerator or a combustion chamber of a special waste boiler.

18.3.1.2 Boiler

There are three major types of waste boiler widely implemented in commercial operation which differ in combustion chambers:

1. Moving grates, which consists of a moving grate that transports waste through the combustion chamber. While on the grate, the waste is dried and burned at high temperature with continuous air supply from below the grate, while the ash and non-combustible parts leave the grate either downwards under normal gravity or at the end of the grate path. Properly designed grates transport and agitate the waste and evenly distribute air to make combustion process efficient (see Fig. 18.5).
2. Rotary kiln combustion chambers, which use a special inclined rotating cylinder transporting the waste through the furnace as illustrated in Fig. 18.6. This cylinder can be up to 5 m in diameter and up to 20 m long. Due to its rotation, the waste mixes and allows proper burning process. Rotary kiln combustion is sometimes used to burn hazardous materials like medical, biological or industrial waste. It can provide higher burning temperatures to achieve full destruction of viruses or microbes.

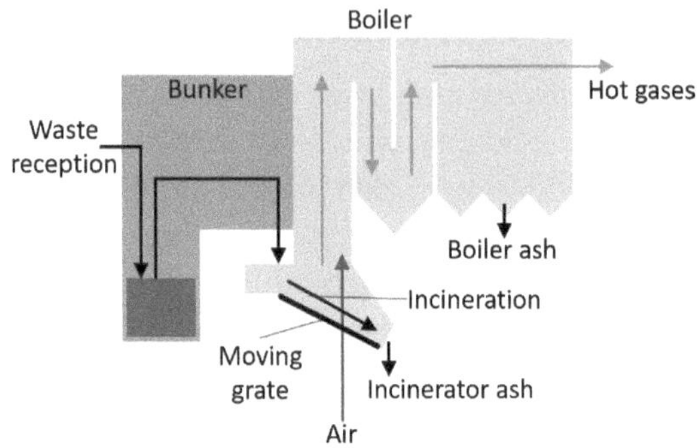

Figure 18.5 ***Simplified schematic of a moving grate waste boiler.*** Waste is delivered to the bunker, pretreated and fed to the moving grate. Air is blown from below to maintain the incineration process. Hot gases follow the complicated hot gas path to transfer its heat to multiple heat exchanger tubes and evaporate water.

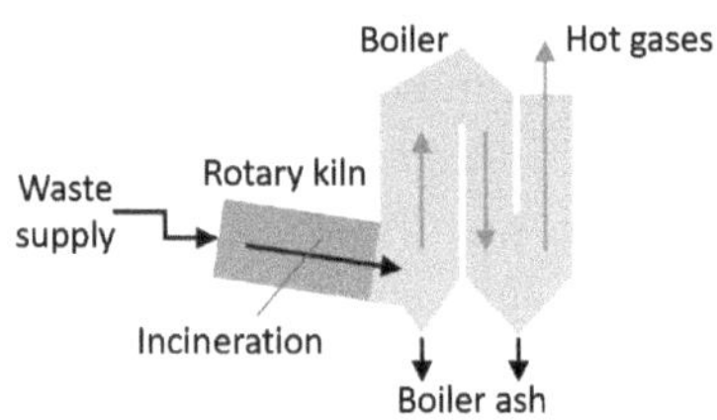

Figure 18.6 ***Simplified schematic of a rotary kiln waste boiler.*** Treated waste is delivered to the rotary kiln that mixes the waste and maintains combustion process. Hot gases follow the hot gas path to transfer its heat to multiple heat exchanger tubes and evaporate water.

3. Fluidized bed combustion chambers, which have the same logic as for coal: fine solid particles of fuel waste are mixed with the air coming from the bottom, fly out of the solid layer (like bubbles) and burn. While the technology seems quite efficient, it requires special pretreatment of waste and its consistency, which is very rarely achievable in practice. The schematic of a fluidized bed boiler within a WTE plant is presented in Fig. 18.7.

Remark 18.4. There some other types of boiler and technology that may be utilized for waste, however, moving grate, rotary kiln and fluidized bed are the top three technologies used in commercially built and operated WTE plants.

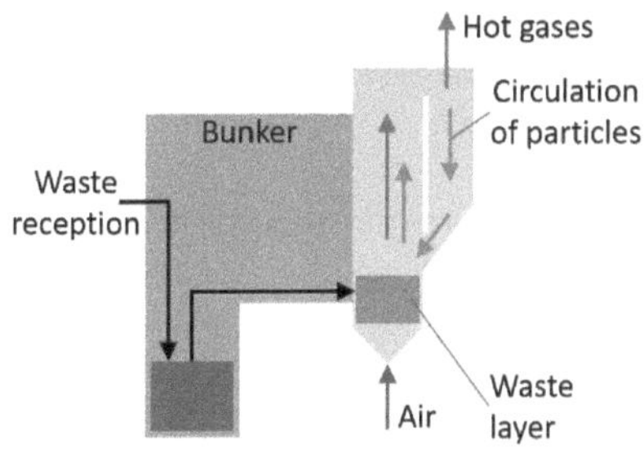

Figure 18.7 ***Simplified schematic of a fluidized bed boiler.*** Treated waste is delivered to the boiler combustion chamber to form a layer of particles. During combustion they start to "bubble" and burn, follow the hot gas path and, if unburnt, return back to the chamber.

For startup purposes and to maintain the furnace temperatures, an auxiliary firing system is implemented. This is usually done with fuel oil or natural gas and is similar to that in a conventional boiler or an HRSG.

Like in a conventional fossil-fired boiler (as discussed in Chapter 8), hot gases of approximately 1000–1200°C from the combustion area go through the system of heat exchangers — water tubes, which absorb the heat of the flowing hot combustion gases and transfer it to water. Depending to the size and configuration, various designs of boilers may be exercised. The most established is the drum technology: evaporation of water takes place in a vessel called drum, which also serves as a water–steam separator to feed steam further to a downstream consumer. If the heat balance allows for higher energies, this steam may be further driven through a heat exchanger called superheater, which further increases is temperature and pressure, and, in turn, overall plant efficiency.

Obviously, superheated steam is far better for further use in a steam turbine as it increases its output and, depending on the flow, allows for a multipressure design, though mostly without reheater due to relatively smaller size of plant compared to conventional coal-fired power plants.

18.3.1.3 Emissions and process waste treatment

Combustion of municipal solid waste also releases waste in the form of solid ash, small particles that flow with the gases, and gaseous emissions. Therefore, any WTE facility must be equipped with modern treatment systems to comply with ecological regulations. These systems are similar to those in a coal-fired power plant and always include:

- Bottom ash handling;
- Fly ash treatment;
- Flue gas handling.

Bottom ash is the noncombustible fraction of the waste that is burnt in the furnace. Sometimes referred to as grate ash, clinkers or slag, is usually generated at the rate of roughly 20–30% of the weight of the waste combusted, but only 5–10% by volume [1], however, it can vary on either side depending on the composition of waste.

A bottom ash handling system has to allow for ash removal during the operation and its delivery to the intermediate storage exposed to the environment for its preliminary natural treatment, which can be simply in a pile on site with further mixture with cement or other material. In many cases bottom ash goes through magnets or eddy current separators to remove ferrous (steel and iron) and other metals (copper, brass, aluminum, nickel, etc.) for further recycling. Then, bottom ash can be utilized or landfilled, depending on the legislation of the country and the need of the local industry. It can be generally assumed that bottom ash generation rate is roughly 10% of the waste's original volume.

Contrary to bottom ash, fly ash particles are carried from the furnace by the flue gases. These are mostly residues of incomplete combustion of waste and therefore can be harmful if released to the atmosphere. Compared to coal combustion, where fly ash particle composition is rather limited due to coal's nature, waste residues may contain heavy metals of antimony (Sb), arsenic (As), cadmium (Cd), chromium (Cr), cobalt (Co), copper (Cu), lead (Pb), manganese (Mn), mercury (Hg), nickel (Ni), thallium (Tl), and vanadium (V) [1]. This variety is due to inorganic nature of waste as a composition of many artificial products.

However, fly ash particles can be removed with the same technology used in coal-fired power plants. These are (see Chapter 8 for detailed description):

- Mechanical systems, both dry (gravitational, cyclone, and bag fabric filters) and wet; and
- Electrostatic precipitators (ESP) that use static charge to catch particles from the flue gas flow.

When captured, fly ash is collected and treated through multiple processes like chemical stabilization, solidification, or thermal treatment and stored in a fly ash silo. Due to high content of various chemical elements resulting from its nature, fly ash from waste combustion is difficult and too expensive to reuse or separate valuable materials, and in most cases it goes to landfills after treatment.

Gaseous pollutants may contain HCl mainly from the combustion of polyvinyl chloride (PVC, the world's third-most widely produced synthetic plastic polymer, after polyethylene and polypropylene), SO_2 from combustion of sulfurous compounds, HF from combustion of fluorine compounds, and nitrogen oxides NO_x [1]. These gases are usually sent through a flue gas scrubber to remove poisonous acidic material right before the flue gases go through the fly ash filters.

The NO_x emissions control is carried out in a conventional way by injecting ammonia or urea in the hot gas path and, exposed to high temperature, the products of these reactions are neutralized up to 99%. This process is implemented, for instance, within SCR or SNCR systems (for more details refer to description presented in Chapter 8).

Other poisonous substances like heavy metals (mercury or cadmium) can be removed from the hot exhaust gases by injecting activated carbon into the flow.

18.3.2 Power island and balance of plant

Similar to any fossil-fired steam power plant, the power island of a conventional WTE power plant accommodates a steam turbine with its generator and related essential systems.

Generally, WTE plants have lower capacity than large scale utilities operating on coal or natural gas, and steam turbines installed at WTE plants are therefore much smaller, with their sizes ranging from a few megawatts up to large units of several hundred megawatts with proven performance for such application. An efficient and reliable steam turbine is the key to harnessing energy from waste. They are usually of condensing type, i.e., employ a condenser to transfer energy from the cold steam and turn it into water. Depending on the capacity, smaller machines are usually single casing with one or two pressures and maybe with a few steam extractions for various plant and external needs.

Electrical generators are rather standard for 50 or 60 Hz, depending on the grid.

Major systems, supplied with the units, always include a lubricating and hydraulic oil system, various steam shut off and control valves, electrical subsystems for safe and reliable operation and control of the generator, etc.

Like any thermal power plant, a WTE facility always has several essential balance of plant systems and equipment required for its overall operation. Mechanical BOP systems of a typical WTE facility include, among others:

- A heat rejection system usually represented by an air cooled condenser or a mechanical draft small modular cooling tower to condense steam back to water;
- A water treatment and recycling system to provide demineralized water for the water–steam cycle;
- Grid water heaters (if any) to transfer heat of the steam to water which is fed to the grid;
- Various pumps and valves to run the cycles and control the fluids.

There are multiple other necessary systems which are present at any power plant facility, for instance, HVAC, compressed air, drainage, fire protection, etc. One can refer to Chapter 9 for details. Their design (and sometimes presence) depends on the configuration and size of a plant.

Electrical balance of plant systems of a WTE facility incorporate, among others:

- A power transfer system to supply electricity from the generator to the grid, which includes the main transformer, generator bus bars, control and protection system;
- LV, MV, and HV electrical networks to feed internal systems;
- Main and auxiliary transformers;
- Generator circuit breakers to protect equipment and grid from anomalies;
- A substation, usually based on a gas insulated switchgear (GIS), due to a smaller footprint and often location in the rural areas where space is limited and does not allow having an open switchyard;
- Overall plant control, monitoring and protection systems for safe and reliable operation.

18.4. Project evaluation and economics

The successful implementation of a WTE project rests primarily upon the following essential building blocks or key elements [11]:

- A reason or need for the project because of a critical community solid waste disposal problem or crisis;
- An implementing government agency or private project developer with political commitment willing and able to undertake the project;

- An adequate supply of solid waste for the project or means to assure waste stream control or attract sufficient quantities from other communities;
- Markets for the recovered energy and recovered materials;
- A project site that is environmentally, technically, socially, and politically acceptable.

Even though WTE power plants are similar to conventional solid fuel fired steam power plants, the underlying investment decision-making process differs significantly. While a conventional power plant is based on the electricity and heat sales, a WTE plant is first aimed at burning waste and delivering heat and electricity, which are considered as an economically positive add-on to the waste-conquering problem and, of course, contribute to the investment calculations together with multiple incentives. Moreover, waste as fuel may have zero or even negative cost which would impact the investment analysis.

So far, a WTE plant is one of the options to answer the growing urbanization population, consumption and consequently waste production, but usually for a developed country. Still, the potential of WTE implementation strongly depends on the specific economic, social, and political conditions of the country where the implementation strategy must be carried out [2].

WTE plants have capacities from several dozens to hundreds of megawatts (electrical capacity) with a bit low efficiency around 30% due to lower heating values of waste, however, with heat production it may increase to more than 85% [12].

WTE projects are capital intensive and require planning that frequently extends over a two to five year period [11]. The investor has to identify the right location within the community as the latter is the "producer" of solid waste. The community must be able to supply or divert enough solid waste to the proposed facility since solid waste serves as the feedstock for plant operations and energy production. The community must be able to guarantee both the quantity and the quality of its solid waste. The level of the wastes to be guaranteed will determine the ultimate size of the facility [11].

Normally, major reports include WTE into biomass data so that it may be difficult to allocate required data precisely. The capital costs of a WTE plant are comparable to biomass plants that operate of biogas or bio-liquid, and the data suggest the range of 450–2500 USD per kW of installed capacity [13] as shown in Table 18.1.

WTE plant investment costs tend to be at the upper limit of the general biomass costs. This is mostly due to the fact that the primary objective

Table 18.1 CAPEX and OPEX components for a waste-to-energy plant.

Technology	CAPEX[a]	OPEX[a]		LCOE[a]
	$/kW	Fixed, $/kW-year	Variable, $/MWh	USD/kWh
WTE	up to 8000	100	5–15	0.04–0.08

Source: Based on [13,14], EIC and IIR databases and calculations.

[a] WTE plant investment costs tend to be at the upper limit of the general biomass costs.

of these projects is not power generation, but waste disposal. Capital costs are often higher, as expensive technologies are used to ensure that local pollutant emissions are reduced to acceptable levels [13].

Importantly, WTE projects are often planned, carried out and financed by the community, which clearly puts them aside of the purely commercial thermal or renewable power plants. This is usually due to the fact that strong political and community support is vital for this type of project to become real.

Moreover, the waste treatment policies can considerably change the feasibility study of a WTE plant project. There are multiple policy measures that can be granted to the WTE project [2]:

- Various subsidies of a local government like renewable certificates, feed-in tariffs or heat incentives;
- Zero-waste policies that require decreasing landfills through any of the options, e.g., a WTE plant, while the landfill fees and taxes are kept high;
- Carbon taxes;
- Renewable targets that also cover WTE plants.

It should be noted that WTE does not mean that it can easily lead to a zero waste future and other options can be omitted. WTE plants are an alternative which allows reprocessing bulk quantities of municipal solid or industrial waste and relieving landfills. But the WTE option has to be executed only together or after all other options are exhausted: reduction, reuse, recycling or recover.

18.5. Sustainability challenges and environmental issues

In many scenarios and discussions regarding the global climate change or sustainability topics, the WTE technology is either neglected or only mentioned within the biomass energy technologies as having a very limited share. However, WTE is one of the solutions to conquer global waste production (at least in the short to medium run) and one of the means to recover materials and energy from waste by thermal or biological treatment.

Many countries use waste-to-energy plants to capture the energy stored in municipal solid waste. According to the EIA, the use of WTE plants in some countries is relatively high with about 70% in Japan, more than 50% in Norway, and more than 45% in Switzerland. This is partly due to the fact that these countries have very limited open space for landfills, and the energy demand is constantly growing.

While reuse, recycling and recovery are the preferred solutions for waste treatment, thermal WTE technology has several important environmental advantages over direct landfilling of the MSW, which can be summarized as follows [1]:

- Greenhouse gas emissions from WTE incineration contribute less to climate change than those from landfills. WTE mostly emit carbon as CO_2 due to the combustion process (most of other pollutants are filtered out), whereas in landfills both methane and CO_2 are generated and partly escape into the atmosphere. The global warming potential of the CO_2 is 21 times lower than that of methane, explaining its lower contribution to climate change. Contrary to landfilling with some gas capture, which emits up to 2750 g/kWh of CO_2 equivalent emissions, incineration has much lower levels of 1600 g/kWh [2].
- It saves land, as significantly more land is needed for a landfill than for a WTE plant.
- It allows recycling of ferrous and nonferrous metals and granulates.
- From biomass in the MSW, renewable energy that can replace fossil fuels is produced. It is generally assumed that 50% of the energy produced is biogenic due to biodegradable fraction and thus does not contribute to climate change.

Despite several advantages, waste incineration is still treated with caution by the public, as some concerns still exist. Like any other technology, WTE does possess some disadvantages, especially in the sense of emissions and the impact to the environment:

- Even though modern filtration and cleaning systems allow capturing most of the harmful substances, there some air emissions associated with waste combustion: metals (mercury, lead, and cadmium), organics (dioxins and furans), acid gases (SO_2, HCl), nitrogen oxides, and carbon monoxides [2].

- The ash from the furnace can still contain toxic chemicals and is usually disposed in landfills, which can potentially contaminate soil and ground water.
- Like any other thermal technology, a conventional WTE plant consumes a lot of water for the water and steam cycle, cleaning, cooling, scrubbing, etc. The discharged water is contaminated with chloride and some soluble heavy metals (the exact composition depends on the nature of MSW). Instead of discharging this water, closed cleaning systems have to be adopted at site.

Contrary to fossil combustion, where the fuel chemical composition is way simpler compared to municipal soil waste and therefore the emissions from fossils are more predictable and hence easier to treat, waste incineration emissions treatment is a challenge, which can be solved at the modern level of technology. The problem here is not technological but rather concerns the attitude of the operator: the cleaning systems will operate properly only if they are properly maintained, which is costly. And these costs are not negligible. This is a significant adoption barrier faced by the industry in the developing countries, along with the lack of trained personnel to successfully handle such a complex and daunting process [2].

References

[1] Kalogirou E. Waste-to-energy technologies and global applications. CRC Press. ISBN 9781351977913, 2017.
[2] World Energy Council. World energy resources 2016. London: World Energy Council. ISBN 9780946121588, 2016.
[3] Hoornweg D, Bhada-Tata P. What a waste. A global review of solid waste management. Urban Development Series, vol. 15. World Bank; 2012.
[4] Branchini L. Waste-to-energy: advanced cycles and new design concepts for efficient power plants. Springer International Publishing. ISBN 9783319136080, 2015.
[5] Bogner J, Abdelrafie A, Diaz C, Faaij A, Gao Q, Hashimoto S, et al. Waste management, vol. Climate change 2007: mitigation. Contribution of Working Group III to the Fourth Assessment Report of the Intergovernmental Panel on Climate Change. Cambridge, United Kingdom, New York, NY, USA: Cambridge University Press; 2007.
[6] World Bank. Solid waste management. URL: http://www.worldbank.org/en/topic/urbandevelopment/brief/solid-waste-management.
[7] Waste atlas. URL: http://www.atlas.d-waste.com/.
[8] United Nations Environment Programme (UNEP). Guidelines for national waste management strategies. Moving from challenges to opportunities. UNEP. ISBN 9789280733334, 2013.
[9] Directive 2008/98/ec of the European Parliament and of the Council of 19 November 2008 on waste and repealing certain directives. Official Journal of the European Union 2008.

[10] Van Ewijk S, Stegemann J. Limitations of the waste hierarchy for achieving absolute reductions in material throughput. Journal of Cleaner Production 2016;132:122–8.
[11] Rogoff M, Screve F. Waste-to-energy: technologies and project implementation. Elsevier Science. ISBN 9781437778724, 2011.
[12] Renewable energy sources and climate change mitigation. Special Report of the Intergovernmental Panel on Climate Change. Cambridge, UK, New York, NY: Cambridge University Press; 2011.
[13] IRENA (International Renewable Energy Agency). Power generation costs in 2017. Abu Dhabi: IRENA; 2017.
[14] Black & Veatch. Cost and performance data for power generation technologies; 2012.

CHAPTER NINETEEN

Bioenergy

Contents

19.1. Bioenergy for electricity generation

Bioenergy is by far the largest renewable contributor to the transport and heating sectors, and also provides an important share of renewable electricity generation [1]. According to [1], in 2016 bioenergy supplied around 500 TWh of electricity, accounting for approximately 2% of global electricity generation with the total installed capacity reaching 110 GW. While over the period of 2010–2016 annual capacity additions were steady in the range of 5–7 GW per year, bioenergy only plays a prominent role in the electricity generation portfolios of a limited number of countries. Thus, in 2016 around 90% of all capacity was located in just 26 countries, which indicates that bioenergy is growing but not expanding strongly into a wider array of countries, in many cases despite biomass resource availability [1].

Looking ahead, Asia is expected to replace Europe as the largest market for bioenergy electricity deployment due to a combination of increasing energy demand, low-cost biomass waste and residue resources, and long term targets in emerging economies such as China, India, and Thailand [1].

One of the major challenges of bioenergy within electricity production sector is relatively high cost of electricity generation and limited scope for lowering these costs due to general maturity of technology implemented. These technologies are mostly based on conventional power plant technologies like direct combustion of solid fuels or use of gaseous products of biomass within gas turbines or gas engines. Even combined in most of the

Sustainable Power Generation
https://doi.org/10.1016/B978-0-12-817012-0.00033-5

cases with heat production, biomass electricity generation is still limited in improvements of project economics or technology.

19.2. Sources of biomass feedstock and their treatment

Biomass varies in its nature and, consequently, requires specific technologies in each case, both in the management of the particular crops and in the energy use, like heat, electricity, etc. [2].

19.2.1 Sources of feedstock

Generally, biomass resources can be classified into three major categories of feedstock, determined by their origin:

- Organic residues and waste, where biomass comes as a byproduct of other processes, activities or product streams;
- Forestry, when biomass is harvested directly;
- Agriculture, when biomass is grown specifically for energy production.

In many cases the term energy farming is used in a very broad sense to mean the production of fuels or immediate energy as a main or subsidiary product of agriculture, forestry, aquaculture and sometimes industrial or social activities leading to organic waste production [3]. Energy farming incorporates all of the above mentioned sources of biomass, and on a broad level has both advantages and disadvantages [3].

The major attractions of biomass as an energy raw material due to energy farming include [3,4]:

- Renewability and domestic availability in many countries. Biomass is often widely dispersed and therefore can be well considered as a distributed energy source rather than as a raw material confined to a few limited sectors of a region or country;
- Biomass has lower content of sulfur and thus produces lower emissions of sulfur oxides during combustion;
- Large potential supply or certain energy crops that can meet the demand;
- Variety of crops which would allow for combinations with food and energy production in various regions and climatic conditions;
- Biomass can provide multiple energy and energy-intensive products like solid, liquid or gaseous fuels, further production of electricity and heat, chemicals and chemical feedstock, as well as paper, lumber, and other finished goods and products;

Table 19.1 Estimated maximum performance energy harvests from dry biomass.

Feedstock	Yield, t/ha	Energy, GJ/ha
Forest wood	7	130
Tropical wood	11	200
Sorghum (3 crops / Yr)	50	850
Corn	25	77
Sugar cane	30	150

Source: Aggregated from [2,3].

- Increased overall efficiency by using by-products, residue, and waste, or combining biomass with other conventional types of fuel, like coal.

Major disadvantages of energy farming include, but are not limited to, the following points [3]:

- By all measures, the energy density of biomass is low compared to that of coal or liquid oil products. The yield of biomass per unit growing area of land or water is small. Table 19.1 shows both the biomass yield (tons per hectare) and the energy equivalent produced per hectare and year for different energy crops [2,3]. The heat content of biomass, even on a dry basis, is, at best, comparable with that of a low-rank coal or lignite, and substantially (25–50%) lower than that of anthracites, most bituminous coals and oil products;
- Most biomass as received has a high burden of physically absorbed moisture, that may be up to 50% of the total raw material on a mass basis. This requires substantial drying that requires energy and shall be preferably done naturally. Feeding high-moisture biomass to the combustion would result in the thermodynamic cycle efficiency as low as 20%;
- Energy crops may substitute for human food production, especially due to large footprint of biomass production, or withdraw land from natural use that may lead to deforestation, destruction on natural habitats of animals and plants, and overall changes in ecosystems;
- Energy crop production requires a large amount of water resources and energy and may catalyze other ecological issues and resource constraints;
- The totality of both intense food and biofuel production in intense farming may lead to soil infertility and its erosion;
- In many cases transportation of bulky solid biomass is limited or requires much effort;
- If poorly controlled, energy crop production can result in increased emissions into air, water, and soil.

Remark 19.1. Biomass can be a potential means to counter CO_2 buildup because it may demonstrate that energy expended to grow, collect, process the biomass (e.g., for fertilizer production, operating of machinery, means of transportation and storage) does not emit greater quantities of CO_2 [4].

There can be various approaches to use energy crops more efficiently, especially when growing plants that can supply both human food and energy from the waste products, or avoiding burning residue biomass in the field.

There are two major categories of energy biomass cultivated worldwide:

1. Forest residue and wood waste for direct combustion are usually obtained from modified forests having a larger tree density. However, wood can be considered as a sustainable energy resource only if it is grown as fast as it is consumed. Moreover, wood resources are consumed also for sawn timber, paper making, cleared for agricultural land (and not harvested but wasted). In the ideal case, wood shall not be used directly for electricity and heat generation on a large scale.
2. Agricultural crops like sugar and starch crops (sugar cane, corn, cereals), oil crops (e.g., palm, canola, sunflower), and various types of waste (like biomass fraction of municipal solid waste, wet waste from agriculture), which are usually used for production of liquid or gaseous biofuels or chemical feedstock. In many cases they are considered as a by-product of agricultural activity and do not substitute food production.

Remark 19.2. Municipal solid waste does not include only biomass waste, but also various other materials like paper, plastics, metals, and other compound materials. Its treatment is subject to special requirements and challenges that are covered in Chapter 18.

19.2.1.1 Forestry residue and wood waste

Forest residues are a byproduct from forest harvesting, which is a major source of biomass for energy. This includes thinning, cutting stands for timber or pulp, clearing lands for construction or other use that also yields tops and branches usable for bioenergy. On top, stands damaged by insects, diseases or fire can be an additional sources of biomass. However, these residues usually have low density and heating values with high transportation costs.

Wood waste is mostly the result of wood processing industries like sawmills, plywood, panels, and other wood products supplies, which may generate significant amount of byproduct. The main types of waste include sawdust, off-cuts, trims, shavings that can be further pressed into wood pellets. Wood pellet producers turn low-value wood fiber into pellets. Common sources are forest residues, thinnings, tree tops, and limbs, and low-quality fiber that is unwanted by other industries or lacks a local market. The other sources are roundwood not suitable for lumber or wood from short rotation woody crops [1].

Wood pellets and wood chips are the predominant solid biofuel commodity on the international market, with Europe being the key region that consumes pellets from major exporters like the Unites States, Canada, Russia, and Brazil. It is important to mention that the major production of wood chips is for pulp to produce paper, so there is a need to separate this production from wood chips for heat and power, which are around 10% of the total production [1].

Global wood pellet consumption for both industrial and heating purposes increased by 60% during 2010–2016. Wood pellet production in 2016 reached 28.5 million tonnes. The principal markets for industrial and heating wood pellets are found in the European Union, supplemented by industrial pellet demand in Japan and Korea and heating demand in North America. Industrial wood pellet demand is still dominated by a relatively small number of large-capacity consumers, e.g., coal power stations converted to biomass, and therefore can undergo notable demand changes as a result of technical, economic or policy factors. Conversely, fuel consumption in heating markets is influenced by weather conditions and biomass fuel costs relative to competing heating fuels and technologies [1].

19.2.1.2 Agricultural residue

Agricultural residues can include straw, stover, cane trash and other green agricultural wastes from various types of agricultural production like sugarcane bagasse, oil crops waste, animal waste (cow manure and poultry litter), and some others.

Large quantities of such agricultural waste are produced worldwide and in most cases underutilized, significant quantities are left in the fields, burnt as a waste without recovering the heat, or used to plough back to the soil as a kind of fertilizer.

Sugar cane is probably one of the best examples of successful combination of food and energy production from crops implemented on a large scale. The cane is usually transported to a central mill, where rollers crush the cane to extract the juice as the main initial product. The juices further pass for sugar extraction, while the residue (molasses) can be used for cattle feeding, fermented on site into bioethanol or used in other chemical processes. The cane's fibrous residue (bagasse) can be burnt in boilers as an alternative to fossil fuels. The boiler ash is further used as a phosphate rich fertilizer.

19.2.2 Processes of treating feedstock

Feedstock use for energy production depends on many circumstances. While some of the feedstock like wood or crop residues can be directly burnt for electricity production, further treatment can increase the overall efficiency of the energy conversion process and decrease the level of emissions. These processes are usually aimed at converting solid biomass into liquid or gaseous combustibles.

Therefore, there are several major production processes that include, for instance:

- Solid fuel preparation (chipping, pelletisation, sorting, and separation, etc.) for further combustion in a boiler;
- Pyrolysis;
- Gasification;
- Anaerobic digestion.

19.2.2.1 Pyrolysis

Pyrolysis is the thermal decomposition of the biomass feedstock into gaseous, liquid, and solid products without any oxidizing agent [5]. Exposed to high temperatures in the absence of oxygen, biomass undergoes thermal degradation when its chemical components transform into a wide range of products that can be separated into gas, condensable oil and solid char [6].

The biomass feedstock needs to be conditioned in order to reduce moisture and particle size, as these aspects affect not only the energy requirements of the pyrolysis process but also the characteristics of the resulting products. The conditioned biomass is fed into the special pyrolysis reactor where the thermal decomposition takes place at temperatures usually between 450°C and 600°C [6]. The main product is bio-oil, which is a mixture of organics (75–80% of weight) and water, accompanied by some by-products like char and gas which can be used within the process so there are no waste streams [7].

19.2.2.2 Gasification

In the gasification techniques, biomass is partially oxidized by means of gasifying agent, for instance, air, oxygen, steam, CO_2, or mixture of these components at high temperatures 600–1000°C to form a synthesis gas or *syngas* — a mixture of several combustible gases like CO_2, CH_4, and H_2 in various properties [2,5]. This gas usually has low-to-medium heating value

Biomass-derived syngas can typically be obtained from gasification of agricultural and forestry residues, along with industrial waste such as black liquor, which is a major biomass-containing waste produced in pulp and paper manufacturing regions worldwide [5].

Syngas can be burnt in conventional boilers, gas turbines or in co-firing applications, but can also be used for secondary purposes other than electricity production — production of fuels and chemicals.

While the overall process of gasification is not yet efficient as it still undergoes research and development stage and not yet implemented on a global scale, the fuel gas is still more versatile in applications and less polluting. The syngas calorific values may vary from 3–7 MJ/Nm^3 for those produced with air and up to 15 MJ/Nm^3 if carried out in oxygen. Within gas turbine combustion, its efficiency can be up to 40% for simple cycle machines. Importantly, most of the modern gas turbines that run on natural gas can admit syngas, though different type of combustion chamber is required. Syngas can be further enriched with methane for more efficient combustion or converted into liquid fuel [2].

Another option consists in the synthesis of hydrogen from syngas for use in integrated fuel cells, with an estimated electric conversion efficiency of 50–55%, but this technology requires further development [2].

19.2.2.3 Anaerobic digestion

Contrary to the above mentioned thermo-chemical conversion processes of biomass, when significant heat is required to run the reactions, anaerobic digestion is the biological degradation of biomass in oxygen-free environment without significant heating. The main product of this process is biogas, which is usually rich in CH_4 and CO_2.

The process takes place in a special sealed and warmed container called digester. If properly mixed, up to 60% of the organic material can be converted into biogas during the process. This biogas can be burnt directly or conditioned (upgraded) to be fed into main gas pipeline system and mixed with natural gas used for further purposes. Anaerobic digestion can operate with any type of biomass that can be digested by animals (that is, except biomass from wood). The process is particularly efficient with wet biomass, sludge from sewage treatment plants, as well as in landfill cells for municipal waste storage (see more details on solid waste combustion in Chapter 18). If the biomass used as raw material is not contaminated, the solid digestate can be also used as fertilizer, adding value to the process [2].

19.3. Biomass power plants

While most of the liquid biofuels are usually utilized as transportation fuels within cars, trucks or other means, solid biomass and biogas or syngas are often used in biomass power plants for electricity and/or heat generation.

There are three major types of biomass power plant:

- Solid biomass direct combustion power plant;
- Integrated gasification power plant;
- Co-firing plant of biomass or biomass products, i.e., using biomass fuels (solid, liquid or gaseous) within conventional power plants operated on fossil fuels. Biomass is used as one of the fuel component mixed with traditional fuel, for example, mix of syngas and natural gas, or mix of wood and coal.

19.3.1 Direct combustion of solid biomass

Technology of direct combustion of solid biomass is similar to conventional solid fuel combustion discussed in Chapter 7, with the power plant diagram represented in Fig. 19.1. The process starts with the biomass, for instance, wood chips, delivered by ground transportation (truck or rail) to the storage bunker. From there, this solid fuel is delivered by the conveyor to the boiler's combustion chamber, where it undergoes exothermic reaction of combustion. Hot gases flow around the heat exchanger tubes and heat up the water that flows inside these tubes, so that it evaporates. Hot exhaust gases are further treated and released to the atmosphere.

Pressurized steam is then fed to the steam turbine, where it expands and transfers its energy into the mechanical energy of the spinning shaft, which is coupled to the electric generator. Cold steam is then guided to the heat rejection system that may be represented by a traditional condenser (e.g., with cooling loop) or heat exchanger for district heating. Finally, cold condensate is pumped back to the boiler to close the cycle.

Therefore, a solid biomass power plant consists of the following equipment areas:

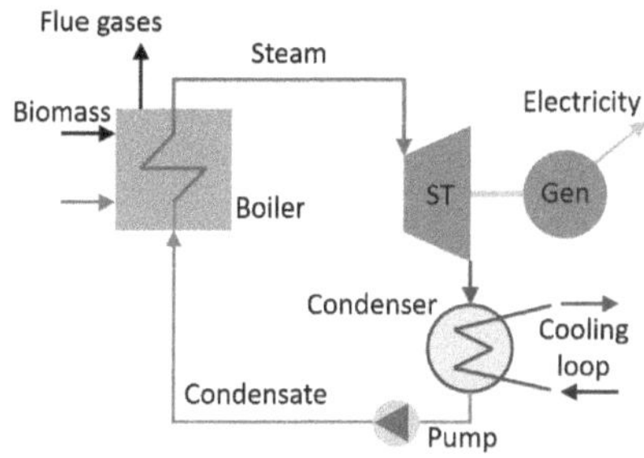

Figure 19.1 ***Process diagram of a solid biomass thermal power plant.*** Biomass from the storage is delivered to the boiler for combustion process. The steam is fed to the steam turbine coupled to electric generator. The process is similar to conventional coal-fired power plant or to a WTE facility.

- Boiler island that includes solid biomass bunker, pretreatment (if required), feeding system, boiler itself with essential waste control and treatment systems like ash handling, ESP, etc.;
- Turbine island, which accommodates steam turbine with generator and essential systems;
- Balance of plant, both mechanical and electrical: heat rejection system, district heating heat exchangers, pumps, valves, piping, power evacuation system, instrumentation and control, etc.

In the majority of solid-fired biomass plants, one of the two types of boiler is employed:

1. Moving grate, or
2. Fluidized bed boiler.

Both types of boiler are equipped with waste handling systems that may contain:

- An ash handling system that evacuates unburnt particles and ash from the bottom of the boiler for further disposal, recycling or use;
- An exhaust air cleaning system to remove fly ash particles and other gaseous components that are considered dangerous or harmful for the environment. These systems at least employ electrostatic precipitators (or bag filters) to minimize the impact.

The mentioned systems are similar to those for coal-fired power plants discussed in detail in Chapter 8.

Steam turbines and BOP systems are similar to those within conventional or WTE power plants.

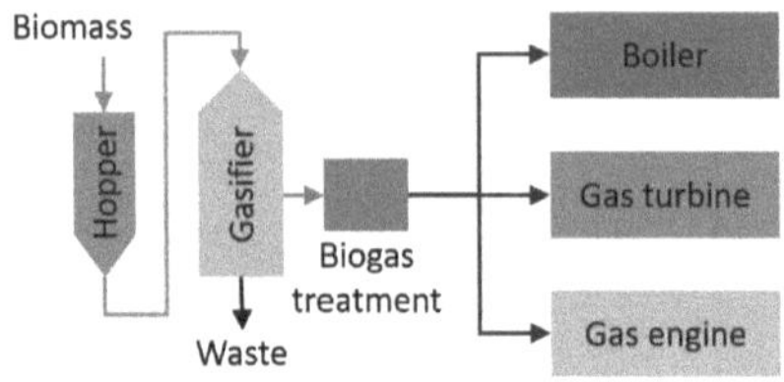

Figure 19.2 ***Simplified process diagram of integrated gasification options.*** Biomass is delivered to the hopper which feeds the gasifier. Biogas is treated to match the requirements of an incineration system: boiler, gas turbine or gas engine.

19.3.2 Integrated gasification

A biomass plant with integrated gasification system may have various configurations, as schematically presented in Fig. 19.2. The process starts with the biomass delivered to the hopper, which feeds the gasifier that treats the feedstock to produce gaseous fuel. After special conditioning within the gas treatment system, ready-to-burn gas can follow several paths depending on the plant type:

1. Gas can be fed to the boiler and burnt there to generate steam that will be used in a steam turbine. This configuration is similar to that operated on solid biomass.
2. Gas can be fed to gas engines, for instance, reciprocating, that would power respective generators and feed electricity to the grid. Usually, such machines have sizes of several megawatts and can operate very flexibly.
3. Finally, gas can be used within utility scale gas turbine, usually with a modified combustion system as biogas or syngas may have lower methane content or much higher hydrogen. This can be further implemented either in simple cycle or even combined cycle configuration to increase efficiency.

Gasification and gas treatment are an important system that converts biomass into synthesis gas and conditions it according to the requirements of the combustion system of the boiler, gas engine or gas turbine. While boiler is usually less critical to gas composition and properties, gas engines and especially gas turbines are sensitive to the fuel and therefore require proper pretreatment before fuel is admitted to the machine.

There are various types of gasifiers that can be used to carry out conversion of biomass into syngas, for instance, fixed bed gasifier or fluidized bed gasifiers. However, all of them run the gasification process in several similar steps, always including:

1. Pyrolysis, which takes place at temperatures around 450–600°C, when volatile components are released from the biomass, and the reaction produces solid residue called char;
2. Oxidation, when char reacts with steam supplied to the gasifier (or reactor) or is burnt in the presence of air or oxygen, which results in release of combustible gases.

The obtained raw gas usually contains various impurities that may negatively affect the equipment and lead to corrosion, erosion or damages. These are usually condensible hydrocarbons, various particles, metal compounds, water, etc. Therefore gas has to undergo a treatment process that may include cooling of the gas to the required temperature, filtering some of the unnecessary solid particles, scrubbing, reduction or increase of pressure, etc.

Further utilization of syngas depends on the available technology. Usually, due to the size of a syngas operated power plant, smaller boiler units are implemented. The same logic applies to gas engines and gas turbines, which are usually smaller compared to large-scale heavy duty gas turbines. In many cases, gas engines, small heavy duty gas turbines and aeroderivative turbines are used. The latter are aircraft turbines implemented for power generation applications and can be up to 100 MW per unit. They normally boast high flexibility, fast start and ramp up rates and mobility (these machines can be easily supplied in container modules).

The efficiencies of electricity generation can reach the levels of 40% for simple cycle applications (e.g., gas turbine driven), while combined heat and power production can even lead to the dramatic numbers of 90%.

19.3.3 Biomass energy economics

Technology options largely determine the cost and efficiency of biomass power generation equipment, although equipment costs for individual technologies can vary significantly. The factors affecting this depend on the region, feedstock type and availability, and how much feedstock preparation or conversion happens on site [8].

Biomass energy is one of the technologies that gives dramatic variation in capital costs, as represented in Table 19.2. Despite a growing number of biomass energy projects, the information is limited and in many cases the type of technology implemented in a project is not disclosed. Hence, the estimates [8,9] suggest that for steam boiler technology with solid biomass combustion the project capital costs may differ from as low

Table 19.2 CAPEX and OPEX components for various types of biomass power plants.

Technology	CAPEX	OPEX		LCOE
	$/kW	Fixed, $/kW-year	Variable, $/MWh	USD/kWh
Steam boiler	450–2000+	100	5–15	0.04–0.08
Gasifier	2000–8000	100	5–15	0.04–0.08

Source: Based on [8,9], EIC and IIR databases and calculations.

as 500 USD/kW for capacity addition projects up to 2000 USD/kW and higher. For integrated gasification biomass plants, the costs are much higher and usually vary from 2000 USD/kW up to 8000 USD/kW. The less expensive projects are executed in Asia and South America, while the more expensive ones are in the developed countries, especially in Europe.

Fixed operations and maintenance costs for bioenergy power plants typically range from 2–6% of total installed costs per year (or can be estimated at an average level of around 100 USD per kW-year), while variable O&M costs are typically relatively low, at around 5–15 USD per MWh. The fixed O&M costs of larger plants are lower per kW due to economies of scale, especially for labor. Variable O&M costs are determined by the output of the system depending on its capacity factor and are usually expressed as USD/kWh [8].

Same problems are applicable to LCOE calculation. Averaged estimates suggest [8] the range from 0.04 USD/kWh in India up to 0.08 USD/kWh in Europe, reflecting more advanced technologies. The lowest cost can be witnessed at sites where biomass is readily available at lowest price (e.g., industrial processing sites of forestry or agricultural production).

Biomass power plants vary dramatically in capacities because of the amount of available fuel, its utilization for residential and rarely utility scale plants and applications. Thus, commercial solid biomass power plants may have capacities around 100 MW with the efficiency ranging from 20–40% for electricity generation up to 70–90% for CHP applications. Large scale gasification may go beyond a 100 MW threshold in capacity, however, usually for heat production. Still, biomass plants can be competitive to conventional thermal power plants in efficiency, and well exceed many emerging technologies in capacity. It is important to mention that many technologies are still active in the research, development and demonstration stages for reaching enhanced efficiencies and reduced costs.

19.4. Biomass energy sustainability questions

While biomass power plants are usually considered CO_2 neutral from an accounting standpoint, this is not exactly the truth. Even though the amount of CO_2 released during the biomass energy conversion process is compensated by the absorption of CO_2 by that same biomass from the atmosphere, it assumes only operation-related emissions. Emissions from construction and decommissioning phases, as well as from biomass harvesting, treating and transporting must also be considered [2]. The levels of these emissions, together with ammonia and nitrogen oxide, which are mostly due to fertilizers' production, are quite significant, while the emissions from construction are relatively low. Thus estimated lifecycle CO_2 emissions for biomass also vary depending on many factors. The values lie between 16 and 74 g per kWh for electricity generation biomass applications [10].

Another important sustainability point to be considered for a biomass power plant is its relative footprint, which includes the power plant site itself together with the energy crop fields. This, as well, can be twofold:

- Pure energy crops for fuel (in solid, liquid or gaseous form) production will be fully attributed to the biomass power plant's footprint;
- Residue feedstock footprint, which is left after food production, shall not be attributed to the biomass plant, which in a sense recycles waste and recovers energy out of it.

It is therefore important to ensure that energy crops do not cannibalize the food production industry by occupying land. This is achieved by means of joint food and energy crops, or by increasing the biomass production based on the selection of appropriate genotypes of crops to enhance their energy production per unit area. Importantly, scientists prefer to base energy crops on those plants that are able to grow in poor soils and adverse climatic conditions, which are not appropriate for food crops. Similarly, another goal is to minimize water consumption of crops, required amount of fertilizers, and care.

Biomass power generation is a large consumer of land, water, and nutrient from the soil. The race towards much higher crop yield, both for energy use or combined food and energy fuel production, may lead to deforestation and loss of natural CO_2 sinks, degradation or complete loss of biodiversity, contamination with fertilizers, potential erosion of soil, and finally loss of agricultural productivity. Even though it is a possible negative scenario, bioenergy deployment should be executed with caution.

However, the overall idea of using natural fuel producers is promising. Various innovative approaches are now under research, for instance, natural "facilities" that consume water, sunlight and CO_2 and produce liquid fuels with the release of by-product — oxygen. In the ideal case, such plants should be suitable for periodic release of this fuel without significant damage.

Having the potential of being CO_2 neutral, biomass power generation still has a way to go, but the application of innovative ideas may well change this part of energy system to make it a full participant of sustainable energy system. The ongoing climate change may affect overall biosphere and hence food and energy crop production in the long run, which shall be closely observed and carefully studied.

References

[1] International Energy Agency (IEA). Technology roadmap. Delivering sustainable bioenergy. France: OECD / IEA; 2017.

[2] Guerrero-Lemus R, Martínez-Duart J. Renewable energies and CO_2: cost analysis, environmental impacts and technological trends – 2012 edition. Lecture notes in energy. Springer. ISBN 9781447143857, 2013.

[3] Twidell J, Weir T. Renewable energy resources. Taylor & Francis. ISBN 9781317660361, 2015.

[4] Tester J, Drake E, Driscoll M, Golay M, Peters W. Sustainable energy: choosing among options. MIT Press. ISBN 9780262304634, 2012.

[5] Richardson Y, Drobek M, Julbe A, Blin J, Pinta F. Biomass gasification to produce syngas. In: Recent advances in thermo-chemical conversion of biomass. Elsevier; 2015. p. 213–50 [chapter 8].

[6] San Miguel G, Makibar J, Fernandez-Akarregi A. New advances in the fast pyrolysis of biomass. Journal of Biobased Materials and Bioenergy 2012;6:1–11.

[7] Bridgwater A, Meier D, Radlein D. An overview of fast pyrolysis of biomass. Organic Geochemistry 1999;30:1479–93.

[8] IRENA (International Renewable Energy Agency). Renewable power generation costs in 2017. Abu Dhabi: IRENA; 2017.

[9] Black & Veatch. Cost and performance data for power generation technologies; 2012.

[10] Renewable energy sources and climate change mitigation. Special Report of the Intergovernmental Panel on Climate Change. New York, NY: Cambridge University Press; 2011.

CHAPTER TWENTY

Geothermal energy

Contents

20.1. Geothermal resources, current use, and potential

As discussed in Chapter 2, the Earth possesses enough thermal energy from the core and mantle, and contributed by radioactive decay within the crust. While the temperature gradient of the Earth's crust can be assumed as 30°C per kilometer of depth, there are still some areas where it can be up to 150°C. The latter are the most promising areas as they contain far more energy that can be utilized by humankind, however, these regions are limited to a few areas around the world.

Most of the accessible geothermal regions lie on the boundaries of the tectonic plates. When magma moves upwards, it brings significant amounts of thermal energy from the lower crust and mantle. However, most of these

Sustainable Power Generation
https://doi.org/10.1016/B978-0-12-817012-0.00034-7

active zones are within ocean basins and hence are unsuitable for geothermal development. A notable exception is Iceland, Kamchatka peninsula in Russia, and some other limited parts of the world, where recent spreading of the plates and associated volcanic activity provided these regions with rich and easily accessible geothermal resources. Still, rifting of the Earth's crust can also occur in other continental blocks, for instance, in East Africa, where they result in young volcanic activity and host multiple geothermal systems [1].

Moreover, continental and oceanic plates converge through a subduction process, when one plate — usually ocean plate — is pushed under a continental one, since the former is denser. This process causes melting near the leading edge of the subducted plate, which results in a series of volcanoes parallel to the boundaries of the interaction of these plates. Consequently, if the two continental plates collide, they crumple and form mountains. On top, plates may slide relative to each other, which can result in extensional troughs like the Salton Trough of Southern California. This process is also associated with volcanic activity. In some areas volcanic activity takes place far away from the plate boundaries at the so-called "hot spots" [1].

Finally, there are some areas with no plate boundaries but the continental crust is thin enough, as it is supposed to undergo tensional spreading, and therefore providing elevated heat flow. This process is usually not accompanied by volcanic activity.

It is therefore evident that commercial use of thermal energy that comes from the Earth shall be done in the areas where the heat is accessible, i.e., not so deep. All modern geothermal production is currently restricted to those areas where the resource is sufficiently hot, accessible, contains enough liquid that is heated up down the Earth (such a liquid is called geothermal fluid), and can support sufficient productivity over a long period of time.

This way, in general terms, geothermal energy is the thermal energy stored mainly in the Earth's crust [2] as it is the most accessible for the present level of technology. In most areas, this heat is generated a long way down and is mostly dissipated by the time it reaches the surface, but in some places the geothermal resources are relatively close to the surface and can be used as non- or low-polluting sources of energy. The available hotspots include the western part of the USA, west and central Eastern Europe, Iceland, multiple place in Asia and New Zealand, and Eastern and Central Africa [3]. The general idea of the geothermal energy utilization is to access this heat and implement the process to transform it to other usable energy like hot water, steam or electricity in an economically efficient way.

At present, the estimated identified geothermal resources using today's technology have the potential of 35–73 TW_e of electrical generating capacity [1].

20.1.1 Types of geothermal resources

While all of the geothermal resources are still based on the heat of the crust, there are several types available for utilization. These include:

- *Hydrothermal* or hot water geothermal resource, which assumes that precipitation from the surface penetrates through the crust faults goes down to the layer of permeable rock where it is heated from the mantle and tends to rise naturally due to its lower density. If this hot water encounters a major fault it rises until it reaches the boiling point and therefore flashes into steam, which may result in hot springs, geysers, etc. For the commercial project it is therefore important to locate artificially drilled wells to control this release. These resources are usually located in seismic areas with multiple natural faults;
- *Hot dry rock* (HDR), which is the deep hot, but dry rock. The HDR based geothermal system would recover the Earth's heat from hot, but dry regions through the closed-loop circulation of pressurized fluid. The pressurized fluid is injected from the surface to open the joints within the rock to increase its permeability and create an artificial reservoir. This fluid absorbs thermal energy from the rock surfaces and is then extracted from the production well for practical use. This approach is now under development within the enhanced geothermal systems (EGS);
- *Geopressured resources*, which are deep reservoirs with geofluids which are trapped is sand layers under an impermeable layer of caprock. This fluid is usually water that contains dissolved natural gas. This closed fluid heats up and makes a high-pressure environment with temperatures more than 200°C. If the reservoir is large enough, it can produce hot geothermal fluid for quite a long time;
- *Low-temperature resources*, which are hydrothermal resources with temperatures usually below 150°C. Their applications include power generation, direct heating, drying, agriculture, etc.

As of 2014, hydrothermal resources are the predominant geothermal systems that have been developed commercially for electricity genera-

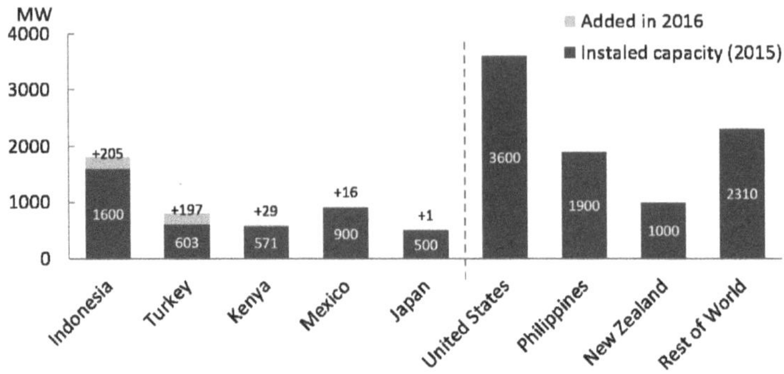

Figure 20.1 ***Geothermal power capacity as of 2015 and the largest capacity additions in 2016.*** The largest commercial geothermal capacities in 2016 were commissioned in Indonesia, Turkey, Kenya, Mexico, and Japan. *Source: adapted from [5].*

tion [4] as they are the most accessible, and the required technology is the most developed.

20.1.2 Status and installed capacities

Geothermal resources provide electricity and thermal energy services (heating and cooling). In 2016, the estimated electricity and thermal output from geothermal sources was 567 PJ (157 TWh), with each providing approximately equal shares. Some geothermal plants produce both electricity and thermal output for various heat applications [5].

An estimated 0.4 GW of new geothermal power generating capacity came online in 2016, bringing the global total to an estimated 13.5 GW. Indonesia and Turkey were in the lead for new installations. Kenya, Mexico, and Japan also completed projects during the year, and several other countries had projects under development. The countries with the largest amounts of geothermal power generating capacity at the end of 2016 were the United States (3.6 GW), the Philippines (1.9 GW), Indonesia (1.6 GW), New Zealand (1.0 GW), Mexico (0.9 GW), Italy (0.8 GW), Turkey (0.8 GW), Iceland (0.7 GW), Kenya (0.6 GW), and Japan (0.5 GW), as presented in Fig. 20.1 [5].

The uses of geothermal energy for electricity generation or heating needs depend on the temperatures. Low and moderate temperature at less than 80–90°C or between 90°C and 150°C can be used for the heat directly, and the highest temperature resources (above 150°C) are suitable only for electric power generation [3]. Today's total global geothermal electricity generation is approximately 10 700 MW, with nearly one-third in the USA (over 3000 MW), and the next biggest share in the Philippines (1900 MW) and Indonesia (1200 MW) [3].

20.2. Types of geothermal power plant

Most high-temperature geothermal fields are liquid-dominated, which means that they provide compressed, hot liquid water filling the fractured and porous rock of the reservoir down the Earth. In liquid-dominated geothermal systems used for electrical production, hot pressurized water flows into the wells from the reservoir, and the pressure decreases as the water moves upwards towards the surface. Once the saturation pressure corresponding to the geofluid temperature is reached, the fluid begins to boil or *flash*, creating a two-phase, liquid–vapor mixture that goes through the upper section of the well [1].

In a few geothermal fields, the wells produce only steam, therefore the extracted pressurized geofluid has only one phase — vapor, so that the field is called vapor-dominated. In such a favorable case steam can be used directly in the process.

There are also multiple cases, when the geothermal field has lower temperatures below 150–175°C. Due to a large concentration of dissolved chemicals within the geofluid, they will deposit at the flash point and restrict the flow within the well and piping. Therefore, these geofluids are artificially kept at higher pressure to avoid saturation, and its thermal energy is absorbed through heat exchangers.

This way, depending on how the geothermal fluid behaves and transfers its thermal energy, there are three major geothermal power plant technologies:

1. *Flash*, when the two-phase fluid is separated into water and steam phases once it reaches the saturation point (as pressure drops at the top of the well), and the vapor is then used further in the process. Flash technology can be implemented in several steps and consequently named as single-flash, double-flash, etc.;
2. *Dry steam* that uses steam from the well directly in the process; and
3. *Binary cycle* that uses heat transfer equipment to transfer heat from the hot geothermal fluid (yet of lower temperature) to other working fluid.

20.2.1 Single flash system

Single-flash technology means that the geothermal fluid coming from the wells undergoes a single flashing process, that is, the process of transitioning

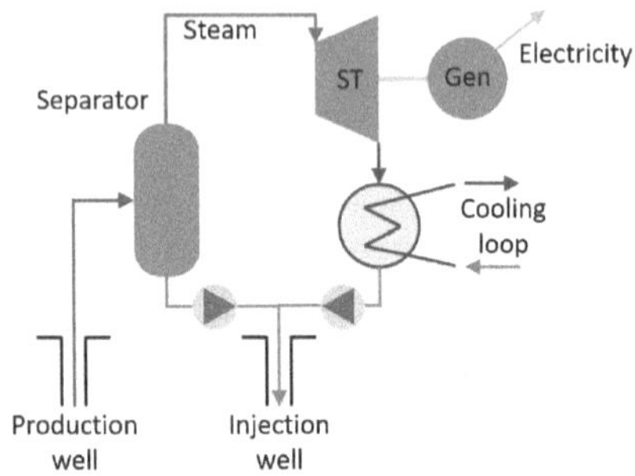

Figure 20.2 ***Simplified cycle of a single-flash geothermal power plant.*** Production well delivers a water–steam mixture to the separator, which separates brine from the steam. Steam is fed to the steam turbine (ST) that runs generator (Gen), and the condensate is pumped back to the injection well.

from a hot pressurized liquid to a two-phase mixture of liquid and vapor as a result of lowering the geofluid pressure below the saturation pressure corresponding to the fluid temperature [4].

A simplified heat balance diagram of a single-flash geothermal power plant is presented in Fig. 20.2. Geofluid from the production well PW is supplied through the gathering system to the steam separator, which separates steam from the two-phase fluid. While the liquid called the *brine* is pumped back to the injection well, dry steam is fed to the steam turbine. Within the steam turbine, steam expands and cools down, while converting its energy into the energy of the rotating shaft of the turbine. The latter is coupled to the generator that converts mechanical energy into electricity.

Cold steam from the steam turbine is further fed to the condenser, where vapor cools down and turns back into liquid due to the continuous supply of cold cooling water from the heat rejection system (e.g., cooling tower). This liquid is finally pumped back to the injection well to close the cycle.

20.2.2 Double and triple flash systems

After the flashing process inside the separator, the fluid can still remain hot and possess enough energy for extraction. *Double* and *triple-flash* plants are a logical extension of the single-flash process. These systems can produce some combinations of steam of various parameters, e.g., high (HP) and low-pressure (LP) as schematically shown in Fig. 20.3 for a double-flash configuration. While thermodynamically a theoretically infinite number of flashes would result in the highest power, the physical and economic

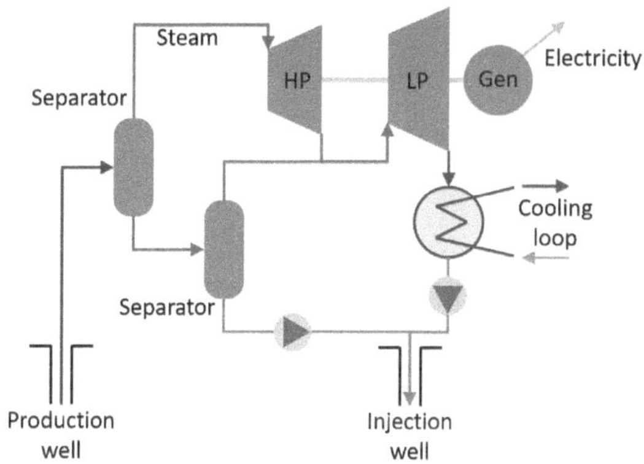

Figure 20.3 ***Simplified cycle of a double-flash geothermal power plant.*** A cascade of two separators is used to separate brine from the mix and feed high and low pressure steam to the respective ST section.

constraints, and the diminishing returns of increasing the number of flashes, limit current plants to no more than three flashes [6].

The process, as schematically shown in Fig. 20.3, starts at the production well and follows the flash cycle. While the HP steam from the first stage of separation is fed to the steam turbine, the high-pressure hot brine is fed to the second separation stage. The second flash stage separates steam that is further fed to the low-pressure stage of the steam turbine. Finally, if the brine after the second flash stage possesses enough energy, it can be further used in a third separation (within a triple-flash process), otherwise the brine is pumped back to the injection well to close the cycle.

Depending on the type of heat rejection system (condenser and cooling tower, which are a part of mechanical balance of plant), the brine available for re-injection is usually limited to 80–85% of the total extracted from the production wells [1].

20.2.3 Dry steam system

If the geothermal resource can supply dry steam of a reasonable temperature and pressure, it can be separated from particles and further fed directly to the steam turbine, as illustrated in Fig. 20.4.

The process of a *dry steam* geothermal system is fairly straightforward: dry steam from the production well is fed through the conditioning systems like particle remover and mist separator (to remove mist and droplets of fluid) directly to the steam turbine, where its energy is converted to

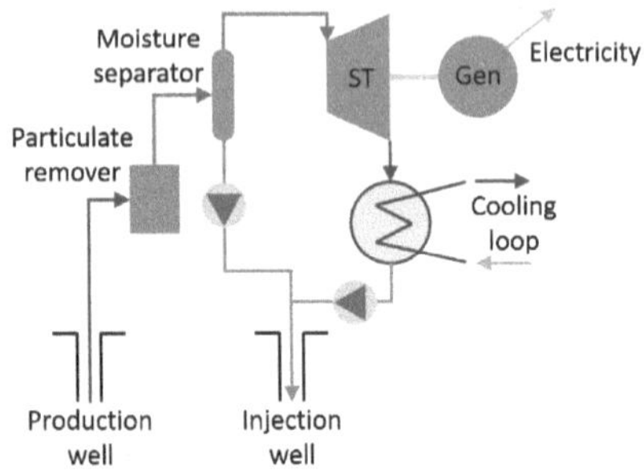

Figure 20.4 ***Simplified cycle of a dry steam geothermal power plant.*** Dry steam from the well is conditioned through particulate remover and moisture separator and fed to the steam turbine.

mechanical energy of the rotating shaft. Cold steam is then fed to the condenser, after being cooled down, is pumped back to the injection well.

It is important to mention that steam quality is crucial to reliable long-term operation of a geothermal steam plant and especially steam turbine. To provide clean, high-quality steam to the machine, dry steam geothermal power plants employ steam treatment facilities and equipment to remove chemicals, compounds, liquids, and solids that could result in scale formation, corrosion, or erosion damage to the steam turbine and other plant equipment [6].

Dry steam systems are much simpler compared to flash-type geothermal power plants as they have less equipment, for instance, no separators, have higher efficiency and better economics of electricity generation. However, due to the very limited dry steam resources, they are very rare. Moreover, dry steam systems have roughly 15–20% of condensed geofluid available for re-injection as most of it is wasted in the heat rejection system (e.g., wet cooling towers) and dissipated to the atmosphere [1].

20.2.4 Binary cycle system

If the geothermal fluid that reaches the surface is not hot enough to produce steam, or it has to be kept in liquid state to avoid deposits of the dissolved minerals, its energy can still be used for electricity generation in a *binary cycle system*, which typically uses geofluid at a temperature below 150–175°C. The hot geothermal fluid is fed into a special heat exchanger called evaporator, as schematically represented in a heat balance in Fig. 20.5, where its heat is transferred to a secondary working liquid, for instance, isopentane, which boils at a much lower temperature. The isopentane "steam" is then

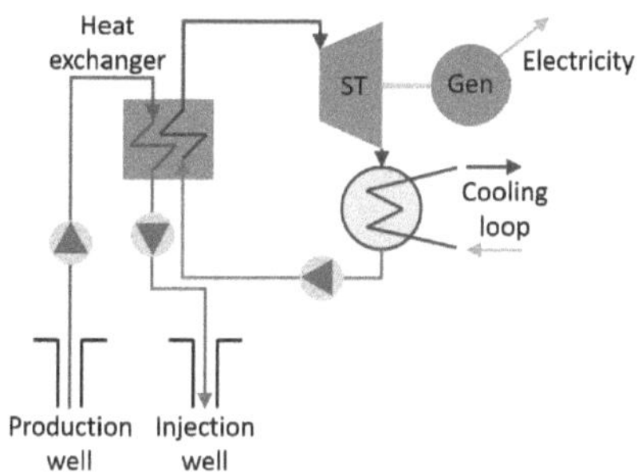

Figure 20.5 ***Simplified cycle of a binary geothermal power plant.*** **Production well supplied geothermal fluid to a heat exchanger that transfers thermal energy to the working fluid of a separate cycle.**

fed to the special type of a steam turbine and then condenses back to its liquid state, which closes the second cycle.

Organic Rankine cycle and Kalina cycle are the most common in binary plants. For the organic Rankine cycle, a high molecular weight organic fluid with a boiling point well below that of water is used. For the Kalina cycle, a working fluid consisting at least of two components (e.g., water and ammonia) is employed, and the ratio between them varies in different parts of the system to obtain higher reversibility and, therefore, to increase the total thermodynamic efficiency of the system [7].

The advantage of such a plant is that the geothermal fluid loop and secondary fluid are separated, allowing for utilization of lower temperature geothermal fluids. Moreover, all geofluid is pumped back to the reservoir and therefore eliminates geofluid emissions to the atmosphere and maintains the fluid supply to the geothermal system with a properly designed injection system [1]. This is extremely important as it maintains the production of reservoir and does not require any additional water supply, which is especially beneficial for areas with scarce water resources.

Similarly to a conventional thermal power plant, in order to optimize plant performance, the heat exchanger can consist of several parts, including:

- Economizer, to preheat the secondary fluid;
- Evaporator, to convert secondary fluid to steam, and
- Superheater, to produce steam of higher pressure and temperature.

Still, binary plants have a higher auxiliary power load to run pumps (which maintain the geofluid cycle) and heat rejection system (to run

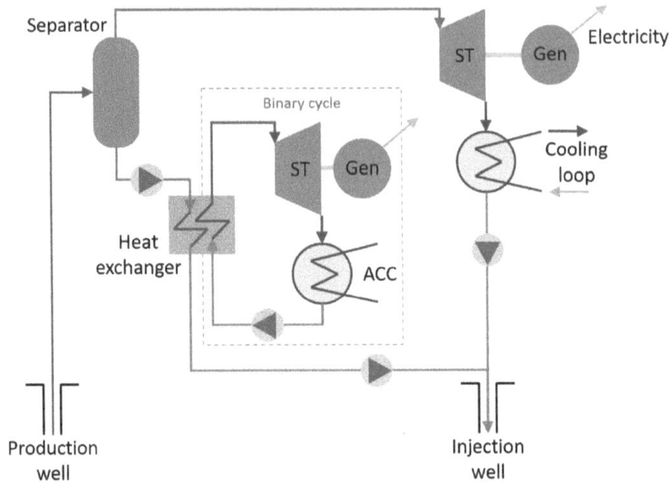

Figure 20.6 ***Simplified cycle of a mixed type geothermal power plant.*** The hot geothermal fluid is fed to the separator where it flashes and the steam runs the steam turbine of the main cycle. The hot fluid from the separator transfers its energy in the heat exchanger to the binary cycle working fluid. Here ACC denotes air cooled condenser.

cooling water pumps or ventilators if mechanical draft cooling towers or air-cooled condensers are used).

20.2.5 Mixed and combined cycles

Depending on the available resources, there have been approaches to combine various cycles together in one geothermal power plant to enhance its performance and efficiency by converting as much energy as possible from the geothermal fluid.

A *mixed cycle* uses a flash cycle together with a binary technology in a manner where the two units can operate independently. The simplest example would be that of a single-flash plant with a steam turbine harnessing the flashed steam, and a binary plant harnessing the hot brine, as shown in Fig. 20.6 [6].

To utilize even more energy, the heat dissipation system (that is, a condenser with an external cooling loop) of the major flash cycle can be replaced with the second binary cycle: cold steam that comes from the steam turbine still has enough energy that can be harnessed, as schematically shown in Fig. 20.7.

Although the mixed cycle may be purpose-built as an original single plant, it is common to encounter these as retrofits to an existing flash plant.

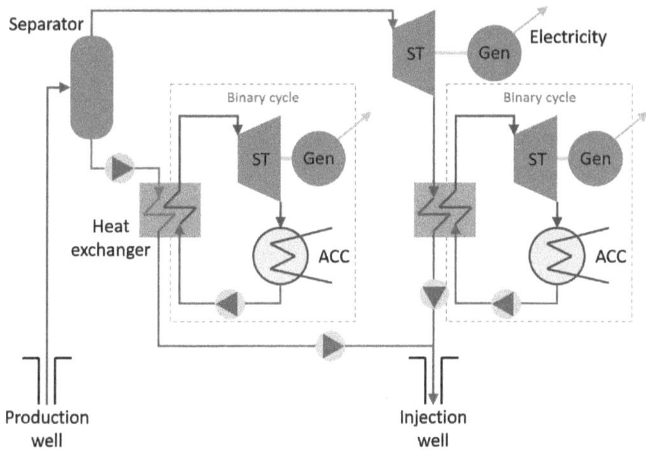

Figure 20.7 ***Simplified cycle of an optimized mixed type geothermal power plant.*** Similarly to the previous example, the current setup uses two binary cycles: the one that uses energy from the hot fluid of the separator, and the other that utilizes heat of hot exhaust steam from the ST of the main cycle.

The advantage of a retrofit is that the operation of the flash plant over a period of time prior to the binary addition allows the operators to characterize the chemistry of the brine from the separator more precisely. A more accurate determination of the appropriate injection temperature reduces the likelihood that scaling will occur and helps optimize the potential output of the binary *bottoming* cycle [6].

Although there are evident advantages in assessing the extra potential at single-flash plants due to the higher brine temperatures, even double-flash plants can have a margin in the brine temperature suitable for a bottoming binary cycle [6].

Combined or mixed cycle geothermal power plants can increase output and efficiency without additional extraction of the geofluid. However, the brine that is re-injected back to the reservoir would have lower temperature, hence this impact on the resource has to be thoroughly studied by geothermal specialists before project implementation.

20.3. Plant configuration

A geothermal power plant can be considered as a further development of a traditional steam power plant, where energy to produce steam is derived from the Earth. As this energy usually comes in the form of hot pressurized geofluid, the plant has to employ special equipment and systems to absorb energy from this hot fluid and convert it into electricity and, if required, heat. Therefore, any currently built and operated geothermal power plant has the following major equipment areas:

- *Gathering system*, to collect hot geofluid, in some cases process it, and deliver it to the major energy conversion equipment;
- *Power island* that consists of a steam turbine with its generator and essential systems;
- *Balance of plant*, which includes mechanical and electrical systems for plant's operation.

Depending on the type of geothermal power plant, these systems may differ in configuration, footprint and functionality.

20.3.1 Steam gathering system

The major component of a steam gathering system is a set of *production wells* that extract geofluid from the reservoir and bring it up to the surface. Depending on the type of the resource, the drilling techniques differ:

- For high-temperature, water-dominated reservoirs, methods similar to those used in oil and gas industry are used: conventional rotary drilling rigs with circulating drilling fluid to bring back cuttings to the surface and cool down the well.
- In steam-dominated or dry steam reservoirs, drilling conditions differ significantly: a well is usually drilled conventionally with drilling fluid to the point above the entry to the steam reservoir, and then high-pressure air is used instead of the drilling fluid (as the drilling muddy fluid would plug the fractures and prevent steam from coming to the well).
- For low-temperature water-dominated reservoirs, technology similar to water well drilling is used: using the same equipment, drilling fluid can be replaced with common water.

After the well is drilled, steel casing pipe is installed and cemented into the well. Cementing the casing properly is important as it provides stability of the production well and prevents mixing of various fluids at different levels and their uncontrolled release.

The configuration of the gathering system, for instance, for a flash type power plant, can take two basic approaches [6] as shown in Fig. 20.8:

- Wellhead separation (A), when each production well has its own separator station for initial separation of steam. Steam and brine are conveyed separately to the plant from the gathering field. This type of separation

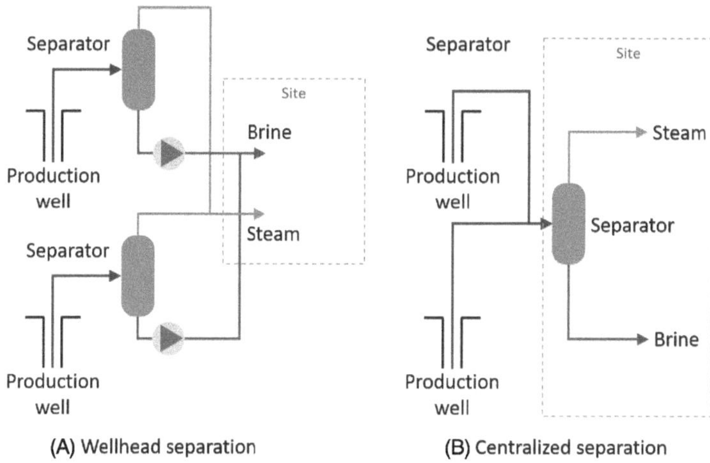

Figure 20.8 ***Major types of steam gathering system configurations.*** In a wellhead separation configuration (A), steam from the production wells is separately fed to the separators, and in centralized separation (B), steam is gathered from multiple wells and fed to a single separator.

results in a large number of separators with their own set of instrumentation and therefore large quantity of equipment in the field and multiple pipelines for steam and brine.

- Centralized separation station (B) relies on conveyance of two-phase fluid flow from the well heads to a joint separator located at site, which may lead to lower capital costs due to smaller amount of field equipment.

The system of pipes brings the geofluid from the production wells to the separation equipment or to special heat exchangers (for binary cycle plants). Multiple instrumentation is always installed to control the processes and maintain the systems.

Re-injection wells, which are located based on the resource model to minimize temperature impact, are drilled in the same way as production wells. They may also have special pumps installed on top or in the well for pumping back the brine.

20.3.2 Power island

Power island of a geothermal power plant accommodates major rotating equipment, namely steam turbines with their related generators and essential systems supplied with the major equipment, usually in the form of small containers or skids. While the power island of a geothermal power plant is similar to that of any conventional steam power plant, it is normally smaller in size and capacity, and in many cases machines differ in construction and material due to lower pressures and more aggressive working fluids.

Depending on the type of a power plant, the power island may include heat exchange equipment of the binary cycle or steam separators of a flash type plant.

20.3.2.1 Steam turbine and generator

Steam turbines implemented in a geothermal power plant use exactly the same principles of operation and energy conversion, but are usually much smaller compared to large units installed in fossil-fired steam power plants. Contrary to large utility units which consume dry superheated steam, geothermal steam turbines have to be able to accept dry steam or saturated steam from the separator. Steam generated from geofluid may have various fractions of minerals and gases such as hydrogen sulfide, so that steam turbine materials have to be properly chosen to prevent erosion, corrosion, and fouling. As a result, geothermal power plants require specially designed and fine-tuned steam turbines: unique high-grade materials to prevent or at least minimize corrosion and to prevent fatigue-related crack formations and fractures.

Depending on the configuration of a power plant, a steam turbine can employ from one up to three different pressure sections, with single and double flows of each section, and single or multiple casings. The turbine is further coupled either directly or through the gear box to the electric generator. Both machines are usually supplied by the same OEM or at least designed to operate together. While steam turbines are usually pre-designed for particular site conditions, generators are mostly taken from the standard product line.

Both steam turbine and its related generator have several essential systems supplied by the manufacturer and include, among other things:

- Lubrication and hydraulic oil system, to lift the rotor on the bearings and minimize friction losses;
- Set of steam safety shut-off and control valves for each pressure section, to control the incoming steam;
- Cooling system for generator and turbine (depending on the design);

- Instrumentation and protections, to control critical parameters like velocity, vibration, rotor displacement, temperature of bearings, and so on.
- Control panel and its connection to the overall plant control.

20.3.2.2 Balance of plant

Balance of plant systems of a geothermal power plant are traditionally divided into two major parts: mechanical and electrical.

Mechanical BOP systems include, among other components:

- Steam separators in a flash power plant, to separate the two-phase geofluid into vapor and brine. Also called as flash evaporators or flash tanks, these vessels may use natural gravity to separate fluid and vapor;
- Evaporators, or surface heat exchangers in the binary cycle, to transfer heat from the geofluid to the secondary working fluid;
- Heat evacuation system that consists of a condenser and a cooling tower. Contrary to traditional steam plants, geothermal plants may implement spray condensers and mix working fluid with the cooling water inside the cooling tower. However, despite slightly higher efficiency of a direct contact condensers, they would lead to increased auxiliary power consumption and higher operating costs, therefore surface condensers can be a preferred option. Both types of condenser are also equipped with a larger non-condensible gas removal system to remove the air in-leakage and the gases contained in the geofluid. This is implemented via steam or mechanical type ejectors;
- Environmental systems, to remove flammable or noxious gases like H_2S, which is usually contained in the geofluid. While these gases are generally vented to the atmosphere, some countries like the USA, Japan, and Italy have implemented regulations to prevent this release. There are several technologies to prevent release of these gases, mostly H_2S: regenerative thermal oxidation (burning), liquid redox methods (converting H_2S to sulfur to be disposed as solid), separation for further re-injection to the reservoir, etc.

Major electrical BOP systems are designed to properly transfer electricity from the plant to the grid and ensure operation of the whole plant (as all systems would require power and control). Therefore, these systems shall include such major components as:

- Power transfer system with the main transformer, switchyard or substation for safe and reliable connection to the grid, conductors in between the generator and transformer, and required protections;

- Instrumentation and control system, including DCS, to ensure safe and reliable operation;
- MV and LV distribution systems for power plant needs, to supply energy to motors, pumps, valves, etc.

20.4. Geothermal project development

Like any other project, the success of a geothermal project is influenced by a variety of factors. Though all of the projects are usually considered as unique due to their nature, they still have a lot in common also with other power plant construction projects.

20.4.1 Exploration and feasibility study

Each geothermal project starts with the preliminary study for geothermal exploration which consists of several major steps:

- Reconnaissance survey, to identify specific areas with potential geothermal resources and assign priority ranking to them. This usually includes deep analysis of available geological, geophysical, geochemical and other surveys (if any), study of maps and satellite and air photos, volcanic activity and other traces of thermal activity of the Earth in the region. This survey usually closes with the detailed report that indicates potential areas for geothermal project development;
- Conduct investigation at the top priority sites, to locate exact places for test drilling to confirm the resource potential. This study include more detailed physical, geological and chemical studies of the top priority sites to develop a model of the geothermal reservoir and allocate possible drilling sites;
- Investigation of drilling sites, which is conducted to localize and rank the reservoirs, their potential and size, preferably with surface survey only to keep the costs down;
- Exploratory drilling combined with geo-scientific surveying, to prove the reservoirs and evaluate their productivity and fluid chemistry. Usually small mobile drilling rigs are used.

The exploration stage finishes with the general understanding of the reservoir, its properties and estimated productivity, which would normally allow starting with the overall feasibility study of the project. This study begins with the appraisal drilling and detailed reservoir evaluation to come up with the revised reservoir model and properly allocated test wells and future production wells, acquire detailed geothermal data, and confirm resource to minimize the risks.

Based on this detailed information, the overall plant capacity and applicable technology type can be determined, and allow for detailed economic evaluation to come up with the final investment decision.

Project exploration and feasibility study is a long process. Generally, it is impossible to give a clear timeline for all sorts of geothermal projects as it depends on reservoir studies and its complexity, which is unique for each site. This process usually takes several years (from 2–3 up to 10 for the whole project) as it is aimed at minimizing future project risks and as it requires significant upfront investment.

20.4.2 Investment decision

Investment decision requires detailed cost and payback analysis based on market surveys and cost estimation from peer projects, as well as forecast of the energy and heat prices. Geothermal electricity development costs vary considerably as they depend on a wide range of conditions, including resource temperature and pressure, reservoir depth and permeability, fluid chemistry, location, drilling market, size of development, number and type of plants (dry steam, flash, binary or hybrid), and whether the project is a greenfield site or expansion of an existing plant. Development costs are also strongly affected by the prices of commodities such as oil, steel, and cement [8].

Traditionally, the CAPEX of a geothermal power plant is dominated by preparation and drilling costs, with a typical cost split presented in Fig. 20.9. Preparation and drilling costs may vary significantly from 30% up to 60% depending on the type of the resource and location, and the facilities and equipment costs depend on commodities prices (like steel), and may be affected by the location (delivery of bulk material, availability of skilled workforce, etc.).

The industry is sensitive to trends in oil and natural gas prices. Low oil and gas prices tend to reduce global demand for drilling rigs for oil and gas exploration, which can have a positive effect on the geothermal industry by reducing the associated costs of geothermal exploration and the development of new fields. Conversely, low fossil fuel prices in general, and natural gas prices in particular, tend to reduce the competitiveness of geothermal heat and power [5].

Total capital costs of a new geothermal power plant development project range from 1800 USD/kW$_e$ to 4800 USD/kW$_e$ for flash plant develop-

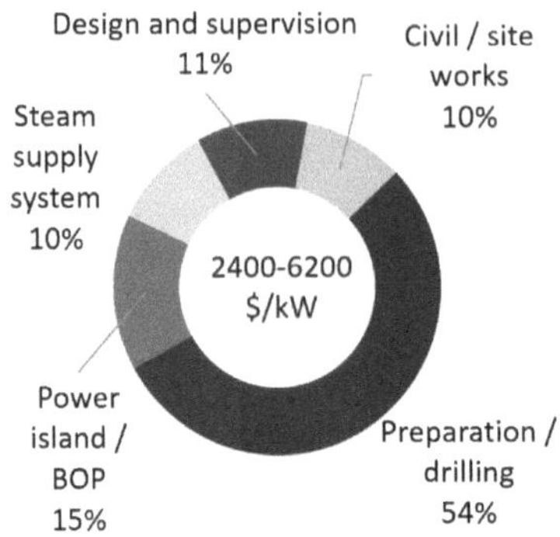

Figure 20.9 ***Capital cost breakdown for binary geothermal power plant.*** A typical binary geothermal power plant with 5 MW_e capacity. *Source: Based on [9] and EIC and IIR databases.*

Table 20.1 Estimated values for CAPEX and OPEX components for various types of geothermal power plants.

Technology	CAPEX, $/kW	OPEX, $/kWh	LCOE, $/kWh
Flash	2000–4800	0.009–0.031	0.07–0.12[a]
Binary	2400–6200	0.009–0.031	0.07–0.12[a]

Source: Based on [8,7,11] and EIC and IIR databases analysis.

[a] LCOE level depends on region and chosen technology.

ments and 2400 USD/kW_e to 6200 USD/kW_e for binary developments, as represented in Table 20.1. The highest investment costs for binary plants can be found in Europe in small binary developments of a few MW_e used in conjunction with low- to medium-temperature resources, with the lower investment costs for larger plants in South-East Asia or Central Africa due to greater capacities. The availability of renewable energy feed-in tariffs and the sale of heat from CHP development (e.g., for district heating purposes) can increase economic viability significantly [8,7]. Expectations of decreasing CAPEX costs are low over the next years and are at the level of 5% by 2020 [10].

Remark 20.1. It is not yet possible to assess reliable investment data for EGS because it is still at the experimental stage [7]. However, some studies estimate capital costs at the level of 9000–10 000 USD/kW of installed capacity [11].

Although the studies are considering geothermal plant average lifetime around 20–30 years, usually its operational life is considerably greater, reaching in some cases more than 50 years [7].

Operation and maintenance costs for a geothermal power plant, as represented in Table 20.1, are limited, as geothermal plants require no fuel. Typical O&M costs depend on location, type and size of the facility. They range from 9 USD/MWh$_e$ for larger flash and binary plants up to 25 USD/MWh$_e$ for much smaller binary plants. These costs, however, exclude well replacement drilling costs [10].

Levelized generation costs of geothermal power plants vary widely. On average, production costs for hydrothermal high temperature flash plants are estimated to range from 0.05 USD/kWh$_e$ to 0.08 USD/kWh$_e$, while for binary plants the range is estimated to be 0.06–0.12 USD/kWh$_e$. While the given numbers are considered as averaged ranges, the real LCOE numbers will differ from project to project and from region to region, as indicated in Table 20.1. While in some cases geothermal development projects give a highly competitive projection of LCO estimation, other plants may have higher upper limits for LCOE up to as high as 200 USD/MWh$_e$ in Europe for small plants and lower-temperature resources. Estimated EGS LCOE values are usually expected at higher levels around 100–300 USD/kWh$_e$ depending on the region [8,7,10].

Project finance modeling is based on the common financial instruments like NPV to calculate the project rate of return. While geothermal energy project is usually more capital intense for the construction due to risks associated with the geothermal resource potential estimation, the cost of electricity from geothermal power plant is usually higher compared to conventional sources of electricity. Therefore, many countries implement incentives to facilitate geothermal projects.

Feed-in tariffs are the most common direct incentive for geothermal investment. Countries in Europe use this mechanism to promote renewable energy projects to reduce greenhouse emissions. While prices represent a significant premium over wholesale power prices, they are subject to change as with the recent removal of solar price supports in Spain and Germany. In contrast, countries such as Indonesia and Kenya rely on feed-in tariffs to stimulate capacity investment by removing price risk and providing a premium to cover resource and country risks. These tariffs, commonly issued under competitive tenders, are embedded into multiple-year power purchase agreements (PPA) with a national utility and so have far less risk and higher bankability than the European renewable tariffs [12].

It is worth noting that tax exclusions on import duties and value-added levies are also important mechanisms to attract geothermal investment in many developing countries [12].

Typically, small renewable energy projects, as well as small geothermal projects, require equity sponsorship of 25–50% of the total value of the project. As the real or perceived risk associated with a project increases, lenders require a larger equity component to finance the project [13]. However, most of the energy projects, and geothermal power plants are among them, usually get debt financing through bank loans or government funding and subsidies. These can be done either through local or regional development banks or, in many cases, through the World Bank, supported by loans provided by the OEMs through export credit agencies, or carbon funds and loans (aimed at reducing carbon footprint). However, it is the finalized feasibility study report with the resource and risk assessment that is essential for proper finance allocation and approval for opening the bids for equipment.

There are multiple economic risk factors, which can be broadly distinguished into the following major categories [13]:

- Project financing options, including loan and grace period, interest rates and loan service charges, international currency trends;
- Geothermal power plant operating risks, including reservoir management and its sustainability, and plant operation and maintenance;
- Energy market factors that may include tariff policies, energy market stability, competition, energy purchase agreements (expiration and renewal terms), market forecasting and understanding of market development;
- Political stability, legislation and policy changes (including environmental).

After the investment modeling for the properly assessed resource has been developed, and the reservoir model and associated risks estimated and financing is confirmed, the project goes to the execution phase.

20.4.3 Project execution phases and lifecycle

After the financing has been confirmed, the project execution starts. It is possible to formally divide a geothermal project into the following key phases:

1. Preliminary design, to come with a key engineering solutions and prepare request for proposal documentation for potential suppliers;
2. Tendering and bidding, to identify key equipment suppliers and negotiate contracts for equipment and material supply, as well as for construction and other services;

3. Detailed engineering done by EPC company, separate engineering consultant, or technology integrator and OEMs (the structure of the deal depends on the project);
4. Construction of the project, which includes:
   - Site mobilization, preparation for civil and drilling works;
   - Production drilling to develop the geothermal resource (including production and injection wells);
   - Construction and erection of equipment, interconnecting piping for geofluid and brine, etc.
5. Commissioning and testing for guaranteed characteristics of the plant. This is usually prerequisite for commercial operation of a plant.

Project execution timeline, starting from the bidding phase till the COD, can normally be assessed accurately and may last around 2–3 years, depending on the complexity and long lead material cycles (e.g., steam turbine or other major equipment), availability of workforce and adequate risk management.

Progress on the development of geothermal energy around the world has been constrained, in part, by a lack of clear resource assessment standards. To help address this challenge, in 2016 new geothermal specifications were completed under the UN Framework for Fossil Energy and Mineral Reserves and Resources. The framework's objective is to harmonize standards for reporting geothermal resources in a manner similar to other extractive industries worldwide, for the benefit of investors, regulators and the general public [5].

Typically geothermal power plants are designed to operate for at least 30 years. The thermal efficiency of many geothermal power plants is rather low — up to 20–30%, as the easily accessible geothermal resources do not reach high temperatures compared to the steam temperatures in conventional thermal power plants. However, as the geothermal resource is almost constant over time, the capacity factor of geothermal systems to produce electricity can be large (up to 96%) [7].

20.5. Challenges and sustainability issues

Geothermal energy contributes a tiny proportion of the world's primary energy consumption. Even in electricity generation, geothermal produces less than 1% of the world's output [12]. Even though on a global scale this type of electricity generation contributes less than many other energy technologies, it still plays a crucial role in some countries' electricity generation mix.

This type of energy is a long-term renewable resource with the strong advantage of being a dependable baseload source, independent of the weather and, in the right circumstances, an energy store for energy delivered to the ground [14]. Contrary to other renewable power sources like solar radiation or wind, geothermal power plant does not depend on climate conditions and therefore can operate reliably with the capacity factor around 95%.

The geothermal industry continues to face challenges, mostly due to high risks of geothermal exploration and project development, the associated lack of procedures and established measures to mitigate these risks, and the financial competition from cheaper gas-fired energy. Yet the progress has been made with the development of new projects in several key markets, and industry leaders continued their partnerships to pursue new opportunities [5].

20.5.1 Effects on climate-relevant emissions

Despite the fact that geothermal energy is considered green and renewable, it still has certain environmental impact, as other energy technologies do.

According to recent studies [15], geothermal power plant lifecycle emissions include those due to constructing the plant itself and to evaporation of carbonic acid and other substances dissolved in geofluid drawn from the Earth's crust. These lifecycle emissions are estimated as approximately 15 g CO_2 equivalent per kWh of generated energy, whereas the evaporative emissions are estimated as 0.1 g of CO_2e per kWh for binary plants and up to 40 g CO_2e per kWh for other types of plant.

These numbers prove that geothermal energy is one of the cleanest forms of energy now available in commercial quantities. Using geothermal energy avoids the problems of acid rain and greatly reduces greenhouse gas emissions and other forms of air pollution [1]. These emissions mostly stem from the noncondensible gases that are carried in the geofluid in dissolved form [4]. In commercially operated plants, these gases would accumulate within the condenser and be removed and released to the atmosphere by standard means, e.g., ejectors or vacuum pumps, which consume certain power and therefore are calculated within auxiliary needs of the plant and decrease the net output.

Carbon dioxide CO_2 and hydrogen sulfide H_2S are the most common and prominent noncondensible gases in geothermal steam, but gases such as methane, hydrogen, sulfur dioxide, or ammonia can also be found, usually in very low concentrations [4].

Because the CO_2 concentrations in geothermal gases can be significant, some experts are still concerned about the potential greenhouse gas impact of open-loop geothermal power generation, although emission rates depend on local geology and operating conditions. For example, in some regions CO_2 emissions from geothermal power plants are significantly lower per kilowatt-hour than those from coal- or natural gas-fired plants and can range from less than 0.2 kilograms per kWh from flash steam geothermal plants to about 0.03 kilograms per kWh for dry steam geothermal plants (all open-loop). However, in some regions, by contrast, studies have found that an open loop geothermal plant can emit up to 1 kg CO_2 per kWh, or approximately 1200 tonnes per day, which can be due to high levels of dissolved calcite in the geothermal reservoirs [5].

While the numbers of CO_2 and CO_2-equivalent emissions differ greatly from region to region, they are generally lower than those of a conventional thermal power plant. If a country has enough accessible geothermal resources, this can be a good and economically feasible alternative to provide clean and reliable baseload electricity generation.

20.5.2 Geologic impact

Geothermal energy requires extensive drilling of production and injection wells, which may be expected to cause seismic activity. According to studies [6], slightly more frequent occurrence of localized microseismic activities, hydrothermal steam eruptions, and ground subsidence may be potentially influenced by geothermal power operation. Because geothermal operations usually take place in areas that are also tectonically active, it is often difficult to distinguish between geothermal-induced and naturally occurring events. However, these small seismic activities generated by geothermal projects are not felt by humans or not reported as hazardous.

While conventional geothermal projects are developed from near surface reservoirs, regions without volcanoes or hot spot areas seek geothermal energy in deeper layers of the crust down to 5 km drilling depths. This could result in some tremors that may be felt on the surface and lead to project closures as shown by example in Switzerland in 2006 and 2013 [6], where man-made earthquakes are related mainly to geothermal energy projects. In 2006, an earthquake in Basel with a magnitude of 3.4 was triggered by water being injected at high pressure into the well, and an earthquake with a magnitude of 3.5 occurred near St. Gallen in 2013 [16].

20.5.3 Water impact and land use

In general, geothermal energy has a limited water footprint compared to other thermal power plants, especially conventional, with 0.38 l/kWh as estimated by the Geothermal Energy Association [6]. This water is normally required for two major reasons: to maintain the cooling loop within the cooling tower and to add water to the brine if required to maintain a higher percentage of re-injection.

One of the most important lessons learned from more than 100 years of geothermal power production is the criticality of re-injection of the geofluid after its energy has been extracted. Ideally, 100% of the produced fluid should be re-injected to maintain reservoir pressure without cooling the production zones in the reservoir. Flash steam plants that use water cooling towers with make-up water supplied by the geofluid condensate cannot maintain 100% injection without augmentation by outside supplies. If flash plants were to be built with air-cooled condensers, 100% re-injection would be possible, but no such geothermal flash plants have ever been designed [1].

Moreover, the location of injection wells relative to production wells must be carefully determined in order to prevent the flow of the cooler injected fluid from prematurely cooling the reservoir fluid. This requires a reliable reservoir model that can predict the behavior of the resource under various scenarios for the expected lifetime of the field [1].

Moreover, there is a concern regarding drilling and using the hydrothermal resource as it may impact ground waters and change the water flows in the crust.

Geothermal energy has a relatively small footprint and does not require any mining activities and other large excavations compared to conventional coal-fired plants. An average geothermal plant is estimated to use only 2.47–19.77 ha of land per megawatt of installed capacity. Since minimal above-ground infrastructure is required for this type of plant, the remaining area of the concession can be shared with stakeholders for other activities, for example, as farming, forestry, national park, etc. Being found in natural habitats, these locations are sometimes converted into ecotourism sites [6].

20.5.4 Perspectives of geothermal power

The discussion about the technology and its positive and negative impact on the environment can lead towards the conclusion that geothermal electricity generation can provide a good alternative to conventional power generation technologies. The global resource base for geothermal energy is large and well distributed. Geothermal power, generated from these resources, has grown steadily for the last 100 years at an average rate of more than 8% per year [2]. This growth has been due to high-temperature hydrothermal resources utilized in flash-type power plants.

These plants can boast multiple advantages: they are relatively simple (as they are still based on an established conventional technology), safe and modular, have certain benefits in reducing greenhouse gas emissions and provide stable and reliable electricity, as well as small footprint.

We can assume that the 21st century can face the use of low-temperature geothermal resources within the binary cycles and further development of high-temperature and geopressured systems for electricity generation, but the speed of geothermal energy implementation is still extremely low, also for direct geothermal energy use for heating purposes.

Research efforts continue in Europe, Japan, Russia, Iceland, the USA, and elsewhere to improve technology, like enhanced geothermal systems for heat mining in deep, hot formations of low permeability, and to demonstrate the viability of operating systems at a higher commercial scale. If these efforts receive sufficient support, share learning, and are successful, they could make geothermal energy universally available within a few decades [2]. It is important to mention that geothermal energy has been an emerging topic along with other green energy technologies like wind and solar within major energy development scenarios, where it plays its role, though rather modest.

References

[1] Letcher TM. Future energy: improved, sustainable and clean options for our Planet. Elsevier Science. ISBN 9780080994222, 2014.

[2] Tester J, Drake E, Driscoll M, Golay M, Peters W. Sustainable energy: choosing among options. MIT Press. ISBN 9780262304634, 2012.

[3] Teske S, European Renewable Energy Council, Greenpeace International. Energy [r]evolution. A sustainable world energy outlook 2015. Greenpeace International; 2015.

[4] DiPippo R. Geothermal power plants: principles, applications, case studies and environmental impact. Elsevier Science. ISBN 9780081002902, 2015.

[5] REN21. Renewables 2017 global status report. Paris: REN21 Secretariat. ISBN 978-3-9818107-6-9, 2017.

[6] DiPippo R. Geothermal power generation: developments and innovation. Elsevier Science. ISBN 9780081003442, 2016.

[7] Guerrero-Lemus R, Martínez-Duart J. Renewable energies and CO_2: cost analysis, environmental impacts and technological trends – 2012 edition. Lecture notes in energy. Springer. ISBN 9781447143857, 2013.

[8] International Energy Agency (IEA). Technology roadmap. Geothermal heat and power. France: OECD / IEA; 2011.
[9] Serdjuk M, Dumas P, Angelino L, Tryggvadóttir L. Geothermal investment guide. GEOELEC project; 2013.
[10] International Energy Agency (IEA). Renewable energy essentials: geothermal. IEA; 2010.
[11] Black & Veatch. Cost and performance data for power generation technologies; 2012.
[12] World Energy Council. World energy resources 2016. London: World Energy Council. ISBN 9780946121588, 2016.
[13] Sukhgerel J. Geothermal project development with a case study from Tsetserleg, Mongolia. Report 2006;11:197–218.
[14] International Energy Agency Geothermal (IEA Geothermal). 2016 annual report. France: OECD / IEA; 2017.
[15] Jacobson MZ. Review of solutions to global warming, air pollution, and energy security. Energy & Environmental Science 2009;2:148–73. https://doi.org/10.1039/B809990C.
[16] Swiss Seismological Service, ETH Zurich. Geothermal energy and induced earthquakes. URL: http://www.seismo.ethz.ch/en/knowledge/things-to-know/geothermal-energy-earthquakes/geothermal-energy-and-induced-earthquakes/.

CHAPTER TWENTY-ONE

Ocean energy conversion

Contents

21.1. Introduction

Oceans can be considered as large accumulators of solar and gravitational energy which are converted into mechanical energy in the form of ocean streams, waves, tides, and other patterns of water movement. The mechanism of this energy conversion, as broadly discussed in Chapter 2, relies on fundamental energy sources, which are infinite on the humankind scale and therefore can be considered as purely renewable.

While ocean energy has been known and somehow utilized for many years, its commercial implementation has not yet been put on a global and considerable scale. With less than 1% of the total electricity generation, ocean energy of all sorts is still a future, yet promising source of green and sustainable energy. Very few commercial ocean energy facilities have been built to date. Of the approximately 536 MW of operating capacities at the end of 2016, more than 90% was represented by two tidal barrage facilities (a 254 MW plant in the Republic of Korea and a 240 MW in France) [1].

In general, ocean energy resources are contained in ocean waves, tidal range, tidal currents, ocean currents, ocean thermal energy, and salinity gradients [2,3]. These resources are vast but not evenly distributed; however,

Sustainable Power Generation
https://doi.org/10.1016/B978-0-12-817012-0.00035-9

there is always some source of ocean energy available at every coast. These various forms of energy can potentially be harnessed and converted into electricity.

Given the relative immaturity of ocean energy technologies versus most other energy technologies, much of the focus in terms of barriers to deployment has been on the level and type of energy innovation support for ocean energy [4].

It is important to mention that marine energy is probably among the least visible and least supported among emerging energy sources. Of course, most countries in which ocean energy technologies are currently being developed have renewable energy targets, although very few of them have specific policies to promote ocean energy. They mostly implement various mechanisms like capital grants or financial incentives to create early-stage opportunities in renewables generally, but almost never use market mechanisms or initiatives specifically for ocean energy development [3].

21.2. Energy from the ocean streams and tidal range

The ocean waters are continuously moving. The global oceanic currents, as discussed in detail in Chapter 2 are caused by several factors: changes in temperature, salinity difference, gravitational forces and some others. These moving masses of water possess enough energy that can potentially be harnessed, at least partially.

Contrary to deep water, the shallow or surface ocean streams are available for harnessing at the present level of technology. There are three major ways the shallow waters are moving [3,5]:

1. These are the cold and warm surface currents like the Gulf Stream or the California current. These ocean currents are the constant flows of water around the oceans and always flow in one direction and are mostly driven by winds, differences in water temperature, salinity, and density.
2. Strong currents are caused by the oceanic tides, which are the result of the gravitational forces of the Sun and the Moon. These forces cause periodic movement of the waters that result in oceanic streams that change their direction and level according to the positions of the Sun and the Moon.
3. Thanks to the gravitational impact, tides cause the vertical rise and fall of the water level which is accompanied by the inflow of the waters to the shoreline and its further outgoing flow. The change in the water

level is known as tidal range, leading to the creation of the water head — stored potential energy.

Overall, the steady movement of water, either continuous like oceanic currents or periodic induced by the tides, causes the water flow it its essence. Therefore a major approach of water stream or tidal range energy facility is to implement such a technology which allows capturing that energy of the flow and converting it into electricity.

21.2.1 Energy potential

The energy potential of tidal currents is typically located in areas with the greatest tidal range [4]. It further increases in the areas with favorable geographic conditions, e.g., narrow or shallow straights that increase the speed of the streams, but the overall tidal energy potential is quite difficult to estimate. The global estimates suggest the value of 3 TW, however, the technically harvestable part of this resource in coastal areas does not exceed 1 TW [6]. This number, however, shall be further proved for each country based on the local geography of the coast line: in some areas tidal range may exceed 15 m, while the others can capture only a few meters, if any.

Tidal stream energy potential is even more complicated to evaluate. It depends on the shape of the cost and the depth profile and is usually located in the areas with the highest tide range, but has not yet been properly mapped and estimated for most of the countries [4,6].

Still major studies suggest [4,6,7] that the amount of energy stored in the ocean streams and tides is high enough to compete with some of the conventional technologies and contribute to the green portfolio.

21.2.2 Major approaches

There are two major approaches to harness the ocean stream and tidal range energy [3–6]:

- Implementing the underwater equivalent of a wind turbine to capture the kinetic energy of the flowing water of the tidal and ocean streams. Because water is more than 800 times denser than air, the potential for power extraction from fast currents is tremendous. There are multiple concepts of more than 1 MW of unit capacity that have been widely tested. Some may look more like an underwater wind turbine installed, for instance, on the seabed or on a floating base moored by cables. For ocean currents these turbines are usually deployed in arrays similar to wind ones. Several different water current energy converter device

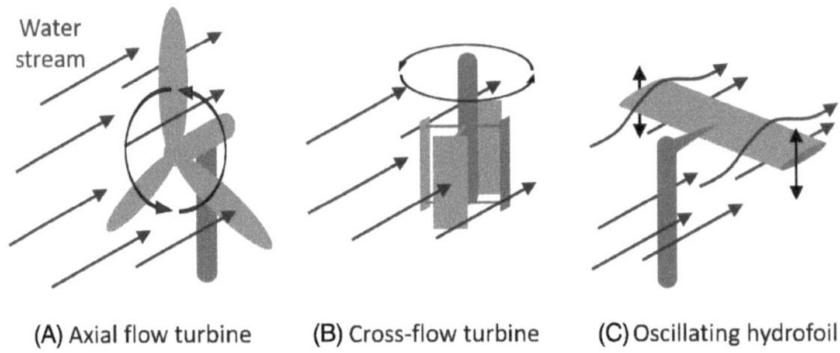

Figure 21.1 ***Tidal and stream flow energy converters.*** There have been three major technologies proposed to capture the energy of water stream: horizontal axis axial flow turbine (A), vertical axis cross-flow turbine (B), and oscillating hydrofoil (C).

technology concepts have been proposed and developed in recent years (see Fig. 21.1):

- Horizontal axis axial flow turbines are similar in design to wind turbines to capture the flow of water, however, their blades are smaller and move slower due to higher density of water. They usually have two or three blades.
- Vertical axis cross-flow turbines usually have two to three blades mounted vertically along the shaft to form a rotor.
- Oscillating hydrofoil or reciprocating device operates like an airplane wing: the control system changes the angle of the wings against the water flow creating lift and drag forces that make the device oscillate. This periodic movement is then fed into the power conversion system.

- To tap the tidal range requires building a dam across a suitable bay to capture the water level: once the tide starts, the high waters from the ocean create the water head with the captured part which can be used to power a hydroturbine of the bulb type. The same process can also operate once the captured water head remains high once the tide goes down, causing the flow through the turbine in the opposite direction. This type of power plant has the same logic as a conventional hydroelectric power plant but has cyclic operation.

All water stream devices require a support structure, either fixed to the seabed or floating. It has to withstand severe oceanic conditions and depends on the depth of the sea, geology of the seabed, and the available competency of construction. These structures can be similar to wind offshore.

Table 21.1 Typical tidal and ocean stream energy costs.

Technology	CAPEX, $/kW	OPEX, $/kW/Yr	LCOE, $/kWh
Tidal stream	3300–14 600	160–1160[a]	0.36–0.63
Tidal range	100–850[b]	–[c]	–[c]

Source: Based on [4–6] and calculations.

[a] For pre-commercial arrays of tidal current technology.
[b] Capital costs are subject to project nature and location, and may be significantly different from the reference numbers.
[c] No reliable data available.

Based on an overview of existing tidal current projects, 76% of all turbines are horizontal axis turbines and 12% are vertical axis turbines [6]. The devices can be either stand alone (limited to small plant capacity) or deployed in arrays similar to wind parks. The connection to the substation is usually done via an underwater cable. Still, the number of sites suitable for tidal power plants is limited by several factors [8], for instance, the length of the dams to close the bay that drives the capital costs up, the size of the head, proximity to infrastructure, and so on.

Contrary to tidal current and water stream technologies, tidal range power plant is similar to a hydroelectric power plant. To tap the range requires building dam which may be quite long to capture enough water potential. These plants usually employ low head two-directional hydroturbines.

21.2.3 Project economics and future perspectives

Tidal energy has been in operation since the 1960s in several countries, including Canada, France, China, and later South Korea. Still, there is little economic data available, mostly because the costs are too site-specific and it is difficult to make generalizations [6].

The development of purely commercial arrays of tidal current technologies is still in the demonstration phase, which drives the LCOE quite high: 0.36–0.46 $ per kWh (see Table 21.1).

Tidal range power plants are similar to hydroelectric power plants in construction with lower CAPEX due to the fact that the water head is relatively low and may be easier to construct. In some cases it is also combined with other infrastructure projects (like roads or safety dams) and therefore cannot be purely considered as a power project.

So far, tidal and ocean current energy utilization approach is under development. There is no shortage of energy conversion proposals for tidal streams, but many are unproven, most are very complex, and almost all are expensive [9]. While tidal range proved its commercial usability, capturing ocean streams still has to be economically proved. It still requires increasing efficiency of the units and decreasing installation costs (which can be driven down due to the development of offshore wind).

Moreover, ecological considerations would also play role, especially for tidal range plants as they close the bays and can significantly damage local ecosystems. Detailed research and investigation are required before project deployment followed by careful monitoring during operation.

21.3. Energy from the ocean waves

There are multiple types of wave that can be found in the ocean, for instance, caused by wind, earthquake, tides, etc., but the wave energy applies only to a certain type of waves — wind waves. *Wind* or *ocean waves* are derived from solar energy: the heating of the atmosphere leads to different pressure zones and global and local wind circulations. The winds blow over the ocean surface and generate surface waves which can propagate over long distances and deep water with very little energy loss until they reach land or shallow waters. Therefore wave energy can be regarded as a concentrated form of solar energy with the global oceans acting as a gigantic energy conversion system that transforms wind energy into wave energy [10,5].

21.3.1 Wave energy potential

As the ocean covers most of the Earth's surface, and global winds circulate over the waters, they generate waves almost everywhere. While the wind waves propagate over these vast distances with very limited energy loss, the substantial amount of their power reaches almost continuously any stretch of coastline open to the ocean [10]. Captured almost everywhere, waves are seen as an unlimited source of renewable and clean energy with a huge potential.

Global estimates for wave energy potential are still relatively uncertain [4,11]. The global theoretical wave energy resource capacity is estimated at the level of 3.7 TW and, excluding areas with poor wave climate and potential ice cover, are around of 3 TW [5,10]. This gives the rough value of 29 500–32 000 TWh per year, however, some estimates gave much lower values [12].

The technical potential is, of course, lower than the global theoretical potential and can be roughly estimated at the level of 500 GW_e or about 146 TWh/yr assuming that wave energy technologies would only be deployed in the 2% of the world's coastline that exceeds a power density of 30 kW/m [11].

Multiple regional estimates have been conducted to evaluate the regional and local wave energy resources, their technical potential and possible installed capacities for the next decades. Thus, Asia and Australasia receive the largest quantity of wave energy (around 6200 and 5600 TWH/yr, respectively), with South and North America also receiving impressive amounts of more than 4000 TWh/yr each. Western and Northern Europe performs moderately well given its relatively small size. Finally, Central America with 1500 TWh/yr and the Mediterranean Sea and Atlantic Archipelagos perform poorly given their mid-latitude position [4,5]. However, these numbers do not take geographical, economic or technical imitations into consideration.

While there is no final number regarding the wave energy potential, as all of the evaluations depend on varying assumptions about technical and economic feasibility and the ways how to measure and calculate the values, the final conclusions of these estimates are all the same: wave energy has great potential for its development and can supply a significant part of world electricity needs and contribute to sustainable power generation as one of the largest renewable energy sources. Still, global installed capacity of wave energy power plants were around 8 MW in 2017 according to [13].

All this poses an important question: what are the available and future technologies that can be economically used to generate green electricity from waves?

21.3.2 Wave energy conversion technologies

The development of wave energy converters (WEC) dates back to the 19th century when the first attempts to harness this source of energy were made, with some further scattered interest in the 20th century [5]. Over the past decades some countries started to deploy certain technologies to capture wind energy, which has been characterized by a number of different ideas and various concepts and technologies how to convert wave energy into electricity.

There are multiple approaches to how categorize WEC, for example, by their location [5,4,14] as illustrated in Fig. 21.2:

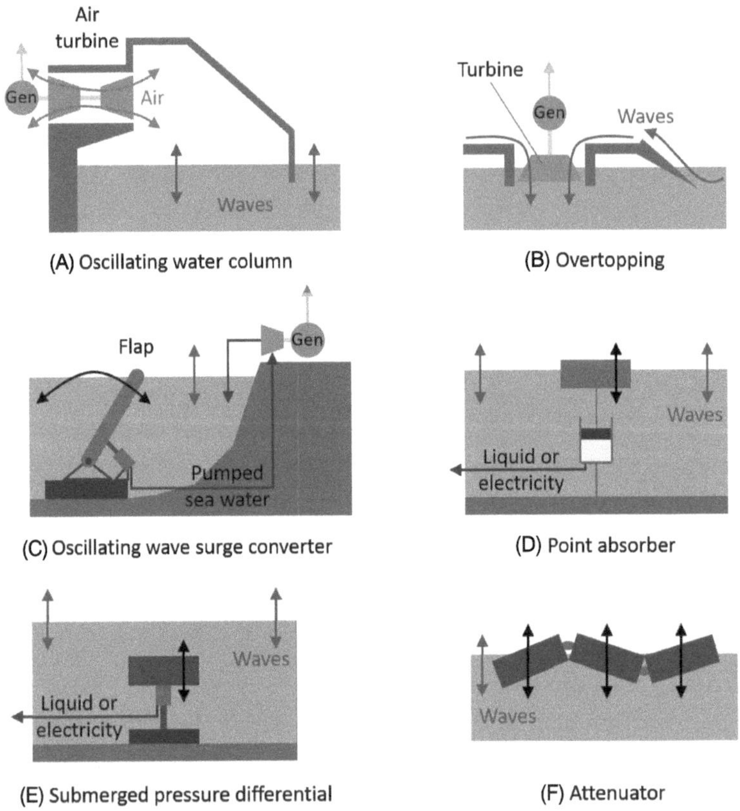

Figure 21.2 ***Major types of wave energy converter.*** Major technologies are: oscillating water columns (A), overtopping devices (B), oscillating wave surge converters (C), point absorbers (D), submerged pressure differential devices (E), attenuators (F).

- Onshore or shoreline devices are integrated into a natural rock face or artificial breakwater structure. They are close to the grid and are relatively easy to maintain. Located at the shallow waters where waves have lower energies, they are less likely to be damaged, however, this reduces the potential resource that could be captured by such plants. There are two major onshore technologies utilized for wave energy conversion:
  - Oscillating water columns (OWC), which use the oscillatory motion of a mass of water induced by a wave in a chamber to compress air to drive an air turbine as shown in Fig. 21.2(A). The water column thus acts as a piston on the air volume, pushing it through the turbine as the waves increase the water level in the chamber, and drawing it

as the water level decreases. This is one of the first types of wave energy converter. Currently, floating OWCs are tested to deploy them offshore;
  - *Overtopping devices* or *terminator WEC* (B), which convert wave energy into potential energy, stored in a reservoir and used to drive low-head turbines. The design of this device facilitates waves breaking on a ramp to be collected in a reservoir located above the free water level, so that it can flow downwards and run a low-head hydroturbine;
- Near-shore devices are located in waters which are shallow enough to fix the equipment to the seabed (e.g., via pinned pile foundation or gravity mass). This gives a suitable base against which an oscillating body can operate. But, similarly to onshore systems, shallow water location limits the potential. Here three major technologies are currently deployed:
  - *Oscillating wave surge converters* (OWSC), as schematically represented in Fig. 21.2(C), which exploit the surging motion of near-shore waves to induce the oscillatory motion of a flap. The moving flap powers the pump to deliver sea water to a water turbine of a Pelton wheel type. Water spins the turbine and is then released back to the ocean. These are mostly bottom-mounted devices, although some prototypes of floating OWSC are already under development;
  - *Point absorbers* (D), which exploit the relative motion between an oscillating body and a fixed structure or component, which can be either moored to the seabed or installed on the seabed through a large foundation mass. They are non-directional devices, as their performance is not affected by the direction of waves. The oscillation movement can pump liquid (hydraulic devices) or generate electricity directly (electric devices);
  - *Submerged pressure differential devices* (E), which are fully submerged devices that employ the hydrodynamic pressure induced by waves to force an upward motion of the device, which then returns to its starting position once the pressure differential is reduced.
- Offshore devices are located in deeper waters and are tethered to the sea bed. They can capture much higher energies but are more difficult to install and maintain, especially because of the severe operating conditions. Offshore WEC employ the following technologies:
  - *Attenuators*, which exploit the incoming wave power to generate an oscillatory motion between adjacent structural components (see

Fig. 21.2(F)). Waves are resisted by the hydraulic rams that hold the components and pump high pressure oil to drive motors. Attenuators are designed to operate offshore, and are commonly surface floating, although fully submerged devices have been proposed as well;

- *Bulge wave devices*, which use wave-induced pressure to generate a bulge wave within a flexible tube. As the bulge wave travels within the device it increases in size and speed. The kinetic energy of the bulge is used to drive a turbine at the end of the tube;
- *Rotating mass converters*, which exploit the relative motion of waves to induce the rotation of an eccentric mass contained within the device. As the mass rotates it drives an electrical generator.

All these technologies employ three major types of power take-off system for wave energy conversion:

1. Fluid power that may run air turbine, hydroturbine or piston-driven motor to power an electric generator;
2. Direct mechanical drive system, to power a generator, and
3. Direct electrical drive system that generates current.

Hence, according to the description of various technologies, each wave energy conversion plant consist of several components [11]:

- The structure and prime mover that captures the energy of the wave;
- Foundation or mooring that keeps the structure and prime mover in place;
- The power take-off (PTO) system by which mechanical energy is converted into electrical energy, and
- The essential control systems to safeguard and optimize performance in operating conditions.

21.3.3 Project economics

Wave energy is still at the early stage of its development, therefore the costs of deployment and electricity generation are indicative and may not properly resemble the potential of mass deployment of wave energy technologies. The present costs are seen to be higher than other emerging technologies. Table 21.2 shows the estimated capital costs for an average 10 MW wave farm: the broad range is due to a number of various technologies, location and varying site conditions, and lack of global experience. The latest estimates for Europe-based wave energy projects suggest that the PTO system accounts for 22%, installation is roughly 18%, O&M is 17%,

Table 21.2 Typical wave energy costs.

Technology	CAPEX, $/kW	OPEX, $/kW/Yr	LCOE, $/kWh
Wave	2700–18 100	64–1500	0.04–0.85

Source: Based on [4,5,11] and calculations.

foundation and mooring occupy 6%, and grid connection is around 5% of the total lifetime project costs [11].

The same logic underlies the high estimated O&M costs and the LCOE. The estimates show that the LCOE for wave energy was at the level of 400–850 USD per MWh in 2013 [11].

Currently many technologies are either at the later stages of the research and development stage, or undergo tests as demonstrators to come up with a final system that has proved its parameters. The estimated capacity of the projects that have been announced as being at various stages of development is estimated at more than 800 GW, but their commercial operation dates are not clear [4]. This means that the project economics is most probably too high and may well be optimized for serial facilities for commercial use.

21.3.4 Challenges and perspectives

One of the major drivers of wave energy is its vast potential across multiple countries and around the globe, its small visual impact on the shoreline and relative stability of the output [11], though some studies consider this resource as stochastic with accurate forecasts limited to around one week in advance [4]. Wave energy is usually positively perceived by the public, and the technology is approaching its commercial implementation. However, these technologies still require initial investments, most probably from the governmental funding, as the attractiveness for private investments is not yet clear. Moreover, as for any new power generation technology the regulation is not yet fully developed. This also addresses the licensing process and evaluation of the environmental impact of wave energy technologies.

Preliminary studies already suggest that this impact is relatively low and usually addressing the following issues:

- Changes in wave climate and patterns along the shoreline;
- Installation of the foundations on the bottom and floating devices may attract new marine life, but longer observations and evaluations are required to conclude if these changes are positive or may negatively affect local ecosystems.

- Some other concerns may include the impact of underwater noise (from installation and operation) on the sea animals, electromagnetic impact from the cables, etc.

Importantly, lifecycle GHG emissions from wave energy technology are estimated as being less than 23 g of CO_2 equivalent emissions [4].

21.4. Ocean thermal energy

Approximately 15% of the solar energy falling onto the ocean is absorbed and retained as thermal energy and stored as heat in the upper layers [4]. Therefore, the surface temperatures of the ocean waters can be significantly warmer than those several hundred meters deep as the thermal conductivity of water is quite low. The temperature gradient would allow employing a heat engine and generating electricity within the *ocean thermal energy conversion* or OTEC facility.

OTEC technology typically requires at least 20°C of temperature differential to work effectively. This means that cold waters of around 5°C are drawn from the depth of 800–1000 m to the surface, where the water temperature equals to 25°C. This difference is usually available within the belt that spans between 20–30° north and south of the equator and may be around 20–25°C. Normally, it is available all the year round with very limited changes.

According to [4,8,15], the estimated total energy resources that could be extracted with the OTEC technology without having a major impact on the thermal characteristics of the oceans range largely between 30 and 90 PWh. This huge potential is the largest among other ocean energy sources, however, the technological potential that could be practically and economically extracted is much lower.

21.4.1 Technology and cycle

There are two major cycles that are employed in an OTEC plant:

1. Closed cycle, as represented in Fig. 21.3(A), which uses fluid with a low boiling point, for instance, ammonia. The warm surface water transfers its heat to the working fluid within the evaporator, and the vapor powers the turbine to drive a generator. Cold water from the deep is used within the heat evacuation system: it is pumped through the heat exchanger (condenser) that condenses vapor back to the fluid.
2. Open cycle, as shown in Fig. 21.3(B), which pumps warm surface waters directly to a low-pressure container, which causes it to boil. The

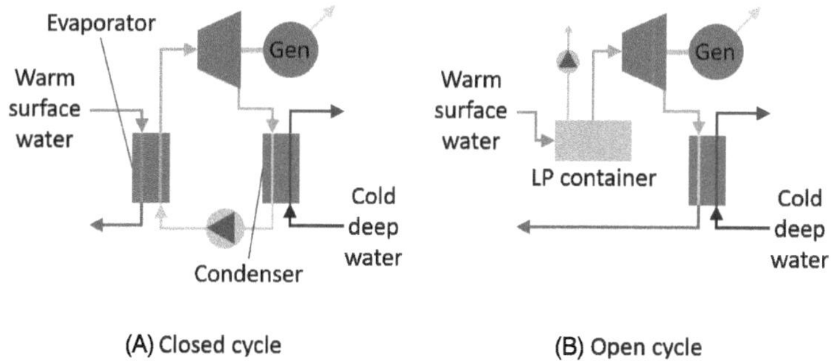

Figure 21.3 ***Schematic diagrams of OTEC cycles.*** There are two major types of OTEC cycle: closed (A) and open (B).

steam can be further fed to a low-pressure steam turbine that powers the generator. Then, it is condensed with the help of cold water from the deep. Moreover, the condensed water is fresh so that this type of OTEC facility can operate as a desalination plant.

There are also some hybrid cycles that use the benefits of the both: flash-evaporates seawater steam transfers its heat to the working fluid within the evaporator.

The design of an OTEC plant is subject to a number of technological challenges. According to [7], the cold water intake pipe that brings water from the deep has to be of a large diameter of around 7–9 m for a 40 MW plant to keep losses in pressure drop tolerably low. If located onshore, the cold water pipes may be much longer to bring cold waters from the depths of around 1 km. In the end, the overall net thermodynamics does not exceed 7% [4].

Moreover, these facilities have to be designed for severe oceanic conditions, especially if the floating approach is used. These platforms have to be properly moored with a dynamic positioning system to maintain its location and safety.

21.4.2 Economic perspectives

OTEC is a stable source of baseload power with the capacity factor up to 95% [4,16], while the heat engine allows varying the load profile, so that it can ramp up rather quickly.

OTEC technologies have been tested since the 1970s and are ready for commercial implementation, but the costs still remain high. Capital expenses for small-scale facilities of less than 10 MW may range around

Table 21.3 Typical OTEC costs.

Technology	CAPEX, $/kW	OPEX, $/kW/Yr	LCOE, $/kWh
Small-scale	16 400–35 400	–[a]	0.19–0.94
Large, installed	5000–15 000	–[a]	0.07–0.32
Large, floating	down to 2500	–[a]	0.07–0.32

Source: Based on [4, 16] and calculations.

[a] No reliable data available.

16 400–35 400 $ per kW of installed capacity, which makes these small project economically viable only if valuable by-products like fresh water are produced [16]. Larger facilities can offer much lower CAPEX of about 5000–15 000 $ per kW for installed facilities and even lower (down to 2500 $/kW) for floating ones as shown in Table 21.3.

According to some of the estimates [4], the LCOE for small-scale OTEC plants up to 10 MW ranges between 190 and 940 $ per MWh, however, if the facility were to be scaled up to several hundred megawatts, the cost would fall dramatically and likely range between 70 and 320 $ per MWh.

Still, the information regarding the costs related to the OTEC technology is limited as it has not yet been deployed on a large commercial basis. This is mostly due to high upfront costs and limited experience, though such facilities can be convenient sources of energy for many countries and locations, e.g., for remote islands.

While the economies of scale clearly show the economic potential of OTEC technologies, there are some issues related to the construction, for instance, large cold water intake piping, connection to the grid or other utilities (for example, for fresh water delivery), etc. This also requires proper maintenance, cleaning, and protection from fouling and corrosion. Moreover, the impact on the ocean life has not yet been properly investigated on a large scale, there may be some impact due to the extraction of cold water and its delivery to the surface that may cause thermal pollution.

21.5. Salinity gradient energy

Seawater has salinities around 30–50 ppt (parts per thousand) which is approximately 200 times more saline than fresh water (which is around 0.5 ppt) that comes from the rivers, ground waters, precipitation and snowmelt [3]. The salinity of seawater differs mostly due to global ocean currents, both deep and surface. This high salinity of ocean water provides a chemical pressure potential with the fresh water which can be further utilized in an energy conversion chain. Therefore salinity gradient power is the energy obtained from the difference in the salt concentrations between the two fluids, for example, sea water and fresh water from the river. Hence, salinity gradient power has the largest potential at the mouths of major rivers, where large volumes of fresh water flow out to sea [3].

Approximately 0.61 kWh of free energy can be harvested by mixing one cubic meter of fresh river water and seawater [17]. The global amount of salinity gradient energy available from oceans and seas is estimated at 3.1 TW [18] or 1650 TWh per year [3]. However, the technical potential depends on many circumstances and is much lower.

The salinity gradient energy conversion technology is based on a membrane approach. Such a membrane is an essential and critical part of most salinity gradient power technologies. Commercially operated facilities require large quantities of membranes: for instance, for a 2 MW power plant at least 2 million square meters of membrane surface every five years is required [18]. The increase of membrane efficiency is an important driver of the technology.

There are two main technologies that are employed to utilize salinity gradients [17–21]:

- *Pressure retarded osmosis* (PRO), which employs the osmosis process: the feed (fresh water) is separated from a solution (salty sea water) with a semi-permeable membrane. The same volume of fresh water has a much lower concentration of salt than the same volume of salty water. This difference in salt concentrations between the feed solvent and the draw solution causes the flow from the solvent side to the solution side through the membrane — the process that tries to equalize the concentrations of salt. This flow creates fresh water pressure over the membrane. Even though the draw solution is partially pressurized, its hydraulic pressure is less than its osmotic pressure, so that there is still a net osmotic driving force to transport water from the feed to the draw solution. This pressure is high enough to run a small hydroturbine, as schematically shown in Fig. 21.4(A). The energy derived from osmotic pressure is comparable to a water potential with a theoretical pressure of 26 bar, or a water column of 270 m height.
- *Reversed electro dialysis* (RED) module, which consists of alternating cation exchange membranes (CEM) and anion exchange membranes (AEM), as illustrated in Fig. 21.4(B). CEM is only permeable for positive ions, and AEM only for negative ones. When the solutions with

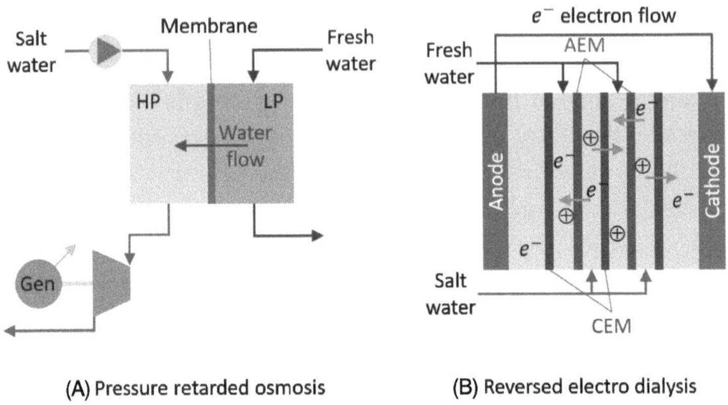

Figure 21.4 ***Salinity gradient energy process.*** Two major technologies that utilize salinity gradient energy: pressure retarded osmosis (A) and reversed electro dialysis (B).

different concentration of salt flow on either side of these ion exchange membranes, electric potentials are created: cations from high concentration migrate to low concentration through CEM, and anions migrate from high concentration to low concentration via AEM. At both ends of the module, electrodes are used to convert the ionic current into the electric current.

Both technologies face a number of technological challenges that prevent them from upscaling: the quality of the membranes, bio-fouling, the development of the modules that contain membranes, water pre-treatment to increase the process efficiency and reliability, and the technology to produce large amounts of durable and affordable membranes [18]. Several research and pilot projects are under development in multiple locations throughout the world, but the commercial applications are expected to be launched not earlier than in the 2020s.

Due to the early stage of the development, the economic side of these technologies is still unclear, and all numbers should be treated as projections. The membrane accounts for around 80% of the total plant costs, as the development is still ongoing. Other capital costs are available at average market rates. The estimated LCOE that is expected for the commercial projects is around 0.15–0.30 $ per kWh for PRO technology and 0.11–0.20 $ per kWh for RED [18]. Again, these numbers are just an indication and may differ from the reality once the projects are commissioned and put into commercial operation.

Salinity gradient energy systems are claimed to have very limited environmental impact [21]. So far, the absence of large capacities does not allow providing detailed analysis of this impact, however, the expected influence is comparable to that of a desalination plant.

References

[1] REN21. Renewables 2017 global status report. Paris: REN21 Secretariat. ISBN 978-3-9818107-6-9, 2017.

[2] Uihlein A, Magagna D. Wave and tidal current energy — a review of the current state of research beyond technology. Renewable and Sustainable Energy Reviews 2016;58:1070–81. https://doi.org/10.1016/j.rser.2015.12.284.

[3] Huckerby J, Jeffrey H, Sedgwick J, Jay B, Finlay L. An international vision for ocean energy — version II. Ocean Energy Systems Implementing Agreement; 2012.

[4] World Energy Council. World energy resources 2016. Marine energy. London: World Energy Council. ISBN 9780946121588, 2016.

[5] Pecher A, Kofoed J. Handbook of ocean wave energy. Ocean engineering & oceanography. Springer International Publishing. ISBN 9783319398891, 2016.

[6] IRENA International Renewable Energy Agency. Tidal energy. Technology brief. Abu Dhabi: IRENA; 2014. URL: http://www.irena.org/-/media/Files/IRENA/Agency/Publication/2014/Tidal_Energy_V4_WEB.pdf.

[7] Tester J, Drake E, Driscoll M, Golay M, Peters W. Sustainable energy: choosing among options. MIT Press. ISBN 9780262304634, 2012.

[8] Charlier R, Justus J. Ocean energies: environmental, economic and technological aspects of alternative power sources. Elsevier oceanography series. Elsevier Science. ISBN 9780080870946, 1993.

[9] Letcher T. Wind energy engineering: a handbook for onshore and offshore wind turbines. Elsevier Science. ISBN 9780128094297, 2017.

[10] Greaves D, Iglesias G. Wave and tidal energy. Wiley. ISBN 9781119014478, 2018.

[11] IRENA (International Renewable Energy Agency). Wave energy. Technology brief. Abu Dhabi: IRENA; 2014. URL: www.irena.org/documentdownloads/publications/wave-energy_v4_web.pdf.

[12] Babarit A. Ocean wave energy conversion: resource, technologies and performance. Elsevier Science. ISBN 9780081023907, 2017.

[13] Ocean Energy Systems (OES), International Energy Agency (IEA). Annual report. An overview of ocean energy activities in 2017. The Executive Committee of Ocean Energy Systems; 2017. URL: https://report2017.ocean-energy-systems.org/documents/OES-Annual-Report-2017/.

[14] Lehmann M, Karimpour F, Goudey C, Jacobson P, Alam MR. Ocean wave energy in the United States: current status and future perspectives. Renewable and Sustainable Energy Reviews 2017;74:1300–13. https://doi.org/10.1016/j.rser.2016.11.101.

[15] Nihous G. Mapping available ocean thermal energy conversion resources around the main Hawaiian islands with state-of-the-art tools. Journal of Renewable and Sustainable Energy 2010;2. https://doi.org/10.1063/1.3463051.

[16] IRENA (International Renewable Energy Agency). Ocean thermal energy conversion. Technology brief. Abu Dhabi: IRENA; 2014. URL: http://www.irena.org/-/media/Files/IRENA/Agency/Publication/2014/Ocean_Thermal_Energy_V4_web.pdf.

[17] Jihye K, Kwanho J, Myoung JP, Ho KS, Joon HK. Recent advances in osmotic energy generation via pressure-retarded osmosis (PRO): a review. Energies 2015;8:11821–45. https://doi.org/10.3390/en81011821.

[18] IRENA (International Renewable Energy Agency). Salinity gradient energy. Technology brief. Abu Dhabi: IRENA; 2014. URL: http://www.irena.org/-/media/Files/IRENA/Agency/Publication/2014/Salinity_Energy_v4_WEB.pdf.

[19] Jihye K, Jijung L, Joon HK. Overview of pressure-retarded osmosis (PRO) process and hybrid application to sea water reverse osmosis process. Desalination and Water Treatment 2012;43(1–3):193–200. https://doi.org/10.1080/19443994.2012.672170.

[20] Ramon G, Feinberg B, Hoek E. Membrane-based production of salinity-gradient power. Energy & Environmental Science 2011;11. https://doi.org/10.1039/c1ee01913a.

[21] Cipollina A, Micale G. Sustainable energy from salinity gradients. Elsevier Science. ISBN 9780081003237, 2016.

PART SEVEN

Future of sustainable power generation

Overview

The world is now living through a number of important transitions that span over all branches of economy and society. Their impact questions the sustainable future through the changing ecology, climate issues, scarcity of resources caused by unreasonable consumption, yet directs the society further into electrification of things, further industrialization, urbanization and further challenges we may not be aware of.

We therefore devote this final part of the book to the questions of the very possibility of the sustainable power generation. While summarizing key findings of the previous chapters, we cover such important topics like regulation, sustainable electricity management through smart grids and storage technologies, and potential alternatives.

Finally, we outline the trends of the industry, power generation world and technology, and societal changes to draw attention to the need of the mindset changes. The latter can be achieved through sustainability education which is also a hot topic covered here on our way towards our common future.

CHAPTER TWENTY-TWO

Can we build a sustainable power generation system?

Contents

22.1. Introductory

A solution to the global climate change problem and the transition towards sustainable future that is based on clean energy generation requires a complex approach and cannot be conducted in a single step. This does not only imply the shift from the GHG emissions that may lead towards the increase of the atmospheric temperature, but also fresh water supply issue, energy insecurity and limited supply to many regions, health issues, and many other problems. All this requires a large-scale conversion to clean, perpetual, and reliable energy at affordable cost and increased level of efficiency [1], and delivered to the consumer.

The number of technologies to convert various sources of primary energy into one of the most promising forms of secondary energy, that is, electricity, is significant. Major traditional technologies like combustion of conventional and cheap fossil fuels, use of energy of the flowing or stored water, or conversion of the nuclear energy has been implemented on a commercial basis worldwide. Recent development in emerging technologies has shown fantastic results which prove the technical feasibility and economic viability of many of novel approaches to energy conversion that broaden the choice of our options to an amazing portfolio, which continues to expand. However, having this broad range of possibilities makes it

Sustainable Power Generation
https://doi.org/10.1016/B978-0-12-817012-0.00037-2

even more complicated to answer an emerging question of sustainability: "What shall we do?"

Over the last decades there have been a number of studies conducted by scholars to identify the transition plan towards cleaner energy production. Thus, for instance, Jacobson and Masters [2] suggest to replace 60% of coal-fired generation by more than 250 000 wind turbines to let the US meet the requirements of the Kyoto protocol. In [3] the authors proposed the course toward the stabilization of the carbon dioxide component of the climate change that would require the development within the coming decades of primary energy sources that do not emit carbon dioxide to the atmosphere, in addition to the ongoing efforts to reduce the end-use energy demand. The proposed portfolio of solutions may well include not only the increased use of renewable and nuclear energy, but also demand decarbonization of fossil fuels. A similar approach has been proposed in [4] by Pacala and Socolow, where they highlighted several options that rely on the available technologies.

Further on, multiple long-tern strategies have been studied and proposed by a number of scholars. They analyzed the technological, geographical, and economic potential of separate technologies and their contribution to the sustainable energy scenarios with the conclusion that completely renewable future, at least for certain regions, is possible [5,6]. In [7] the authors conclude that wind, water, and solar technologies can provide 100% of the world's energy demand, therefore eliminating all fossil fuel consumption. This research has been further detailed to provide reasoning behind the promising conclusion [1,8].

Major agencies, governmental organizations, and NGOs have also contributed to the development of the forecasts on how to target global warming issues and mitigate climate change, focusing on energy production and consumption scenarios and introduction of novel technologies. One can name the scenarios proposed by the World Energy Council [9]. The cited "Grand Transition" targets the next half of the century till 2060 and provides an open, transparent, and inclusive framework to think about a very uncertain future.

Similarly, various scenarios have been built by the International Energy Agency, for instance, "World Energy Outlook" [10], "Energy and Air Pollution" [11], "Status of Power System Transformation" [12], etc. The US Energy Information Administration has significantly contributed to the development of scenarios and modeling of the sustainable future by issuing the International Energy Outlook (see, for instance, [13]) mostly for the period up to 2040. Such reports address the question of the temperature increase, climate change, and the required actions that have to be taken by the governments to mitigate these changes.

Some reports have been issued by multinational companies like BP with their statistical report [14], or by NGOs like Greenpeace with their famous "Energy [Re]volution" report [15].

Thus, in the next sections we try to aggregate and summarize the results aggregated and analyzed within the current book, together with those derived by major scholars and institutions, and provide a general vision towards sustainable energy future, by comparing the existing technologies, their pluses and drawbacks, and the complexity of the scenario modeling. This means that, unfortunately, there is no "one size fits all" solution to the climate change issue, or a single recipe to come up with such solution. However, we do try to attract attention to multiple sides of the problem to ensure we target its versatility with care and address all sides of sustainability: environmental, social, and economic.

22.2. Long-term sustainable energy system

The implementation of novel technologies that are currently going through the research and development phase will most likely take long to be adopted on a commercial scale and provide significant and tangible impact towards combating the global climate issues. With this in mind, we have drawn our attention to those technologies which have already proved to be at least technically viable and are approaching commercial parity with conventional ones.

Current sustainability issues caused by temperature increase, various emissions, multiple sources of pollution and consequent energy insecurity require immediate to short-term actions that shall likely bring straightforward changes to the current situation. This means that the current energy system with its portfolio of traditional and quite reliable scope of technologies — thermal, nuclear, and hydroelectric — has to be modified in the way that it employs the existing best practices and proven innovations that can be scaled up as part of the global energy system without further major technology development [1]. Moreover, the technological capability and the ability to implement the emerging solution on an industrial scale prevails over the economic feasibility of such a change. This means that the shift towards sustainable energy system may be physically possible, but the cost of such a shift can extend well beyond the profitability limits of the energy companies, which actually pulls us back to the sustainability paradigm. We may even conclude that this is all about the motivation of the society to perform such a shift and go beyond the short-sighted, yet natural approach of thinking of only a couple of generations ahead.

Thus, we shall ensure that the transition goes in the direction of a clean and reliable energy system that is capable of supporting the growth of population and energy demand in the longer run.

22.2.1 Available energy resources

One of the most important requirements for continuous development is the availability of resources required to power the energy system of the future. Table 22.1 shows an estimated theoretical potential of the available energy and the currently installed capacities that produce electricity out of this source of primary energy. It also indicates the currently installed capacity for each energy source [5].

- Coal is one of a few fossil primary energy sources that is readily available throughout the world. The total proved reserves of coal equal to 891.5 billion tons and may last for more than 100 years with the same rate of production. As coal supplies around 40% of total electricity generation, with 2400 kW per ton, the available energy can be estimated at the level of 20 PW per year.
- Natural gas reserves have the approximate volume of more than 186 trillion cubic meters which are expected to last at least for the next 50 years. The estimated energy theoretically available is quite large, but only 39% of natural gas is used for electricity and heat generation.
- Oil is the primary source for chemical industry as its use in power generation is less than 1%.
- Nuclear fuel is expected to last for at least the next 200 years with the same level of consumption, but its reserves are scattered and most of them are costly to extract. The implementation of closed fuel cycles may well extend this number to more than 1000 years.
- Hydroenergy can boast a quite large "green" energy potential of around 42 PWh per year with the technical potential estimated at around 15 PWh/yr, or about 35% of the theoretical potential derived from the total annual runoff of precipitation.
- Wind energy potential is usually estimated at the level of 1700 TW, which is around 14 900 PWh per year at the heights of about 100 m if all wind at all speeds is used to power wind turbines. This corresponds

Table 22.1 Worldwide theoretically available energy, installed capacity and world electricity generation for the end of 2016.

	Total available energy	Capacity		Electricity generation	
	PWh per year	GW	%	TWh	%
Coal	20[a]	3954.0[b]	61.1[b]	9594	38.4
Oil	193[c]	–[b]		931	3.7
Natural Gas	131[d]	–[b]		5794	23.2
Nuclear	30[e]	391.0	6.0	2606	10.4
Hydro	42[f]	1200.0	18.5	4170	16.7
Wind	14 900[g]	487.0	7.5	958	3.8
Solar PV	14 000[h]	303.0	4.7	328	1.3
Other	–[i]	138.0	2.1	592	2.4

Source: Based on [7,10,13,16–19] and calculations.

[a] With the total proven reserves estimated at 895.1 billion tons and estimated 2400 kWh per ton, coal could last for around 110 years. The number is given for total coal reserves, but only around 60% is used for electricity generation and combined heat and power.

[b] Capacity is given for all conventional thermal power generation technologies together. The estimated shares of coal and natural gas are around 2500 and 1400 GW, respectively. No precise data is available.

[c] With the total proven oil reserves of 1700.1 billion barrels and estimated at around 1700 kWh per barrel, cheap oil could last for the next 52 years. The number is given for total oil reserves, but only less than 1% is used for electricity generation and combined heat and power.

[d] Total proven reserves are estimated at 186.5 trillion m^3 with 10.5 kWh per m^3, cheap natural gas would last for the next 50 years. The number is given for total natural gas reserves, but only less 39% is used for electricity generation and combined heat and power.

[e] Nuclear fuel reserves are estimated at more than 10 million tons, but most are expensive to extract. The reserves are estimated to last more than 200 years unless closed fuel cycle is implemented.

[f] The technical potential for hydropower is usually estimated at around 15 000 TWh/yr, or about 35% of a theoretical potential derived from the total annual runoff of precipitation.

[g] Theoretically, around 1700 TW of wind energy are available over the world's land plus ocean surfaces at the heights of about 100 m.

[h] The amount of the Sun energy that reaches the Earth and can potentially be captured is estimated to be from 15 750 to 49 837 EJ.

[i] The final number comprises geothermal, biomass, ocean energy, biomass, waste, and is difficult to estimate.

to the average hub height of modern wind turbines. However, the wind power in locations over land and near the shoreline with the wind speeds faster than 7 m/s is around 72–170 TW, with only around 50% of this number available for practical development at the current level of technology.

- Solar radiation has the global theoretical power potential of approximately 14 000 PWh per year. This number has to be further tuned to exclude areas of competing land or high latitudes where solar insolation is low. According to [5], the technical potential can be estimated at the level of around 20%, which is still a huge number to cover all human needs.

- Other sources of energy include geothermal heat with up to 640 PWh/y, tidal with around 25 PWh/y, wave energy estimated at 30 PWh, and ocean thermal that ranges from 30 up to 90 PWh per year. This also includes energy from waste and biomass, which are difficult to estimate precisely.

The fossil reserves provide limited a amount of energy in time, but have historically been the most developed source of electricity. Conventional fossil-fired technologies now account for more than 65% of total electricity generation, with the largest renewable source of energy — hydro — occupying more than 16% only. While solar and wind energy provide by far the largest theoretical amount of energy, their utilization is really low compared to conventional sources.

22.2.2 Climate-relevant emissions

One of the major parameters to use for comparison of the power generating technologies that can potentially be used for future energy system is the amount of lifecycle CO_2 equivalent emissions. They imply not only direct carbon dioxide emissions from the process, but also other greenhouse gases and particles according to their warming potential. Such a metric estimates the total emissions that are released during the whole lifecycle of the power plant, including its construction, operation, and possibly decommissioning, as well as production of equipment, mining of fuel (if applicable), etc.

Table 22.2 summarizes ranges of lifecycle CO_2 equivalent emissions for emerging technologies that are currently available on the market for large-scale commercial implementations versus the existing conventional power generating plants.

Wind energy offers the lowest level of lifecycle CO_2 emissions compared to other technologies both conventional and emerging. This is due to the nature of the energy conversion chain and limited amount of material required for the construction (including concrete works and steel production). Hydroelectric power can range significantly depending on the project size, but most of the emissions associated with the project come from concrete works. Solar PV technology has a much higher level that comes from the PV panel production process, but these levels are, of course, much lower than for conventional thermal technologies.

It is important to mention that nuclear energy can be as low as some of the emerging technologies, but the calculations are likely to miss those

Table 22.2 Equivalent carbon dioxide lifecycle emissions, g of CO_2 per kWh.

Technology	Emissions
Coal	650–1200
Coal with CCS	255–442[a]
Natural gas fired	350–500
Nuclear	9–70[b]
Hydroelectric	4–22
Wind	2.8–7.4
Solar PV	19–80
Geothermal	15–55
CSP	8.5–32
Waste to energy	up to 1600
Biomass	16–75
Wave	8–50
Tidal	14

Source: Based on [5,20,21] and calculations.

[a] Based on the estimates of the pilot projects.

[b] These numbers do not include proper estimation of decommissioning phase and related emissions, as well as potential threats associated with nuclear power generation.

emissions associated with decommissioning and other risks dealing with nuclear weapons proliferation, possible accidents, and waste management.

There are some other parameters that are important for choosing among options:

- The footpring of a power plant, which comprises the land for the plant and associated facilities. Conventional thermal power plants are therefore impacted by the associated mines, especially open pit, and the nuclear facilities have a safety zone around. Solar fields require large territories to concentrate enough energy (but can be combined with the buildings on the rooftops), but biomass has to compromise with the food production. Wind parks would require spacing in between, however, the land around onshore windmills can be further used for agriculture.
- Water consumption per kWh defines the amount of water consumed by the process. Water consumption for wind power and PV can be considered negligible, but all thermal technologies are very water intense, and, together with nuclear, require a lot of cooling means: more than 2 l per kWh for coal-fired power generation. By far the largest wa-

ter consumption relates to biomass energy, where the bulk amount of water is used for watering the energy crops. Hydroelectric plants with reservoirs consume water since it evaporates from the reservoir surface and is somehow "lost".

22.2.3 Economic comparison

Economic evaluation of the technology is still the main tool for decision making, though may not be perfect for choosing the most sustainable technology as the trade-off between economic profitability and environmental impact has not been widely supported, yet. Thus, the major tool to compare various technologies is the levelized cost of electricity, or LCOE, which defines the cost of kWh generated and levelized for the whole plant's lifecycle.

Table 22.3 represents the capital costs and the LCOE analysis of various technologies. Indubitably, the lowest levelized electricity costs come from the established conventional technologies like large hydroelectric power (with as low as 0.02 USD per kWh) and traditional coal and gas fired power generation. This is due to the years of their development, efficient, and proved technology. However, the implementation of incentives and introduction of carbon taxes may change the picture and make these costs much higher for thermal technologies, while hydroelectric power is limited to other factors, such as resource location.

Anyway, the trend is that LCOE costs of emerging technologies like wind (both on- and offshore) go down and are already competitive to conventional technologies without incentives. It is likely that the trend continues so that the novel energy technologies will gradually replace conventional technologies within the new construction sector, which would be a major step towards sustainable development.

22.3. The importance of climate change regulation

Current energy markets, institutions, and policies have been developed to support the production and use of fossil fuels [8]. The fossil-based economy has dominated the production and transformation of the primary energy into the secondary due to its economic benefits that have been proved over time, the regulation has eventually followed the same way. The policies that have been developed over time provide a solid basis for fossil-fuel energy, its production, transformation, distribution, and consumption.

Table 22.3 Capital costs, levelized cost of electricity, and energy payback ratio for major technologies.

	CAPEX, USD/MW	LCOE, USD/kWh
Coal	1300–3000	0.04–0.14[a]
Coal with CCS	5500–6950	–[b]
Natural gas fired	450–1400	0.04–0.12
Nuclear	3500–6520	0.09–0.15[c]
Hydroelectric	940–8400	0.02–0.27
Wind	900–6500	0.06–0.21
Solar PV	1000–3500	0.07–0.31
Geothermal	2000–6200	0.05–0.12
CSP	2500–13 200[d]	0.14–0.35[e]
Biomass and waste	450–8000[f]	0.04–0.08[f]
Wave	2700–18 100	0.12–0.85
Tidal	100–14 600	0.36–0.63

Given costs vary significantly based on region, technology, capacity, local regulations, fuel supply, proximity to consumer, resource or grid, and available market incentives.

Source: Based on [20–30], EIC and IIR databases analysis and calculations.

[a] Mostly excluding costs related to emissions taxes that are being implemented in many countries.
[b] No commercially reliable data available.
[c] Does not include reliable decommissioning cost and full lifecycle costs of spend fuel storage and treatment. Decommissioning projects always overrun the budget and timeline.
[d] The installed cost depends largely on the capacity of the thermal storage: without a storage, the cost may be 2550–11 200, and adding 4–8 hours of storage drives it up to 6050–13 200 USD/kW.
[e] Auction prices show much lower values of around 0.06 $/kWh, but they have to be treated with caution due to auction prices being dependent on a set of obligations and terms in the contract that can be very market and project specific.
[f] WTE plant investment costs tend to be at the upper limit of the general biomass costs.

Similarly, hydropower generation has induced the second large pillar of regulatory requirements that have already been worked out to the significant level of completeness.

While the mentioned two have been the pioneering technologies to become the basis for energy regulation, nuclear energy can be studied as an example of a difficult and sometimes quite bumpy way to build the regulatory system from scratch in many countries. Despite major experience in other traditional technologies, nuclear energy is an example of a different world that requires special approach and therefore is highly impacted and strictly regulated by the international bodies like IAEA.

The focus on fossil-fuel energy system has been until the first attempts to realize the impact of the fossil energy on the environment and the international gatherings to formulate some generic opinion on that impact. When the UN Framework Convention was negotiated in 1992 in Rio de Janeiro, its goal seemed to be well achievable. However, the Kyoto Protocol of 1997 that became the major instrument to achieve this goal failed miserably [31]. Followed further by the Paris agreement that has been ratified by many and diverse countries that are planning (and are normally taking actions) to decrease their GHG emissions and minimize the temperature increase, it is supposed to bring some effect by 2020. This has been mostly driven through two broad ways:

- Increasing the efficiency and decreasing the impact of the existing conventional facilities, mostly coal-fired, and
- Replacing the existing capacities and introducing new from the renewable emerging technologies' pool: wind, solar, new hydro, and others.

In turn, such a change requires hard work on the regulation level. Because fossil-fuel energy systems have different production, transmission, and end-use costs and characteristics than emerging renewable energy systems, new policies are needed to ensure that such novel systems develop as quickly and broadly as is socially desirable and, still, economically feasible [8].

While the countries promised to come back every five years to estimate the progress and reconsider their goals set by the Paris agreement, the situation seems to be somehow falling apart at the seams, and the world clock is still ticking.

22.3.1 Current status of regulation

Presently, according to [32,33] there are more than 1300 national climate laws worldwide, broadly defined as laws that address all issues related to climate change, its mitigation, and adaptation. Since the time of the Kyoto protocol was signed in 1997, when there were only around 60 relevant laws and policies, the next 20 years of collaborative work increased the number of regulatory documents around 20 times. As their number has doubled roughly every 4–5 years, this speaks for the growing attention that legislators, governments and the society in general are devoting to the issue and for their ambition to battle climate change [32].

There are several ways to address climate change issues on a global scale. While in some countries the primary approach is driven by the acts of the parliament (in the form of the formal laws passed), in other countries the direction of the climate policy is defined through executive policies (for instance, executive orders, decrees, strategies, and development plans). Both are equally important and have relatively the same share of more than 40% each [32].

Next, climate change laws and policies also differ in scope and ambition [32,33]. Thus, some of them are specifically focused on climate change, advancing explicitly greenhouse gas emissions reduction and adaptation to climate change, but they usually represent around a quarter of the total number of documents. The remaining laws and policies address climate change and transitions to low-carbon economies through other areas indirectly, for example, via wider frameworks like economic development strategies or green growth plans.

Contrary to the first attempts of creating climate change regulations, the presently introduced documents are aimed at having bottom-up planning with less ambitious goals to make sure they are manageable and achievable. For example, this underlies the nature of the new COP with its approach to reduce GHG emissions: it is bottom-up in nature, flexible and supported by top-down elements [34]. However, in order to meet the still challenging goals of decreasing the GHG emissions and hold the temperature increase, the policies have to target the mentioned above two approaches directly:

- Decreasing the support of the fossil energy and pushing it to the cleaner limits, and
- Promoting the deployment of renewable energy sources.

Both support the overall goal of decarbonization, which has been one of the priorities in many international and regional policies.

According to [15], there is a clear agenda for the policy changes which need to be made to encourage a shift to renewable sources. The main demands from the new regulations include, among other things:

- Using established and proven technologies with the highest efficiency during the transition period in short to medium term;
- Phasing out subsidies for fossil-fired power generation to ensure their continuous replacement by other technologies or, similarly, the introduction of fossil-energy special taxation regimes;
- Establishing legally binding targets for renewable energy and combined heat and power generation to increase the efficiency of the systems and decrease waste;
- Starting the reform of the electricity markets to guarantee access to the grid for renewable power generators;
- Providing defined and stable returns for investors, for example, through feed-in tariff payments that will give the possibility for long-range investment planning of the renewable projects;
- Increasing research and development budgets for renewable energy and energy efficiency.

Regulation and policies can be considered as a key tool for enabling the change, which shall still be in combination with the favorable market conditions. This can be achieved by not only issuing multiple decrees or other regulatory documents, but also with the help of various instruments that may stimulate the shift. This means that the market would rather demand a mix of various technologies than the use of a single one, so that there is no single pathway that can be defined on a regulatory level.

The established conventional energy system has led to an established energy market. The direct introduction and application of strict regulatory measures towards renewables will shake it and may worsen the situation. Of course, at all times emerging technologies required acceptance of the public and support from the governments and other organizations on their way to become a common reality, and the current situation is not completely new or any different. However, the regulations do need to cover the environment part of sustainability paradigm first to ensure that we no longer continue the same magnitude of the climate change. This resulted in the introduction of the financial incentives to support wind or solar parks (which is now not needed anymore in many locations due to the natural evolution of emerging technologies and their competitiveness), opening of the carbon market to mitigate the costs, implementation of the feed-in tariffs related to renewable technologies, renewable output quotas, funding of the corporate research programs, etc.

According to [8], most studies find that feed-in tariffs that subsidize the difference between generation cost and wholesale market electricity price are especially effective at stimulating generation from renewable energy sources [6,35]. Eventually, the feed-in tariffs are an instrument that is required now but shall be gradually decreased since the production costs of new technologies continue to decrease, so that the innovations will finally become competitive with the conventional power generation.

But the changes of the power generation system regulation is only one part. The policies must address the complete energy system, include the optimization of the infrastructure to increase its efficiency, as well as the increase of the end use efficiency and decrease the overall energy demand. The latter seems quite challenging in the face of the population increase and hence continuous energy consumption growth, but shall be serious considered to optimize the per capita consumption, replace inefficient processes, activities and technologies.

Last but not least, even the introduction of the most efficient and overwhelming and smart regulation may fail due to the social side of the issue.

Therefore, the buy-in of the society on all levels is an important and essential step. Society, in its broad sense, has to be involved in discussions which would require the socially challenging tasks also defined within the SDGs, for instance, quality education, peace ad justice, and partnership for the goals.

22.3.2 Climate engineering

Some serious voices are now saying that, barring a technological revolution, a seismic shift in global policies, or a worldwide economic depression, the only way to advert global climate catastrophe may be climate engineering [3,31]. Even though it is debatable whether to cover the topic of geoengineering within the regulation topic, we believe that it is more impacted by the social acceptance, the global nature of the risks and hence regulatory framework rather than pure technology.

The idea of climate engineering dates back to the first discoveries in meteorological science, even suggesting burning fossil fuels to prevent global cooling and possible ice ages, and further do weather modifications to cause precipitation artificially or destroy clouds. The visual representation of the climate change and weather control projects and technologies can be found at Weather Modification History portal[1] showing that the potential of such changes grew tremendously and caused many countries to seriously consider the possible consequences including the use of such climate alteration for military purposes. Therefore, in 1977 the Environmental Modification Convention prohibited the military and other hostile use of the environment modification technologies that may have severe and long-lasting impact on the society. The treaty went into force in 1978 and finally banned weather warfare.

Climate engineering, sometimes also referred to as *geoengineering*, is the deliberate large-scale intervention in the Earth's climate system, in order to moderate global warming. Such an intervention always has two major elements involved [31]:

- Large scale to have this intervention big enough to impact the climate of the whole planet or at least on a regional scale, and
- Specific intent to alter the climate.

Generally, the climate engineering approaches that are now under discussion worldwide, can be broadly divided into two major proposals [31]:

[1] https://weathermodificationhistory.com/.

- Solar radiation management (or albedo management), which attempts to decrease the amount of solar energy reaching or staying on the Earth's surface. There have been several technological approaches to scattering or reflecting some of this energy back to space, for instance, with aerosol sprays at the upper layers of the atmosphere that would form highly reflective particles. Another idea was to launch a number of mirrors to the near-Earth orbit. This solutions seems quite cheap, manageable, and also reversible, but has number of flaws: it does not decrease the overall GHG level (and therefore will not mitigate health issues, for instance), may disrupt weather patterns, impact ecosystems in the way no one can predict, and may probably be required to last very long.
- Greenhouse gas or carbon dioxide removal, which is another possible strategy of climate engineering. In its essence, it removes carbon dioxide as the major GHG from the atmosphere, which is a high level case of a specific carbon capture and storage. There have been multiple technologies proposed by scholars and engineers: machines to directly remove CO_2 from the air, fertilizing the oceans to grow CO_2 consuming algae, capture it back in charcoal, plant more trees, etc. This approach is less risky and seems to have much fewer side effects, it is really slow to achieve the required impact. Moreover, there is no guarantee that all captured carbon would be held for a long period of time.

Naturally, there is a philosophical debate about whether there is anything ethically wrong with the climate engineering approach. While some believe that the advancement we have already achieved is the beginning of the end of Nature, other believe that the climate and geoengineering may be an essential step to shape the world for the future generations. Of course, anything other than a preservation of the current structure and function demands, a definition of "improvement" of the natural system is required. And this judgment will be very different across diverse cultures and over generations [36].

22.4. How should we power the world?

The amount of total primary energy supply required by the humankind is growing and is estimated to be at the level of roughly 700–900 EJ by 2050 [9,37]. This energy is expected to be the major driving force of the transition towards sustainable future of the humankind, which shall be compliant with the sustainability paradigm. This means that the combination of the environmental, social and economic benefits have to be considered in conjunction.

According to the findings summarized briefly in the previous sections, a large-scale wind, water, and solar energy system can reliably supply all of the world's energy needs, with significant benefit to climate, air and water quality, ecological systems, and energy security, at a reasonable cost [8]. Of course, this cannot be achieved within the short period of time, hence one of the possible ways to mitigate this is to split the action plan into three major periods:

1. Transition period, when the rapid energy demand of the developing countries has to be fulfilled by conventional hydro, clean coal, nuclear and with the largest possible share of renewables, e.g., wind and solar. At the same time, old conventional capacities have to be either replaced with renewables only (and this may be executed through regulation), or upgraded or repowered to increase their efficiency and decrease ecological footprint.
2. The pre-sustainability period shall initiate the process of decommissioning old conventional thermal and nuclear power plants to replace them with wind and solar technologies, and potentially advanced nuclear plants with a closed fuel cycle. All new capacities have to be of renewable and green nature.
3. Sustainability phase should start when the last conventional thermal plant has been decommissioned to ensure there is no source of real-time GHG emissions. While this seems to be the ideal case, it may be considered as a common sustainability goal for a power generation system.

There is an ongoing debate about whether nuclear energy shall be considered as a sustainable source of energy for the future. In our opinion, advanced nuclear energy, if properly treated, can be an extremely useful source of cheap, reliable, and safe electricity that can be used for multiple needs beyond pure power generation, for example, desalination or production of hydrogen. Moreover, floating plants and micro nuclear stations can be implemented in local grids to deliver power to remote regions. It may be too optimistic to mention that nuclear fusion can become an impressive step forward, but this is not likely to take place in the time of the next couple of generations.

So can we really power the whole world with wind energy, solar radiation and the movement of water? Or shall we further use the advanced nuclear cycle and other energy sources like geothermal heat or energy crops? Current electricity generation required for the humankind in 2016 was at the level of around 25 PWh [17], and the world total final consumption of energy was slightly more than 111 PWh. Compared to the amount of energy theoretically available from all sources, especially solar radiation and wind, as well as from other sources, we can easily see that the current values are easily manageable. Even the forecast increase of energy consumption up to the level of 18 mtoe [38] equals to 209 PWh, which is still a manageable amount of energy to extract from renewable sources. With the technical potential of wind estimated at the level up to 400 PWh and solar PV will less than 3000 PWh, these numbers do not seem even extraordinary.

In reality, however, a lot has to be done to manage this challenge. The obstacles to realizing this transformation of the energy system are primarily social and political, not technological [8]. There are readily available and mostly commercially feasible solutions to provide all required energy with the renewable sources based on water, wind, and solar energy, backed with a combination of modern incentives (like feed-in tariffs) and modern transmission and distribution system that is capable of managing the change.

Thus, the solution sounds simple: make the electricity production clean and electrify everything in our lives.

References

[1] Jacobson MZ, Delucchi MA. Providing all global energy with wind, water, and solar power, part I: technologies, energy resources, quantities and areas of infrastructure, and materials. Energy Policy 2011:1154–69. https://doi.org/10.1016/j.enpol.2010.11.040.

[2] Jacobson MJ, Masters GM. Exploiting wind versus coal. Science 2001;293:1438.

[3] Hoffert MI, Caldeira K, Benford G, Criswell DR, Green C, Herzog H, et al. Advanced technology paths to global climate stability: energy for a greenhouse planet. Science 2002;298:981–7.

[4] Pacala S, Socolow R. Stabilization wedges: solving the climate problem for the next 50 years with current technologies. Science 2004;305:968–72.

[5] Jacobson MZ. Review of solutions to global warming, air pollution, and energy security. Energy & Environmental Science 2009;2:148–73. https://doi.org/10.1039/B809990C.

[6] Sovacool B, Watts C. Going completely renewable: is it possible (let alone desirable)? The Electricity Journal 2009;22:95–111. https://doi.org/10.1016/J.TEJ.2009.03.011.

[7] Jacobson M, Delucchi M. A path to sustainable energy by 2030. Scientific American 2009;301:58–65. https://doi.org/10.1038/scientificamerican1109-58.

[8] Delucchi MA, Jacobson MZ. Providing all global energy with wind, water, and solar power, part II: reliability, system and transmission costs, and policies. Energy Policy 2011:1170–90. https://doi.org/10.1016/j.enpol.2010.11.045.

[9] World Energy Council (WEC), Accenture Strategy, Paul Scherrer Institute. World energy scenario 2016. The grand transition. London: WEC; 2016.

[10] International Energy Agency (IEA). World energy outlook. Paris: OECD/IEA; 2017.
[11] International Energy Agency (IEA). Energy and air pollution. Paris: OECD/IEA; 2016.
[12] International Energy Agency (IEA). Status of power system transformation 2017. System integration and local grids. Paris: OECD/IEA; 2017.
[13] Information Administration (EIA). International energy outlook 2016. US Energy Information Administration; 2016.
[14] BP. BP Statistical review of world energy. BP; 2015.
[15] Teske S, European Renewable Energy Council, Greenpeace International. Energy [r]evolution. A sustainable world energy outlook 2015. Greenpeace International; 2015.
[16] REN21. Renewables 2017 global status report. Paris: REN21 Secretariat. ISBN 978-3-9818107-6-9, 2017.
[17] IEA. Key world energy statistics 2018. International Energy Agency; 2018.
[18] International Energy Agency (IEA). Technology roadmap. Hydro. France: OECD / IEA; 2012.
[19] United Nations Development Programme (UNDP), World Energy Council. World energy assessment: energy and the challenge of sustainability. New York: UNDP; 2000.
[20] Letcher TM. Future energy: improved, sustainable and clean options for our Planet. Elsevier Science. ISBN 9780080994222, 2014.
[21] Guerrero-Lemus R, Martínez-Duart J. Renewable energies and CO_2: cost analysis, environmental impacts and technological trends – 2012 edition. Lecture notes in energy. Springer. ISBN 9781447143857, 2013.
[22] Black & Veatch. Cost and performance data for power generation technologies; 2012.
[23] World Energy Council. World energy perspective. Cost of energy technologies. London: World Energy Council. ISBN 9780946121304, 2013.
[24] Andrews J, Jelley N. Energy science: principles, technologies, and impacts. Oxford: OUP. ISBN 9780199592371, 2013.
[25] IFC. Hydroelectric power. A guide for developers and investors. International Finance Corporation; 2015.
[26] IEA-ETSAP (International Energy Agency – Energy Technology Systems Analysis Programme), IRENA (International Renewable Energy Agency). Wind power. Technology brief. IRENA, IEA-ETSAP; 2016. URL: https://www.irena.org/-/media/Files/IRENA/Agency/Publication/2016/IRENA-ETSAP_Tech_Brief_Wind_Power_E07.pdf.
[27] International Energy Agency Wind (IEA Wind). Report available at: https://community.ieawind.org/publications/ar, 2016. ISBN 978-0-9905075-3-6.
[28] IRENA (International Renewable Energy Agency). Renewable power generation costs in 2017. Abu Dhabi: IRENA; 2017.
[29] International Energy Agency (IEA). Technology roadmap. Geothermal heat and power. France: OECD / IEA; 2011.
[30] International Energy Agency (IEA). Renewable energy essentials: geothermal. IEA; 2010.
[31] Gerrard M, Hester T. Climate engineering and the law. Cambridge University Press. ISBN 9781107157279, 2018.
[32] Averchenkova A, Fankhauser S, Nachmany M. Trends in climate change legislation. Edward Elgar Publishing. ISBN 9781786435781, 2017.

[33] Nachmany M, Fankhauser S, Setzer J, Averchenkova A. Global trends in climate change legislation and litigation. Grantham Research Institute on Climate Change and the Environment; 2017.
[34] Rosenzweig R. Global climate change policy and carbon markets: transition to a new era. Energy, climate and the environment. UK: Palgrave Macmillan. ISBN 9781137560513, 2016.
[35] Bürer M, Wüstenhagen R. Which renewable energy policy is a venture capitalist's best friend? Empirical evidence from a survey of international cleantech investors. Energy Policy 2009;37:4997–5006.
[36] Schneider S, Rosencranz A, Niles J. Climate change policy: a survey. Island Press. ISBN 9781559638814, 2002.
[37] World Energy Council (WEC), Paul Scherrer Institute. World energy scenarios. Composing energy futures to 2050. London: WEC; 2013.
[38] BP. BP energy outlook. BP; 2018. URL: https://www.bp.com/content/dam/bp/en/corporate/pdf/energy-economics/energy-outlook/bp-energy-outlook-2018.pdf.

CHAPTER TWENTY-THREE

Sustainable electricity management beyond generation

Contents

23.1. Energy system and future challenges

The historical development of the present energy system dates back to the 19th century, when the dim, unreliable, smoky gas lights in growing cities started to be replaced by an innovative light bulb technology. Street lighting spread through the cities and, accompanied by the advancement of the electric machines, drove the growth of generating capacities. These capacities were mostly private and differed dramatically in size and parameters.

While some of the large coal-fired plants emerged in cities to cover urban development needs, industry demanded power as well. This led to the blossom of concentrated power generating facilities of large capacities to power factories, metallurgy, pulp and paper, chemical and other industries. Therefore the initial infrastructure for electrical energy emerged around large centers of gravity: urban islands and concentrated industrial centers. These were mostly large fossil-fired power plants with base load output, further extended by freshly built nuclear power plants. They were strategically

Sustainable Power Generation
https://doi.org/10.1016/B978-0-12-817012-0.00038-4

located near the fossil fuel reserves or large-scale transportation channels like water or rail. They fed electricity to the transmission network that further brought energy to the centers of distribution, either residential or industrial. Such local grids grew over time, and in the beginning of the 20th century they eventually interconnected into larger grids for economic reasons: a larger and diversified grid was much more reliable to supply power to its customers.

Further on, the emergence of the large hydroelectric power plants influenced the overall topology of the grid: located in the areas with favorable hydrology, these plants were not always near the consumers and therefore required longer transmission lines. The same was relevant to nuclear power plants that were located near large sources of cooling water.

The ongoing spread of the electrical home appliances, needs for heating, ventilation and air conditioning, and growth of businesses had led to the increased electricity consumption and, consequently, caused disturbances in the grid: variable consumption started to become more random and followed only generic time patterns, individual electricity consumers grew larger, and the traditional grid was not able to fully meet the demand. Multiple approaches have been proposed to balance the grid, including the use of the hydroelectric power generation resources, which are capable of varying the output and supporting the grid, and gas turbine based generation that boasts short ramp-up time and great operational flexibility. The control of the market allowed meeting the supply and demand, and the energy management has become an essential function for almost real-time regulation.

With that being said, modern electrical power systems have been developing since the 19th century and taking different forms around the world [1]: some of the systems are advanced and have multiple opportunities for optimization and smart control, some span over large territories and interconnect large consumer centers, other operate in island mode and are isolated. In emerging markets, energy systems are rapidly evolving to expand over the regions, for example, the Asian grid project emerging from China.

Growing concerns regarding global warming warn about large utility-scale fossil-fired plants (mostly traditional coal-fired), nuclear accidents put extreme pressure, and hydro has been limited to some geographical locations. In many countries, there are sufficient renewable energy resources to meet a major share of local demand. At the same time, the characteristics of many renewables include their natural unpredictability and variability over time scales ranging from seconds to years. These can constrain the ease of integration and result in additional system costs, particularly when reaching higher RE shares of electricity, heat, or gaseous and liquid fuels [1].

Renewable power generation systems can be integrated into various electricity supply systems: from large, interconnected, continental-scale supergrids to on-site generation and utilization in small, island mode microgrids. Technically and economically feasible levels of renewable energy penetration depend on the unique characteristics of a system: the status of infrastructure development and interconnections, mix of generation technologies, control and communication capability, demand pattern, and geographic location in relation to the renewable resources available, market designs, and institutional rules [1].

Achieving higher renewable energy share compared to the present level within the current power generation system will, of course, require additional integration efforts starting now and continuing over the longer term. These include not only deeper understanding of the renewable energy resource characteristics and availability, but mostly investments in enabling infrastructure and research, development and demonstrations, modifications to institutional and governance frameworks, innovative thinking, attention to social aspects, markets and planning, and capacity building in anticipation of renewable energy growth [1]. The development of modern technologies to harness renewable sources of primary energy opens a new era of power generation with the extreme challenges to the traditional grid and established infrastructure:

- Emerging renewable energy technologies like wind and solar have intermittent electricity generation pattern, which does not fit into the current market energy consumption profile;
- Newly built large base load capacities like hydroelectric plants are located in remote areas and require long transmission lines to feed the grid with certain backup;
- Small private generating capacities like rooftop PV cells or single wind turbines are entering the global grid and make the energy patter even less predictable and difficult to manage;
- Emerging large electricity consumers like data centers now tend be scattered in multiple areas and shift towards urban islands. The overall urbanization leads to energy-intense consumers.
- Decisions to stop large capacities of coal-fired on nuclear plants in developed countries would lead to significant challenges in managing the grid.

- Energy market is transforming into a more complicated system: incentives for renewables, GHG taxes, new financing models, shift of large project geographies, the introduction of new technologies, entrance of smaller players with start-up mentality — the list can be continued — are among the drivers towards the energy system change. The centralized grid will not be able to manage it in the long run.

As in fashion, energy systems are changing — fundamentally and fast.[1] The importance of centralized power generating capacity is decreasing, and the ways of electricity transmission and distribution have to follow. This leads to an increasing complexity of the modern decentralized grid, and its management is also way more complicated. On top, electricity consumption continues to steadily grow.

There are several approaches for addressing these challenges. One of them is to further expand and diversify the existing grid so that it is more likely to balance supply and demand effects. The second strategic approach is to transform the grid in a way that it becomes smarter with the help of novel technological solutions. Further development may well introduce the third strategic approach of introducing a fully smart grid with a grid-scale storage. This may create new options for integrating new sources of energy supply, emerging consumption patterns, quickly balance changes in customer behavior and respond to the events that may affect supply.

The demand for such changes is already there and has created a type of environment that made it possible for new technologies and approaches to energy system emergence, which are usually referred to as the mentioned smart grid. They are targeting the transmission and distribution sides of the energy system and the overall infrastructure, various energy storage technologies that are likely to blossom in the face of renewables integration into the grid-scale storage, ongoing talks of hydrogen as a promising energy carrier, and so on. These are a few among the disruptive innovations that are happening around the power generation industry to support it become sustainable.

23.2. The concept of a smart grid

The concept of the new type of grid, which is smarter than the present one in the way of efficient, reliable and sustainable electricity delivery, is an emerging topic. There have been multiple specialized books on

[1] https://www.siemens.com/global/en/home/products/energy.html.

smart grids, technology behind and the ways to deploy them (see, for example, [2–5]). We here limit ourselves to the conceptual view of the smart grid as an essential part of the sustainable energy system development.

The term *smart grid* is used as the name for the new generation of the energy delivery network, from the generating facilities to the end user. A major difference of the smart grid from the traditional one is that it allows a two-way communication between the supplier and consumer. Smart grids enable real time data collection concerning electricity supply and demand during the transmission and distribution process, making monitoring, generation, consumption, and maintenance more efficient [6].

The term smart grid has been introduced publicly in the early 2000s, with the Energy Independence and Security Act of 2007 [7] providing a comprehensive definition of the smart grid through its ten major attributes:

1. Increased use of digital information and controls technology, to improve reliability, security, and efficiency of the electric grid;
2. Dynamic optimization of grid operations and resources, with full cyber-security;
3. Deployment and integration of distributed resources and generation, including renewable resources;
4. Development and incorporation of demand response, demand-side resources, and energy-efficiency resources;
5. Deployment of "smart" technologies (real-time, automated, interactive technologies that optimize the physical operation of appliances and consumer devices) for metering, communications concerning grid operations and status, and distribution automation;
6. Integration of "smart" appliances and consumer devices;
7. Deployment and integration of advanced electricity storage and peak-shaving technologies, including plug-in electric and hybrid electric vehicles, and thermal-storage air conditioning;
8. Provision to consumers of timely information and control options;
9. Development of standards for communication and interoperability of appliances and equipment connected to the electric grid, including the infrastructure serving the grid;
10. Identification and lowering of unreasonable or unnecessary barriers to adoption of smart grid technologies, practices, and services.

This means that a smart grid is an evolution of the existing grid technology to optimize its performance, conservation and delivery of power. Smart grids use technologies to instantly relay information to match supply with demand, support well-informed decisions, and keep systems operating at optimal efficiency [8]. The smart grid therefore becomes a more information management system that allows the real-time coordination and optimization of the electricity supply and demand, especially in the face of diversified consumption patterns and intermittent energy sources.

The European energy regulators defined the smart grid as an electricity network that can cost-efficiently integrate the behavior and actions of all users connected to it — generators, consumers, and those that do both — in order to ensure an economically efficient, sustainable power system with low losses and high levels of quality, and security of supply and safety [9].

The smart grid therefore should include a complex network of technologies and systems, hardware and software, communications, and controls that taken together would provide both producers and consumers a high level of visibility and control [3]. Such a vision of the grid of the future means that the grid has to meet the demands of the society on the way towards affordable, clean, and reliable electricity supply for the sustainable development.

There have been other definitions of the smart grid, but the general idea stayed the same: ensure sustainability, security of supply, and competitiveness of the electricity market to meet the ongoing demand and decarbonization goals. The smart grid should be an essential step forward to accommodate large-scale renewable energy sources, including intermittent generation, integrate small-scale distributed energy sources, develop a more intelligent, competitive and accessible energy market, and improve security.

The implementation of the smart grid concept on a global scale would bring multiple unbeaten advantages and help maintain the transition towards sustainable future. In brief, it would provide a number of key attributes of sustainable energy system:

- Efficiency of the system with minimized energy losses, e.g., from the energy transportation;
- Flexibility to interchange the components of the system and use various resources for best performance results without disrupting the whole system;
- Modularity that would allow pluging in and switching off both consumers and energy providers without disturbing others;
- Economic efficiency for all stakeholders. As the costs of renewables are likely to continue their fall, there will be an increased implementation of more cost-effective sustainable technologies;
- Reliability of the system, to ensure energy security and affordability of energy for all stakeholders. This can be achieved by introduction of

energy storage systems to store the intermittent energy from renewables and dispatch it when and where necessary;
- Environmental responsibility, to decrease the GHG emissions and provide all global energy with the renewable sources.

Despite a number of obvious advantages, the implementation of the smart grid concept is not possible without the development of the proper regulation. A key principle of good regulation is to concentrate on the outputs of the regulated entity and effects of a given activity. This can be achieved through the performance-based incentives that provide financial support and benefits to those entities that implement the smart grid concept, with the penalties related to some parameters. This, in turn, requires a sound definition of goals and measurable performance indicators. This shall provide a solid basis for longer investments and the ability of the companies to properly do economic evaluation and plan their returns.

Secondly, the global implementation would require cooperation among all necessary stakeholders, including generating companies, grid operators, regulators, consumers, with a clear focus on standardization.

Thirdly, the implementation of the smart grid is not possible without innovation, development of new solutions and the required support of the regulators. This shall be extended from pure R&D financing also to the investments into demonstrator projects to understand the cost structure, optimize it, and further roll-out for commercial use.

Evidently, the next era of the energy system will not any longer be controlled by large generating companies. Instead, it is the global cooperation between customers and utilities that would allow working together to create a sustainable energy system. Generally speaking, the technology is already here: emerging generating capacities and systems to harness renewable energies and to store extra energy for further use have already been implemented or are undergoing development phase. Thus, with a global effort, a new sustainable energy system structure is possible.

23.3. Energy storage

A decentralized grid employs several electricity generating facilities that depend on intermittent primary energy sources. These are the most promising solar and wind energy sources: they both depend on climatic conditions (e.g., cloudy days or low wind speeds) and therefore cannot meet the changing or peak demand and may cause curtailment energy pricing during the day. Thus, the best approach is to store excess energy of the renewable sources and then dispatch it during the peak times.

Unfortunately, we do not yet possess a technology that would allow storing electrical energy in large quantities and for an unlimited period of time without significant losses. This is commercially possible only through the conversion of electricity into one of the other forms of energy [10]. The idea of storing excess energy has been under development for many years, and currently there are several options to store electricity and other forms of energy:

- Mechanical energy storage systems:
  - Flywheel,
  - Pumped hydro,
  - Compressed air;
- Thermal energy storage:
  - Sensible heat,
  - Latent heat,
  - Thermochemical;
- Electrical energy storage in supercapacitors;
- Electrochemical technologies (batteries);
- Hydrogen for energy storage.

There are some other energy storage technologies that have been deployed or are still under development, including biological or magnet, but they are not used in utility scale power generation applications.

23.3.1 Flywheel

The flywheel as a means of energy storage has been in place for many centuries. The material of the wheel has changed from stone and wood to metals and other advanced materials, but the concept remained the same: having a massive rotor spin with certain speed so that, when required, it can transfer its rotational energy. The industrial revolution employed flywheels on steam engines in transportation or in the factories to store energy and smooth the performance of the machines. With the development of electric machines, the need of the flywheel on a large industrial scale had declined, but recent developments in new materials, bearing technology, and power electronics opened a new era of more flywheel energy storage applications.

Flywheel energy storage is based on a simple rotating mass principle as each moving mass possesses a certain amount of kinetic energy. This kinetic energy of the rotating mass can be expressed as

$$E_r = \frac{1}{2} I\omega^2,$$

where ω is the angular velocity and I is the moment of inertia of the mass. The useful energy of a flywheel within a speed range from ω_{min} to ω_{max} can be obtained by [11]

$$E_r = \frac{1}{2} I \left(\omega_{max}^2 - \omega_{min}^2\right) = \frac{1}{2} I\omega_{max}^2 \left(1 - \frac{\omega_{min}^2}{\omega_{max}^2}\right).$$

Typically, industrial flywheels normally operate between the maximum and minimum speeds to avoid voltage variation and to limit the maximum motor and generator torque for a given power rating [11].

In most cases flywheel rotors are made in the form of hollow or solid cylinders or disks. In the case of the solid cylindrical or disk rotor of mass m and radius R, the moment is

$$I = \frac{1}{2} mR^2,$$

which delivers

$$E_r = \frac{1}{4} mR^2\omega^2. \tag{23.1}$$

The obtained equation (23.1) shows the importance of the mass of the flywheel, its angular speed of rotation and the size that directly affect the amount of energy that can be stored in a flywheel. The maximum speed limit at which a flywheel may operate is determined by the strength of the rotor material, and the stress limit must be always met to keep the spinning rotor safe.

Therefore the stored energy can be optimized by either increasing the angular speed or by increasing the moment of inertia. Due to material limitations, low speed flywheels (up to 10 000 rpm) are made out of metals and are supported by mechanical or magnetic bearings, while the high speed ones (up to 100 000 rpm) use lighter but stronger composite materials and require magnetic bearings to minimize friction losses [11].

Modern flywheel energy storage devices have a long lifecycle of more than 20 years, high cycle life (hundreds of thousands), very fast response of seconds, high efficiency of 90–95%, high charge and discharge rates, high power density, high energy density, and low environmental impacts [11].

Despite the fact that they self-discharge at a much higher rate than other storage options, they provide an extremely cost-effective solution with low O&M requirements, modularity, and capacities of kilowatts to megawatts. Therefore flywheel energy storage systems (FESS) are gaining certain popularity.

Modular flywheel energy storage system now have multiple applications for a modern power grid:

- Grid balance with the FESS electric load leveling and pulse dampening, as well as providing short bursts of additional power [12];
- Renewables integration, which requires storage capacities to instantly balance their short-term intermittency, to help restore local or grid-wide power quality;
- Managing isolated or island grids by providing grid stabilizing solution that provides frequency control and voltage support through active and reactive power injection and absorption.

23.3.2 Pumped hydro

The concept of a pumped hydroenergy storage has been well-known and widely used in a pumped storage hydroelectric power plant, which is discussed in Chapter 14. The potential energy of water is stored in two reservoirs, the upper one and the lower one so that there is a water head between them. These reservoirs are connected with the waterway, where hydroturbines, which may also operate as pumps, are installed.

The operating principle of a pumped storage hydropower facility is simple:

1. During the low demand period, when the grid has the excess of energy, the electricity is used to run the pumps and bring water to the upper reservoir from the lower — transforming electrical energy into the potential energy of the stored water;
2. During the peaking or high demand periods when energy is limited in the grid, water from the upper reservoir flows to the lower one due to gravity and runs hydroturbines that generate electricity.

The overall efficiency of this storage depends on the characteristics of the turbine and pumps, but can reach the level of 60–80%, and the life expectancy is similar to a traditional hydropower plant and can be up to 50 years.

This type of storage is a mature technology that is commercially available on the market. It offers enough operation flexibility and efficiency. It can support the grid in multiple ways: provide energy management, frequency control, have emerging capacity that can be implemented quickly. Being historically established, it can provide capacities of hundreds of megawatts up to the level of several gigawatts.

However, this type of energy storage is limited to a number of geographical locations suitable for two relatively large reservoirs and additional water source, high capital costs (as discussed in Chapter 15) and long construction cycles.

23.3.3 Compressed air

Compressed air energy storage (CAES) is the way to store excess energy in the reservoir with compressed air in the form of potential energy that can be further released to do some work. For a utility scale, it usually amounts to air being pumped into a large underground natural formation of special storage tanks. These energy storage applications are generally similar to pumped hydro in the range of applications and limitations as most of them are subject to suitable geographical locations.

The CAES works in a simple way. When there is excess energy in the grid, it is used to run the motors that power air compressors to bring air into the reservoir. The special point about the compressed air is that it heats up during the compression from the atmospheric pressure to a storage pressure that may be around 50–80 bar.

When the energy demand is high, the stored air runs a gas turbine–generator unit. During this discharge, the expansion removes heat so that if no extra heat is added to the process the discharged air will be much colder after expansion and release from the storage. Without proper heating, such released air flow cannot be used in existing turbines. This is usually achieved by mixing this air with fuel (mostly natural gas) within a combustion chamber to warm it up and increase efficiency. The heat can then be further recovered from the exhaust gases. Combustion of fuels, however, leads to CO_2 emissions.

In order to minimize the overall temperature issue and increase the round trip efficiency of this technology, two major technologies have been exercised [13]:

1. Diabatic storage, which dissipates much of the energy released as heat during compression into the atmosphere as waste. Therefore the temperature of the released air can be considered as the measure of the stored energy and, with further heating for combustion, it drives the efficiency further down and costs up;

2. Adiabatic process, which assumes that the heat, which is released during the compression, is recovered, stored, and then further utilized to reheat the compressed air during the turbine operation. The expected efficiency of such a technology may be well around 70%. However, this technology is still going through the pilot project phase.

Compressed air energy storages have many great advantages: they are quite cheap and flexible in size from small to a large-scale application that can easily support bulk power management similar to pumped hydro. This mature technology has proved to have a safe and reliable design and may leverage existing gas turbine technology.

On the contrary, the CAES requires stable and suitable geology and therefore is difficult to modularize. Moreover, the efficiency is mainly influenced by the requirement to warm up the air during the expansion, which is considered the most significant loss. If done with fossils like natural gas, it is then subject to emissions regulations and the volatility of fuel prices.

23.3.4 Thermal energy storage

As briefly discussed in Chapter 17, there are three main types of thermal energy storage system: sensible or single-phase, latent heat, and thermochemical storage.

Sensible heat storage The most widely used form of thermal energy storage in the power generation sector is sensible heat storage [14]. In a sensible heat storage system, a liquid or solid medium — such as water, molten salt (nitrates, carbonates, chlorides), or solids (sand, ceramic, graphite, concrete, rocks) — is heated or cooled to store thermal energy. The thermal energy is stored in a medium simply by transferring heat to that medium to increase its temperature (charging). This energy may later be drawn from the same medium when lowering the temperature (discharging) [15].

Generally, the first choice for the CSP plant project with a thermal energy storage system is molten salts, which can withstand extremely high temperatures to maintain high levels of cycle efficiency. They can offer up to 10–15 hours of energy storage depending on the size and type of material and technology. This is the least expensive option, but it requires large volumes of the medium due to their low energy density.

Latent heat storage Latent heat storage relies on the phase changes of the medium, for instance, from solid to liquid, that use latent heat to store energy. These materials normally include [16]:

- Organic materials, paraffin- or non-paraffin-compound. They are mostly used for heating and cooling applications in buildings as their melting point ranges around 20–32°C. They are chemically stable, non-corrosive and non-toxic with heat capacity, but are flammable.
- Inorganic phase changing materials like metallic alloys and salt compositions, with the latter mostly used in solar thermal applications. They offer high heat capacity, better thermal conductivity and rather low material cost. In contrast, they are corrosive and cool down quickly.
- Eutectic mixtures can be of three types: organic–organic, organic–inorganic, and inorganic–inorganic. They are generally used in building applications.

For solar power generation, mostly inorganic materials are used thanks to their melting point range of 115–897°C [16].

Thermochemical storage Thermochemical storage uses chemical reactions to store energy. It is based on the reversible chemical energy which absorb heat for storage and then releases it. The storage is "charged" by the endothermic reaction and is later discharged by an "exothermic" one.

The basic principle can be illustrated by simplified reactions:

$$AB + \text{energy} \leftrightarrows A + B.$$

Here the decomposition process is endothermic, meaning that the energy is absorbed, while the reverse synthesis reaction is exothermic and results in giving stored heat back [17]. The most common reaction used for this process is the hydration of salts.

During the energy storage phase, the chemical compounds are stored separately at ambient temperature, there is no thermal loss except for initial cooling, however, some other losses may be associated with the degradation of materials.

Thermochemical energy storage systems provide high energy storage density and are considered as a promising technology. Still, this technology remains expensive: the cost of the equipment is still very high compared to other technologies and the medium itself, even though the latter is also relatively expensive.

Overall, thermal energy storage, in general, offers relatively low cost and flexibility in sizing up to large scale for solar thermal power plant operation.

23.3.5 Supercapacitors

Supercapacitors are double-layer capacitors that have much higher capacitance but low voltage limits. They consist of two metal plates which are usually

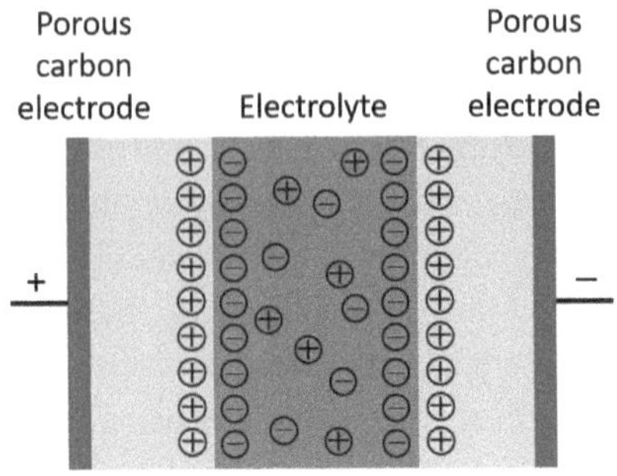

Figure 23.1 ***Simplified diagram of a supercapacitor.*** The applied voltage creates two separate charged layers on the surfaces.

covered by a porous activated carbon and immersed into the electrolyte. When a voltage is applied (as shown in Fig. 23.1), two separate charged layers are formed of the surfaces with a small separation distance.

Supercapacitors can withstand a long cycle life with minimal change in performance and their life can be around 20 years. However, they have quite low energy density and therefore cannot be employed as a continuous power supply. Therefore, they are mostly used in energy management applications which may require multiple rapid discharge and charge cycles for short-term power needs, for instance:

- Voltage stabilization during start-up and shut-down;
- Primary frequency response to compensate inertia of many renewables;
- Microgrid distribution system management for voltage, frequency, and power stabilization;
- Fast short-term backup power between a major blackout and a stable source like diesel generator start-up;
- Integrated with batteries, they can provide almost real time frequency, voltage and power smoothing for island mode operation;
- They can be used together with other energy storage systems for stable operation.

Overall, supercapacitors are an essential system for short-time energy storage and intermittency support for a modern grid management, especially in the face of its decentralization and implementation of large and small scale renewables.

23.3.6 Batteries

Another commonly used energy storage technology is the chemical one. Such energy systems are composed of one or more chemical compounds that absorb released energy when they react with another chemical compound. One of the simplest examples is the battery, which can be frequently referred to as electrochemical energy storage as the chemical reactions there are caused by electrical energy and subsequently produce electrical energy [18].

Battery storage can be probably named the next disruptive technology in the power sector [19]. Battery storage prices are dropping very fast due to the growth in electronic and electric vehicles. Many countries are investing into the increase of production of lithium-ion batteries to serve the growing demands in transportation and power.

Recent research has found [20] that storage is already economical for many commercial customers to reduce their peak consumption levels. Nowadays, battery storage is starting to play a much broader role in power generation and distribution sectors, moving from niche uses such as grid balancing to broader ones such as replacing some of the conventional power generators for reliability, providing power-quality services, and supporting renewables integration.

Storage batteries that allow storing energy generated by the wind or solar plants, are similar in chemistry to the commonly used car batteries but differ in design. Car batteries have to give short bursts of very high current to start the engine but are not designed to operate as storage batteries because they are made to be fully charged and have very short life if subjected to the deep recharge cycles [18].

There are several major types of battery that are applicable for large scale grid use:

- Lithium-ion which mostly dominate the market;
- Sodium-ion batteries are cheaper than lithium-ion but offer lower power density and are still under development;
- Molten-state batteries, which are composed of two molten metal alloys separated by an electrolyte. They are relatively simple to manufacture but have to be operated at high temperatures to keep alloys in liquid state. For instance, the sodium–sulfur batteries from NAS[2] are implemented for grid management renewables stabilization on a utility scale with the capacity of up to 50 MW.
- Rechargeable flow batteries (also known as redox batteries) is a type of battery where chemical energy is provided by two chemical components dissolved in liquids and separated by a membrane. The ion

[2] https://www.ngk.co.jp/nas/why/.

exchange occurs through the membrane, while both liquids separately circulate through the system. This technology offers a long lifecycle, flexible capacity and much higher power densities than the most popular Li-ion battery. This makes them better for applications where rapid charge and discharge takes place over a short time period, such as in frequency response services and demand management.

The rapid cost reduction of the battery technology is expected to continue, mostly due to massive investments of the car manufacturers. This may be important for the renewable energy systems and their integration. For instance, for solar PV systems, the cost of the balance of system becomes less critical as the cost of the core battery technology declines.

At the moment, utility-scale battery systems are mostly used for uninterrupted power supply and power quality control, as well as for transmission and distribution grid support and rapid load shifting [13].

23.3.7 Comparison and trends

There are multiple technologies that can be implemented on a grid scale for energy storage and grid management. Table 23.1 shows a brief comparison of major technologies regarding storage capacity, advantages, and costs.

Energy storage applications are becoming more and more important due to the introduction of new energy storage and their connection to the grid. The ongoing grid expansion, shift to smart grids and overall shift towards sustainable energy future requires energy storage technologies to support three major functions [13]:

Table 23.1 Comparison of major energy storage technologies.

	Capacity, up to, MW	**Life, years**	**Cycle life**	**Round-trip efficiency, %**	**Energy cost[a], $/kWh**
Flywheel	1	15–30	1 000 000	80–90	1500–6000
Pumped hydro	100+	up to 100	100 000	65–80	100–260
Compressed air	150+	up to 100	100 000	65–75	100–650
Thermal storage	100+	20–30	–[b]	90+	15–80[c]
Batteries	100+	10+	15 000[d]	70–96	250–3000[d]

Source: Based on [21,13].

[a] Average installation cost per generated kWh.
[b] The cycle lifetime is usually until the plant is decommissioned.
[c] Cost per thermal kWh.
[d] Battery number of cycles and cost highly depend on the size and technology used.

1. Uninterrupted power supply and power quality control for smaller capacities up to 100 kW and short periods of time is usually delivered by supercapacitors, flywheels, and various batteries. These are, for example, the standard sources of emergency power in utilities to address possible blackouts safely and quickly.
2. Transmission and distribution grid support and load shifting can be managed by batteries, larger flywheels, and supercapacitors that can provide capacities of dozens of megawatts for various time durations.
3. Bulk power management requires high capacity technologies of hundreds of megawatts and longer discharge times. This is usually executed through two mature utility-scale technologies: pumped hydro and CAES. Their relatively low installation cost per kWh allows them to be a great balancing mechanism to counterweight intermittent sources. Here thermal storage and hydrogen fuel cell technologies can also be mentioned as a source to provide bulk power.

It is likely that the energy storage technologies will continue to develop towards higher round-trip efficiency, lower energy production cost and flexibility of operation. This is important in the face of new grid challenges.

23.4. Hydrogen as an emerging energy carrier

Hydrogen has been used as a feedstock for many areas for decades, and the hydrogen industry is well established [22]. According to IRENA [22], the largest share of hydrogen demand comes from the industries such as chemicals (production of ammonia, refining for hydrocracking and desulphurization of fuels), iron and steel, glass, electronics, and others.

Recent trends showed a growing interest in innovative hydrogen applications, mostly in transportation sector. Several companies have released hydrogen-powered cars, however, the infrastructure still remains a bottleneck. Hydrogen as a gas is not that easy to transport, feed into devices, and take care of securely.

23.4.1 Production and storage

Hydrogen is abundant in the environment. It is stored in water, hydrocarbons (for instance, in methane CH_4), and other organic matter. Currently, over 95% of hydrogen production is based on the hydrocarbons of the fossil fuels: natural gas, oil, and coal. Steam–methane reforming — a reaction that

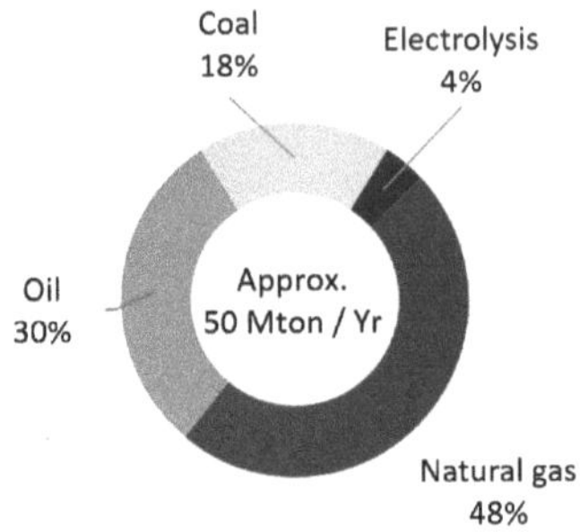

Figure 23.2 ***Sources of world production of hydrogen.*** The estimated world production of hydrogen is around 50 million tons per year. Most of hydrogen is produced out of fossils: natural gas, oil, and coal. *Source: Based on [22,24,25].*

reacts steam at high temperature around 1000°C with methane and catalyst — is the most common way to produce hydrogen, with the gasification of coal and oil being widely used for industrial scale production. Electrolysis accounts for only 4% of the total utility-scale hydrogen production, as represented in Fig. 23.2. There are also some other methods to produce hydrogen, e.g., reforming of alcohols, pyrolysis of the biomass, oxidation of hydrocarbons, etc. [23], which are not employed on a large industrial scale.

Electrolysis is an energy intensive process, but it can take the advantage of inexpensive excess renewable energy of wind or solar so that this energy can be stored in the chemical energy of hydrogen.

The industrial use of hydrogen is dominated by the chemical industry for ammonia production and fertilizers, which account for much of its industrial consumption, petrochemical industry where hydrogen is used for hydrocracking and desulphurization. Some other applications include metal production, food processing, electronics manufacturing, etc. [23]. This requires large amounts of hydrogen to be stored and transported to long distances. There are several major ways to store hydrogen [26,27] that have been globally employed:

- Liquid hydrogen LH_2 storage is currently used in industrial hydrogen demand centers such as refineries. It is normally produced in a liquefaction plant (which consumes considerable energy) and delivered by trucks or railroad transport to the consumer, which is the most economic way of its delivery. The equipment for liquid hydrogen storage and handling is even more expensive compared to LNG: the latter has a higher boiling point of −160°C compared to hydrogen's −253°C. Moreover, a boil-off of hydrogen of 0.06–0.5% per day (depending on the volume) on average would always occur no matter what type of vessel insulation is used.

- Compressed gas storage is widely implemented in chemical plants and refineries, as well as deployed in hydrogen-powered cars. While it requires less energy and can be scaled down, the volumetric energy density is much lower compared to LH_2 and the storage tanks can be costly. There are also certain safety concerns with highly pressurized and flammable gas storage.
- Bulk underground storage is widely used for hydrogen in the same way as for natural gas or oil. Subject to geological conditions, it requires special porous cavern with impermeable cap on top. Its extraction is made with another fluid that is pumped into the cavern to displace hydrogen.
- Hydrogen can be stored as chemical hydride of such elements as calcium, magnesium, lithium, or in sodium borohydride. Mixed with mineral oil in the form of slurry, hydrides can be stored in conventional tanks like liquid fuels. They have higher energy density compared to LH_2, and hydrogen can be easily liberated through the hydrolysis reaction when the compound is exposed to water. However, this reaction is highly exothermic, and the excess heat can be recovered for other needs. The used fuel — metal hydroxides — can be further sent for regeneration, but require a lot of energy for that.
- Another possibility is a metal hydride. Here absorption of hydrogen is exothermic and requires cooling, while the release of hydrogen is endothermic and needs a heating source. It also has a higher energy density compared to LH_2, is safe, and does not have leakages. But such energy storages are heavy and may require high temperatures for hydrogen release, for example, 300°C for magnesium hydride. This type of storage has been applied in facilities where weight is not an issue.

Remark 23.1. There has also been an ongoing discussion regarding ammonia. Due to its structure of three hydrogen and one nitrogen atoms, it contains more hydrogen by volume that pure compressed or liquefied hydrogen. Its energy density is much higher, and the ability of this gas to become liquid at a lower pressure makes it easier for transportation and storage. Thus, some approaches suggest converting renewable energy of wind and solar into this energy-rich gas, which can be easily liquefied, transported, and then further used within ammonia-powered fuel cells, hydrogen production, or even burnt directly.

23.4.2 Conversion of hydrogen and fuel cells

Hydrogen as an energy carrier can be converted into multiple useful forms of energy [27,23]:

- Hydrogen can be used in internal combustion or jet engines. Compared to gasoline engines, the efficiency of using hydrogen is around 20% higher with limited amount of emissions. While burning gasoline releases CO_2 and NO_x, the only products of hydrogen combustion in air are water vapor and small amounts of nitrogen oxides. The latter can be lowered by excess air or optimized cooling with water injection or using liquid hydrogen.
- Combustion with pure oxygen results in pure steam generation with the temperatures in the flame zone of around 3000°C. With the injection of water the temperature of the steam can be regulated, and the process can deliver both saturated and superheated steam.
- Hydrogen and oxygen may be combined at the temperature significantly lower than the flame combustion (from ambient to around 500°C) in the presence of a suitable catalyst (for example, platinum). This principle is used in special catalytic burners, were catalyst can be dispersed in a porous structure to increase the reaction surface, and the reaction is controlled by the hydrogen flow. This reaction cannot take place outside the reaction zone due to a high concentration of hydrogen (above flammable limit of 75%) and absence of flame. There are also no nitrogen oxide emissions thanks to low temperatures.
- Hydrogen can be combined with oxygen without combustion in an electrochemical reaction and produce DC electricity. This process takes place in a device called *fuel cell* (also see Chapter 1 for details).
- Metal hydride conversion uses the property of hydrogen to form hydrates with metals as a way of energy conversion.

Fuel cells are considered most promising technologies of converting hydrogen into electricity for commercial use. Evidently, the simplified formula

$$2H_2 + O_2 \Rightarrow H_2O \quad + \mathit{Electricity}$$

looks great for sustainable future: no CO_2 emissions, converts chemical energy into the most useful energy carrier — electricity. Contrary to batteries, it requires continuous supply of hydrogen fuel and oxidizing agent (oxygen being simple). Depending on the type of the electrolyte used in the process, there are several major types of fuel cell [23,27,28]:

- *Alkaline fuel cells* (AFC), which use concentrated KOH as an electrolyte, retained in a matrix, and a wide range of electrocatalysts can be used (e.g., Ni, Ag, metal oxides, noble metals). This fuel cell is sensitive to the CO_2 content and requires pure hydrogen;
- *Polymer electrolyte membrane* or *proton exchange membrane fuel cells* (PEMFC), which use a thin polymer membrane as the electrolyte and an excellent proton conductor. The catalyst is typically platinum, or Pt–Ru alloy if hydrogen supply contains small amounts of CO_2 — both catalysts are costly. They normally operate at temperatures below 120°C;
- *Phosphoric acid fuel cells* (PAFC), which use concentrated phosphoric acid to serve as electrolyte. The electrocatalyst is platinum which is expensive. Another disadvantage is the use of acidic electrolyte;
- *Molten carbonate fuel cells* (MCFC), which have the electrolyte composed of a combination of alkali (Li, Na, K) carbonates in a ceramic matrix. The operating temperatures are around 600–700°C when carbonates form a highly conductive molten salt. These fuel cells have low power density and long startup times;
- *Solid oxide fuel cells* (SOFC), which use a solid, nonporous metal oxide as the electrolyte. The cell operates at high temperatures of around 700–1000°C where ionic conduction by oxygen ions takes place. However, high operating temperatures require longer startup time and put certain operating limits.

Table 23.2 gives an overview of fuel cells technologies that have been under development these days. Their electrical efficiency is around 60% with the trend now targeting at the decrease of their cost of production and scalability.

23.4.3 The future of hydrogen

Hydrogen is likely to gain a broader use in the coming decades if the cost of its production, transportation, and distribution goes down. Due to the lack of modern infrastructure beyond traditional hydrogen use (for chemical or petrochemical industries) and the absence of mature policies and regulations (including safety) regarding hydrogen use, it is even more complicated to roll it out for mass consumption.

For a while, the idea of hydrogen economy has been promoted vigorously by scientists, politicians, business and environmentalists as one of the most promising solutions [26], but so far, the present day role of hydrogen

Table 23.2 Comparison of fuel cell technologies.

Technology	Electrolyte	Operating temperature, °C	Stack size	Electrical efficiency, %
AFC	KOH	< 100	1–100 kW	60
PEMFC	Polymer membrane	< 120	up to 100 kW	40–60
PAFC	Phosphoric acid	150–200	up to 100 kW	40
MCFC	Molten alkali carbonates in a porous matrix	600–700	300 kW per module, up to 3 MW	50
SOFC	Solid metal oxide	700–1000	up to 2 MW	60

Source: Based on [23,27,28].

for energy production is still negligible. There have been plenty of demonstrator projects, prototypes, and even commercial applications, but they have not yet been rolled-out on a large scale. One of the major approaches to unlock hydrogen's potential is to integrate the renewables, hydrogen storage, and fuel cell technology into a future decentralized energy system.

The long-term benefits of hydrogen are compelling, and it provides a promising pathway for the energy transition. Continuous improvements in cost and performance of hydrogen related technologies are being made along the entire value chain [29], but the challenges are still there:

- Integrating large scale renewables, to ramp up the generation of hydrogen, compensate for the variability of power generation by converting wind or solar energy into chemical energy of hydrogen;
- Increasing storage capacity and developing proper infrastructure to bring to life its potential to store energy for longer periods at the levels of gigawatts;
- Improving the economic performance of the hydrogen utilization of a large scale, that is, making fuel cells more cost-effective.

The basic technology of the fuel cells has been in place already for a while, initially starting for some applications in the space industry. While fuel cells can be and are used for multiple applications such as commercial, industrial and residential primary and backup power generation, serve as power sources in remote locations, and in some specific applications, they are still a costly solution. However, the power generation sector is still in search of a profitable niche [30]. This may well be driven by the introduction of carbon policies that may well effect the business model of hydrogen use. Such assumptions are used in some of the scenarios to model the future energy system [25,31].

An important factor that has to be always considered is hydrogen safety. Since hydrogen has the smallest molecule, it has a great tendency to escape through even tiny openings, which may not allow other fuel gases or liquids leak. This requires the use of special materials, i.e., certain grades of steel that can prevent hydrogen penetration. However, prolonged exposure to hydrogen, especially at high temperatures and pressures, can cause the steel to loose its strength that may lead to material failure [27].

Even though hydrogen is highly diffusive, its leakage may lead to an explosion: a hydrogen–air mixture can burn in a wide range of concentrations, though its flammability limit is pretty high. Moreover, liquid hydrogen may cause cold burns due to its extremely low liquefaction temperature, but its leakage would dissipate relatively fast. Moreover, a failure in a relief valve of a pressure vessel may lead to a violent explosion of a boiling hydrogen [27].

Over the past decades hydrogen has been widely used in many industries with high safety standards implemented in production, storage, transportation and its further utilization [29]. Multiple reports can prove the capabilities to treat hydrogen in a safe way, though its mass use would need further development of extra safety procedures.

References

[1] Renewable energy sources and climate change mitigation. Special report of the intergovernmental panel on climate change. Cambridge, UK, New York, NY: Cambridge University Press; 2011.

[2] Gabbar H. Smart energy grid engineering. Elsevier Science. ISBN 9780128092323, 2016.

[3] Sioshansi F. Smart grid: integrating renewable, distributed and efficient energy. Elsevier Science. ISBN 9780123864529, 2012.

[4] Skopik F, Smith P. Smart grid security: innovative solutions for a modernized grid. Elsevier Science. ISBN 9780128023549, 2015.

[5] Borlase S. Smart grids: infrastructure, technology, and solutions. Electric power and energy engineering. CRC Press. ISBN 9781351833721, 2017.

[6] Department of Energy (DOE). The smart grid: an introduction. URL: https://www.energy.gov/sites/prod/files/oeprod/DocumentsandMedia/DOE_SG_Book_Single_Pages%281%29.pdf.

[7] 110th Congress. Energy independence and security act of 2007. URL: https://www.congress.gov/110/plaws/publ140/PLAW-110publ140.pdf.

[8] IRENA (International Renewable Energy Agency). Smart grids and renewables. A guide for effective deployment. Abu Dhabi: IRENA; 2013. URL: https://www.irena.org/DocumentDownloads/Publications/smart_grids.pdf.
[9] European Regulators' Group for Electricity and Gas (ERGEG). Position paper on smart grids. URL: https://www.ceer.eu/documents/104400/-/-/c481db2a-3cfb-6d6f-4b58-da3dee68de4a, 2009.
[10] Burheim O. Engineering energy storage. Elsevier Science. ISBN 9780128141014, 2017.
[11] Amiryar M, Pullen K. A review of flywheel energy storage system technologies and their applications. Applied Sciences 2017;7. https://doi.org/10.3390/app7030286.
[12] Tester J, Drake E, Driscoll M, Golay M, Peters W. Sustainable energy: choosing among options. MIT Press. ISBN 9780262304634, 2012.
[13] IRENA (International Renewable Energy Agency). Electricity storage and renewables: costs and markets to 2030. Abu Dhabi: IRENA. ISBN 978-92-9260-038-9, 2017. URL: http://www.irena.org/-/media/Files/IRENA/Agency/Publication/2017/Oct/IRENA_Electricity_Storage_Costs_2017.pdf.
[14] Harvey A. The latest in thermal energy storage. Power 2017. URL: https://www.powermag.com/the-latest-in-thermal-energy-storage/.
[15] Bergan P, Greiner C. A new type of large scale thermal energy storage. Energy Procedia 2014;58:152–9. https://doi.org/10.1016/j.egypro.2014.10.422.
[16] Anisur M, Kibria M, Mahfuz M, Saidur R, Metselaar I. Latent heat thermal storage (LHTS) for energy sustainability. In: Energy sustainability through green energy. Green energy and technology. New Delhi: Springer; 2015.
[17] Kato Y. Chemical energy conversion technologies for efficient energy use. In: Paksoy H, editor. Thermal energy storage for sustainable energy consumption. Springer; 2007. p. 377–91.
[18] Dincer I. Thermal energy storage: systems and applications. Wiley. ISBN 9781119956624, 2011.
[19] Frankel D, Wagner A. Battery storage: the next disruptive technology in the power sector. McKinsey & Company; 2017. URL: https://www.mckinsey.com/business-functions/sustainability-and-resource-productivity/our-insights/battery-storage-the-next-disruptive-technology-in-the-power-sector.
[20] D'Aprile P, Newman J, Pinner D. The new economics of energy storage. McKinsey & Company; 2016. URL: https://www.mckinsey.com/business-functions/sustainability-and-resource-productivity/our-insights/the-new-economics-of-energy-storage.
[21] Lazard. Lazard's levelized cost of storage analysis — version 3.0. URL: https://www.lazard.com/media/450338/lazard-levelized-cost-of-storage-version-30.pdf.
[22] IRENA (International Renewable Energy Agency). Hydrogen from renewable power: technology outlook for the energy transition. Abu Dhabi: IRENA. ISBN 978-92-9260-077-8, 2018. URL: http://www.irena.org/-/media/Files/IRENA/Agency/Publication/2018/Sep/IRENA_Hydrogen_from_renewable_power_2018.pdf.
[23] International Energy Agency (IEA). Hydrogen production and storage. R&D priorities and gaps. Paris: IEA; 2006.
[24] Santhanam K, Press R, Miri M, Bailey A, Takacs G. Introduction to hydrogen technology. Wiley. ISBN 9781119265542, 2017.
[25] International Energy Agency (IEA). Technology roadmap. Hydrogen and fuel cells. France: OECD / IEA; 2015. URL: https://www.iea.org/publications/freepublications/publication/TechnologyRoadmapHydrogenandFuelCells.pdf.

[26] Ekins P. Hydrogen energy: economic and social challenges. Earthscan. ISBN 9781849774949, 2010.
[27] Sherif S, Goswami D, Stefanakos E, Steinfeld A. Handbook of hydrogen energy. Mechanical and aerospace engineering series. Taylor & Francis. ISBN 9781420054477, 2014.
[28] Office of Energy Efficiency & Renewable Energy. Comparison of fuel cell technologies. URL: https://www.energy.gov/eere/fuelcells/comparison-fuel-cell-technologies.
[29] Hydrogen Council. How hydrogen empowers the energy transition. Hydrogen Council; 2017. URL: http://hydrogencouncil.com/wp-content/uploads/2017/06/Hydrogen-Council-Vision-Document.pdf.
[30] Overton T, Maize K. Whatever happened to fuel cells?. Power 2017. URL: https://www.powermag.com/whatever-happened-to-fuel-cells.
[31] REN21. Renewables 2017 global status report. Paris: REN21 Secretariat. ISBN 978-3-9818107-6-9, 2017.

CHAPTER TWENTY-FOUR

Transitions towards a sustainable power generation system of the future

Contents

24.1. What is going on?

The world is living through a number of important transitions that span over the borders and concern the whole society. The first industrial revolution initiated the rapid development of world economies, and many countries emerged as global industrial leaders harnessing innovation and implementing best practices to go further and quicker. Such growth inevitably caused growing consumption of resources to increase the manufactured capital in an exponential manner. Production of steel, generation of electricity, oil production, manufacturing of machines and appliances — these a few examples of the evident consequences of the exponential economic growth. In turn, economic growth would close the feedback loop to further increase the rates of production and consumption.

By the beginning of the 21st century it became evident to the open public beyond scholars that exponential growth like this is not possible to

Sustainable Power Generation
https://doi.org/10.1016/B978-0-12-817012-0.00039-6

last forever. The presently known closed system of the Earth has limited resources: cheap fossil reserves are likely to be depleted in some years (with the forecast ranging from a few decades to a few hundred years), fresh water is going to be limited (and no-one ever thought of it as an issue a couple of centuries ago), the population is beating another record (and it has to be fed and provided with sanitation), and so on.

Finally, the ongoing observations of the climate warn us about the gradual changes: the average temperature of the Earth is likely to rise and continue doing so, GHG emissions have never been properly taken care of, and the overall future is vague. Multiple discussions have taken place, research publications, reports, conferences, governmental documents, and policies have been discussed, issued and ratified without a clear and solid answer what to do on a global scale.

According to the UN [1]:

- Almost 800 million people live below the international poverty line;
- Globally, one in nine people in the world today are undernourished;
- 3 in 10 people lack access to safely managed drinking water services and 6 in 10 people lack access to safely managed sanitation facilities;
- 13% of the global population still lacks access to modern electricity;
- Half of humanity lives in cities today, and 5 billion people are projected to live in cities by 2030;
- Arable land loss is estimated at 30 to 35 times the historical rate.

The list can be continued but the overall picture suggests that urgent steps shall be taken to ensure the transition towards sustainable development.

As discussed in detail in the previous parts of this book, energy is probably the major "prime mover" of the development, the "blood" of the system, and also one of the major limiting factors: the use of fossils seems to contribute to the negative change the most. Power generation here is not the last to mention. Of course, multiple measures have already been taken to increase the efficiency of the power generation system, but they were mostly driven by economic considerations. One merely can mention any of the countless developments that have been introduced during the last decades, including such breakthroughs like:

- Conventional steam power plant efficiency breaking the record of 47% with the modern ultrasupercritical technology;
- Modern combined cycle power plant efficiency breaking the psychological limit of 60% with the value of 62.22% and being expected to grow further;

- Remote control and monitoring allows predicting the state of the machinery and optimizing maintenance plans to further increase flexibility, performance, and efficiency;
- Renewables are steadily increasing their capacity factor, breaking the value of 40% alongside their booming capacity growth;
- A floating nuclear power plant has been released for its performance tests to become the source of power for remote isolated locations;
- Mobile and small nuclear plants have been under development already for a while with the first commercial specimen to come soon;
- Wind and solar are gradually increasing their share and capacity per unit.

The list can be further extended with countless developments and R&D breakthroughs in power generation industry on a broader level. There is, probably, only one promising technology that has been dreamed of since more than 60 years without major success — the nuclear fusion. Every year since almost seven decades there have been announcements that in 50 years there will be a commercial plant commissioning, but we may expect to hear the same promising slogan in the next 50 years as well. Thus, this illustrates the non-stop process not only in power generation industry, but the whole economy. There are both internal transitions within the energy and power generation industries and external influencing factors from the neighboring areas of economy and society, philosophical concepts and mindset changes. The former include such major trends as:

- Decentralization of electricity generation,
- Switch to emerging green technologies like wind, water, and solar,
- Digitalization and implementation of analytics; and
- The efficiency race.

Among dozens of changes that are currently taking or will inevitably take place in the next decades in the minds of people, corporations and NGOs, there are several trends, or rather, conceptual changes that may be crucial for the sustainable power generation industry and the world:

- Global electrification of things, including one of the fastest growing areas — electric mobility;
- Digital transformation of industries;
- Urbanization and sustainable city design;
- The emerging concept of justice applied to energy;
- Sustainability education that emerges in business environment;

- Corporate social responsibility concept that has been applied on a regular basis but which shall transform to a normal business practice and style of doing business.

24.2. Major transitions in power generation industry

It has been proclaimed by many of the world leading energy-related companies that the world is likely to go through a number of hurdles which would challenge the system and pose the question of choice: whether to stay within the unlimited consumption paradigm or to draw our direction into the sustainability. However, since we still live in coal dominated world with the figures suggesting slightly less than 40% of electricity coming from it, it is necessary to start the transition and manage its impact all together. Even though the directions of change are the same in all regions, the speed and intensity towards clean energy would be different and may require various scale of efforts by societies.

24.2.1 Centralized to decentralized power

We may say that we are now witnessing an exciting crossroad of technologies when the world is tending to shift from a traditional centralized grid — a grid that connects generating capacities and delivers electricity to the consumers connected to the same grid. As discussed in Chapter 23, the concept of the energy network has been evolving around the idea of centralized electricity generation, transmission, storage, and distribution. Such a system had to manage the waves of industrial revolutions and support the exponential growth, and in most cases it managed, though not always very well.

The present power generation system, plugged into the centralized network, faces an increasing challenge of the mismatch between supply and demand, which is an overwhelming economic problem. This is due to the layout that is not capable of supporting the emerging needs, already old-fashioned engineering solutions of the traditional grid, and overall efficiency.

Distributed energy is the utilization of smaller power generation capacities and local storage systems located in the proximity of consumers. These systems employ emerging renewable energy sources like solar, wind or small hydro, where batteries should play an increasing role of storing energy and dispatching it when necessary.

However, such a transition will not take place overnight and completely deplete centralized power generating facilities like large hydrodams, which are located in remote areas, or large offshore wind parks. This is mostly important for massive industrial consumers that require constant and high energy supply, but the increased electrification demands power at the right time and place, so that decentralization will be the perfect match.

Recent projects that come from conventional and historically "large and base load" technologies like nuclear suggest that their path has not yet been fully exercised: a floating nuclear power plant that has been recently tested in Russia may be a significant player in powering island countries or remote areas, and the development of modular and autonomous nuclear reactors may contribute to many local grids.

24.2.2 Fossil to renewable power in the long run

While the deployment of renewable energy technologies has been agreed as important and essential, they are still very limited compared to their potential, despite governmental incentives and support thanks to obvious reasons: their capacities are much smaller compared to large utility-scale conventional technologies, while the economic side of new energies becomes more and more attractive [2,3]. On top, the renewable energy reserves represent enough primary energy potential and therefore can theoretically support all needs of the growing economies [2]. However, the overall contribution of non-hydro renewable energy technologies is still very limited compared to their potential, and there is an urgent need to design, develop, and deploy these technologies with minimum dependence on external sources [4].

The problem here lies in choosing the right strategy of the gradual replacement of the currently operating capacities to ensure a smooth transition without interruptions and, more importantly, to avoid the possible negative effect of a poorly planned action. The development of the roadmap shall be conducted on a global scale and take into account such topics as:

- The required level of electricity for a specific region or country;
- The available resources of the specific country, to provide or not decrease its energy independence and security;
- The projected growth trajectory;
- The economic burden of the transition that has to be addressed by the current and, importantly, future generations.

The transition period, therefore, would require the implementation of conventional technologies and still on a global scale, thus continuing the process of their substitution with a more sustainable sources of power. The speed of the transition will be largely governed by the possibility of governments to unite on this path and the choice of the sustainability paradigm.

To sum up, despite multiple skeptical claims that renewables will not be capable of powering the society with the required electricity, there are a number of example countries that have come up to an extremely high level of renewable energy generation from wind, solar, biomass. The list is likely to grow year after year, and the green club will become something common in the nearest future.

24.2.3 Digitalization and analytics

The Industry 4.0 that is taking off these days promises to become a breakthrough in optimizing performance, lowering operation and maintenance costs, and providing new solutions that are more efficient, environment-friendly and economically more beneficial.

Digitalization is a key compound to glue together the whole industry and make energy systems more connected, efficient, intelligent, and sustainable, increase their safety and productivity, and make them more accessible for the society.

The amount of data has grown dramatically since the deployment of the Internet, and the present day requires efficient and productive algorithms and analytics to get most out of this data. Grid management is probably the most developed from the digital point of view, as this part of the energy system has been utilizing digital solutions and data analytics for the efficient energy distribution and market operation.

Similarly, the industry has been implementing digital control systems, process automation solutions, and systems to manage industrial processes autonomously with minimal involvement of the operator. This has now been extended towards smart meters, fully automated control systems, augmented reality solutions to simplify inspections, AI solutions to predict process parameters and ensure the quality of product is stable. Software solutions and built-in analytics allow getting information from almost any point of the process and aggregating it, analyzing remotely (for instance, in the technical center of the equipment manufacturer), and issuing commands from the remote, if required.

Similarly, modern digital systems can aggregate condition monitoring systems of the equipment, for instance, turbines, generators, compressors, or motors, analyze them based on historical data (e.g., about the maintenance), plan the outages and lifecycle, and optimize warehouses to order spare parts in advance or 3D print them on the premises.

The stories are endless with one simple idea in common: making most of the modern digital innovations in data collection, transmission, analytics to make the processes more efficient, safer, more reliable, and profitable in the end. We shall still draw attention to the fact that digitalization is not a goal by itself, yet a solution for some of the process bottlenecks, a way to save money and, finally, increase efficiency in energy consumption and material use.

24.2.4 Efficiency game

To direct our way towards sustainability, we may start with the increase of efficiency as the simplest means each and every person can contribute to. With the energy generation in mind, efficiency means not only efficient use of energy, but also responsible consumption of other resources:

- The use of energy efficient and energy saving appliances, like LED lights, smart or passive heating in the houses, unplugged appliances, etc.;
- Responsible food consumption, to minimize waste production as food industry is super energy intense;
- Responsible and careful fresh water consumption, to ensure that water is not wasted (for example, from the dipping tap);
- Waste management that results in proper sorting, recycling, and further re-use of materials.

These are a few examples of everyday efficiency increase that, if done on a regular and global basis, can significantly support the sustainability journey.

From the industrial scale, there is a trend to repower older capacities with new and more efficient equipment. One can name the retrofit projects when older inefficient coal-fired facilities have been upgraded to USC boilers, when their efficiency almost doubles, and the installation of innovative cleaning solutions allows minimizing the impact of GHG emissions to a much lower scale.

Another trend is to replace some of the old materials with the newly developed ones. In many cases the oil associated gas that is released during the oil production has been burnt, which then contributs to the global warming. But, being a valuable resource for the chemical industry, this gas is now extracted and supplied to the factories for chemical production of plastics like polyethylene, polypropylene, and others. These novel materials are a step forward against glass (which is too energy consuming to manufacture), paper (which is still water-intense to manufacture and requires significant effort and energy for proper recycling), and metals. The list is likely to be continued, and the plastics will continue to dominate the production. On top, they can be either recycled or, in the worst case, burnt to recover energy in WTE plants as they are carbon-based.

24.3. Transitions beyond energy system

The energy system is an integral part of the global economy, which comprises all production and services' areas and social sphere. The society has always been evolving under multiple forces: political, social, environmental. The information flows, disruptive innovation, social changes impact the whole economy and energy system. There are, however, several major trends that directly affect the power generation system or, in other words, are dependent on energy production.

24.3.1 Electrification of things

Modern society is highly dependent on electricity, and this trend is going to advance even further. Electricity is the most convenient form of energy to use in small or large devices, transmit over short and long distances, convert efficiently to other forms of energy. Its overwhelming usability has proved to be the cornerstone of modern material world: electricity brings light, runs machines, powers our appliances, provides sanitation, manages communication and information flows.

Even though there are still areas where other sources of energy are utilized for doing work, for instance, powering transportation or producing materials, electricity is gradually growing its share of secondary energy use. This global trend towards the electrification of things and ongoing global digitalization require much more efficient electricity production, transportation, and consumption, which has to be distributed through the globe. Electricity generation is at the center of the global electrification transition [5], and the electric systems of the future shall include both central and distributed generation sources with a mix of flexible and non-dispatchable capacities [6]. Contrary to historically localized industrial centers (like iron and aluminum metallurgy, pulp and paper, chemical plants) which had consumed most of the energy, today's electrification requires scattered centers of electricity generation through deployment of distributed power managed by smart grids. At the same time, the grid management has become more and more complicated not only due to specific consumer patterns, but also because of various generating means and capacities connected to the grid.

One of the best examples is the growth of electric car industry. The switch towards electric cars would lead to a decrease of GHG emissions but only in case of reliable and on-demand supply of "green" electricity where required and at the required quantities, and the development of sustainable production and recycling processes.

And, by the way, no-one can ever guarantee that humanity would never discover another innovative form of energy that may become even more convenient to use than electricity.

24.3.2 Digital transformation of industry

We are living in the times of a major transformation of industry on a global scale — the fourth one in human history. The first three industrial revolutions are well-known:

1. The adoption of mechanization, to replace labor with machines, and the employment of water and steam energy to run these machines, considerably increased productivity of the factories;
2. The introduction of mass production through assembly lines and the beginning of the electrification;
3. The wide use of computers and automation in the manufacturing process, which increased efficiency and quality.

The fourth one, usually known as the Industry 4.0, is aiming at complete digitalization that changes technologies, processes, and business models. The approach is all about connecting the real and virtual worlds, to drive productivity and efficiency through a number of disruptive innovations. According to BCG analytics,[1] there are nine technology trends that form the building blocks of Industry 4.0:

1. Big data collection from various sources and analytics, which would become an integrated tool for real-time decision making;
2. Autonomous robots, which would gradually start to replace people or semi-autonomous machines and cooperate with each other for safer operation;

[1] https://www.bcg.com/en-ru/capabilities/operations/embracing-industry-4.0-rediscovering-growth.aspx.

3. Simulation, which is expected to be used more extensively to leverage real-time data, create models, and conduct scenario analysis for development and optimization purposes;
4. Horizontal and vertical system integration, which would tend towards companies, departments, functions, and capabilities becoming much more cohesive through various (even automated) value chains;
5. The Industrial-Internet-of-Things (IIoT), which advises that more and more devices will be enriched with embedded computing to enable communication and interaction between them, so that decision-making can be decentralized;
6. Cybersecurity, which is an important measure to ensure security of multiple connected devices. Safety of critical data, reliability of connections, and access management are becoming coherent to the modern industry;
7. The cloud, which can provide universal data storage and access, as more and more data sharing is required by almost all spheres of activity. This would further stimulate the data driven services that can be performed on the cloud;
8. Additive manufacturing like 3D printing, which is an emerging technology for prototyping and possible mass production. The development of a variety of technologies and materials allows speeding up the development process and can completely transform business models in manufacturing.
9. Augmented reality, which can support the design process, selecting parts in warehouses or repair activities. These technologies are still in their infancy but are expected to come into much broader use in the years to come.

Evidently, the new industrial revolution is reshaping the whole industry to bring it to the next level. And the change requires resources, with energy being at the first place. The disruptive innovations all are energy-intensive: data centers need power for cooling, machines operate on electricity, networks are not able to run without electricity. Electric energy is required in large quantities, at the right place and at a right capacity in time. This puts a lot of pressure on the power generation system, which shall be the backbone of the implementation of the Industry 4.0.

24.3.3 Sustainable cities

The ongoing urbanization results in the situation that about half of the world population lives in urban areas, and the trend is likely to continue.

Though urban areas normally provide a higher level of living, they also require more resources in turn to maintain them. The rapid growth of cities and high rate of migration from rural to urban areas during the last 100 years have led to the boom of mega-cities.

Such concentration of people causes social problems and, unfortunately, leads to the concentration of the poor, increased crime rate, limited sanitation, and lowers overall living conditions. The cities may fail to accommodate the inflow of people if the trend continues without any precautions.

The creation of sustainable cities and communities, which is a Sustainable Development Goal 11, aims at providing favorable living conditions for the growing urban population:

- Providing safe and affordable housing;
- Deploying and extending safe and affordable public transportation for all categories of inhabitants;
- Protecting cultural heritage of the communities against rapid (and sometimes brutal) growth;
- Reducing the mortality rate by paying special attention to the air and water quality, developing more green areas and parks.

Multiple approaches to sustainable cities have been discussed, ending up with great and promising ideas to create safe, comfortable, sustainable environment for working, living, and cultural and leisure activities:

- Developing eco-industrial parks where multiple companies work together to decrease their environmental footprint by lower energy consumption, efficient material and water use;
- Promoting urban farming to provide food almost on the premises;
- Either increasing the density of population in urban centers to utilize economies of scale or, to the contrary, promoting wider public spaces and walkable areas, to extend the boundaries of the urban location;
- Implementing car-free cities' strategy with pedestrian areas and public transportation, to avoid carbon footprint of cars.

All these and many other trends and ideas of urbanization and sustainable city development would require more energy, which shall also fit into these models: be sustainable, reliable, flexible, and affordable.

24.4. Sustainability education

The whole society is now exposed to the potential impact of both the favorable transitions and emerging problems, and the latter seem to be unlikely to fade away without immediate measures and actions. However, only a few understand the basic causes of the problems and the rest consequently fail to realize that fundamental social and technological changes are required to overcome them [7]. It is obvious that existing ways of thinking, living and consuming need to be changed in order to enter a process of transition towards sustainable development [8]. The present world educational system, in the global sense, turns out to be incapable of providing a complete picture regarding the modern energy, sources of its supply, ways of consumption and related social, technological, and environmental issues.

The lack of specialists having the required knowledge, skill set and competences can be seen as one of the major challenges towards sustainable development [8–10]. With the boundaries of managerial competences expanding rapidly, present day leaders must necessarily be aware of the subjects that continue to become more and more complex and crucial. In the general business focus on the sustainability and future development, however, their knowledge seldom progresses beyond the level of generic reports and some ideas of the forecasts regarding major energy trends, as well as the level of basic vision of the green energies being the driver of the future. More detailed discussions or the history and underlying problems and developments, especially those that made the present situation happen, are usually left outside of the focus areas.

While some business schools and universities started offering programs or specially designed courses in the fields of sustainability and green energy technologies, they are not always considered as an essential skill of the future leader unless being part of a specific program chosen by an individual. Therefore a conceptual approach is now required for the consolidation and representation of the key features of the managerial integrative competences [8] that would address the challenges and the needs of the society in sustainable energy supply, including the ability to cooperate with scholars from various disciplines and practitioners from the business side. The latter are an extremely important part of the knowledge base as they question the theory in real-life cases and therefore become a valuable source of up-to-date information and examples.

24.4.1 The need for sustainability education

During the past decades there have already been multiple approaches to emphasize the need and introduce educational programs and courses in the field of sustainability [8,11–13], sustainable development [14] and renewable energy [15], and address the needs of the industry and the public [4], however, in the overall abundance of energy-related research works the topics of competence build up and education have been limited [16]. Present research directions aim at multiple advancements in the field of global warming challenge, problem of clean energy and new technologies, especially those that can potentially serve as the replacement for fossil sources like coal, oil, and gas.

While education itself is one of the most powerful means for providing solutions to the problems faced by the society [4], it:

1. Has to possess a number of relevant attributes, and
2. Should be supported by relevant experience.

Intrinsically, education is supposed to be beneficial, especially when it is shaped and delivered in a way to fulfill the needs of the person and cover multiple topics to be more versatile. This leads to the fact that the educational side of the competence shall be integrative [17,18], hence it shall be subject to the following requirements:

- Education has to be properly shaped and planned to ensure the common thread that undergoes the complete educational program till the final specialization. This has to be designed and delivered by the professionals of the respective field and challenged by the industry and academia.
- The fundamental basis of education has to be deep enough to equip the future leader with the solid problem solving skills and let this person be open to innovation and change.
- The specialization which is built on top of the foundation is supposed to be way more dynamic and exposed to the rapid changes of technology and society, therefore it has to be verified and tested by the industry.

This is especially important with regard to the competence, where education forms an important, yet only one pillar. The second one is built by real-time skills and experience gained by a person throughout the practical interactions within the industry, i.e., along the career path, either business-related or academic. While education equips the future leader with the necessary knowledge and tools, the experience has to be gained on top to ensure the expertise is established and developed to the required level. Importantly, conceptions of sustainability and their associated learning and teaching are context driven and vary considerable [12]. Thus, education has multiple perspectives on the way to interpret the concepts and their implications, but still has a vital role to play in the development of a sustainable society. It is a powerful agent of social change, raises awareness of new developments, provides training for the professionals and trains researchers who will develop the next generation of systems and devices [7]. Therefore, the idea of competence in the area of sustainable development build-up approach should be educative, informative, and innovative. The specific objectives of such an approach shall extend over pure renewable energy education or sustainability [4,7,8,10] and may include:

- Drawing attention to the way the environment has been impacted by the human activity on the way towards economic growth and increase of population. This requires understanding of energy as one of the major drivers of the humankind development and an essential resource for human well-being;
- Developing awareness about the nature and causes of energy related challenges being faced by the humankind, such as the scarcity and depletability of fossil fuels, impact to the environment and the climate, fresh water problem, etc.;
- Forming awareness of the traditional sources of energy and the economic, political, and social reasoning behind as major historical drivers, and building the bridge towards novel and innovative approaches of the unconventional sources of power like wind, solar energy, geothermal or tidal power, and others. It shall be represented on the comparative landscape with revealed positive and negative sides of each possible solution;
- Motivating and preparing future leaders to take actions in the direction of development and implementation of alternative strategies to face current and future challenges of the energy sector as the key driver and enabler of growth;
- Developing responsibility and social awareness of the leaders towards sustainable development as this requires deep understanding of the longevity of the goals that would impact not only current but future generations. It should build up competences related to responsible consumption and awareness of the consequences that may follow an unsustainable way of living.

24.4.2 Desirable features of the sustainability education

While the renewable education follows the trend of being provided at mass level on a global scale [4], it has to be a part of the global sustainability educational framework approach to grow responsible leaders. Recent discussions in the field of management education have revealed changes needed for responsible education of future managers and entrepreneurs [18–20], however, the implementation of these changes is still ongoing within the limited number of business schools and other educational institutions.

The sustainable development competence is by far more complicated and demanding, hence it has to be implemented through formal and informal education and further developed through the extensive hands-on experience. For instance, the traditional way to solve the energy supply problem shall result in answering two basic questions: the "what?" and the "how?". Even though answering them will allow a leader to get a solution (the how) for the energy problem (the what), it may not necessarily be optimized in the sense of sustainability. The sustainability competence shall be founded on the three major pillars:

1. Energy and material value chains, the "what?";
2. Innovation and change management, the "how?";
3. Sustainability paradigm, "to which extent?" and "why?".

Energy and material value chains competency has to provide an answer to the question what the resource we have are, and how we can utilize them. It consequently addresses the key topics of energy resources, their production, transformation, transportation, and consumption. On top of the general physical and engineering knowledge, it has to imply estimation of reserves, associated lifecycle costs, uneven distribution, political and environmental risks, and other related issues. On the transformation side, it shall at least focus on electricity generation as the most useful form of energy, fuel production and its implications, and new ways to mitigate the risks against sustainability.

The latter are supposed to cover the intersection between the value chain competence and innovation and change management, which sets its goal in equipping the responsible leader with necessary knowledge and skills to provide a suitable and economically feasible solution to meet the demand and supply, and, in the best case, contribute to the global development through developing new ideas, technologies, business models, and processes.

In turn, knowledge of the sustainability paradigm shall be aimed at mastering the topics dealing with, among others [21]:

- Global environmental challenges and ecological considerations of the environment, the development of the human habitat and the nature, its processes and the potential anthropogenic impact;
- Conservation of resources, including those extracted from the earth, water, and air, as well as the overall resource lifecycle and their impact on our development;

- Definition of sustainable development in various areas of human activity and the technological opportunities and challenges;
- Increase of process efficiency and elimination of humankind activity waste in a broad sense;
- Economics of sustainable development, implications to politics and governance, and the questions of justice.

It is not much of exaggeration to say that the identified three competence areas are all interrelated and equally important on the way to set up the goals and make decisions based upon their full environmental, social, and economic implications [12]. It is valid to emphasize that the sustainability paradigm part shall not be limited only to the environmental sustainability (e.g., global warming or climate change issues in a simple way). While there has emerged a new scientific paradigm called sustainability science, the concept of sustainability shall not be considered its result: to the contrary, it is primarily an ethically grounded concept [22]. Generally, sustainability has more dimensions than nature (or the environment), society, and the economy, and may include cultural or political domain as well. However, sustainability can not be ensured on the way towards the goal of the development without innovating and implementing changes. That is why the concept of innovation is sometimes considered as a process when an organization learns to do something it did not know how to do before, and then proceeds to do it in a sustained way [23].

The mentioned three-pillar framework for the competence build up assumes that the level of knowledge of each part is determined by the needs of the person and specifics of the occupation. Nevertheless, the present industry and business society require more generalization and broader outlook than ever due to more complicated and versatile economic and business activity. A timely response to the challenges of the rapidly evolving technologies and social changes requires a certain skill set, which becomes far more important with regards to the goals of sustainable development: safeguarding human life, maintaining society's productive potential, and preserving the scope of development [22].

So the managerial competence in the energy for sustainable development should, among others:

- Cover traditional energy resources with the clear understanding of their historical importance, ways of utilization, limiting factors and potential future, at least during the transition period. So far, traditional power generating technologies are forecast and expected to play a declining, yet dominant role within the world energy mix [24];

- Cover all renewable energy resources with particular emphasis on new advancements in technology as they continue to increase their hold [25];
- Provide a complete framework to assess the resource potential, available technology and its limitations, economic implications, and efficiency levels of the existing and innovative approaches in energy production, transportation, and consumption;
- Address key socio-economic and cultural peculiarities of various approaches to energy value chain that would lead towards sustainable society and energy justice. In turn, it would offer an opportunity to develop new cross-cutting social science agendas on exploring where injustices occur, developing new processes of avoidance and remediation, and provide a new framework for bridging existing and future research on energy production and consumption [26];
- Provide balance between theory and practice through setting up the sustainability goal as the overall driver of change. This directly follows the Brundtland Commission's definition of sustainable development, which means to "ensure that it meets the needs of the present without compromising the ability of future generations to meet their own needs" [27];
- Prepare the leader towards changes aimed at delivering sustainable development in all spheres of activity, not only in responsible energy production and consumption;
- Guide the manager to share knowledge, skills, and expertise throughout the community and to other members of the society to ensure global acceptance of sustainability ideas.

24.4.3 Concluding remarks

The idea of sustainability in its modern understanding is a relatively recent initiative, with its need being realized a few decades ago and still revealing its full potential. It originates from the sustainable consumption of resources and is mostly dealing with the issues of the environment that is challenged by the humankind and therefore endangered by the past and present activity. The organic growth limitations we face are due to the very nature of the environment and therefore have to be critically assessed and reasonably mitigated on our way of growth. However, with the current goal of material growth it is unlikely that the humankind can sustain the same pace in the long run.

It is the area of sustainable development that is concerned with aspects of learning that enhance the transition towards sustainability [28], where reliable and clean energy would still play one of the dominant roles. However, education in the area of energy sustainability shall spread further over the problems of climate change and renewable energy, dealing with the changes in mindset and overall business attitude. The existing approaches to develop energy for sustainable development competence are not yet mature enough and would require more and more attention both from the scholars and business society. While the former shall provide scientific background and identify new ways towards sustainable future, which is largely the aim of the sustainability science, the latter are responsible for implementing these ideas in practice.

The identified three-pillar competence model shall be further investigated to define clear pathways for all three stages of education, as well as come up with the implementation road map. It can be worth outlining the link between academia and industry within the energy sustainability in the sense of managerial competence, as its urgency and need have been proved to be clear: there is a growing demand for highly educated responsible leaders with the competences of sustainability in energy who are capable of making the change.

24.5. Our common future

The title of this section has been encouraged by the UN conference's Brundtland Report [27] which gives one of the clearest visions of sustainability. This is because the challenges that underline the sustainability topic and determine the ways towards sustainable development cannot be addressed by only a part or group of the society, a single country or even several countries, and therefore have to be addressed globally by all of us.

The changes of the climate, depletion of resources, health issues, food and fresh water scarcity are among the hottest topics which may still give indirect signals. The humankind can respond in three major ways to these signals that consumption and its impact have grown beyond their sustainable limits [29]:

- Deny, disguise, or confuse the signals of these challenges, which may result in many forms. Some claim there is no need to worry at all about the limits which have been largely discussed, and the natural development of technology will automatically solve all issues. Others claim that the problem requires an additional study before taking any action. But most often people seek to shift the costs of their mistakes to others, especially those who are far away in space or time, e.g., build higher stacks to release emissions to upper atmosphere, dispose toxic chemicals in distant regions, use more fertilizers to compensate for the falling soil fertility, use force to secure the use of resources instead of buying them.

- Alleviate the pressure from limits by technical or economic fixes, for instance, gradually reduce the rate of pollution per energy generation and use, increase efficiency of the use of resources and implement recycling procedures, substitute renewable sources for nonrenewable ones.
- Work on the underlying causes and admit that the currently employed socio-economic system is unmanageable from the sustainability point of view, and therefore seek to change the structure of the system.

Evidently, the change of the structure does not mean any negative or unlawful change of the society, political systems, economic crisis, revolution, and so on. In systems language, changing structure means changing the feedback structure, that is, the information links in a system: the content and timeliness of the data that system components have to work with, and the ideas, goals, incentives, costs, and feedbacks that motivate or constrain behavior [29].

All these considerations add some credence to the point of view which says that we should not worry about the spaceman economy at all, and that we should just go on to maintain our growth at a high rate. This means that we leave the problems of the future to the future, so that when scarcities arise, the needs of the then present society will have to determine the solutions [30].

Still, the Earth is our common and the only home which supports the humankind in all aspects of living. Nature does not really care about the artificial borders between the countries, but it does care about the impact the humans put on the environment. Unfortunately, this impact has been negative for many centuries. Extinction of many species of mammals, birds or plants due to unaccountable or even wanton consumption, societal overconsumption and longing for material development without really considering the effect on the environment, irresponsible disposal of waste in "remote" locations that — a short-sighted behavior that can be well characterized by a French saying "Après moi, le déluge".[2]

[2] After all, who cares?

The impact of our activity has already been witnessed throughout the world, the climate change sprouts have already grown in multiple corners of the environment, and these changes may increase their pace so that the impact will be dramatic already for the next couple of generations or even for the current generation. But to understand the problem the society has to change the global mindset, which is not really achievable if the basic human needs have not been fulfilled. Conquering hunger, fresh water, clean air, and safe living problems worldwide is prerequisite but not sufficient. This can be the first step towards starting the educational change in the way of human responsible consumption to form a moderate pattern and promote human capital development — a high-level goal of development.

The paradigm of sustainability that combines human, manufactured, and natural capital is an idealization and might be seen as a utopia, but it gives a clear vision. This vision cannot be achieved by a few people, even by half of the population. The Earth is our common home and therefore we have our common future to learn from the mistakes we have done in the past. This might be the ideal way towards sustainable development and, to start with, start with yourself.

We are living through amazing, yet challenging times. The number of potential dealbreakers that may well impact the society, businesses, and established political structures is enormous, but it is all about the way the new generation of managers and sustainability leaders would perceive them: a challenge or an opportunity to apply the change towards our common and better future.

References

[1] United Nations (UN). Transforming our world: the 2030 agenda for sustainable development. Resolution adopted by the General Assembly on 25 September 2015, A/RES/70/1. New York: UN General Assembly; 2015.

[2] Jacobson MZ, Delucchi MA. Providing all global energy with wind, water, and solar power, part I: technologies, energy resources, quantities and areas of infrastructure, and materials. Energy Policy 2011:1154–69. https://doi.org/10.1016/j.enpol.2010.11.040.

[3] Delucchi MA, Jacobson MZ. Providing all global energy with wind, water, and solar power, part II: Reliability, system and transmission costs, and policies. Energy Policy 2011:1170–90. https://doi.org/10.1016/j.enpol.2010.11.045.

[4] Kandpal T, Broman L. Renewable energy education: a global status review. Renewable and Sustainable Energy Reviews 2014:300–24.

[5] Petit V. The energy transition. An overview of the true challenge of the 21st century. Springer; 2017.

[6] The future of the grid. Evolving to meet America's needs. Final report. An industry-driven vision of the 2030 grid and recommendations for a path forward; 2014.

[7] Jennings P. New directions in renewable energy education. Renewable Energy 2009:435–9.
[8] Adomßent M, Fischer D, Godemann J, Herzig C, Otte I, Rieckmann M, et al. Emerging areas in research on higher education for sustainable development — management education, sustainable consumption and perspectives from Central and Eastern Europe. Journal of Cleaner Production 2014:1–7.
[9] Al-Rawahy KH. Engineering education and sustainable development: the missing link. Procedia 2013:392–401. https://doi.org/10.1016/j.sbspro.2013.10.754.
[10] Perkins JH, Middlecamp C, Blockstein D, Rivers Cole J, Knapp RH, Saul KM, et al. Energy education and the dilemma of mitigating climate change. Journal of Environmental Studies and Sciences 2014:354–9. https://doi.org/10.1007/s13412-014-0189-5.
[11] Vare P, Scott W. Learning for a change: exploring the relationship between education and sustainable development. Journal of Education for Sustainable Development 2007;1:191–8. https://doi.org/10.1177/097340820700100209.
[12] Sidiropoulos E. Education for sustainability in business education programs: a question of value. Journal of Cleaner Production 2014;85:472–87. https://doi.org/10.1016/j.jclepro.2013.10.040.
[13] Roorda N, van Son H. Education for sustainable development. In: Sustainability science. An introduction. Springer; 2016.
[14] United Nations Economic Commission for Europe (UNECE). UNECE strategy for education for sustainable development. URL: http://www.unece.org/fileadmin/DAM/env/documents/2005/cep/ac.13/cep.ac.13.2005.3.rev.1.e.pdf, 2005.
[15] Jennings P, Lund C. Renewable energy education for sustainable development. Renewable Energy 2001:113–8.
[16] Sovacool BK. What are we doing here? Analyzing fifteen years of energy scholarship and proposing a social science research agenda. Energy Research and Social Science 2014;1:1–29. https://doi.org/10.1016/j.erss.2014.02.003.
[17] Barth M, Godemann J, Rieckmann M, Stoltenberg U. Developing key competencies for sustainable development in higher education. International Journal of Sustainability in Higher Education 2007;8:416–30. https://doi.org/10.1108/14676370710823582.
[18] Lambrechts W, Mula I, Ceulemans K, Molderez I, Gaeremynck V. The integration of competences for sustainable development in higher education: an analysis of bachelor programs in management. Journal of Cleaner Production 2013;48:65–73. https://doi.org/10.1016/j.jclepro.2011.12.034.
[19] Kleef J, Roome N. Developing capabilities and competence for sustainable business management as innovation: a research agenda. Journal of Cleaner Production 2007;15:38–51. https://doi.org/10.1016/j.jclepro.2005.06.002.
[20] Hesselbarth C, Schaltegger S. Educating change agents for sustainability — learnings from the first sustainability management master of business administration. Journal of Cleaner Production 2014;62:24–36. https://doi.org/10.1016/j.jclepro.2013.03.042.
[21] Heinrichs H, Martens P, Michelsen G, Wiek A, editors. Sustainability sciences: an introduction. Springer; 2016.
[22] Michelsen G, Adomßent M, Martens P, von Hauff M. Sustainable development — background and context. In: Heinrichs H, Martens P, Michelsen G, Wiek A, editors. Sustainability science. An introduction. Springer; 2016. p. 5–29.
[23] Shepard H. Innovation-resisting and innovation-producing organizations. The Journal of Business 1967;40:470–7.

[24] Teske S, European Renewable Energy Council, Greenpeace International. Energy [r]evolution. A sustainable world energy outlook 2015. Greenpeace International; 2015.
[25] REN21. Renewables 2017 global status report. Paris: REN21 Secretariat. ISBN 978-3-9818107-6-9, 2017.
[26] Jenkins K, McCauley D, Heffron R, Stephan H, Rehner R. Energy justice: a conceptual review. Energy Research and Social Science 2016;11:174–82.
[27] World Commission on Environment and Development (WCED). Our common future. Oxford: Oxford University Press; 1987.
[28] Barth M. Teaching and learning in sustainability science. In: Sustainability science. An introduction. Springer; 2016.
[29] Meadows D, Meadows D, Randers J. Limits to growth the 30-year update. Earthscan. ISBN 9781849775861, 2012.
[30] Boulding KE. The economics of the coming spaceship Earth. Environmental Quality in a Growing Economy: Essays from the Sixth RFF Forum 2013;3:1–20.

Index

FSC
www.fsc.org
MIX
Papier | Fördert
gute Waldnutzung
FSC® C083411

Kolibri 360 GmbH
Ferdinand-Jühlke-Straße 7
99095 Erfurt, Deutschland
produktsicherheit@kolibri360.de